ADVANCES IN CONTINUUM MECHANICS
AND THERMODYNAMICS OF MATERIAL BEHAVIOR

Advances in Continuum Mechanics and Thermodynamics of Material Behavior

In Recognition of the 60th Birthday of Roger L. Fosdick

Edited by

DONALD E. CARLSON
University of Illinois at
Urbana-Champaign, U.S.A.

YI-CHAO CHEN
University of Houston, U.S.A.

Reprinted from *Journal of Elasticity*, Vol. 59, Nos. 1–3 (2000)

Springer-Science+Business Media, B.V.

Library of Congress Cataloging-in-Publication Data

A C.I.P. Catalogue record for this book is available from the Library of Congress.

ISBN 978-94-010-3837-9 ISBN 978-94-010-0728-3 (eBook)
DOI 10.1007/978-94-010-0728-3

Printed on acid-free paper

Originally published by Kluwer Academic Publishers in 2000
Softcover reprint of the hardcover 1st edition 2000

TABLE OF CONTENTS

Journal of Elasticity 59: 1, 2000

In Recognition of the Sixtieth Birthday of Roger L. Fosdick

A symposium entitled **Advances in the Continuum Mechanics and Thermodynamics of Material Behavior** in honor of Roger Fosdick was held as part of the 1999 Joint ASME Applied Mechanics and Materials Summer Conference at Virginia Tech on June 27–30, 1999. The symposium was organized by Yi-Chao Chen as a vehicle for recognizing Roger Fosdick for his many significant contributions to this area on the occasion of his (then) recent sixtieth birthday. Contributors to the symposium were invited to submit full-length paper versions of their presentations for possible publication in a special extended issue of the *Journal of Elasticity*, edited by Donald E. Carlson. The results appear here in this complete volume. We are grateful to Roger's friend and former student K.R. Rajagopal for the following remarks.

As Editor-in-Chief of the *Journal of Elasticity*, Professor Fosdick has not been entirely comfortable with several issues of the *Journal* being comprised of contributions solicited in his honor. We wish to make it clear that he had nothing to do with the conception of the symposium nor with the solicitation and the editing of the articles, all of which were refereed.

DONALD E. CARLSON
University of Illinois at Urbana-Champaign

YI-CHAO CHEN
University of Houston

Guest Editors

Roger L. Fosdick

Roger L. Fosdick

This volume in honor of Professor R.L. Fosdick's sixtieth birthday is a fitting tribute to his wide ranging contributions to continuum mechanics. In addition to his research in various aspects of continuum mechanics, Prof. Fosdick has had a strong influence in shaping the direction that the field has taken in the past few decades through the numerous students who have learnt continuum mechanics from him, both at the Illinois Institute of Technology and at the University of Minnesota. Many are the students who decided to pursue their doctoral work in mechanics after having taken his meticulously prepared courses delivered with pellucid clarity.

After obtaining his undergraduate degree in Mechanical Engineering at the Illinois Institute of Technology, he decided to pursue his doctoral degree in mechanics in what was considered at that time as the Mecca of Mechanics, the Division of Applied Mathematics at Brown University. His initial foray into mechanics concerned the small bending of a circular bar that is finitely elongated or compressed. The systematic and methodical analysis of this problem portended the contributions that were to follow. On completing his doctoral studies, he joined the faculty of the Department of Mechanics at the Illinois Institute of Technology as an assistant professor, and in the short span of three years rose to the rank of associate professor. During his sojourn there, his interests widened, and he started working in the area of non-Newtonian fluid mechanics with Prof. Barry Bernstein. His first doctoral student Schuler studied the plane deformation variant of an interesting problem in elasticity proposed by Ericksen.

In 1969 Prof. Fosdick moved to the Department of Aerospace Engineering and Mechanics at the University of Minnesota and has stayed there for the past three decades. His work with Serrin on secondary flows of non-Newtonian fluids in pipes of non-circular cross-section answered once and for all a conjecture due to Ericksen as to when purely rectilinear flows are possible in such fluids. He followed this work with an analogous study of secondary deformations in non-linear solids with Kao. His detailed, deliberate and careful study of the thermodynamics and stability of the fluids of second grade with Dunn, which was followed by other studies on fluids of second and third grade with me, helped dispel certain misconceptions that had ossified into false tenets in the field and instigated a critical re-evaluation of the status of the fluids of the differential type.

It was about this time that Prof. Fosdick turned his attention to the study of the equilibrium of coexistent phases. His work with Dunn on the morphology and stability of such coexistent mixtures is a thoughtful and thorough study of the problem. His interest in the mechanics of coexistent phases has never waned, and he has co-authored several papers with Hertog, MacSithigh, Mason, Zhang, Royer and Weinberger on a variety of interesting problems in the area. His penchant for theoretical studies has never come in the way of his attacking interesting applied problems; examples which immediately come to mind are his study with Villagio on the bending stiffness of belts, the analysis of free surface flows of non-Newtonian fluids in channels with Castro, and the study of thermoviscoelastic vibration absorbers with Ketema and Yu. Most recently with Aguiar he has studied singular deformations in elasticity that have very interesting applications.

Professor Fosdick's research contributions span a broad spectrum of areas in continuum mechanics, elasticity, viscoelasticity, thermodynamics, mechanics of coexistent phases, and shock physics, to name some. The papers in these issues reflect this breadth and depth of Prof. Fosdick's own research contributions.

His being asked to serve as the Editor-in-Chief of the *Journal of Elasticity* and as a member of the Editorial Board of the *SIAM Journal of Applied Mathematics* are a testament to the high regard in which he is held by his peers. A Fellow of the American Academy of Mechanics, he has had the rare distinction of serving twice as the Chairman of the Society of Natural Philosophy.

K.R. RAJAGOPAL
Texas A&M University

Publications of Roger L. Fosdick

Books

1. *Shock Induced Transitions and Phase Structures in General Media*, J.E. Dunn, R. Fosdick and M. Slemrod (eds). The IMA Volumes in Mathematics and its Applications, Vol. 52. Springer, Berlin (1993).

Publications

1. Extremum principles in the theory of small elastic deformations superposed on large elastic deformations. In: *Progress in Applied Mechanics*, The Prager Anniversary Volume, MacMillan, New York (1963). With R.T. Shield.
2. Small bending of a circular bar superposed on finite extension or compression. *Arch. Rational Mech. Anal.* **12** (1963) 223–248. With R.T. Shield.
3. On the vector potential and the representation of a polyharmonic function in n-dimensions. *J. Math. Mech.* **4** (1965) 573–588.
4. Remarks on compatibility. In: *Modern Developments in the Mechanics of Continua*. Academic Press, New York (1966) pp. 109–127.
5. Extremum principles in heat conduction for rigid heat conductors. *Acta Mech.* **4** (1967) 241–272.
6. Visco-plastic flow of a thin circular plate. In: *Proc. of the 10th Midwestern Mechanics Conf.*, 1967. With E.J. Appleby, J.J. Carey and P.G. Hodge, Jr.
7. On the displacement boundary-value problem of static linear elasticity theory. *ZAMP* **19** (1968) 219–233.
8. A note on non-uniqueness in linear elasticity theory. *ZAMP* **19** (1968) 906–912. With W.S. Edelstein.
9. Dynamically possible motions of incompressible, isotropic, simple materials. *Arch. Rational Mech. Anal.* **29** (1968) 272–288.
10. On general measures of deformation. *Acta Mech.* **6** (1968) 275–295. With A.S. Wineman.
11. A class of dynamically possible steady motions of incompressible, isotropic simple materials. *Internat. J. Non-Linear Mech.* **4** (1969) 79–73.
12. On Ericksen's problem for plane deformations with uniform transverse stretch. *Internat. J. Engrg. Sci.* **7** (1969) 217–233. With K.W. Schuler.
13. On four rheological relations. *Rheol. Acta* **4** (1970) 186–193. With B. Bernstein.
14. Non uniqueness of second order fluids under steady radial flow in annuli. *Internat. J. Engrg. Sci.* **7** (1969) 555–569. With B. Bernstein.
15. On the complete representation of biharmonic functions. *SIAM J. Appl. Math.* **19** (1970) 243–250.
16. A note on the C-N inequality. *ZAMP* **21** (1970) 1083–1084.
17. Elastic stability and the zero moment condition. *J. Elasticity* **1** (1971) 19–28.
18. Statically possible radially symmetric deformations in isotropic, incompressible elastic solids. *ZAMP* **22** (1971) 590–607.
19. Rectilinear steady flow of simple fluids. *Proc. Roy. Soc. London A* **332** (1973) 311–333. With J. Serrin.

20. The free surface on a liquid between cylinders rotating at different speeds, Part I. *Arch. Rational Mech. Anal.* **49** (1973) 321–380. With D.D. Joseph.
21. The free surface on a liquid between cylinders rotating at different speeds, Part II. *Arch. Rational Mech. Anal.* **49** (1973) 381–401. With G.S. Beavers and D.D. Joseph.
22. Thermodynamics, stability and boundedness of fluids of complexity 2 and fluids of second grade. *Arch. Rational Mech. Anal.* **56** (1974) 191–252. With J. Dunn.
23. Global properties of continuum thermodynamic processes. *Arch. Rational Mech. Anal.* **59** (1975) 97–109. With J. Serrin.
24. Global properties of continuum thermodynamic processes. In: *Proc. of the 12th Annual Meeting of Soc. Engrg. Sci.*, University of Texas at Austin (1975) pp. 185–191.
25. Universal flows in the simplest theories of fluids. *Annali Della Schola Normale Superiore Di Pisa IV* **2** (1977) 323–341. With C. Truesdell.
26. Transverse deformations associated with rectilinear shear in elastic solids. *J. Elasticity* **8** (1978) 117–142. With B. Kao.
27. Uniqueness and drag for fluids of second grade in steady motion. *Internat. J. Non-Linear Mech.* **13** (1978) 131–137. With K. Rajagopal.
28. Lemmas in thermodynamic stability: A priori inequalities and induced global bounds on thermodynamic states. *Arch. Rational Mech. Anal.* **67** (1978) 183–224. With J.E. Dunn.
29. On an inequality in thermodynamic stability. In: G. De LaPenha and L. Medeiros (eds), *Contemporary Developments in Continuum Mechanics and Partial Differential Equations*, North-Holland Math. Studies, Vol. 30 (1978) pp. 143–170.
30. On the impossibility of linear Cauchy and Piola–Kirchhoff constitutive theories for stress in solids. *J. Elasticity* **9** (1979) 83–89. With J. Serrin.
31. Anomalous features in the model of second order fluids. *Arch. Rational Mech. Anal.* **70** (1979) 145–152. With K.R. Rajagopal.
32. The morphology and stability of material phases. *Arch. Rational Mech. Anal.* **74** (1980) 1–99. With J.E. Dunn.
33. Thermodynamics and stability of fluids of third grade. *Proc. Roy. Soc. London A* **339** (1980) 351–377. With K.R. Rajagopal.
34. Shearing motions and the formation of shocks in an elastic circular tube. *Quart. Appl. Math.* **38** (1980) 191–207. With G.P. MacSithigh.
35. The elastica and the problem of pure bending for a non-convex stored energy function. *J. Elasticity* **11** (1981) 165–186. With R.D. James.
36. Steady flow of a simple fluid around a rotating sphere. *Rheol. Acta* **19** (1980) 675–697. With B.G. Kao.
37. Catastrophic instabilities and related results in a fluid of third grade. *Internat. J. Non-Linear Mech.* **16** (1981) 191–198. With B. Straughan.
38. On the existence of a manifold for temperature. *Arch. Rational Mech. Anal.* **81** (1983) 317–332. With K.R. Rajagopal.
39. Helical shear of an elastic, circular tube with a non-convex stored energy. *Arch. Rational Mech. Anal.* **84** (1983) 31–53. With G. MacSithigh. Reprinted in: C.M. Dafermos, D.D. Joseph and F.M. Leslie (eds), *The Breadth and Depth of Continuum Mechanics*. Springer, Berlin (1985).
40. Structure and dynamical stability of Gibbsian states. In: J. Serrin (ed.), *New Perspectives in Thermodynamics*. Lecture Notes in Mathematics Proceedings of the Laws and Structures of Continuum Thermomechanics, Minneapolis, MN, June 1983. Springer, Berlin (1985).
41. Elementary implications of viscometry on the free surface profile for the problems of climbing and channel flow. *J. Non-Newtonian Fluid Mech.* **15** (1984) 127–144. With J. Castro.
42. On the creeping motion of a simple fluid past a submerged sphere. *Latin American J. Chem. Engrg. Appl. Chem.*, in press. With J. Castro and C. Filipich.
43. Minimization in incompressible nonlinear elasticity theory. *J. Elasticity* **16** (1986) 267–301. With G. MacSithigh.

44. A plane non-linear shear for an elastic layer with a non-convex stored energy function. *Internat. J. Solids Struct.* **22**(10) (1986) 1129–1135. With K. Rajagopal and G. MacSithigh.
45. The influence of the bending stiffness in the motion of a belt. *J. Appl. Mech.* **53** (1986) 266–270. With P. Villaggio.
46. On the Gibbsian thermostatics of mixtures. *Arch. Rational Mech. Anal.* **93** (1986) 203–221. Reprinted in: B.D. Coleman, M. Feinberg and J. Serrin (eds), *Analysis and Thermomechanics*. Springe, Berlin (1987). With J. Patino.
47. Steady, structured shock waves. Part 1: Thermoelastic materials. *Arch. Rational Mech. Anal.* **104** (1988) 295–365. With J.E. Dunn.
48. A dissipation principle and its consequences for structures shock waves in thermoelastic materials. In: S.C. Schmidt and N.C. Holmes (eds), *Shock Waves in Condensed Matter*, 1987. Elsevier Science, Amsterdam (1988) pp. 215–218. With J.E. Dunn.
49. A variational proof of the stress theorem of Cauchy. *Arch. Rational Mech. Anal.* **105** (1989) 95–103. With E.G. Virga.
50. Steady, structured shock waves of arbitrary intensity in thermoelastic materials. In: M.F. McCarthy and M.A. Hayes (eds), *Elastic Wave Propagation*, 1988. Elsevier Science, Amsterdam (1989) pp. 133–140. With J.E. Dunn.
51. The Maxwell relation and Eshelby's conservation law for minimizers in elasticity theory. *J. Elasticity* **22** (1989) 193–200. With B. Hertog.
52. Material symmetry and crystals. *Arch. Rational Mech. Anal.* **110** (1990) 43–72. With B. Hertog.
53. Minimization in nonlinear elasticity theory for bodies reinforced with inextensible cords. *J. Elasticity* **26** (1991) 239-289. With G. MacSithigh.
54. Steady, structured shock waves in a viscoelastic solid of differential type. *Internat. J. Engrg. Sci.* **28** (1990) 469–483. With W.H. Warner and J.H. Yu.
55. Normality and convexity of the yield surface in nonlinear plasticity. *Quart. Appl. Math.* **21** (1993) 117–127. With E. Volkmann.
56. The torsion problem for a nonconvex stored energy function. *Arch. Rational Mech. Anal.* **122** (1993) 291–322. With Y. Zhang.
57. The penalty method and the numerical solution of plane problems in nonlinear elastostatics. MSI Report (September 1993). With A. Aguiar.
58. The Weierstrass condition for a special class of elastic materials. *J. Elasticity* **34** (1994) 167–184. With J.E. Dunn.
59. Coexistent phase mixtures in the anti-plane shear of an elastic tube. *Z. angew. Math. Phys.* **45** (1994) 202–244. With Y. Zhang.
60. Stress and the moment-twist relation in the torsion of a cylinder with a nonconvex stored energy function. *Z. angew. Math. Phys.* **46** (1995) 146–171. With Y. Zhang.
61. A structured phase transition for the anti-plane shear of an elastic circular tube. *Quart. J. Mech. Appl. Math.* **48** (1995) 189–210. With Y. Zhang.
62. Single phase energy minimizers for materials with nonlocal spatial dependence. *Quart. Appl. Math.* **24** (1996) 161–195. With D. Mason.
63. The static state of a two-phase solid mixture in a stressed elastic bar. *Internat. J. Solids Struct.* **33** (1996) 2267–2281. With G. Royer.
64. Thermodynamics, stability and nonlinear oscillations of a viscoelastic solid Part 1: Differential type solids of second grade. *Internat. J. Nonlinear Mech.* **31** (1996) 495–516. With J.-H. Yu.
65. A note on the optimal state of a binary solid mixture in a stressed elastic bar. *Meccanica* **31** (1996) 519–525. With G. Royer and H. Weinberger.
66. Numerical solution of singular problems in nonlincar clastostatics. Research Report (1995). With A. Aguiar.
67. Alloy separation of a binary mixture in a stressed elastic sphere. *J. Elasticity* **42** (1996) 49–77. With G. Royer.

68. Thermodynamics, stability and nonlinear oscillations of a viscoelastic solid, Part II: History type solids. *Internat. J. Nonlinear Mech.* **33** (1997) 165–188. With J.-H. Yu.
69. Vibration damping through the use of materials with memory. *Internat. J. Solids Struct.* **35** (1998) 403–420. With Y. Ketema and J.-H. Yu.
70. A nonlinear oscillator with history dependent force. *Internat. J. Nonlinear Mech.* **33** (1998) 447–459. With Y. Ketema and J.-H. Yu.
71. On a model of nonlocal continuum mechanics, Part I: Existence and Regularity. *SIAM J. Appl. Math.* **58** (1998) 1278–1306. With D. Mason.
72. On a model of nonlocal continuum mechanics, Part II: Structure, Asymptotics and Computations. *J. Elasticity* **48** (1997) 51–100. With D. Mason.
73. Dynamics of a viscoelastic spherical shell with a nonconvex strain energy function. *Quart. Appl. Math.* **26** (1998) 221–244. With Y. Ketema and J.-H. Yu.
74. A thermoviscoelastic dynamic vibration absorber. *J. Appl. Mech.* **65** (1998). With Y. Ketema.
75. A singular problem in incompressible nonlinear elastostatics. Accepted for publication. *Math. Models Methods Appl. Sci.* (1999). With A. Aguiar.
76. A rheological model for materials which support coexistent shear rates. *Internat. J. Nonlinear Mech.* **35** (2000) 1023–1043. With C. Balan.
77. The state of pure shear. *J. Elasticity* **52** (1998) 91–98. With P. Belik.
78. Shape memory alloys for passive vibration damping. Accepted for publication. *J. Intelligent Material Systems Struct.* (1999). With Y. Ketema.
79. Self-intersection in elasticity. Accepted for publication. *Internat. J. Solids Struct.* (2000). With A. Aguiar.
80. The constraint of local injectivity in linear elasticity theory. Submitted for publication (1999). With G. Royer.
81. The Lagrange multiplier in incompressible elasticity theory. *J. Elasticity* **55** (1999) 193–200. With G. Royer.
82. On the rigid motion of a body. Draft available (1999). With J.E. Dunn.
83. Rank 1 convexity for a class of incompressible elastic materials. Draft available (1999). With J.E. Dunn and Y. Zhang.
84. Temperature effects on the vibration damping properties of shape memory alloys. Draft available (2000). With Y. Ketema.
85. Multiple natural states for an elastic isotropic material with polyconvex stored energy. Submitted for publication (2000). With G. Royer.
86. On the stability of Oldroyd upper and lower convected models of viscoelastic behavior. In progress (1999).
87. An electrodynamic basis for minimization in the static theory of elastic ferroelectrics and ferromagnetics. In progress (2000).
88. The Lagrange multipliers and hyperstress constraint reactions in incompressible multipolar elasticity theory. In progress (2000). With G. Royer.
89. Penalty method, modified Lagrangian and constraint norm in multipolar elasticity theory. In progress (2000). With G. Royer.

Journal of Elasticity **59:** 9–22, 2000.

On Invariance Groups for Equilibrium Theories

J.L. ERICKSEN
5378 Buckskin Bob Rd., Florence, OR 97439, U.S.A.

Received 27 October 1998

Dedicated to Roger Fosdick, on the occasion of his 60th birthday.

Abstract. As continuum theory evolves to meet new demands, we see an increasing number of different kinds of theories. Our understanding of how to treat invariances associated with these should similarly grow, by better understanding concepts involved in these for well established theories, and exploring lines of thought that have not been used for these. The purpose of this paper is to present some of my thoughts about this.

1. Introduction

Theories of invariance, such as those of material symmetries, form an important part of the theory of constitutive equations used for the various kinds of continuum theories. It is a part concerning which I have often found myself in disagreement with some other workers in this area. Also, in various communications with younger workers, I have found some confusion about matters of disagreement, owing to the fact that they had not been exposed to relevant facts and reasoning, at least in part. This led me to believe that it would be useful to make public a discussion of some of the things I have discussed with them. This is what I shall do here, for equilibrium theories. Also, I'll discuss a possible strategy for finding transformations that can be used in constitutive theory, when one is dealing with theories for which this might not be an obvious matter.

2. Enantiomorphism

Many years ago, when I was graduate student, I had an interesting conversation with an old geometer, who told me that there were two different definitions of congruence in the literature on Euclidean geometry, and he showed me examples of writing confirming this. Unfortunately, I lost track of his references, although I recall a later conversation with a philosopher, who showed me an old discussion of this by the philosopher Kant, in his *Prolegomena.* What is the difference? Well, some allowed reflections and others did not. So, if you like, your right and left hands are (approximately) congruent according to one group of persons, but not to the other. A bit later, when I first read Cauchy's [1, 2, 3] macroscopic derivation of equations of linear elasticity theory, I was motivated by this to see what he did about it, in the invariance arguments he used. I noted that he did not assume invariance under reflections, thinking more in terms of superposing (infinitesimal) rigid displacements. What now goes under the name "principle of objectivity" and other names generally also involves time dependent orthogonal transformations. This was first used in the derivation by Stokes [4] of what we now call the Navier–Stokes equations and some nonlinear constitutive equations. He did not consider reflections, merely the idea that superposing rigid body motions should have no effect. Then, some later workers began to incorporate improper orthogonal transformations, asserting that constitutive equations should also be invariant under reflections. For the two kinds of theories mentioned, and various others, the final equations are invariant under reflections, so the difference does not matter. The latter workers seem to think more in terms of changing frames of reference. Other workers did not go along with the stronger assumption and my [5, Section 9] discussion of dynamical theory of cholesteric liquid crystals puts me on record as being in the latter camp. So, workers in continuum theory are now in two camps, very much like those geometers, with one important difference. With the geometers, neither group is wrong, by the rules of mathematics, but they are developing inequivalent theories. It is my view that workers in continuum theory should take pains to try to ensure that assumptions made are consistent with what we know about nature. To assess this, one needs to consider theories where the indicated difference in assumptions has some consequences.

One of the types of theories which seems to me to be important for this is the kind describing optical activity. For some indication of how old these are, one can turn to my favorite reference for such matters, the book by Whittaker [6, Chapter X]. He mentions two theories dealing with early nineteenth century observations of sugar solutions, the older one predating what we now know as the electromagnetic theory of light. I will ignore the older one, believing that it would take too much explanation for readers unfamiliar with ancient theories. The newer one, attributed by Whittaker to Goldhammer in 1892, is intended to apply to a fluid at rest in an inertial frame, with light passing through it. It involves a constitutive equation of the form

$$\mathbf{E} = \frac{1}{\varepsilon}\mathbf{D} + k \operatorname{curl} \mathbf{D}, \tag{2.1}$$

with the usual interpretations of $\mathbf{D}$ and $\mathbf{E}$, ε and k being material constants. As I interpret it, this is invariant under time-independent rotations and translations, as is appropriate for what is, essentially, a static theory. However, it is not invariant under improper orthogonal transformations, so it is not invariant under the static subgroup of the Galilean group, if you regard the latter as including reflections. As far as I am concerned, this theory is acceptable from the viewpoint of invariance, and it describes the observations well enough, by the standards by which such optical theories are judged. Whittaker makes no disparaging remarks about it. I do not know of a constitutive equation that is invariant under reflections which is as good, and I doubt that one exists. Of course, there are fluids that are not optically active, often described by (2.1). with $k = 0$ and, for these, (2.1) is invariant under reflections. As I see it, whether such constitutive equations are invariant under reflections is a matter of material symmetry.

Somewhat newer, but still quite old are the equilibrium theories of cholesteric liquid crystals, with similar invariance properties. Like most liquids, they are, to a good approximation, incompressible, the theory being used to describe a unit vector field $\mathbf{n}$ identified as the optic axis of these optically uniaxial liquids. This means that $\mathbf{n}$ and $-\mathbf{n}$ are physically equivalent, so the transformation relating them describes a kind of material symmetry. For the energy per unit volume W, a constitutive equation of the form*

$$2W = K_1(\operatorname{div}\mathbf{n})^2 + K_2(\mathbf{n}\cdot\operatorname{curl}\mathbf{n} + q_0)^2 + K_3(\mathbf{n}\wedge\operatorname{curl}\mathbf{n})^2, \tag{2.2}$$

is used where the K's and q_0 are material constants, the former being positive. Again, this is invariant under static rotations and translations, but not reflections when, as is usually the case, $q_0 \neq 0$. This theory has done well in describing a variety of phenomena in these fluids, and I consider the invariance assumptions to be sound. Observations indicate that, in particular, minimum energy configurations for cholesterics are well described by (2.2), the orientation being twisted in a simple way. For a pair of enantiomorphic materials, one will have a right-hand twist, the other being of the left-hand kind.

Analogous constitutive equations for nematic liquid crystals are invariant under reflections, being otherwise similar ($q_0 = 0$ for these). Again, I conclude that invariance under reflections is a matter of material symmetry, being appropriate for some but not all materials.

Here, we have seen two examples of kinds of materials which, as some physicists describe them** distinguish between the right and left. My experience is consistent with the view that such materials occur in pairs such that one can be viewed as a mirror image of the other, the basic idea of enantiomorphic materials. That is, it seems that if one is found, a good chemist can find a way to produce its enantiomorph. For (2.1), such enantiomorphs will share the same value of ε, but the value of k for one will be the negative of that for the other, for an observer using the same frame of reference for both. Similarly, for (2.2), enantiomorphs will share the same values of the K's, but the value of q_0 for one will be the negative of that for the other. One can get different values of q_0 by mixing two with values of q_0 opposite sign, including $q_0 = 0$ for the mixture, at some particular concentration. If they are enantiomorphs, mixing equal amounts gives $q_0 = 0$, as one would expect.‡ For these examples, it is clear that the pair of constitutive equations for enantiomorphs is invariant under reflections, and one can reasonably expect this to

* Some workers include another term which is a null Lagrangian but, here, this is unimportant.

** See, for example, the discussion by de Gennes [7, Section 1.3.2.2].

‡ For more information on such mixtures, cf. [8, Chapter 4].

be true generally. By viewing these pairs of functions as a double-valued one, one could regard them as invariant under reflections. So, those who abhor constitutive equations that are not invariant under reflections could use this alternative. Of these, those that I have mentioned this to did not much like the idea.

For nonlinear elasticity theory, for example, the equations are so well established by common usage that many of us, myself included, become uncritical about arguments leading to them. It is rather generally accepted that these constitutive equations are invariant under reflections, as are those involved in various other commonly used theories of continuum mechanics. In my opinion, this is not quite correct, and some discussions of this that I have encountered employ faulty logic. This might not cause real difficulty if one is dealing only with elasticity theory because, as will be discussed, one can argue that, here, constitutive equations for enantiomorphs can be regarded as the same, because of the special nature of these constitutive equations. However, if one is interested in various other kinds of theories, one needs to understand these matters better. My experience is that the number of workers in continuum theory who don't is not so small, so I think it worthwhile to elaborate this.

3. Invariance of Functions

To treat the invariance of a function under a group, it is clearly necessary to decide how its arguments and values are to be transformed by elements of the group. Beyond this, the following statements seem to me to be self-evident.

(a) By common consent in mathematics, a function has some domain and by definition, a domain is a connected set.

(b) For a group G of transformations to leave a function invariant, it is necessary that each of the transformations map its domain onto itself. For a function $f(x)$, if Tf and Tx denote the transforms of f and x for any $T \in G$, the statement that $f(x)$ is invariant under G means that

$$Tf(x) = f(Tx) \quad \forall T \in G. \tag{3.1}$$

(c) If some group of transformations leaves a function invariant, and one considers the restriction of the function to a subdomain, this will be invariant under the subgroup mapping the subdomain onto itself.*

For elasticity theory, what some call hyperelasticity theory, the basic idea is that it suffices to specify one constitutive equation, for the strain energy per unit

* Mathematically, a restriction could be invariant under some transformations which do not leave invariant the function defined on the larger domain. I don't recall seeing an equilibrium theory well exemplifying this, but a theory covering both fluid and solid phases might well do so, if it accounts for shear stresses in solid phases. Things like sandwich plates could be viewed as examples, but these seem to me artificial.

mass W, as a function of the deformation gradient $\mathbf{F}$, perhaps also depending on the position in the reference configuration, denoted by $\mathbf{X}$. So, we have

$$W = \widehat{W}(\mathbf{F}, \mathbf{X}), \tag{3.2}$$

say. For present purposes, I think it best to interpret the symbols in a simplistic way, using rectangular Cartesian coordinates. Then $\mathbf{X}$ denotes the triple of material coordinates and, if $\mathbf{x}$ labels the triple of spatial coordinates,

$$\mathbf{F} = \left\| \frac{\partial x^i}{\partial X^K} \right\|. \tag{3.3}$$

For us to be able to use it to calculate stresses, $\widehat{W}$ must be differentiable. Physically, this is likely to fail for values of $\mathbf{F}$ such that $\det \mathbf{F} = 0$ and, since the domain is connected, I use the

ASSUMPTION. Throughout the domain of $\widehat{W}$, one has either

$$\det \mathbf{F} > 0 \quad \text{or} \quad \det \mathbf{F} < 0. \tag{3.4}$$

Commonly, two different kinds of groups are considered as invariance groups for $\widehat{W}$. One is considered to apply generally to theories of finite deformations, although two different versions are preferred by different groups of workers. According to one group, $\widehat{W}$ is to be invariant under $O(3)$, in the sense that

$$\widehat{W}(\mathbf{F}, \mathbf{X}) = \widehat{W}(\mathbf{QF}, \mathbf{X}), \quad \forall \mathbf{Q} \in O(3). \tag{3.5}$$

To me, this is unacceptable since, with the above assumption, this includes transformations mapping the domain of $\widehat{W}$ to a different, disjoint domain, any $\mathbf{Q}$ with $\det \mathbf{Q} = -1$. Of course, one can use such a transformation as a change of variables, and for any such change, the usual rules assign the same values of a function at corresponding points. This hardly means that the function is invariant under all changes of variables. Here, it is important not to confuse functions with their values although, generally, I tend to be casual about such matters. I conclude that $\widehat{W}$ cannot be invariant under reflections in this sense, so I and some others modify this to exclude improper transformations, i.e.

$$\widehat{W}(\mathbf{F}, \mathbf{X}) = \widehat{W}(\mathbf{QF}, \mathbf{X}), \quad \forall \mathbf{Q} \in SO(3), \tag{3.6}$$

with the usual assumption that the domain of $\widehat{W}$ is chosen to be invariant under this group.* Of course, $\mathbf{F}$ is associated with deformations from some reference configuration, and different choices of it generally give different functions $\widehat{W}$ for the same material. There is the possibility that some of these are the same function, leading to invariances described by equations of the form

$$\widehat{W}(\mathbf{F}, \mathbf{X}) = \widehat{W}(\mathbf{FH}, \mathbf{X}), \quad \forall \mathbf{H} \in G_{\mathrm{M}}, \tag{3.7}$$

* I exclude the linear theories using approximations to (3.6).

where G_M is some group depending on the material. In this paper, except for some considerations of reflections, I deal only with symmetries involving just one material point, not with descriptions of homogeneity or uniformity. Commonly, $\mathbf{H}$ is considered to be the Jacobian matrix associated with a change of reference configuration, but we will find reasons to generalize this. It seems to be generally accepted that, for any reasonable choice of this group,

$$\mathbf{H} \in G_M \quad \Rightarrow \quad \det \mathbf{H} = \pm 1, \tag{3.8}$$

and I concur. For Bravais lattices, consideration of their atomic arrangements supports the notion that $-\mathbf{1} \in G_M$ for them. However, for elasticity theory, elements with $\det \mathbf{H} = -1$ won't map the domain of $\widehat{W}$ onto itself, so we can only have

$$\widehat{W}(\mathbf{F}, \mathbf{X}) = \widehat{W}(\mathbf{FH}, \mathbf{X}), \quad \mathbf{H} \in G_M^+, \tag{3.9}$$

where

$$G_M^+ = \{\mathbf{H} \in G_M \mid \det \mathbf{H} = 1\}. \tag{3.10}$$

Here, I have used the simplest rule for transforming values of $\widehat{W}$, that they are unchanged by these transformations. Arguments given by Wang and Truesdell [9, Chapter III] support the idea that this assumption applies, with G_M also being the invariance group for the Cauchy stress, when G_M is a subgroup of the orthogonal group, which covers most theories of solids. For crystals, this applies when $\widehat{W}$ is restricted to one of Pitteri's neighborhoods. However, when the domain includes all $\mathbf{F}$ with $\det \mathbf{F} > 0$, considerations of the atomic structure of Bravais lattices suggest an infinite, discrete group.* Then, the arguments presented by Wang and Truesdell indicate that (3.9) might not apply, and, in general terms, they describe possible modifications. However, for these crystals, the indicated group simply describes ambiguities in describing a fixed configuration,** with (3.9) interpretable as the statement that the energy depends on the configuration, not on which description one uses. It then seems to me clear that one should continue to use (3.9) for these theories.

Some writings on continuum theory might leave the impression that the theory of material symmetry is the same as that of the groups G_M described above. Certainly, the transformation $\mathbf{n} \to -\mathbf{n}$ used for liquid crystals does not fit this mold. Neither do some others to be considered, along with others not mentioned here.

I have not yet mentioned another possibility. The domain of $\widehat{W}$ can be mapped onto itself by a pair of improper transformations and, obviously, for any $\mathbf{F}$ in the domain

$$(-\mathbf{1})\mathbf{F}(-\mathbf{1}) = \mathbf{F}. \tag{3.11}$$

* Further discussion of these matters is given by Pitteri and Zanzotto [10, Chapters 5, 7], for example.

** Note that this does not exactly agree with the common idea that $\mathbf{H} \in G_M$ describes a change of reference configuration.

Let G^E_{M}denote the group generated by G_{M} and $\mathbf{H}_0 = -\mathbf{1}$. Then, $G_{\mathrm{M}} \subseteq G^E_{\mathrm{M}}$, and even if it is a proper subgroup, $\widehat{W}$ will be invariant under the transformations

$$\mathbf{F} \to \mathbf{QFH}_0\mathbf{H}, \quad \text{with} \quad \{\mathbf{Q} \in O(3) \mid \det \mathbf{Q} = -1\}, \quad \mathbf{H} \in G^+_{\mathrm{M}}, \tag{3.12}$$

because of (3.11). At least in cases I have explored, this is reasonably interpreted as implying that, for elasticity theory, the constitutive equations for enantiomorphic materials are really the same. For example, quartz crystals commonly occur in right- and left-handed forms, apparently as an accidental happening when they are grown, the difference being detectable by X-ray methods. These can even occur in the same sample, associated with the so-called Brazil twins. The experience is that it is impossible to convert one to the other by applying loads to the crystal so, in this sense, they are different materials. However, if one considers elastic phenomena and an observer uses a right-handed (left-handed) reference configuration for the right-handed (left-handed) kinds, one can use (3.12) to relate the two strain energy functions, as being the same. Given the usual idea that enantiomorphs are mirror images of each other, I expect that this type of relation holds generally. I do not object to regarding this as a statement of invariance under reflections, for elasticity theory. However, I do not believe that this is what most other writers have in mind by this. Usually, workers consider $W$to be a function of $\mathbf{C} = \mathbf{F}^{\mathrm{T}}\mathbf{F}$. Then, the points I raise are of little or no consequence for elasticity theory, but they can be important for other kinds of continuum theory.

4. Elastic Dielectrics

As an example of a kind of theory for which the previously described ideas are important, let us consider the theory of elastic dielectrics proposed by Toupin [11]. This changes the constitutive equation for W to be of the form

$$W = \widehat{W}(\mathbf{F}, \mathbf{p}, \mathbf{X}), \tag{4.1}$$

where $\mathbf{p}$ is the electric polarization vector, a spatial vector. The previous considerations apply to the dependence on $\mathbf{F}$ so, in particular, (3.4) applies, etc. Under $O(3)$, $\mathbf{p}$ transforms as indicated by

$$\mathbf{p} \to \mathbf{Qp}, \quad \mathbf{Q} \in O(3), \tag{4.2}$$

and, under G_{M}, it transforms in the trivial manner,

$$\mathbf{p} \to \mathbf{p} \quad \forall \mathbf{H} \in G_{\mathrm{M}}. \tag{4.3}$$

Suppose that we try to force $\widehat{W}$ to be invariant under reflections, in the commonly considered cases where $\widehat{W}$ does not depend explicitly on $\mathbf{X}$, by assuming that (3.12) applies. Then, using (3.12) with $\mathbf{Q} = -\mathbf{1}$ and $\mathbf{H} = -\mathbf{1}$, we get

$$\widehat{W}(\mathbf{F}, \mathbf{p}) = \widehat{W}(\mathbf{F}, -\mathbf{p}). \tag{4.4}$$

Of course, this requires that $-\mathbf{p}$ be in the domain of $\widehat{W}$ whenever $\mathbf{p}$ is, a mild condition. However, constitutive equations for piezoelectric crystals never satisfy (4.4) so, by forcing this invariance, we lose the ability to describe piezoelectric effects, with the common understanding that this is a phenomenon which can be described by linear theory, at least qualitatively. For such crystals, G_{M} is a proper subgroup of G^E_{M},and we have gone astray by using the latter. A similar possibility occurs in the more complicated theory of crystal multi-lattices I [12] proposed, which does not explicitly involve electromagnetic fields.

For materials for which (4.4) does not hold, but when $-\mathbf{p}$ is in the domain of $\widehat{W}$ when $\mathbf{p}$ is, I think it reasonable to regard $\widehat{W}(\mathbf{F}, \mathbf{p})$ and $\widehat{W}(\mathbf{F}, -\mathbf{p})$ as constitutive equations for a pair of enantiomorphic materials. Granted this, workers who think it imperative to have invariance with respect to reflections could, as before, regard these as values of a double-valued function. Otherwise, the idea that all constitutive equations should be invariant under reflections seems to me to be inconsistent with experience.

5. Time Reversals

For equilibrium theories, physicists long used the idea that constitutive equations for thermodynamic potentials should be invariant under time reversals. Obviously, this is associated with the common belief that equilibria are reversible processes. I accepted this until I [13] noted that this is inconsistent with some observations. As I now see it, such constitutive equations are invariant under these transformations for some but not all materials, so they are in the realm of material symmetry transformations. Again, these are conceptually rather different from those included in the group G_{M} discussed previously.

The phenomenon of piezoelectricity has been known for a long time. Consider using (4.1) to describe this. One can have $\widehat{W}$ invariant under time reversals and do this by assuming that, under time reversals,

$$\mathbf{p} \to \mathbf{p}, \tag{5.1}$$

and this or an equivalent has long been used by physicists in such discussions. For a long time, it was believed that the magnetic analog, piezomagnetism, did not occur. As an analogous theory, one can use constitutive equations of the form

$$W = \widehat{W}(\mathbf{F}, \mathbf{m}, \mathbf{X}), \tag{5.2}$$

where $\mathbf{m}$ is the magnetization vector, an axial vector. So, under the central inversion $\mathbf{Q} = -\mathbf{1}, \mathbf{m} \to \mathbf{m}$. However, for time reversals, the accepted rule analogous to (5.1) is

$$\mathbf{m} \to -\mathbf{m}, \tag{5.3}$$

so requiring (5.2) to be invariant under this gives

$$\widehat{W}(\mathbf{F}, \mathbf{m}, \mathbf{X}) = \widehat{W}(\mathbf{F}, -\mathbf{m}, \mathbf{X}). \tag{5.4}$$

In this discussion, I used the rather obvious and accepted assumption that, under time reversals, $\mathbf{F} \to \mathbf{F}$ and $\mathbf{X} \to \mathbf{X}$. This then excludes the phenomenon of piezomagnetism. Then, experimentalists began to discover that piezomagnetic effects do occur in some crystals, contradicting (5.4). Curiously, this was noted in a revision of a volume in the Landau–Lifschitz series [14], without changing the presentation leading to (5.4). Apparently, whoever added the new material did not notice the contradiction.

What bothered me a bit is that the rules noted for transforming $\mathbf{p}$ and $\mathbf{m}$ under time reversals differ from those used in relativistic theories. As is discussed by Truesdell and Toupin [15, Chapter F], for example, these imply that

$$\mathbf{p} \to -\mathbf{p}, \qquad \mathbf{m} \to \mathbf{m}. \tag{5.5}$$

If one uses the same kind of reasoning, replacing the previous rules by these, it would exclude piezoelectric effects, an unacceptable conclusion. I believe that what is wrong is the assumption that these two kinds of constitutive equations are always invariant under time reversals.

There is a simple way of understanding the difference between the two sets of transformation rules. Reasonably, $\mathbf{p}$ should transform like $q\mathbf{r}$, where q denotes a charge, $\mathbf{r}$ a relative position vector. For $\mathbf{m}$, one can similarly use the elementary description of Amperian currents. The first set of rules emerges if one assumes that, under time reversals, $q \to q$, while (5.5) involves the different assumption that $q \to -q$. If, by the first rules, one combines a charge reversal with a central inversion and a time reversal, to get what is commonly called a CPT (charge-parity-time) transformation, we get

$$\mathbf{F} \to \mathbf{F}, \qquad \mathbf{p} \to \mathbf{p}, \qquad \mathbf{m} \to \mathbf{m}, \tag{5.6}$$

so the constitutive equations considered are invariant under this combination, when $\widehat{W}$ does not depend on $\mathbf{X}$. From this view, (5.5) combines a time reversal with a charge reversal. Perhaps this is what should replace the faulty assumption of invariance under time reversals for equilibrium theories describing homogeneous materials. However, I think it wise to look hard for evidence of failure of this before accepting it, something I have not done. Often, workers consider extending the list of arguments in (5.2) to include the gradient of $\mathbf{m}$. A CPT transformation converts this to its negative, so the invariance assumption does restrict the possible forms of $\widehat{W}$. I have not noticed in the literature any constitutive equations of this kind that fail to have this invariance, but I have not made a careful search. For many materials, the assumption illustrated by (5.4) is used and agreeable with experience. Obviously, it is easier to interpret if you use (5.1) and (5.3) instead of (5.5) and, for that reason, I prefer them for considerations of material symmetry.

The invariance described by (5.4) is like the transformation $\mathbf{n} \to -\mathbf{n}$ in that both involve reversing a vector field, something that cannot be properly described using the groups $O(3)$ and G_{M} used in elasticity theory. However, the ideas that motivated use of them are obviously quite different.

Earlier, I mentioned a possible connection between invariance under time reversals and the idea that equilibria are reversible processes. As far as I know, everyone believes that piezoelectricity and piezomagnetism involve no dissipation. Since, for the latter, $\widehat{W}$ is not invariant under time reversals, this seems a bit curious, suggesting that some irreversible processes are inherently non-dissipative. I believe that it would be good to question the common practice of regarding "irreversible" and "dissipative" as essentially synonymous. However, this is so loosely related to considerations of invariance, that I don't think it worth pursuing here.

6. General Invariance Groups

Suppose that we are considering generalizing some familiar kind of equilibrium theory, for which we understand how to describe relevant invariances. We wish to add some different kinds of variables, as we just did, in going from elasticity theory to theories dealing with some electromagnetic effects. We have seen that this can involve changing the basic nature of the relevant transformation groups, exemplified by the inclusion of the time reversals that are not considered in elasticity theory. I think it desirable to have some way of finding a list of all of the kinds of transformations that might be relevant. One possibility is to try to use some of the techniques employed in the large literature dealing with applications of group theory to differential equations. Adeleke [16, 17, 18] has explored this for some types of theories, using special types of generalized coordinate transformations of the kind used by Noether [19], for example, in her derivation of conservation laws implied by invariances in variational problems. As I am doing, he considers only what Truesdell [20, Chapter IV] calls material isomorphism of a point with itself, not with the notions of homogeneity or uniformity of a body, in considering examples. However, his general theory deals with these. For the theories he treats, he derives the set of possible invariance transformations compatible with his assumptions. Some are rather obvious, perhaps after some thought, but others are not. One might well encounter some mathematical possibilities that are not physically relevant for our purposes. Often, applications of group theory deal with one equation, sometimes for a subset of solutions of it. With theories of material symmetry, for example, the interest is in invariances common to a fairly large set of equations, and one concedes that some might be invariant under additional transformations. So, one needs to exercise some judgment about this. For simplicity, I exclude theories of constrained materials, which involve special subtleties. With the common use of Lagrange multipliers, one can use different functions $\widehat{W}$ that agree when the constraints are satisfied. For incompressible elastic materials, one could use a function of $\mathbf{F}/\sqrt[3]{|\det \mathbf{F}|}$ in this way, for example, and there are some advantages to doing so. Then $\widehat{W}$ is invariant under the transformations $\mathbf{F} \to k\mathbf{F}$, for any positive scalar k. It would hardly be reasonable to assume that these are in the invariance group for a compressible material. Similar possibilities occur for some other types of constraints. However, I believe that it is wise to think hard, before rejecting possibilities. So far, at least, I have not encountered any relevant transformations that are not found by this procedure.

Consider constitutive equations of the form (5.2), for example. We start with the idea that they describe some set of fields of the form

$$\mathbf{x} = \hat{\mathbf{x}}(\mathbf{X}), \qquad \mathbf{m} = \widehat{\mathbf{m}}(\mathbf{X}). \tag{6.1}$$

Here, Adeleke's idea is to consider invertible generalized coordinate transformations of the form

$$\overline{\mathbf{x}} = \tilde{\overline{\mathbf{x}}}(\mathbf{x}, \mathbf{X}, \mathbf{m}), \qquad \overline{\mathbf{m}} = \widetilde{\overline{\mathbf{m}}}(\mathbf{x}, \mathbf{X}, \mathbf{m}), \qquad \overline{\mathbf{X}} = \widetilde{\overline{\mathbf{X}}}(\mathbf{x}, \mathbf{X}, \mathbf{m}), \tag{6.2}$$

restricted by mild smoothness conditions. Noether's practice [19] would also include $\mathbf{F}$ in the list of generalized coordinates, treating the arguments of $\widehat{W}$ as coordinates for a manifold. I think that this kind of generalization is worth exploring, but it does make analyses more difficult. Obviously, this includes the aforementioned $\mathbf{F} \to k\mathbf{F}$. However, so does (6.2). Substituting a field of the form (6.1) into (6.2) then gives relations denoted by

$$\overline{\mathbf{x}} = \overline{\overline{\mathbf{x}}}(\mathbf{X}), \qquad \overline{\mathbf{m}} = \overline{\overline{\mathbf{m}}}(\mathbf{X}), \qquad \overline{\mathbf{X}} = \overline{\overline{\mathbf{X}}}(\mathbf{X}). \tag{6.3}$$

Then, by inverting $(6.3)_3$, we can reduce $(6.3)_1$ and $(6.3)_2$ to the form

$$\overline{\mathbf{x}} = \hat{\overline{\mathbf{x}}}(\overline{\mathbf{X}}), \qquad \overline{\mathbf{m}} = \widehat{\overline{\mathbf{m}}}(\overline{\mathbf{X}}), \tag{6.4}$$

defining the transformed fields. The next step is to introduce Galilean invariance in the form

$$\widehat{W}(\mathbf{QF}, \mathbf{Qm}, \mathbf{X}) = \widehat{W}(\mathbf{F}, \mathbf{m}, \mathbf{X}) \quad \forall \mathbf{Q} \in SO(3). \tag{6.5}$$

The idea is to require that the transform of $\widehat{W}$ also satisfy this. For present purposes, the rule used by Adeleke for transforming values of $\widehat{W}$ by any of the generalized coordinate transformations is equivalent to

$$\widehat{W} \to \widehat{\overline{W}} = \widehat{W} + \alpha, \tag{6.6}$$

where α depends on the transformation, but not on the fields (6.1). For elasticity theory, this is what is used by Wang and Truesdell [9, Chapter IV] to get generalizations of (3.9). References to elevant work by others are given by them and Adeleke [16–18]. Wang and Truesdell present analyses of how α depends on $\mathbf{H}$, and mathematical examples of theories for which $\alpha \neq 0$.

The following discussion is made to make clear my opinions on some details relating to Adeleke's procedure, as it applies to theories discussed in my Sections 4 and 5. First, he [16, Chapter 3] shows that the requirement that (6.5) be preserved under the transformations restricts the transformations to be of the form

$$\begin{aligned} \overline{\mathbf{x}} &= a\mathbf{Q}_0\mathbf{x} + f\big(|\mathbf{m}|, \mathbf{X}\big)\mathbf{Q}_0\mathbf{m} + \mathbf{b}, \\ \overline{\mathbf{m}} &= g\big(|\mathbf{m}|, \mathbf{X}\big)\mathbf{Q}_0\mathbf{m}, \\ \overline{\mathbf{X}} &= \widetilde{\overline{\mathbf{X}}}\big(|\mathbf{m}|, \mathbf{X}\big), \end{aligned} \tag{6.7}$$

where the scalar a and vector $\mathbf{b}$ are constants, $\mathbf{Q}_0$ is some particular rotation, and $\overline{\overline{\mathbf{X}}}$ as well as the scalar functions f and g are restricted only by the requirements of invertibility and smoothness. The next and more difficult step is to decide what is to be the domain of $\widehat{W}$, and to determine which of these transformations maps this onto itself. Also, I propose using a replacement for (3.8), which is equivalent there. It is

$$\det \overline{\mathbf{F}} \stackrel{\text{def}}{=} \det \left\| \frac{\partial \overline{\hat{x}}^i}{\partial \overline{X}^K} \right\| = \det \mathbf{F} = \left\| \frac{\partial \hat{x}^i}{\partial X^K} \right\|. \tag{6.8}$$

Essentially, this excludes the possibility of regarding two reference configurations as equivalent when their mass densities are different.

Before exploring these matters, note that (6.7) includes the central inversions and time reversals considered before, and it can be interpreted as including the proper Galilean transformations. If $\widehat{W}$ is defined for all $\widehat{\mathbf{m}}$, it also includes the transformations

$$\overline{\mathbf{x}} = a\mathbf{x}, \qquad \overline{\mathbf{X}} = a\mathbf{X}, \qquad \overline{\mathbf{m}} = b\mathbf{m}, \tag{6.9}$$

where a and b are positive constants, the transformations used in dimensional analysis to describe changing the units of measurement. Of course relations like (4.1) or (5.2) only make sense if such units are considered to be fixed, and one should include appropriate dimensional constants for such analyses. Alternatively, with $b = 1$, (6.9) can be interpreted as a point transformation, stretching the reference and present configurations in the same way. For the theories considered, these leave $\widehat{W}$ invariant in the commonly considered cases where it does not depend explicitly on $\mathbf{X}$. I expect that most readers have used this. However, I consider it to be among the transformations describing material symmetry, and it does not fit the common notion that all such transformations are well described just by the Jacobian matrix associated with a change of reference configuration, as is the case for (3.12).

Now consider $(6.7)_3$, interpreted as a point transformation, at some particular material point $\mathbf{X}_0$. Since we regard this as fixed, we must have

$$\overline{\overline{\mathbf{X}}}(|\mathbf{m}|, \mathbf{X}_0) = \mathbf{X}_0 \quad \Rightarrow \quad \frac{\partial \overline{\overline{\mathbf{X}}}}{\partial \mathbf{m}}(|\mathbf{m}|, \mathbf{X}_0) = \mathbf{0}, \tag{6.10}$$

implying that the gradient of the function given by (6.3), i.e.,

$$\mathbf{K} \stackrel{\text{def}}{=} \left\| \frac{\partial \overline{\overline{X}}^K}{\partial X^L}(\mathbf{X}_0) \right\|, \tag{6.11}$$

can depend on $|\widehat{\mathbf{m}}|$, but not on the gradient of this with respect to $\mathbf{X}$. Using this and (6.7), one can calculate $\overline{F}$ as a function of $\mathbf{X}$, $|\widehat{\mathbf{m}}|$ and $\partial\widehat{\mathbf{m}}/\partial\mathbf{X}$, for any given fields (6.1). For given $\mathbf{X}_0$ and $\widehat{\mathbf{m}}(\mathbf{X}_0)$, we can assign arbitrary values to $\partial\widehat{\mathbf{m}}/\partial\mathbf{X}$ at $\mathbf{X}_0$. Putting this into (6.8) and doing some analysis, one finds that

$$f = 0, \qquad \det \mathbf{H} = 1, \quad \mathbf{H} \stackrel{\text{def}}{=} a\mathbf{K}^{-1}, \tag{6.12}$$

implying that

$$\overline{\mathbf{F}} = \mathbf{FH}. \tag{6.13}$$

Formally, this agrees with the common interpretation of what is used in elasticity theory, with two differences. One is that it is not always appropriate to think of $\mathbf{H}$ as the Jacobian matrix associated with a change of reference configuration, as we have seen. The other is that there is a possibility that $\mathbf{H}$ depends on $|\widehat{\mathbf{m}}|$. There are some trivial possibilities for this, among the continuous groups. For example, for the group describing transverse isotropy, one could use the angle of rotation θ as a parameter labeling group elements. One could also use $\theta + k|\widehat{\mathbf{m}}|$, $k =$ const, which shuffles the group elements differently for different values of $\widehat{\mathbf{m}}$, without changing the group. I have not thought of an example that is likely to be physically relevant, for which the $\mathbf{H}$'s form a group that is different for different values of $\widehat{m}$, and I doubt that any exist.

As to the possibility that the function g in $(6.7)_2$ depends on $|\widehat{\mathbf{m}}|$, my inclination is to exclude it by the requirement that

$$|\overline{\mathbf{m}}| = |\mathbf{m}| \quad \Leftrightarrow \quad g = \pm 1 \tag{6.14}$$

when, as is assumed here, changes of units are excluded. The same analyses apply to the theory of dielectrics discussed earlier: just replace $\mathbf{m}$ by $\mathbf{p}$. With some rather subtle conceptual changes, this brings us back to pretty much the same transformations we discussed before exploring this procedure.

Theories of these kinds have been studied by numerous workers with different backgrounds, so I did not expect to find relevant transformations which have been overlooked, and did not find any that are likely to be important for assessing possible forms of constitutive equations. Rather, they serve as test cases. I believe that the procedure delivers all of the relevant transformations for these theories. This gives us some reason to be confident that this will also be true for theories for which it is less clear what they are. Among these, I would include various theories involving surface energies, in particular theories of shells and membranes more or less like those that Adeleke [17, 18] has analyzed, using these ideas. As was suggested earlier, the procedure also produced some transformations I considered to be irrelevant, excluded by the additional assumptions (6.8) and (6.14).

Finally, I note that, here again, we have used, in an implicit way, the requirement that invariance groups must map the domain of $\widehat{W}$ onto itself. Really, it is this that produces $(6.12)_1$ and, in (6.13), possible choices of $\mathbf{H}$ are subject to this condition, for example. I emphasize this because some workers seem to have overlooked it, as I have done at some times in the past.

References

1. A.-L. Cauchy, De la pression ou tension dans un corps solide. *Ex. de Math.* **2** (1827) 42–56. *See Oeuvres* (**2**)7, pp. 60–78.
2. A.-L. Cauchy, Sur la condensation et la dilatation des corps solides. *Ex. de Math.* **2** (1827) 60–69. *See Oeuvres* (**2**)7, pp. 82–83.
3. A.-L. Cauchy, Sur les équations qui expriment les conditions d' équilibre ou les lois du mouvement intérieur d'un corps solide élastique, ou non élastique. *Ex. de Math.* **3** (1828) 160–187. *See Ouevres* (**2**)8, pp. 124–136.
4. G.G. Stokes, On the theories of internal friction of fluids in motion, and of the equilibrium and motion of elastic solids. *Trans. Cambridge Phil. Soc.* **8** (1845) 75–129.
5. J.L. Ericksen, Liquid crystals with variable degree of orientation. *Arch. Rational Mech. Anal.* **113** (1991) 97–120.
6. E.T. Whittaker, *A History of the Theories of Aether and Electricity*. Vol. 1, revised edn Thomas Nelson and Sons, London (1951).
7. P.G. de Gennes, *The Physics of Liquid Crystals*. Clarendon Press, Oxford (1974).
8. S. Chandrasekhar, *Liquid Crystals*. Cambridge Univ. Press, Cambridge (1997).
9. C.-C. Wang and C. Truesdell, *Introduction to Rational Elasticity*. Noordhoff International, Leyden (1973).
10. M. Pitteri and G. Zanzotto, *Continuum Models for Phase Transitions and Twinning in Crystals*. To be published by CRC Press.
11. R.A. Toupin, The elastic dielectric. *J. Rational Mech. Anal.* **5** (1956) 849–916.
12. J.L. Ericksen, Equilibrium theory for X-ray observations of crystals. *Arch. Rational Mech. Anal.* **139** (1997) 181–200.
13. J.L. Ericksen, Reversible and non-dissipative processes. *Quart. J. Mech. Appl. Math.* **45** (1992) 545–554.
14. L.D. Landau and E. Lifshitz, *Course of Theoretical Physics*. Vol. 8, 2nd revised edn, Pergamon Press, Oxford (1984).
15. C. Truesdell and R.A. Toupin, The classical field theories. In: S. Flügge (ed.), *Handbuch der Physik*, III/1, Springer, Berlin (1960) pp. 226–793.
16. S.A. Adeleke, On material symmetry in mechanics. *Internat. J. Solids Structures* **16** (1980) 199–215.
17. S.A. Adeleke, On transformations of generalized coordinates as symmetry transformations. *J. Elasticity* **14** (1984) 373–386.
18. S.A. Adeleke, On symmetry of shells. *J. Elasticity* **13** (1983) 111–119.
19. E. Noether, Invariante Variationsprobleme. *Nachr. Akad. Wiss. Goettingen, Math.-Phys. Kl.* **2** (1918) 235–257.
20. C. Truesdell, *A First Course in Rational Continuum Mechanics*, Vol. 1. Academic Press, New York (1977).

Journal of Elasticity **59:** 23–50, 2000.

Derivation of Plate and Rod Equations for a Piezoelectric Body from a Mixed Three-Dimensional Variational Principle

S. VIDOLI[1] and R.C. BATRA[2]
[1]*Dipartimento di Ingegneria Strutturale e Geotecnica, Universitá di Roma "La Sapienza", 00184 Roma, Italy*
[2]*Department of Engineering Science and Mechanics, MC 0219, Virginia Polytechnic Institute and State University, Blacksburg, VA 24061, U.S.A.*

Received 16 December 1998; in revised form 24 January 2000

Dedicated to Roger Fosdick on the occasion of his 60th birthday.

Abstract. We use a mixed 3-dimensional variational principle to derive 2-dimensional equations for an anisotropic plate-like piezoelectric body and one-dimensional equations for an anisotropic beam-like piezoelectric body. The formulation accounts for double forces without moments which may change the thickness of the plate and deform the cross-section of the rod. The dependence of the bending rigidities of a transversely isotropic plate upon the angle between the normal to the midsurface and the direction of transverse isotropy is exhibited. The plate equations are used to study the cylindrical deformations of a transversely isotropic plate due to equal and opposite charges applied to its top and bottom surfaces. It is also found that a piezoelectric circular rod with axis of transverse isotropy not coincident with its centroidal axis and subjected to electric charges at the end faces is deformed into a non-prismatic body.

Mathematics Subject Classifications (1991): 74E10, 74E30.

Key words: electroelasticity, approximate theories, anisotropic bodies, double forces.

1. Introduction

Mindlin [1] derived a 2-dimensional theory for plates by expanding displacements as power series in the transverse direction, and integrating the 3-dimensional balance laws and linear constitutive relations in the thickness direction. He proposed correction factors in order for the response predicted from the 2-dimensional theory to match with that deduced from the 3-dimensional linear elasticity theory. He [2] generalized these to piezoelectric plates by also expanding the electric potential as a power series in the transverse direction. Reissner [3] derived a plate theory from the 3-dimensional elasticity theory by using a mixed variational principle, now known as the Hellinger–Prange–Reissner principle. This principle admits different hypotheses on the stress and displacement fields, and has also been employed by Ciarlet and Destuynder [4] to derive the Kirchhoff–Love plate theory. These and other attempts to systematically derive plate and rod theories have been reviewed by Koiter and Simmonds [5], Naghdi [6], Antman [7] and Leissa [8] among others. Naghdi [6] and Antman [7] have also discussed the Cosserat [9] direct theories of plates and rods wherein the bodies are modeled as 2-dimensional surfaces or one-dimensional curves with a suitable number of directors attached to each point; these directors are not tangent to the surface or the curve. The deformation of the directors accounts for the transverse deformations of the plate and of the area of cross-section of the rod. Ericksen [10] has studied plane waves in plates by using such a theory. Because of the immense literature on plates and rods, it is nearly impossible to cite all of the significant contributions.

Podio-Guidugli [11] has used the method of internal constraints to deduce equations of thin plates from the 3-dimensional elasticity equations. He accounted for the material symmetry and the indeterminacy in the stresses induced by the kinematic constraints. Nicotra and Podio-Guidugli [12] have employed the principle of virtual work to derive equations for a plate-like transversely isotropic piezoelectric body by assuming that the displacements and the electric potential vary affinely in the transverse direction. They assumed that the axis of transverse isotropy coincides with the normal to the midsurface of the plate. Their plate equations exhibit uncoupling between the membrane and bending effects. Teresi and Tiero [13] deduced plate theories by seeking stationary points in suitable subspaces of the functional spaces in which the potential energy, the complementary energy and the Hellinger–Prange–Reissner functionals are defined. For an isotropic plate they showed that these methods give different values of the flexural and shear rigidities.

Yang and Batra [14] have given variational principles for piezoelectric bodies. Here we use one of their mixed variational principles (i.e., a generalization of the Hellinger–Prange–Reissner principle to piezoelectric bodies) to derive equilibrium equations and constitutive relations for anisotropic plate-like and rod-like piezoelectric bodies. These equations are used to analyze cylindrical deformations of a transversely isotropic plate with equal and opposite charges applied to its top and bottom surfaces. We also investigate the dependence of the deformations of the plate on the angle between the axis of transverse isotropy and the unit normal to the midsurface of the plate. A similar problem is studied for a transversely isotropic rod. The axial elongations of the rod are significantly increased when the deformations of its cross-sections are considered.

2. Formulation of the Problem

We consider a prismatic anisotropic linear piezoelectric body occupying the region $\mathcal{P} = \mathcal{S} \times \mathcal{I}$ where $\mathcal{S} \subset \mathbb{R}^2$ is a plane surface and $\mathcal{I} = (-h, h)$. For a plate, $\mathcal{S}$ can be identified with its midsurface and $2h$ equals the thickness of the plate. For a prismatic bar, $\mathcal{S}$ is a cross-section and $2h$ the length. The boundary $\partial\mathcal{P}$ of $\mathcal{P}$ can be written as $(\partial\mathcal{S}\times\mathcal{I})\cup(\mathcal{S}^+\cup\mathcal{S}^-)=\mathcal{M}\cup\mathcal{B}$ where $\partial\mathcal{S}$ is the periphery of $\mathcal{S}$, and $\mathcal{S}^+$ and $\mathcal{S}^-$ the upper and lower surfaces of $\mathcal{P}$, i.e., $\mathcal{S}^+=\mathcal{S}\times\{h\}$, and $\mathcal{S}^-=\mathcal{S}\times\{-h\}$. $\mathcal{M}$ is called the mantle of $\mathcal{P}$ and $\mathcal{B}$ is the union of upper and lower bases.

The mixed variational principle of Yang and Batra [14, equation (29)] for quasistatic deformations of a linear piezoelectric body can be written as follows. Find a saddle point of the functional

$$\begin{aligned}\mathcal{H}(\mathbf{u},\mathbf{S},\psi,\mathbf{D}) &= \int_{\mathcal{P}}\mathbf{b}\cdot\mathbf{u}+\int_{\partial_a\mathcal{P}}\mathbf{Sn}\cdot(\mathbf{u}-\bar{\mathbf{u}})+\int_{\partial_b\mathcal{P}}\mathbf{f}\cdot\mathbf{u}-\int_{\mathcal{P}}\mathbf{S}\cdot\mathbf{E}(\mathbf{u})\\ &\quad+\int_{\mathcal{P}}q\,\psi+\int_{\partial_\alpha\mathcal{P}}\mathbf{D}\cdot\mathbf{n}(\psi-\bar{\psi})+\int_{\partial_\beta\mathcal{P}}\chi\,\psi-\int_{\mathcal{P}}\mathbf{D}\cdot\mathbf{W}(\psi)\\ &\quad+\frac{1}{2}\int_{\mathcal{P}}\left[\mathbf{S}\cdot(\mathbb{F}\,\mathbf{S}+\mathbb{M}\mathbf{D})+\mathbf{D}\cdot\left(\mathbb{N}\mathbf{D}-\mathbb{M}^{\mathrm{T}}\mathbf{S}\right)\right],\end{aligned} \tag{1}$$

where $\mathbf{E}(\mathbf{u})=\textsc{Sym}\,\textsc{Grad}\,\mathbf{u}$, $\mathbf{W}(\psi)=\textsc{Grad}\,\psi$, and

$$\begin{aligned}&\partial_a\mathcal{P}\cup\partial_b\mathcal{P}=\partial\mathcal{P}, &&\partial_a\mathcal{P}\cap\partial_b\mathcal{P}=\emptyset,\\ &\partial_\alpha\mathcal{P}\cup\partial_\beta\mathcal{P}=\partial\mathcal{P}, &&\partial_\alpha\mathcal{P}\cap\partial_\beta\mathcal{P}=\emptyset.\end{aligned} \tag{2}$$

Here $\mathbf{u}$ is the displacement of the body point $\mathbf{x}$, $\textsc{Grad}$ is the 3-dimensional gradient operator with respect to $\mathbf{x}$, $\textsc{Sym}\,\textsc{Grad}\,\mathbf{u}=(\textsc{Grad}\,\mathbf{u}+(\textsc{Grad}\,\mathbf{u})^{\mathrm{T}})/2$, $\mathbf{b}$ the density of the body force, $\mathbf{f}$ the surface traction, $\mathbf{S}$ the stress tensor, $\mathbf{n}$ a unit outward normal to $\partial\mathcal{P}$, q the volume charge density, ψ the electric potential, χ the surface charge density, $\mathbf{D}$ the electric displacement, $\mathbf{W}$ the electric field, $\bar{\mathbf{u}}$ the displacements prescribed on $\partial_a\mathcal{P}$, $\bar{\psi}$ the electric potential prescribed on $\partial_\alpha\mathcal{P}$, and

$$\mathbf{E}=\mathbb{F}\mathbf{S}+\mathbb{M}\mathbf{D},\qquad \mathbf{W}=\mathbb{N}\mathbf{D}-\mathbb{M}^{\mathrm{T}}\mathbf{S}, \tag{3}$$

are the constitutive relations for a general anisotropic piezoelectric body. $\mathbb{F}$ is the compliance matrix, $\mathbb{M}$ describes the coupling between the electrical and mechanical effects, and $\mathbb{N}$ is the matrix of dielectric permittivities. We note that electric quantities analogous to mechanical quantities $\mathbf{b}$, $\mathbf{S}$, $\mathbf{f}$, $\mathbf{u}$ and $\mathbf{E}$ are q, $\mathbf{D}$, χ, ψ and $\mathbf{W}$, respectively.

For a prismatic body, a natural decomposition of the position vector, the displacement field and the outward unit normal is

$$\mathbf{x}=\mathbf{r}+\zeta\mathbf{e},\qquad \mathbf{u}=\mathbf{v}+w\,\mathbf{e},\qquad \mathbf{n}=\boldsymbol{\nu}+n\mathbf{e}, \tag{4}$$

where $\mathbf{e}$ is a unit vector perpendicular to $\mathcal{S}$. Thus $\mathbf{v}$ and w denote the displacements of a point within and perpendicular to $\mathcal{S}$. The other field variables can now be written as

$$
\begin{aligned}
\mathbf{E}(\mathbf{u}) &= \text{SYM GRAD}\mathbf{v} + \frac{\mathbf{v}' + \text{GRAD}w}{2} \otimes \mathbf{e} + \mathbf{e} \otimes \frac{\mathbf{v}' + \text{GRAD}w}{2} + w'\mathbf{e} \otimes \mathbf{e}, \\
&= \widehat{\mathbf{E}} + \boldsymbol{\gamma} \otimes \mathbf{e} + \mathbf{e} \otimes \boldsymbol{\gamma} + \varepsilon \mathbf{e} \otimes \mathbf{e}, \\
\mathbf{W}(\psi) &= \text{GRAD}\psi + \psi'\mathbf{e} = \widehat{\mathbf{W}} + \omega \mathbf{e}, \\
\mathbf{S} &= \widehat{\mathbf{S}} + \mathbf{s} \otimes \mathbf{e} + \mathbf{e} \otimes \mathbf{s} + \sigma \mathbf{e} \otimes \mathbf{e}, \\
\mathbf{D} &= \widehat{\mathbf{D}} + \delta \mathbf{e}, \qquad \mathbf{b} = \hat{\mathbf{b}} + \beta \mathbf{e}, \qquad \mathbf{f} = \hat{\mathbf{f}} + \varphi \mathbf{e},
\end{aligned} \tag{5}
$$

and the constitutive relations (3) become

$$
\begin{Bmatrix} \widehat{\mathbf{E}} \\ \boldsymbol{\gamma} \\ \varepsilon \\ \widehat{\mathbf{W}} \\ \omega \end{Bmatrix} = \begin{bmatrix} \mathbb{F}_{ES} & \mathbb{F}_{Es} & \mathbb{F}_{E\sigma} & \mathbb{M}_{ED} & \mathbb{M}_{E\delta} \\ * & \mathbb{F}_{\gamma s} & \mathbb{F}_{\gamma\sigma} & \mathbb{M}_{\gamma D} & \mathbb{M}_{\gamma\delta} \\ * & * & \mathbb{F}_{\varepsilon\sigma} & \mathbb{M}_{\varepsilon D} & \mathbb{M}_{\varepsilon\delta} \\ -* & -* & -* & \mathbb{N}_{WD} & \mathbb{N}_{W\delta} \\ -* & -* & -* & * & \mathbb{N}_{\omega\delta} \end{bmatrix} \begin{Bmatrix} \widehat{\mathbf{S}} \\ \mathbf{s} \\ \sigma \\ \widehat{\mathbf{D}} \\ \delta \end{Bmatrix}, \tag{6}
$$

where $w' = \partial w / \partial \zeta$, a $*$ in row i and column j represents the transpose of the entry in row j and column i, and the tensor product $\otimes$ between two vectors $\mathbf{a}$ and $\mathbf{b}$ is defined by $(\mathbf{a} \otimes \mathbf{b})\mathbf{c} = (\mathbf{b} \cdot \mathbf{c})\mathbf{a}$ for an arbitrary vector $\mathbf{c}$. In terms of fields (5), equation (1) becomes

$$
\begin{aligned}
&\mathcal{H}\big(\mathbf{v}, w, \widehat{\mathbf{S}}, \mathbf{s}, \sigma, \psi, \widehat{\mathbf{D}}, \delta\big) \\
&= \int_{\mathcal{P}} (\hat{\mathbf{b}} \cdot \mathbf{v} + \beta w) \\
&\quad + \int_{\partial_a \mathcal{P}} \big[(\widehat{\mathbf{S}}\boldsymbol{\nu} + \mathbf{s}n) \cdot (\mathbf{v} - \bar{\mathbf{v}}) + (\mathbf{s} \cdot \boldsymbol{\nu} + \sigma n)(w - \bar{w})\big] \\
&\quad + \int_{\partial_b \mathcal{P}} (\hat{\mathbf{f}} \cdot \mathbf{v} + \varphi w) - \int_{\mathcal{P}} (\widehat{\mathbf{S}} \cdot \widehat{\mathbf{E}} + 2\mathbf{s} \cdot \boldsymbol{\gamma} + \sigma \varepsilon) \\
&\quad + \int_{\mathcal{P}} q\psi + \int_{\partial_\alpha \mathcal{P}} \big[(\widehat{\mathbf{D}} \cdot \boldsymbol{\nu} + \delta n)(\psi - \bar{\psi})\big] \\
&\quad + \int_{\partial_\beta \mathcal{P}} \chi\psi - \int_{\mathcal{P}} \big(\widehat{\mathbf{D}} \cdot \widehat{\mathbf{W}} + \delta\omega\big) \\
&\quad + \frac{1}{2}\Big[\int_{\mathcal{P}} \widehat{\mathbf{S}} \cdot \big(\mathbb{F}_{ES}\widehat{\mathbf{S}} + \mathbb{F}_{Es}\mathbf{s} + \mathbb{F}_{E\sigma}\sigma + \mathbb{M}_{ED}\widehat{\mathbf{D}} + \mathbb{M}_{E\delta}\delta\big) \\
&\quad + \int_{\mathcal{P}} \mathbf{s} \cdot \big(\mathbb{F}_{Es}^{\mathrm{T}}\widehat{\mathbf{S}} + \mathbb{F}_{\gamma s}\mathbf{s} + \mathbb{F}_{\gamma\sigma}\sigma + \mathbb{M}_{\gamma D}\widehat{\mathbf{D}} + \mathbb{M}_{\gamma\delta}\delta\big) \\
&\quad + \int_{\mathcal{P}} \sigma\big(\mathbb{F}_{E\sigma}^{\mathrm{T}}\widehat{\mathbf{S}} + \mathbb{F}_{\gamma\sigma}^{\mathrm{T}}\mathbf{s} + \mathbb{F}_{\varepsilon\sigma}\sigma + \mathbb{M}_{\varepsilon D}\widehat{\mathbf{D}} + \mathbb{M}_{\varepsilon\delta}\delta\big) \\
&\quad + \int_{\mathcal{P}} \widehat{\mathbf{D}} \cdot \big(\mathbb{N}_{WD}\widehat{\mathbf{D}} + \mathbb{N}_{W\delta}\delta - \mathbb{M}_{ED}^{\mathrm{T}}\widehat{\mathbf{S}} - \mathbb{M}_{\gamma D}^{\mathrm{T}}\mathbf{s} - \mathbb{M}_{\varepsilon D}^{\mathrm{T}}\sigma\big) \\
&\quad + \int_{\mathcal{P}} \delta\big(\mathbb{N}_{W\delta}^{\mathrm{T}}\widehat{\mathbf{D}} + \mathbb{N}_{\omega\delta}\delta - \mathbb{M}_{E\delta}^{\mathrm{T}}\widehat{\mathbf{S}} - \mathbb{M}_{\gamma\delta}^{\mathrm{T}}\mathbf{s} - \mathbb{M}_{\varepsilon\delta}^{\mathrm{T}}\sigma\big)\Big],
\end{aligned} \tag{7}
$$

where $\mathbf{A} \cdot \mathbf{B} = \mathrm{tr}\,(\mathbf{A}\mathbf{B}^{\mathrm{T}})$ for tensors $\mathbf{A}$ and $\mathbf{B}$.

3. Plate-like Bodies

For a plate-like body, displacements and electric potentials are not prescribed on $\mathcal{S}^+$ and $\mathcal{S}^-$. However, displacements and/or surface tractions, and electric potentials and/or charges may be prescribed on the mantle of the body. Thus,

$$\begin{aligned} \partial_a\mathcal{P} &= \mathcal{M}_a, & \partial_\alpha\mathcal{P} &= \mathcal{M}_\alpha, \\ \partial_b\mathcal{P} &= \mathcal{M}_b \cup \mathcal{B}, & \partial_\beta\mathcal{P} &= \mathcal{M}_\beta \cup \mathcal{B}. \end{aligned} \tag{8}$$

The displacement fields $\mathbf{v}$ and w and electric potential ψ are taken to be affine functions of ζ. That is,

$$\begin{aligned} &\mathbf{v}(\mathbf{r},\zeta) = \mathbf{v}_0(\mathbf{r}) + \zeta\mathbf{v}_1(\mathbf{r}), \qquad w(\mathbf{r},\zeta) = w_0(\mathbf{r}) + \zeta w_1(\mathbf{r}), \\ &\psi(\mathbf{r},\zeta) = \psi_0(\mathbf{r}) + \zeta\psi_1(\mathbf{r}). \end{aligned} \tag{9}$$

The representation $(9)_2$ for the transverse displacement allows for a change in the thickness of the plate. Vel and Batra [15] have used the 3-dimensional equations of elastostatics to analyze deformations of a linear elastic orthotropic plate. They found that the elongation of the normal to the midsurface of the plate is in general nonzero and depends upon the boundary conditions at the plate edges and the loading. Representations (9) for $\mathbf{u}$ and ψ are the same as those assumed by Nicotra and Podio-Guidugli [12], and the representation $(9)_2$ for w is more general than that of Teresi and Tiero [13]. For displacements and the electric potential given by (9), we have

$$\begin{aligned} \widehat{\mathbf{E}} &= \mathrm{SYM\,GRAD}\mathbf{v}_0 + \zeta\,\mathrm{SYM\,GRAD}\mathbf{v}_1 =: \widehat{\mathbf{E}}_0 + \zeta\widehat{\mathbf{E}}_1, \\ 2\boldsymbol{\gamma} &= \mathbf{v}_1 + \mathrm{GRAD}\, w_0 + \zeta\,\mathrm{GRAD} w_1 =: \boldsymbol{\gamma}_0 + \zeta\boldsymbol{\gamma}_1, \\ \varepsilon &= w_1 =: \varepsilon_0, \qquad \widehat{\mathbf{W}} = \mathrm{GRAD}\psi_0 + \zeta\,\mathrm{GRAD}\psi_1 =: \widehat{\mathbf{W}}_0 + \zeta\widehat{\mathbf{W}}_1, \\ \omega &= \psi_1 =: \omega_0. \end{aligned} \tag{10}$$

$\widehat{\mathbf{E}}_0$ is called the membranal strain tensor, $\widehat{\mathbf{E}}_1$ the curvature tensor, and $\boldsymbol{\gamma}$ the shear strain vector. Note that $\boldsymbol{\gamma}$ varies affinely through the thickness. The functional $\mathcal{H}$ given in (7) now becomes

$$\begin{aligned} \mathcal{H} = &\int_{\mathcal{S}} (\mathbf{B}_0\cdot\mathbf{v}_0 + \mathbf{B}_1\cdot\mathbf{v}_1 + \Xi_0 w_0 + \Xi_1 w_1 + Q_0\psi_0 + Q_1\psi_1) \\ &+ \int_{\partial_b\mathcal{S}} (\mathbf{F}_0\cdot\mathbf{v}_0 + \mathbf{F}_1\cdot\mathbf{v}_1 + \Phi_0 w_0 + \Phi_1 w_1) + \int_{\partial_\beta\mathcal{S}} (X_0\psi_0 + X_1\psi_1) \\ &+ \int_{\mathcal{S}} (\mathrm{DIV}\mathbf{N})\cdot\mathbf{v}_0 + (\mathrm{DIV}\mathbf{M} + \mathbf{T})\cdot\mathbf{v}_1 + (\mathrm{DIV}\mathbf{T})w_0 \\ &+ (\mathrm{DIV}\mathbf{C} + \Sigma)w_1 + (\mathrm{DIV}\boldsymbol{\Delta}_0)\psi_0 + (\mathrm{DIV}\boldsymbol{\Delta}_1 + d)\psi_1] \\ &- \int_{\partial_b\mathcal{S}} \left[(\mathbf{N}\boldsymbol{\nu})\cdot\mathbf{v}_0 + (\mathbf{M}\boldsymbol{\nu})\cdot\mathbf{v}_1 + (\mathbf{T}\cdot\boldsymbol{\nu})w_0 + (\mathbf{C}\cdot\boldsymbol{\nu})w_1\right] \\ &- \int_{\partial_\beta\mathcal{S}} \left[(\boldsymbol{\Delta}_0\cdot\boldsymbol{\nu})\psi_0 + (\boldsymbol{\Delta}_1\cdot\boldsymbol{\nu})\psi_1\right] + \mathcal{R}\big(\widehat{\mathbf{S}}, \mathbf{s}, \sigma, \widehat{\mathbf{D}}, \delta\big), \end{aligned} \tag{11}$$

where

$$
\begin{aligned}
&\mathbf{B}_0 = (\hat{\mathbf{f}}^+ - \hat{\mathbf{f}}^-) + \int_{\mathcal{I}} \hat{\mathbf{b}}, \qquad \mathbf{B}_1 = h(\hat{\mathbf{f}}^+ + \hat{\mathbf{f}}^-) + \int_{\mathcal{I}} \zeta \hat{\mathbf{b}},\\
&\mathbf{F}_0 = \int_{\mathcal{I}} \hat{\mathbf{f}}, \qquad \mathbf{F}_1 = \int_{\mathcal{I}} \zeta \hat{\mathbf{f}},\\
&\Xi_0 = (\varphi^+ - \varphi^-) + \int_{\mathcal{I}} \beta, \qquad \Xi_1 = h(\varphi^+ + \varphi^-) + \int_{\mathcal{I}} \zeta\beta,\\
&\Phi_0 = \int_{\mathcal{I}} \varphi, \qquad \Phi_1 = \int_{\mathcal{I}} \zeta\varphi,\\
&Q_0 = (\chi^+ - \chi^-) + \int_{\mathcal{I}} q, \qquad Q_1 = h(\chi^+ + \chi^-) + \int_{\mathcal{I}} \zeta q,\\
&X_0 = \int_{\mathcal{I}} \chi, \qquad X_1 = \int_{\mathcal{I}} \zeta\chi,\\
&\mathbf{N} = \int_{\mathcal{I}} \widehat{\mathbf{S}}, \qquad \mathbf{M} = \int_{\mathcal{I}} \zeta\widehat{\mathbf{S}}, \qquad \mathbf{T} = \int_{\mathcal{I}} \mathbf{s}, \qquad \mathbf{C} = \int_{\mathcal{I}} \zeta\mathbf{s},\\
&\Sigma = \int_{\mathcal{I}} \sigma, \qquad \mathbf{\Delta}_0 = \int_{\mathcal{I}} \widehat{\mathbf{D}}, \qquad \mathbf{\Delta}_1 = \int_{\mathcal{I}} \zeta\widehat{\mathbf{D}}, \qquad d = \int_{\mathcal{I}} \delta,\\
&\hat{\mathbf{f}}^{\pm} = \mathbf{s}n^{\pm}, \quad \varphi^{\pm} = \sigma n^{\pm}, \quad \chi^{\pm} = \delta n^{\pm}, \quad \text{on } \mathcal{S}^{\pm},
\end{aligned}
\tag{12}
$$

and $\mathcal{R}(\widehat{\mathbf{S}}, \mathbf{s}, \sigma, \widehat{\mathbf{D}}, \delta)$ is the part of $\mathcal{H}$ that does not depend upon the displacements and the electric potential. Superscripts $+$ and $-$ on a quantity signify its values on surfaces $\mathcal{S}^+$ and $\mathcal{S}^-$ respectively. Here $\mathbf{N}$ is a 2×2 symmetric matrix giving the in-plane (within the plane $\mathcal{S}$) forces sometimes also called the membranal stress tensor, DIV is the 2-dimensional (within the plane $\mathcal{S}$) divergence operator, $\mathbf{M}$ is the 2×2 symmetric matrix of bending moments also called the moment tensor, $\mathbf{T}$ is the resultant shear force or the shear stress vector, $\mathbf{C}$ is the moment of internal double forces (i.e., of forces acting along $\mathbf{e}$), Σ is the resultant transverse normal stress, $\mathbf{\Delta}_0$ is the resultant in-plane electric displacement, $\mathbf{\Delta}_1$ represents the resultant moment of order 1 of the in-plane electric displacement, and d is the resultant transverse electric displacement. $\mathbf{B}_0 + \Xi_0\mathbf{e}$ is the resultant body force per unit area of $\mathcal{S}$, $\mathbf{B}_1$ the body couple per unit area, Q_0 the resultant applied charge/area, Q_1 the resultant moment of the electric charge, Ξ_1 the double force per unit area, and $\mathbf{F}_0$, $\mathbf{F}_1$, Φ_0, Φ_1 and X_0 have meanings similar to those of $\mathbf{B}_0$, $\mathbf{B}_1$, Ξ_0, Ξ_1 and Q_0 but are defined on the boundary $\partial\mathcal{S}$ of $\mathcal{S}$. The variation of $\mathcal{H}$ with respect to $\mathbf{v}_0$, $\mathbf{v}_1$, w_0, w_1, ψ_0 and ψ_1 gives

$$
\begin{aligned}
&\mathrm{DIV}\mathbf{N} + \mathbf{B}_0 = \mathbf{0}, \quad \mathrm{DIV}\mathbf{M} + \mathbf{T} + \mathbf{B}_1 = \mathbf{0}, \quad \mathrm{DIV}\mathbf{T} + \Xi_0 = 0, \quad \text{on } \mathcal{S},\\
&\mathrm{DIV}\mathbf{C} + \Sigma + \Xi_1 = 0, \quad \mathrm{DIV}\mathbf{\Delta}_0 + Q_0 = 0, \quad \mathrm{DIV}\mathbf{\Delta}_1 + d + Q_1 = 0, \text{ on } \mathcal{S},\\
&\mathbf{N}\boldsymbol{\nu} = \mathbf{F}_0, \quad \mathbf{M}\boldsymbol{\nu} = \mathbf{F}_1, \quad \mathbf{T}\cdot\boldsymbol{\nu} = \Phi_0, \quad \mathbf{C}\cdot\boldsymbol{\nu} = \Phi_1, \quad \text{on } \partial_b\mathcal{S},\\
&\mathbf{\Delta}_0\cdot\boldsymbol{\nu} = X_0, \quad \mathbf{\Delta}_1\cdot\boldsymbol{\nu} = X_1, \quad \text{on } \partial_\beta\mathcal{S}.
\end{aligned}
\tag{13}
$$

We assume that the stress fields $\widehat{\mathbf{S}}$, $\mathbf{s}$ and σ, and the electric displacements $\widehat{\mathbf{D}}$ and δ are polynomials in ζ of degree 1, 3, 2, 1 and 2, respectively. This is guided by the forms of equations $(12)_{11}$–$(12)_{17}$, and the boundary conditions on $\mathbf{s}$, σ and δ at the top and bottom surfaces $\mathcal{S}^+$ and $\mathcal{S}^-$. Recall that tractions and electric charges applied on $\mathcal{S}^+$ and $\mathcal{S}^-$ are included in $(12)_1$, $(12)_2$, $(12)_5$, $(12)_6$ and $(12)_9$. The assumption that $\widehat{\mathbf{S}}$, $\mathbf{s}$, σ, $\widehat{\mathbf{D}}$ and δ are polynomials in ζ of degrees higher than 1, 3, 2, 1 and 2 respectively leads to an indeterminate set of equations. Of course, one could postulate criteria for the determination of the higher-order terms and endow more structure to the 2-dimensional model. Our choice uses the lowest degree polynomials for satisfying $(12)_{11}$–$(12)_{17}$ and the boundary conditions on $\mathcal{S}^+$ and $\mathcal{S}^-$. Thus, we take

$$
\begin{aligned}
\widehat{\mathbf{S}}(\mathbf{r},\zeta) &= \frac{\mathbf{N}(\mathbf{r})}{2h} + \zeta\frac{\mathbf{M}(\mathbf{r})}{J},\\
\mathbf{s}(\mathbf{r},\zeta) &= \frac{3}{4h}\left(1-\frac{\zeta^2}{h^2}\right)\mathbf{T}(\mathbf{r}) + \frac{5}{2J}\left(\zeta-\frac{\zeta^3}{h^2}\right)\mathbf{C}(\mathbf{r})\\
&\quad -\frac{1}{2}\left[\hat{\mathbf{f}}_1\left(1-\frac{3\zeta^2}{h^2}\right)+\hat{\mathbf{f}}_0\frac{\zeta}{h}\left(3-5\frac{\zeta^2}{h^2}\right)\right],\\
\sigma(\mathbf{r},\zeta) &= \frac{3}{4h}\left(1-\frac{\zeta^2}{h^2}\right)\Sigma(\mathbf{r}) - \frac{1}{2}\left[\varphi_1\left(1-\frac{3\zeta^2}{h^2}\right)-2\varphi_0\frac{\zeta}{h}\right],\\
\widehat{\mathbf{D}}(\mathbf{r},\zeta) &= \frac{\mathbf{\Delta}_0(r)}{2h} + \frac{\zeta\,\mathbf{\Delta}_1(\mathbf{r})}{J},\\
\delta(\mathbf{r},\zeta) &= \frac{3}{4h}\left(1-\frac{\zeta^2}{h^2}\right)d - \frac{1}{2}\left[\chi_1\left(1-\frac{3\zeta^2}{h^2}\right)-2\chi_0\frac{\zeta}{h}\right],
\end{aligned}
\tag{14}
$$

where $J = \int_{\mathcal{I}} \zeta^2 = (2h^3)/3$,

$$
\begin{aligned}
\hat{\mathbf{f}}_1 &= \frac{1}{2}(\hat{\mathbf{f}}^+ + \hat{\mathbf{f}}^-), & \hat{\mathbf{f}}_0 &= \frac{1}{2}(\hat{\mathbf{f}}^+ - \hat{\mathbf{f}}^-),\\
\varphi_1 &= \frac{1}{2}(\varphi^+ + \varphi^-), & \varphi_0 &= \frac{1}{2}(\varphi^+ - \varphi^-),\\
\chi_1 &= \frac{1}{2}(\chi^+ + \chi^-), & \chi_0 &= \frac{1}{2}(\chi^+ - \chi^-).
\end{aligned}
\tag{15}
$$

The variation of $\mathcal{H}$ with respect to $\mathbf{N}$, $\mathbf{M}$, $\mathbf{T}$, $\mathbf{C}$, Σ, $\mathbf{\Delta}$ and d gives

$$
\begin{Bmatrix} \widehat{\mathbf{E}}_0 \\ \boldsymbol{\gamma}_0 \\ \varepsilon_0 \\ \widehat{\mathbf{W}}_0 \\ \omega_0 \end{Bmatrix} = \frac{1}{2h}
\begin{bmatrix}
\mathbb{F}_{ES} & \mathbb{F}_{Es} & \mathbb{F}_{E\sigma} & \mathbb{M}_{ED} & \mathbb{M}_{E\delta}\\
* & \frac{6}{5}\mathbb{F}_{\gamma s} & \frac{6}{5}\mathbb{F}_{\gamma\sigma} & \mathbb{M}_{\gamma D} & \frac{6}{5}\mathbb{M}_{\gamma\delta}\\
* & * & \frac{6}{5}\mathbb{F}_{\varepsilon\sigma} & \mathbb{M}_{\varepsilon D} & \frac{6}{5}\mathbb{M}_{\epsilon\delta}\\
* & * & * & \mathbb{N}_{WD} & \mathbb{N}_{W\delta}\\
-* & -* & -* & * & \frac{6}{5}\mathbb{N}_{\omega\delta}
\end{bmatrix}
\begin{Bmatrix} \mathbf{N} \\ \mathbf{T} \\ \Sigma \\ \mathbf{\Delta} \\ d \end{Bmatrix}
$$

$$
-\frac{1}{5}\left\{\begin{array}{c} \mathbf{0} \\ \mathbb{F}_{\gamma s}\hat{\mathbf{f}}_1 + \mathbb{F}_{\gamma\sigma}\varphi_1 + \mathbb{M}_{\gamma\delta}\chi_1 \\ \mathbb{F}^{\mathrm{T}}_{\gamma\sigma}\hat{\mathbf{f}}_1 + \mathbb{F}_{\varepsilon\sigma}\varphi_1 + \mathbb{M}_{\varepsilon\delta}\chi_1 \\ \mathbf{0} \\ \mathbb{M}^{\mathrm{T}}_{\gamma\delta}\hat{\mathbf{f}}_1 + \mathbb{M}_{\varepsilon\delta}\varphi_1 + \mathbb{N}_{\omega\delta}\chi_1 \end{array}\right\},
$$

$$
\left\{\begin{array}{c} \hat{\mathbf{E}}_1 \\ \gamma_1 \\ \widehat{\mathbf{W}}_1 \end{array}\right\} = \frac{1}{J_S}\begin{bmatrix} \mathbb{F}_{ES} & \mathbb{F}_{Es} & \mathbb{M}_{ED} \\ * & \frac{10}{7}\mathbb{F}_{\gamma s} & \mathbb{M}_{\gamma D} \\ -* & -* & \mathbb{N}_{WD} \end{bmatrix}\left\{\begin{array}{c} \mathbf{M} \\ \mathbf{C} \\ \mathbf{\Delta}_1 \end{array}\right\} + \left\{\begin{array}{c} \mathbf{0} \\ -\frac{3}{7h}(\mathbb{F}_{\gamma s}\hat{\mathbf{f}}_0 + \mathbb{F}_{\gamma\sigma}\varphi_0 + \mathbb{M}_{\gamma\delta}\chi_0) \\ \mathbf{0} \end{array}\right\}, \tag{16}
$$

$$
\begin{aligned} &\mathbf{v}_0 = \bar{\mathbf{v}}_0, \quad \mathbf{v}_1 = \bar{\mathbf{v}}_1, \quad w_0 = \overline{w}_0, \quad w_1 = \overline{w}_1, \quad \text{on } \partial_a\mathcal{S}, \\ &\psi_0 = \bar{\psi}_0, \quad \psi_1 = \bar{\psi}_1, \quad \text{on } \partial_\alpha\mathcal{S}, \end{aligned} \tag{17}
$$

where

$$
\begin{aligned} &\bar{\mathbf{v}}_0 = \frac{1}{2h}\int_{\mathcal{I}}\bar{\mathbf{v}}, \quad \bar{\mathbf{v}}_1 = \frac{1}{J}\int_{\mathcal{I}}\zeta\bar{\mathbf{v}}, \quad \overline{w}_0 = \frac{3}{4h}\int_{\mathcal{I}}\left(1 - \frac{\zeta^2}{h^2}\right)\overline{w}, \\ &\overline{w}_1 = \frac{5}{2J}\int_{\mathcal{I}}\left(\zeta - \frac{\zeta^3}{h^2}\right)\overline{w}, \quad \bar{\psi}_0 = \frac{1}{2h}\int_{\mathcal{I}}\bar{\psi}, \quad \bar{\psi}_1 = \frac{1}{J}\int_{\mathcal{I}}\zeta\bar{\psi}. \end{aligned} \tag{18}
$$

Balance laws $(13)_1$–$(13)_6$, constitutive relations (16), boundary conditions $(13)_7$–$(13)_{12}$ and (17), and congruence relations (10) form a complete set of equations for a plate problem. Note that for an anisotropic piezoelectric plate-like body, the problems for the determination of in-plane deformations, shear deformations, changes in thickness that are uniform over $\mathcal{S}$ and the in-plane electric field are coupled together through the constitutive relations. The bending problem is coupled with the problem of determining the thickness variation that is not uniform over $\mathcal{S}$.

Due to the presence of $\mathbf{T}$ in the balance equation $(13)_2$ for $\mathbf{M}$, the transverse shear induces bending of the plate. Also all of the fields are coupled together. A charge applied to $\mathcal{S}^+$ and $\mathcal{S}^-$ will induce bending, shearing, in-plane deformations and variations in the thickness of the plate.

4. Rod-like Bodies

In order to simplify equations, we assume that the ζ-axis coincides with the centroidal axis of the rod. Thus,

$$
\int_{\mathcal{S}}\mathbf{r} = \mathbf{0}. \tag{19}
$$

Furthermore, displacements and electric potentials are assumed to be prescribed only on a part of the upper and lower end faces; however, surface tractions may be prescribed anywhere on the boundary. Thus,

$$\begin{aligned} &\partial_a \mathcal{P} = \mathcal{B}_a, &&\partial_\alpha \mathcal{P} = \mathcal{B}_\alpha, \\ &\partial_b \mathcal{P} = \mathcal{M} \cup \mathcal{B}_b, &&\partial_\beta \mathcal{P} = \mathcal{M} \cup \mathcal{B}_\beta. \end{aligned} \tag{20}$$

For a rod-like body we assume that $\mathbf{v}$ and w are affine in $\mathbf{r}$, and ψ is uniform over the cross-section. That is

$$\begin{aligned} &\mathbf{v}(\mathbf{r}, \zeta) = \mathbf{v}_0(\zeta) + \mathbf{V}(\zeta)\mathbf{r}, \qquad w(\mathbf{r}, \zeta) = w_0(\zeta) + \boldsymbol{\varpi}(\zeta) \cdot \mathbf{r}, \\ &\psi(\mathbf{r}, \zeta) = \psi_0(\zeta), \end{aligned} \tag{21}$$

where $\mathbf{V}$ is a 2×2 matrix. Thus plane sections remain plane during the deformation process. However, the section is in general not rigid, and can undergo in-plane deformations. The assumption $(21)_3$ for the electric potential is motivated by the common applications of piezoelectric rods where charges are not specified on the rod's mantle. Thus the electric potential can be taken to be uniform over a cross-section. For the displacement and electric potential fields given by (21), we have

$$\begin{aligned} \widehat{\mathbf{E}} &= \operatorname{SYM}\mathbf{V} =: \widehat{\mathbf{E}}_0, \\ 2\boldsymbol{\gamma} &= \mathbf{v}_0' + \boldsymbol{\varpi} + \mathbf{V}'\mathbf{r} =: \boldsymbol{\gamma}_0 + \boldsymbol{\Gamma}\mathbf{r}, \\ \varepsilon &= w_0' + \boldsymbol{\varpi}' \cdot \mathbf{r} =: \varepsilon_0 + \boldsymbol{\varepsilon}_1 \cdot \mathbf{r}, \\ \widehat{\mathbf{W}} &= \mathbf{0}, \quad \omega = \psi_0' =: \omega_0. \end{aligned} \tag{22}$$

The functional $\mathcal{H}$ given in (7) simplifies to

$$\begin{aligned} \mathcal{H} = &\int_{\mathcal{I}} (\mathbf{B}_0 \cdot \mathbf{v}_0 + \mathbf{B}_1 \cdot \mathbf{V} + \Xi_0 w_0 + \boldsymbol{\Xi}_1 \cdot \boldsymbol{\varpi} + Q\psi_0) \\ &- \int_{\mathcal{I}} \big(\mathbf{T}' \cdot \mathbf{v}_0 + (\mathbf{C}' + \boldsymbol{\Sigma}) \cdot \mathbf{V} + N' w_0 + (\mathbf{M}' + \mathbf{T}) \cdot \boldsymbol{\varpi} + \Delta' \psi_0\big) \\ &+ \int_{\mathcal{B}_b} (\mathbf{s} - \hat{\mathbf{f}}) \cdot \mathbf{v}_0 + \int_{\mathcal{B}_b} [(\mathbf{s} - \hat{\mathbf{f}}) \otimes \mathbf{r}] \cdot \mathbf{V} + \int_{\mathcal{B}_b} (\sigma - \varphi) w_0 \\ &+ \int_{\mathcal{B}_b} [(\sigma - \varphi)\mathbf{r}] \cdot \boldsymbol{\varpi} + \int_{\mathcal{B}_\beta} (\delta - \chi)\psi_0 + \mathcal{R}(\widehat{\mathbf{S}}, \mathbf{s}, \sigma, \widehat{\mathbf{D}}, \delta), \end{aligned} \tag{23}$$

where

$$
\begin{aligned}
\mathbf{B}_0 &= \int_{\mathcal{S}} \hat{\mathbf{b}} + \int_{\partial\mathcal{S}} \hat{\mathbf{f}}, & \mathbf{B}_1 &= \int_{\mathcal{S}} \hat{\mathbf{b}} \otimes \mathbf{r} + \int_{\partial\mathcal{S}} \hat{\mathbf{f}} \otimes \mathbf{r}, \\
\mathbf{F}_0 &= \int_{\mathcal{S}} \hat{\mathbf{f}}, & \mathbf{F}_1 &= \int_{\mathcal{S}} \hat{\mathbf{f}} \otimes \mathbf{r}, \\
\Xi_0 &= \int_{\mathcal{S}} \beta + \int_{\partial\mathcal{S}} \varphi, & \boldsymbol{\Xi}_1 &= \int_{\mathcal{S}} \beta \mathbf{r} + \int_{\partial\mathcal{S}} \varphi \mathbf{r}, \\
\Phi_0 &= \int_{\mathcal{S}} \varphi, & \boldsymbol{\Phi}_1 &= \int_{\mathcal{S}} \varphi \mathbf{r}, \\
Q &= \int_{\mathcal{S}} q + \int_{\partial\mathcal{S}} \chi, & X &= \int_{\mathcal{S}} \chi, \\
N &= \int_{\mathcal{S}} \sigma, & \mathbf{M} &= \int_{\mathcal{S}} \sigma \mathbf{r}, & \mathbf{T} &= \int_{\mathcal{S}} \mathbf{s}, \\
\mathbf{C} &= \int_{\mathcal{S}} \mathbf{s} \otimes \mathbf{r}, & \boldsymbol{\Sigma} &= \int_{\mathcal{S}} \widehat{\mathbf{S}}, & \Delta &= \int_{\mathcal{S}} \delta.
\end{aligned}
\tag{24}
$$

Note that $\mathcal{R}(\widehat{\mathbf{S}}, \mathbf{s}, \sigma, \widehat{\mathbf{D}}, \delta)$ does not depend upon the displacements and the electric potential. By taking the variation of $\mathcal{H}$ with respect to $\mathbf{v}_0$, $\mathbf{V}$, w_0, $\boldsymbol{\varpi}$ and ψ_0, we obtain

$$
\begin{aligned}
&\mathbf{T}' + \mathbf{B}_0 = \mathbf{0}, \quad \mathbf{C}' + \boldsymbol{\Sigma} + \mathbf{B}_1 = \mathbf{0}, \quad N' + \Xi_0 = 0, \quad \text{on } \mathcal{L}, \\
&\mathbf{M}' + \mathbf{T} + \boldsymbol{\Xi}_1 = \mathbf{0}, \quad \Delta' + Q = 0, \quad \text{on } \mathcal{L},
\end{aligned}
\tag{25}
$$

and

$$
\begin{aligned}
&\int_{\mathcal{B}_b} (\mathbf{s} - \hat{\mathbf{f}}) = \mathbf{0}, \quad \int_{\mathcal{B}_b} (\mathbf{s} - \hat{\mathbf{f}}) \otimes \mathbf{r} = \mathbf{0}, \\
&\int_{\mathcal{B}_b} (\sigma - \varphi) = 0, \quad \int_{\mathcal{B}_b} (\sigma - \varphi)\mathbf{r} = \mathbf{0}, \quad \int_{\mathcal{B}_\beta} (\delta - \chi) = 0.
\end{aligned}
\tag{26}
$$

Here $\mathbf{T}$ is the shear force on a cross-section and $\mathbf{B}_0$ the in-plane body force per unit length. The skew-symmetric part of the 2×2 matrix $\mathbf{C}$ represents the torsional moment and its symmetric part the moment of double forces. The 2×2 symmetric matrix $\boldsymbol{\Sigma}$ gives the in-plane double forces, the skew-symmetric part of the 2×2 matrix $\mathbf{B}_1$ equals the applied torque per unit length, and the symmetric part of $\mathbf{B}_1$ equals the double force/length. The total axial force on a cross-section and the axial body force/length are represented by N and Ξ_0 respectively. Furthermore, $\mathbf{M}$ is the bending moment, $\boldsymbol{\Xi}_1$ the applied bending moment/length, Δ the resultant axial electric displacement, and Q the resultant applied charge per unit length. For $\mathcal{B}_a = \emptyset$, the boundary conditions $(26)_1$ and $(26)_2$ can be expressed in terms of resultant fields defined on the end faces of the rod, a similar remark applies to the case when $\mathcal{B}_\alpha = \emptyset$. Equations (26) describe the boundary conditions on that part of the end faces where mechanical and electric displacements are not prescribed. For $\mathcal{B}_b = \mathcal{B}$, and $\mathcal{B}_\beta = \mathcal{B}$, these equations will give boundary conditions for $\mathbf{T}$, $\mathbf{C}$, N, $\mathbf{M}$ and Δ.

In order for the rod theory to mimic the plate theory developed above, we must choose $\widehat{\mathbf{S}}$, $\mathbf{s}$ and $\widehat{\mathbf{D}}$ such that

$$\widehat{\mathbf{S}}\boldsymbol{\nu} = \hat{\mathbf{f}}, \quad \mathbf{s} \cdot \boldsymbol{\nu} = \varphi, \quad \widehat{\mathbf{D}} \cdot \boldsymbol{\nu} = \chi \quad \text{on } \partial\mathcal{S}. \tag{27}$$

This requires that $\widehat{\mathbf{S}}$, $\mathbf{s}$ and $\widehat{\mathbf{D}}$ be functions of $\mathbf{r}$, the shape of the cross-section, and the distributions of $\hat{\mathbf{f}}$, φ and χ on the boundary of a cross-section. Note that the effects of $\hat{\mathbf{f}}$, φ and χ are included in the balance equations (25) through $\mathbf{B}_0$, $\mathbf{B}_1$, Ξ_0, $\boldsymbol{\Xi}_1$, Φ_0, $\boldsymbol{\Phi}_1$, Q and X. Here we assume that $\widehat{\mathbf{S}}$, $\mathbf{s}$, σ, $\widehat{\mathbf{D}}$ and δ are polynomials in $\mathbf{r}$. For a rod made of a linear elastic material, σ and ε, and δ and ω must be polynomials of the same degree in $\mathbf{r}$. However, the degree of polynomials of $\widehat{\mathbf{S}}$, $\mathbf{s}$ and $\widehat{\mathbf{D}}$ needs to be determined so that $(24)_{11}$–$(24)_{16}$ and the following boundary conditions (28) on $\partial\mathcal{S}$ are satisfied:

$$\int_{\partial\mathcal{S}} \widehat{\mathbf{S}}\boldsymbol{\nu} = \mathbf{0}, \qquad \int_{\partial\mathcal{S}} \mathbf{s} \cdot \boldsymbol{\nu} = 0, \qquad \int_{\partial\mathcal{S}} \widehat{\mathbf{D}} \cdot \boldsymbol{\nu} = 0. \tag{28}$$

Equations $(28)_1$ and $(28)_2$ imply that the resultant of tractions acting on the boundary of every cross-section of the rod vanishes. Similarly, the resultant of normal electric displacement on the boundary of a cross-section of the rod is null. In order for equation $(28)_2$ to be satisfied, $\mathbf{s}$ must be a polynomial of degree 5 in $\mathbf{r}$. However, most classical rod theories assume that $\mathbf{s}$ is an affine function of $\mathbf{r}$ and ignore boundary condition $(28)_2$. We discuss the classical choice below, and the other one in the Appendix. We take

$$\begin{aligned}
\widehat{\mathbf{S}}(\mathbf{r}, \zeta) &= \frac{\boldsymbol{\Sigma}(\zeta)}{A}, \\
\mathbf{s}(\mathbf{r}, \zeta) &= \frac{\mathbf{T}(\zeta)}{A} + \mathbf{C}(\zeta)\mathbf{J}^{-1}\mathbf{r}, \\
\sigma(\mathbf{r}, \zeta) &= \frac{N(\zeta)}{A} + \mathbf{J}^{-1}\mathbf{r} \cdot \mathbf{M}(\zeta), \\
\widehat{\mathbf{D}}(\mathbf{r}, \zeta) &= \mathbf{0}, \qquad \delta(\mathbf{r}, \zeta) = \frac{\Delta(\zeta)}{A},
\end{aligned} \tag{29}$$

where A is the area of cross-section and

$$\mathbf{J} = \int_{\mathcal{S}} \mathbf{r} \otimes \mathbf{r} \tag{30}$$

is the inertia tensor.

The variation of $\mathcal{H}$ with respect to $\boldsymbol{\Sigma}$, $\mathbf{T}$, N, Δ, $\mathbf{M}$ and $\mathbf{C}$ gives

$$\begin{aligned}
&\begin{Bmatrix} \widehat{\mathbf{E}}_0 \\ \boldsymbol{\gamma}_0 \\ \varepsilon_0 \\ \omega_0 \end{Bmatrix} = \frac{1}{A} \begin{bmatrix} \mathbb{F}_{ES} & \mathbb{F}_{Es} & \mathbb{F}_{E\sigma} & \mathbb{M}_{E\delta} \\ * & \mathbb{F}_{\gamma s} & \mathbb{F}_{\gamma\sigma} & \mathbb{M}_{\gamma\delta} \\ * & * & \mathbb{F}_{\varepsilon\sigma} & \mathbb{M}_{\varepsilon\delta} \\ -* & -* & -* & \mathbb{N}_{W\delta} \end{bmatrix} \begin{Bmatrix} \boldsymbol{\Sigma} \\ \mathbf{T} \\ N \\ \Delta \end{Bmatrix}, \\
&\boldsymbol{\varepsilon}_1 = \mathbf{J}^{-1}\mathbb{F}_{\varepsilon\sigma}\mathbf{M} + \mathbf{J}^{-1}\mathbf{C}^{\mathrm{T}}\mathbb{F}_{\gamma\sigma}, \\
&\boldsymbol{\Gamma} = \mathbb{F}_{\gamma s}\mathbf{C}\mathbf{J}^{-1} + (\mathbb{F}_{\gamma\sigma} \otimes \mathbf{M})\mathbf{J}^{-1},
\end{aligned} \tag{31}$$

and

$$\int_{\mathcal{B}_a} (\mathbf{v}_0 - \bar{\mathbf{v}}) = \mathbf{0}, \qquad \int_{\mathcal{B}_a} (\mathbf{V} - \mathbf{J}^{-1}\bar{\mathbf{v}} \otimes \mathbf{r}) = \mathbf{0},$$
$$\int_{\mathcal{B}_a} (w_0 - \bar{w}) = 0, \qquad \int_{\mathcal{B}_a} (\boldsymbol{\varpi} - \bar{w}\mathbf{r}) = \mathbf{0}, \qquad \int_{\mathcal{B}_a} (\psi_0 - \bar{\psi}) = 0. \tag{32}$$

Balance laws (25), constitutive relations (31), boundary conditions (26) and (32), and congruence relations (22) form a complete set of equations for a rod problem. Note that there is no constitutive coupling between the two groups of fields $(\boldsymbol{\Sigma}, \mathbf{T}, N, \Delta)$ and $(\mathbf{M}, \mathbf{C})$. However, they are coupled through the balance laws (25). For $\mathcal{B}_b = \emptyset$ and $\mathcal{B}_\beta = \emptyset$, equations (32) reduce to boundary conditions for $\mathbf{v}_0$, $\mathbf{V}$, w_0, $\boldsymbol{\varpi}$ and ψ_0.

5. Examples

5.1. PLATE-LIKE BODY

5.1.1. *Bending Rigidities of a Transversely Isotropic Plate*

We consider transversely isotropic bodies with the axis of transverse isotropy along the unit vector **a**, and calculate the bending and torsional rigidities of a plate as a function of the angle θ between the unit vectors **a** and **e**. Recall that the unit vector **e** is along the normal to the midsurface of the plate. For simplicity we use rectangular Cartesian coordinate axes with the x_3-axis coincident with the unit vector **e** or the transverse direction, the vector **a** in the $(x_1$–$x_3)$-plane and surfaces $\mathcal{S}^+$ and $\mathcal{S}^-$ free of surface tractions and surface charges. The inversion of equation $(16)_2$ gives **M** and **C** in terms of $\widehat{\mathbf{E}}_1$ and $\boldsymbol{\gamma}_1$. When **a** is parallel to **e**, the matrix $\mathbb{F}$ has the following representation:

$$[\mathbb{F}] = \begin{bmatrix} -\dfrac{1}{E_1} & 0 & -\dfrac{\nu_{12}}{E_1} & 0 & 0 & -\dfrac{\nu_{13}}{E_3} \\ 0 & \dfrac{(1+\nu_{12})}{E_1} & 0 & 0 & 0 & 0 \\ -\dfrac{\nu_{12}}{E_1} & 0 & \dfrac{1}{E_1} & 0 & 0 & -\dfrac{\nu_{13}}{E_3} \\ 0 & 0 & 0 & \dfrac{1}{G_{13}} & 0 & 0 \\ 0 & 0 & 0 & 0 & \dfrac{1}{G_{13}} & 0 \\ -\dfrac{\nu_{13}}{E_3} & 0 & -\dfrac{\nu_{13}}{E_3} & 0 & 0 & \dfrac{1}{E_3} \end{bmatrix}. \tag{33}$$

Here E_1, E_2 and E_3 are the Young's moduli along the principal material axes, ν_{13} and G_{13} are the Poisson's ratio and the shear modulus, respectively. When **a** is parallel to $\mathbf{e}$, the principal material axes coincide with the coordinate axes. The 2×2 symmetric matrix $\widehat{\mathbf{E}}$ has been written as a 3×1 matrix with entries $\widehat{E}_{11}$, $\widehat{E}_{12}$ and $\widehat{E}_{22}$ in columns 1, 2 and 3, respectively. When the unit vector $\mathbf{a}$ is not parallel to the unit vector $\mathbf{e}$, i.e., $\mathbf{a}$ is not in the transverse direction, then the elements of the 3×3 matrix $\mathbb{F}$ are functions of $\cos\theta$, $\sin\theta$, E_1, E_3, ν_{12}, ν_{13} and G_{13}, and

$$\begin{aligned}\mathbb{F}(\mathbf{S}) &= \alpha_1 \mathbf{I}_\pi \mathbf{S} \mathbf{I}_\pi + \alpha_2 (\mathbf{I}_\pi \cdot \mathbf{S}) \mathbf{I}_\pi + \alpha_3 (\mathbf{PS} + \mathbf{SP}) \\ &\quad + \alpha_4 (\mathrm{tr}(\mathbf{S})\mathbf{P} + (\mathbf{P} \cdot \mathbf{S})\mathbf{I}) + \alpha_5 (\mathbf{P} \cdot \mathbf{S})\mathbf{P},\end{aligned} \tag{34}$$

where $\mathbf{P} = \mathbf{a} \otimes \mathbf{a}$, $\mathbf{I}_\pi = \mathbf{I} - \mathbf{P}$, $\mathbf{I}$ is the identity matrix, and

$$\begin{aligned}\alpha_1 &= \frac{1+\nu_{12}}{E_1}, \qquad \alpha_2 = -\frac{\nu_{12}}{E_1}, \qquad \alpha_3 = \frac{1}{G_{13}}, \qquad \alpha_4 = -\frac{\nu_{13}}{E_3}, \\ \alpha_5 &= \frac{G_{13} - 2E_3 + 2G_{13}\nu_{13}}{G_{13}E_3}.\end{aligned} \tag{35}$$

Writing

$$M_{11} = J\widehat{E}_{11} f(E_1, E_3, \nu_{12}, \nu_{13}, G_{13}, \theta), \quad J = \frac{2}{3}h^3, \tag{36}$$

we have plotted in Figure 1 the normalized f, i.e., $f(\theta)/f(0)$, as a function of θ for $E_3/E_1 = \eta = 0.3$, 0.5 and 0.9, and $\nu_{12} = \nu_{13} = 0.3$, $G_{13} = E_1/2(1+\nu_{13})$. It is clear that for η far away from 1.0, $f(\theta)$ depends strongly upon θ because of the inclusion in the plate model of double forces without moments. For $\theta = 0$, the flexural rigidities D_{11}, D_{22} and D_{12} and the torsional rigidity G_{12} are given by

$$D_{11} = D_{22} = \frac{JE_1}{(1-\nu_{12}^2)}, \qquad D_{12} = \frac{JE_1\nu_{12}}{1-\nu_{12}^2}, \qquad G_{12} = \frac{JE_1}{1+\nu_{12}}, \tag{37}$$

and for $\theta = 90°$,

$$\begin{aligned}D_{11} &= \frac{JE_3^2}{(E_3 - E_1\nu_{13}^2)}, \qquad D_{22} = \frac{JE_1E_3}{(E_3 - E_1\nu_{13}^2)}, \\ D_{12} &= \nu_{13}D_{22}, \qquad G_{12} = JG_{13}.\end{aligned} \tag{38}$$

Figure 2 depicts the value of θ as a function of η for which D_{11} is maximum. For a rather large range of values of η, the value of θ that optimizes D_{11} varies between 30° and 34°. For $\theta \neq 0$, several components of the 3×3 flexural rigidity matrix are nonvanishing.

5.1.2. *Cylindrical Deformations of a Transversely Isotropic Plate*

We now use equations (13) to analyze cylindrical deformations of a homogeneous transversely isotropic plate with only equal and opposite charges applied to the top and bottom surfaces. The plate is assumed to be infinite in one of the in-plane directions, and either both edges are clamped and grounded, or one edge is clamped

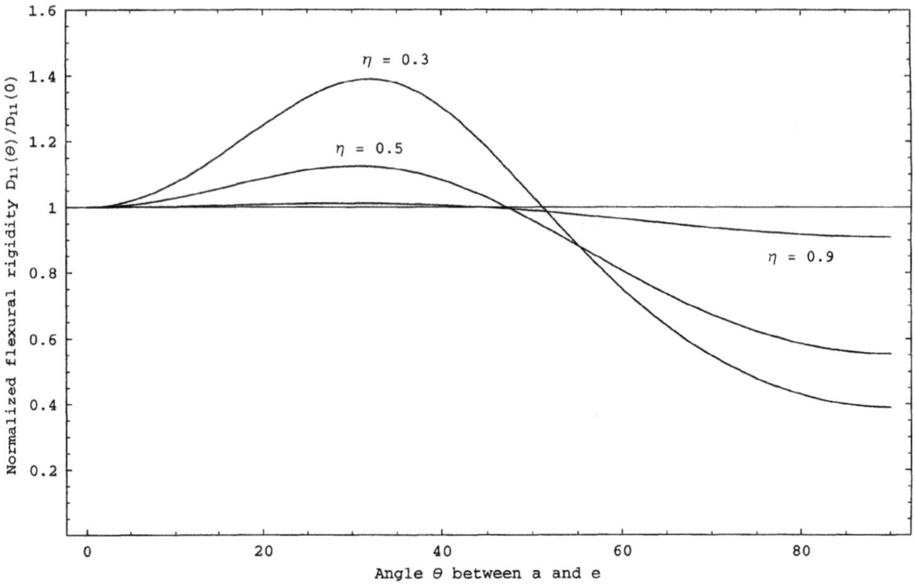

Figure 1. Normalized bending rigidity D_{11} vs. the angle θ between the axis of transverse isotropy and the normal to the midsurface of the plate.

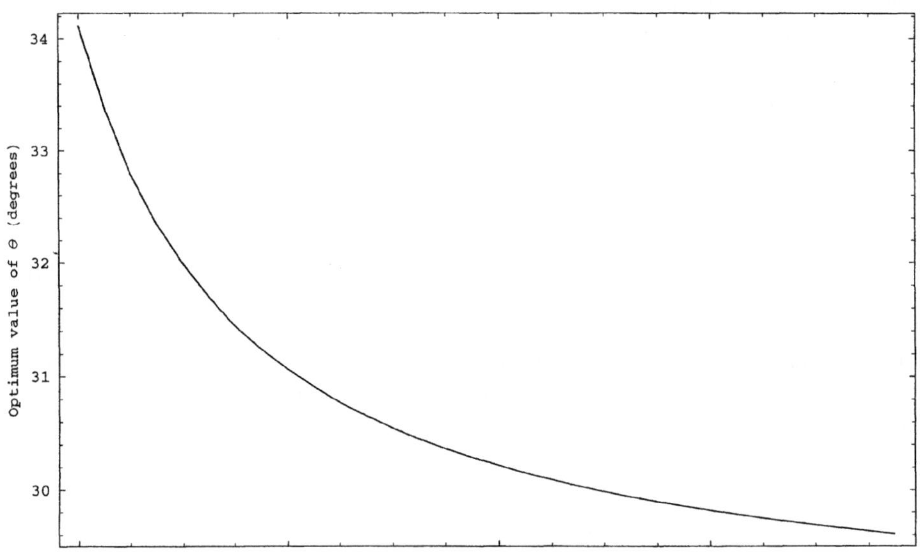

Figure 2. Optimum value of the angle θ between the axis of transverse isotropy and the normal to the midsurface of the plate.

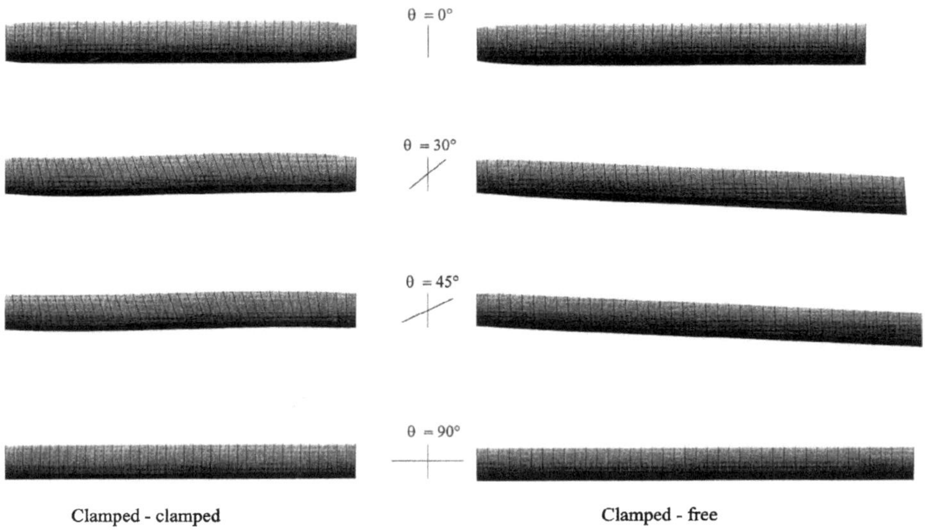

Figure 3. Deformed shapes, for $\theta = 0°, 30°, 45°$ and $90°$, of the clamped–clamped and the clamped-free plate subjected to equal and opposite charges on the top and bottom surfaces.

and grounded and the other is mechanically and electrically free. There are no body forces and body charges. For this problem, equations (9) simplify to

$$\begin{aligned} v(r,\zeta) &= v_0(r) + \zeta v_1(r), \qquad w(r,\zeta) = w_0(r) + \zeta w_1(r), \\ \psi(r,\zeta) &= \psi_0(r) + \zeta \psi_1(r), \end{aligned} \tag{39}$$

where r is the distance measured from the clamped edge of the plate. A numerical solution of equations (13), (16), (17), (10) and (39) was found by the finite element method. Values of material parameters used to compute results are given below in equations (48) for $\theta = 45°$; those for $\theta \neq 45°$ can be obtained by using the appropriate transformation rules. For a surface charge density of 10^{12} C/m^2 and the span to thickness ratio of 5 Figure 3 exhibits the deformed shapes of the plates for $\theta = 0$, 30°, 45° and 90°. The vertical lines indicate the deformed normals to the midsurface of the plate. Figures 4–9 depict the variations of v_0, v_1, w_0, w_1, ψ_0' and ψ_1 along the span of the plate. For $\theta = 0°$ and 90° we recover classical results showing coupling between the electrical and the mechanical effects. That is, the thickness of the plate changes and there is no transverse shear deformation of the plate. For $\theta = 30°$ and 45°, there is significant bending and shearing of the plate but the changes in the thickness are small as compared to those for $\theta = 0°$ and 90°. A considerable part of the energy used to change the thickness of the plates for $\theta = 0°$ and 90° is consumed to bend and shear the plates when $\theta = 30°$ and 45°. For $\theta = 30°$ and 45°, the in-plane electric field exhibits a boundary-layer effect near the clamped edges. However, for $\theta = 0°$ and 90°, the in-plane electric field is essentially zero all over the midsurface of the plate. The electric field in the

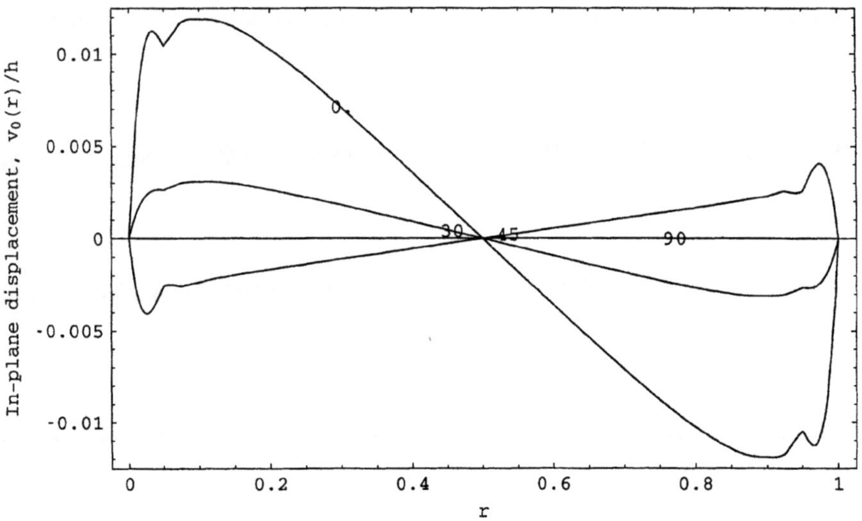

(a)

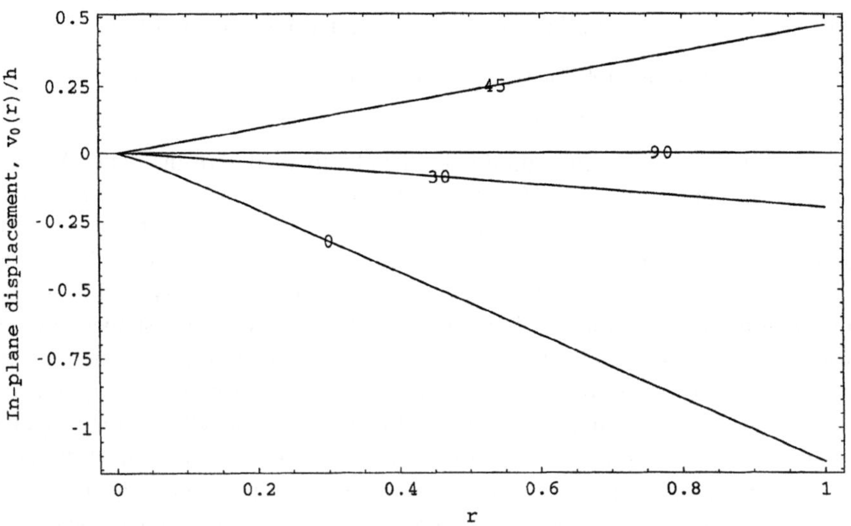

(b)

Figure 4. Variation of the in-plane displacement, v_0, along the span of the plate; (a) clamped–clamped, (b) clamped-free.

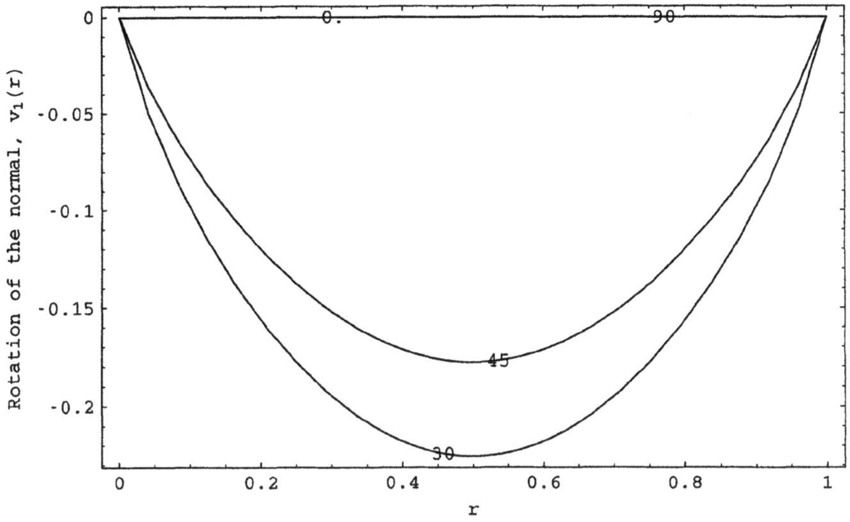

(a)

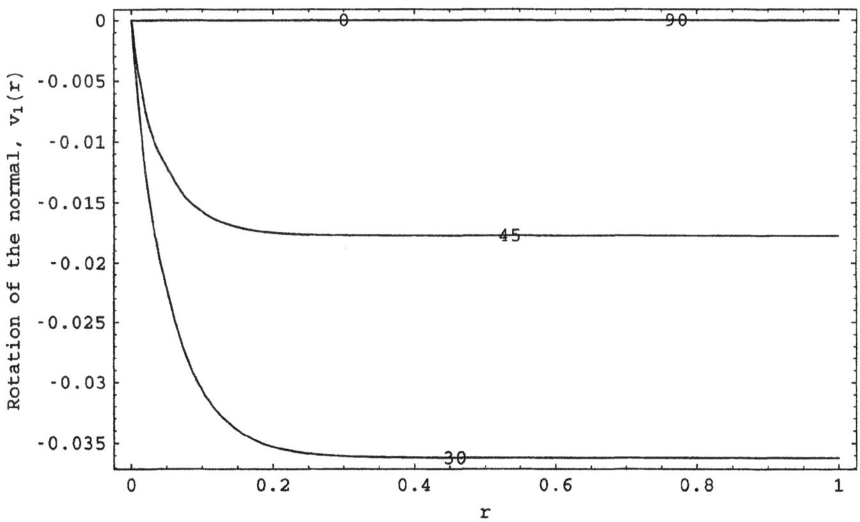

(b)

Figure 5. Variation along the span of the rotation about the long edge of the plate; (a) clamped–clamped, (b) clamped-free.

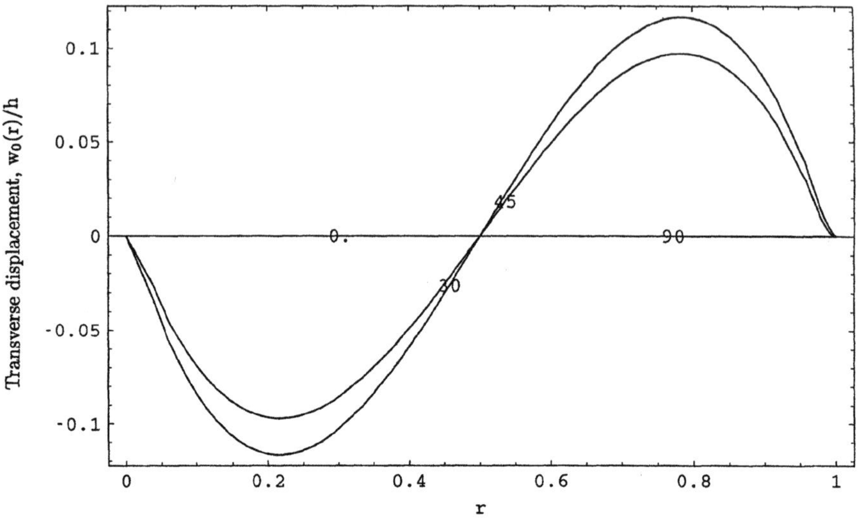

(a)

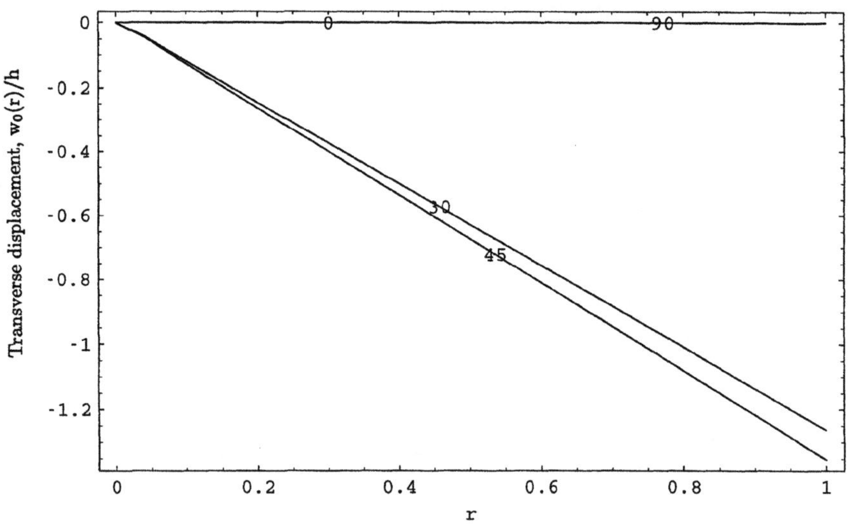

(b)

Figure 6. Variation along the span of the transverse displacement of points on the midsurface of the plate; (a) clamped–clamped, (b) clamped-free.

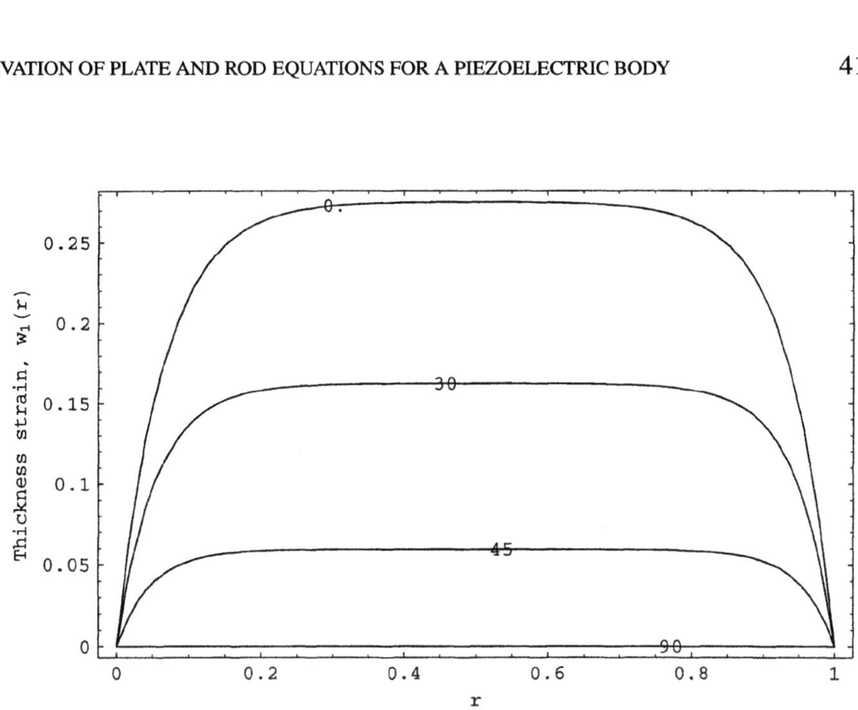

(a)

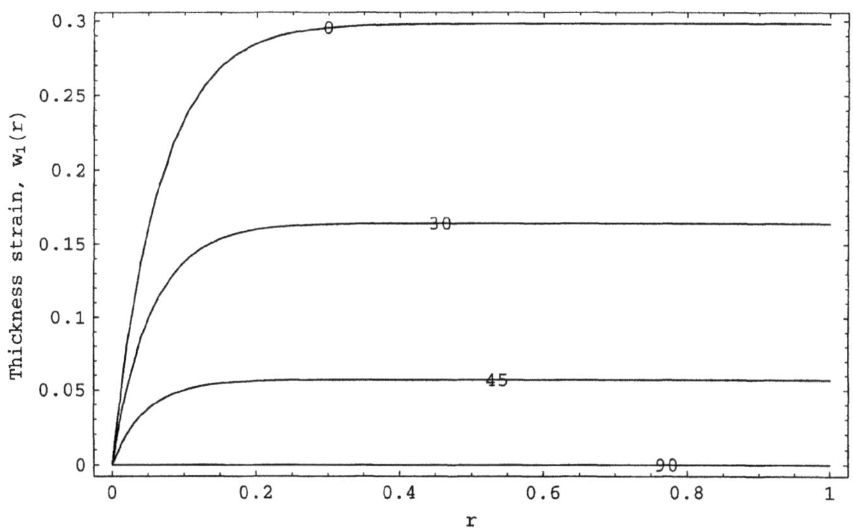

(b)

Figure 7. Variation of the transverse normal strain along the span of the plate; (a) clamped–clamped, (b) clamped-free.

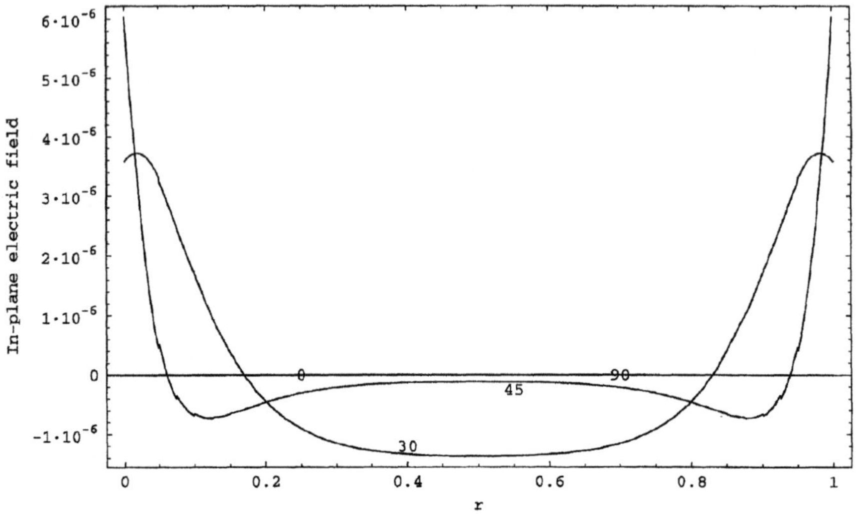

(a)

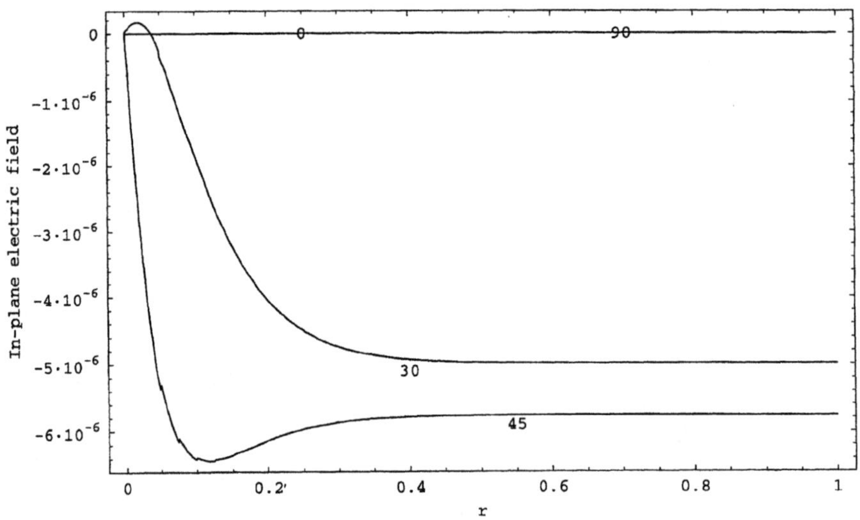

(b)

Figure 8. Variation along the span of the in-plane electric field on the midsurface of the plate; (a) clamped–clamped, (b) clamped-free.

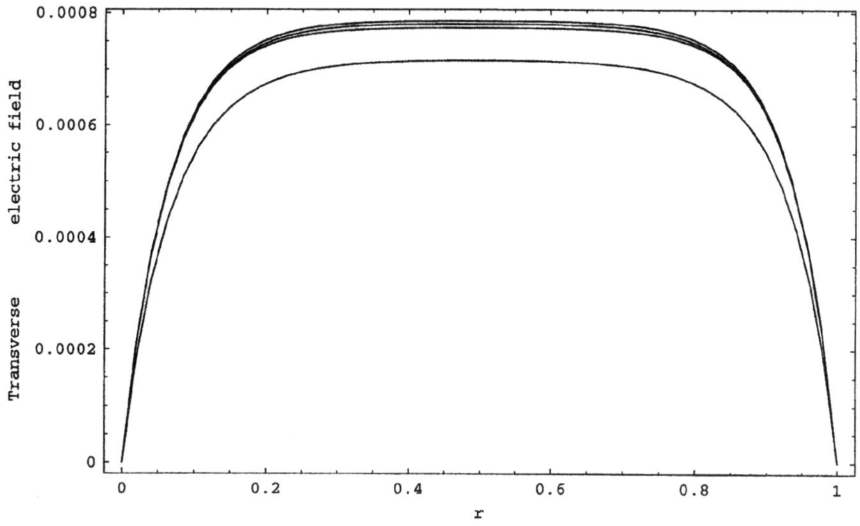

(a)

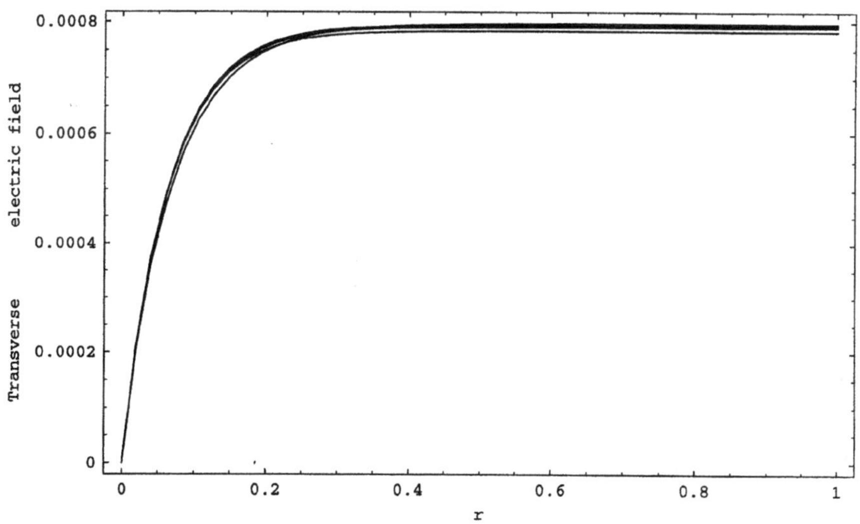

(b)

Figure 9. Variation of the transverse electric field along the span of the plate; (a) clamped–clamped, (b) clamped-free.

thickness direction exhibits a boundary layer effect near the clamped end for all values of θ. There is no boundary layer effect near the free edge of the plate.

5.2. ROD-LIKE BODY

We study electromechanical deformations of a circular cylindrical rod subjected to only electric charges at the end faces. Thus, we set

$$\begin{array}{lllll} \mathcal{B}_a = \mathcal{S}^-, & \mathcal{B}_b = \mathcal{S}^+, & \mathcal{B}_\alpha = \mathcal{S}^-, & \mathcal{B}_\beta = \mathcal{S}^+, & \\ \hat{\mathbf{b}} = \mathbf{0}, & \beta = 0, & q = 0 & \text{in } \mathcal{P}, & \\ \hat{\mathbf{f}} = 0, & \varphi = 0, & \text{on } \mathcal{M} \cup \mathcal{B}_b, & \chi = 0, & \text{on } \mathcal{M}, \\ \chi = \bar{\chi}, & \text{on } \mathcal{B}_\beta, & & & \\ \bar{\mathbf{v}} = \mathbf{0}, & \bar{w} = 0, & \bar{\psi} = 0, & \text{on } \mathcal{S}^-. & \end{array} \tag{40}$$

That is, the end face $\mathcal{S}^-$ is rigidly clamped and kept at zero electric potential while the end face $\mathcal{S}^+$ is traction free and carries an electric charge. The mantle of the rod is free of surface tractions and surface charges. Also, the body force and the charge density vanish in $\mathcal{P}$. Corresponding to (40) the boundary conditions become

$$\mathbf{T}(h) = 0, \quad \mathbf{C}(h) = 0, \quad N(h) = 0, \quad \mathbf{M}(h) = 0, \quad \Delta(h) = \int_{\mathcal{S}^+} \bar{\chi} = X. \tag{41}$$

The solution of the equilibrium equations (25) under the boundary conditions (41) for $\mathbf{T}$, N, $\mathbf{M}$ and Δ is

$$\mathbf{T}(\zeta) = \mathbf{0}, \qquad N(\zeta) = 0, \qquad \mathbf{M}(\zeta) = \mathbf{0}, \qquad \Delta(\zeta) = X. \tag{42}$$

However, the fields $\mathbf{C}$ and $\mathbf{\Sigma}$ can not be found from the equilibrium equations $(25)_2$ because the number of unknowns exceeds the number of equations. The constitutive relations (31) and equations (42) give

$$\begin{aligned} \widehat{\mathbf{E}}_0 &= \mathrm{SYM}\mathbf{V} = \frac{1}{A}(\mathbb{F}_{ES}\mathbf{\Sigma} + \mathbb{M}_{E\delta}X), \\ \mathbf{\Gamma} &= \mathbf{V}' = \mathbb{F}_{\gamma s}\mathbf{C}\mathbf{J}^{-1} \end{aligned} \tag{43}$$

which when substituted in $(25)_2$ with $\mathbf{B}_1 = \mathbf{0}$, and in $(26)_2$, yield the following equations for the determination of $\mathbf{V}$:

$$\left(\mathbb{F}_{\gamma s}^{-1}\mathbf{V}'\mathbf{J}\right)' + A\mathbb{F}_{ES}^{-1}\mathrm{SYM}\mathbf{V} = \mathbb{F}_{ES}^{-1}\mathbb{M}_{E\delta}X, \qquad \mathbf{V}'(h) = 0, \qquad \mathbf{V}(-h) = \mathbf{0}. \tag{44}$$

Equation $(44)_3$ follows from $(32)_2$.

We now assume that the rod is of uniform circular cross-section, the material properties do not vary in the axial direction, $\mathbf{J}$ is the spherical tensor, $\mathbf{J} = j\widehat{\mathbf{I}}$, and introduce the non-dimensionalization $x = (\zeta + h)/2h$. Here $\widehat{\mathbf{I}}$ is the 2×2 identity tensor. With these assumptions, equations (44) become

$$\mathbf{V}^{**} + \frac{4Ah^2}{j}\mathbb{F}_{\gamma s}\mathbb{F}_{ES}^{-1}\mathrm{SYM}\mathbf{V} = \frac{4h^2}{j}\mathbb{F}_{\gamma s}\mathbb{F}_{ES}^{-1}\mathbb{M}_{E\delta}X, \quad \mathbf{V}(0) = 0, \quad \mathbf{V}^*(1) = 0, \tag{45}$$

where * (not to be confused with $*$ in (6)) denotes the derivative with respect to x. Knowing $\mathbf{V}$, and using equations (22) and (31), we can find fields $\boldsymbol{\varpi}$, $\mathbf{v}_0$, w_0 and ψ_0 from

$$
\begin{aligned}
\boldsymbol{\varpi}(x) &= \mathbf{V}^{\mathrm{T}}(x)\mathbb{F}_{\gamma s}^{-1}\mathbb{F}_{\gamma\sigma}, \\
\mathbf{v}_0(x) &= \int_0^x \Big[\mathbb{F}_{Es}^{\mathrm{T}}\mathbb{F}_{ES}^{-1}\mathrm{SYM}\mathbf{V}(\xi) - \mathbf{V}^{\mathrm{T}}(\xi)\mathbb{F}_{\gamma s}^{-1}\mathbb{F}_{\gamma\sigma} \\
&\quad + \frac{1}{A}(\mathbb{M}_{\gamma\delta} - \mathbb{F}_{Es}^{\mathrm{T}}\mathbb{F}_{ES}^{-1}\mathbb{M}_{E\delta})X\Big]\mathrm{d}\xi, \\
w_0(x) &= \int_0^x \Big[\mathbb{F}_{E\sigma}^{\mathrm{T}}\mathbb{F}_{ES}^{-1}\mathrm{SYM}\mathbf{V}(\xi) + \frac{1}{A}(\mathbb{M}_{\varepsilon\delta} - \mathbb{F}_{E\sigma}^{\mathrm{T}}\mathbb{F}_{ES}^{-1}\mathbb{M}_{E\delta})X\Big]\mathrm{d}\xi, \\
\psi_0(x) &= \int_0^x \Big[\frac{1}{A}(\mathbb{N}_{\omega\delta} + \mathbb{M}_{E\delta}^{\mathrm{T}}\mathbb{F}_{ES}^{-1}\mathbb{M}_{E\delta})X - \mathbb{M}_{E\delta}^{\mathrm{T}}\mathbb{F}_{ES}^{-1}\mathrm{SYM}\mathbf{V}(\xi)\Big]\mathrm{d}\xi.
\end{aligned}
\tag{46}
$$

Using the following representation for $\mathbf{V}$

$$
[\mathbf{V}] = V_0\begin{bmatrix} 1 & 0 \\ 0 & -1 \end{bmatrix} + V_1\begin{bmatrix} 1 & 0 \\ 0 & 1 \end{bmatrix} + V_t\begin{bmatrix} 0 & 1 \\ 1 & 0 \end{bmatrix} + V_x\begin{bmatrix} 0 & 1 \\ -1 & 0 \end{bmatrix}, \tag{47}
$$

for a circular rod, we have plotted in Figure 10 the variation of V_0, V_1, V_x and V_t with $x \in [0, 1]$. The wavelengths of V_x and V_t depend upon the angle between the centroidal axis of the bar and the axis of transverse isotropy, and also on the aspect ratio (i.e., diameter of the cross-section/length of the bar). In order to plot results in Figure 10, various material (PZT-5) and geometric parameters were assigned the following values in MKS units (e.g., see [16])

$$
\begin{aligned}
[\mathbb{F}] &= 10^{-12} \\
&\times \begin{bmatrix}
24.299 & 0 & -5.32533 & -11.8846 & 0 & -0.925055 \\
0 & 34.762 & 0 & 0 & 9.53798 & 0 \\
-5.32533 & 0 & 14.4336 & -4.76207 & 0 & -5.32533 \\
-5.9423 & 0 & -2.38104 & 29.2625 & 0 & -5.9423 \\
0 & 9.53798 & 0 & 0 & 34.762 & 0 \\
-0.925055 & 0 & -5.32533 & -11.8846 & 0 & 24.299
\end{bmatrix}, \\
[\mathbb{M}] &= 10^{-18}\begin{bmatrix}
-4750 & -38100 & 4750 \\
-19100 & 0 & -19100 \\
8080 & 0 & -8080 \\
12800 & 0 & -12800 \\
19100 & 0 & 19100 \\
-4750 & 38100 & 4750
\end{bmatrix}, \\
[\mathbb{N}] &= 10^{-18}\begin{bmatrix}
65.9 & 0 & -0.576 \\
0 & 65.3 & 0 \\
-0.576 & 0 & 65.9
\end{bmatrix}
\end{aligned}
\tag{48}
$$

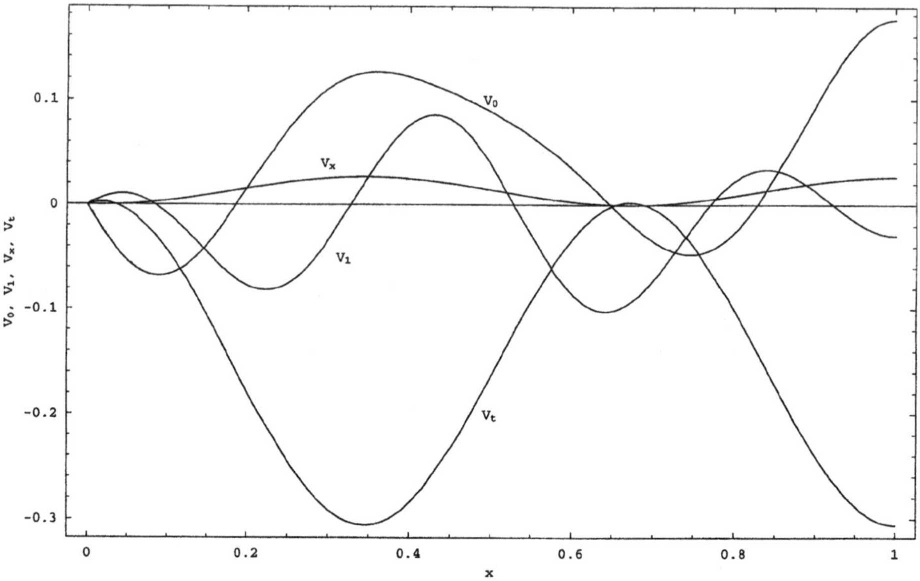

Figure 10. Variation of the twisting and in-plane deformations of a cross-section of the rod along its centroidal axis.

- aspect ratio = 0.4,
- angle between the centroidal axis and the axis of transverse isotropy = 45°,
- total charge applied at one end face = 10^{12}.

In the representation (47) of $\mathbf{V}$, the deformation $\mathbf{V}_0$ transforms a circle into an ellipse, $\mathbf{V}_1$ is a spherical deformation or dilatation of the cross-section, $\mathbf{V}_t$ accounts for the shearing deformation of the cross-section, and $\mathbf{V}_x$ gives the twisting of the cross-section. It is clear from the results plotted in Figure 10 that the deformation of the cross-section due to its being twisted is small, and that due to shearing is quite large.

Figure 11 depicts the deformed shapes of the cross-section at twenty locations along the axis of the bar. Due to different deformations of these cross-sections, a prismatic bar is deformed into a non-prismatic one. By varying the material properties along the axis of the bar, one should be able to optimize a desired mode of deformation. These material properties could be varied by changing the symmetry group of the material.

We have plotted in Figure 12 the axial displacement of points on the centroidal axis of the bar. The dotted curve is for the case when each cross-section of the bar is assumed to be rigid, i.e., $\mathrm{sym}\mathbf{V} = \mathbf{0}$, and the solid curve for deformable cross-sections. It is clear that the axial displacement of a point increases significantly when deformations of a cross-section are accounted for. The difference between the axial displacements of a point for deformable and rigid cross-sections increases noticeably for points away from the fixed end-face. Thus, the electrostrictive effect

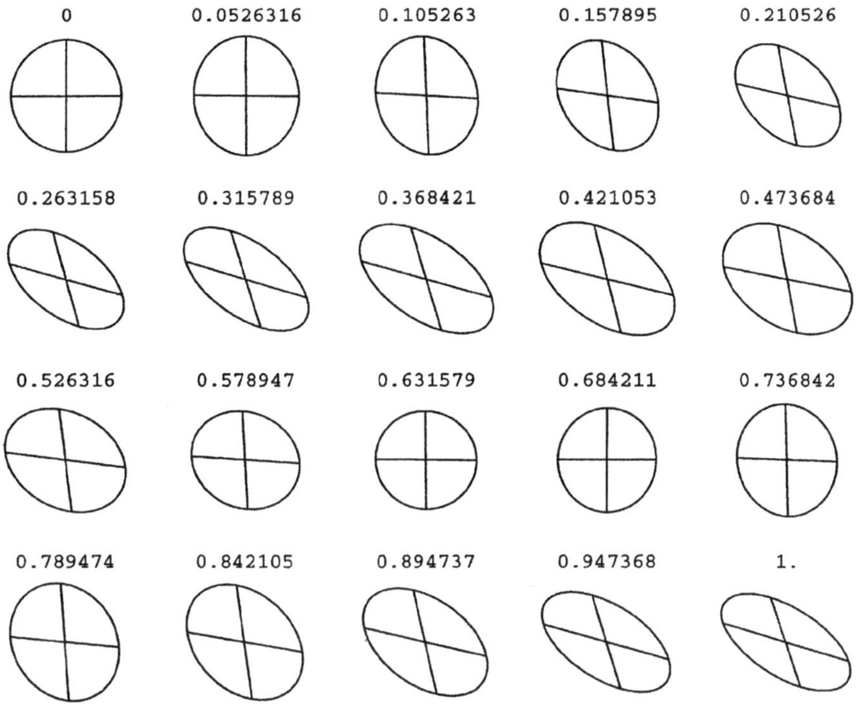

Figure 11. Deformed shapes of a cross-section at different locations on the centroidal axis of the rod.

can be significantly modified by adjusting the angle between the direction **a** of transverse isotropy and the axis **e** of the bar. However, this increase in elongation is accompanied by the simultaneous bending and twisting of the bar.

6. Conclusions

We have derived governing equations, constitutive relations, and boundary conditions for quasistatic deformations of an anisotropic plate-like and an anisotropic rod-like body by using a mixed variational principle of Yang and Batra for a piezoelectric body. The transverse shear deformations of a plate and the changes in its thickness are accounted for. Similarly, deformations of a cross-section of the rod due to double forces without moments applied at its mantle are considered.

We have also computed the bending rigidities of a homogeneous transversely isotropic plate as a function of the angle θ between the axis **a** of transverse isotropy and the unit vector **e** normal to the mid-surface of the plate. It is found that for a large range of the values of E_T/E_L, the value of θ that optimizes D_{LL} varies between 30° and 34°. Here subscripts L and T signify longitudinal and transverse directions, respectively. The plate equations are employed to study the cylindrical

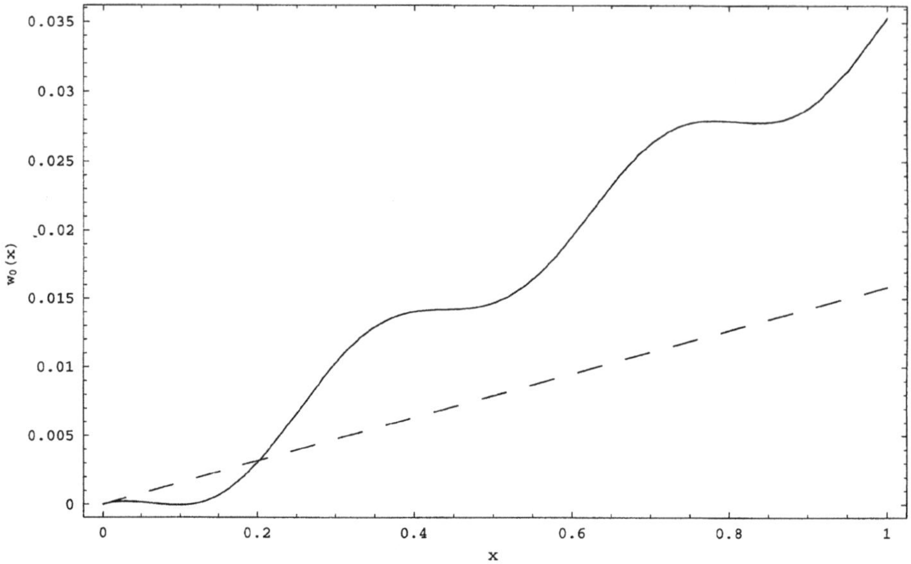

Figure 12. Axial displacements of points on the centroidal axis of the rod.

deformations of a transversely isotropic plate with either both edges clamped and grounded or one edge clamped and grounded and the other mechanically and electrically free. Only electric charges are applied to the top and bottom surfaces of the plate. The angle between the axis of transverse isotropy and the unit normal to the midsurface of the plate strongly influences the transverse deformations of the plate.

We have used the rod equations to analyze deformations of a homogeneous and transversely isotropic bar loaded on one end face by surface charges only with the other end face rigidly clamped and kept at zero electric potential. The mantle of the rod is assumed to be free of surface tractions and surface charges. When the axis of transverse isotropy makes an angle of 45° with the centroidal axis of the rod, the deformations of the cross-section significantly influence the axial elongation of points on the centroidal axis. The axial elongation is simultaneously accompanied by the bending and twisting of the rod.

Acknowledgement

This work was partially supported by the ARO grant DAAG55-98-1-0030 and the NSF grant CMS9713453 to Virginia Polytechnic Institute and State University.

Appendix

The lowest order polynomial in $\mathbf{r}$ that simultaneously satisfies equations $(24)_{14}$ and $(28)_2$ is

$$\mathbf{s} = \mathbf{s}_0 + \mathbf{C}\big[a_1\mathbf{H}_2\mathbf{r} + a_2\mathbf{H}_3(\mathbf{r}\otimes\mathbf{r}) + a_3\mathbf{H}_4(\mathbf{r}\otimes\mathbf{r}\otimes\mathbf{r}) \\ + a_4\mathbf{H}_5(\mathbf{r}\otimes\mathbf{r}\otimes\mathbf{r}\otimes\mathbf{r}) + a_5\mathbf{H}_6(\mathbf{r}\otimes\mathbf{r}\otimes\mathbf{r}\otimes\mathbf{r}\otimes\mathbf{r})\big], \tag{A.1}$$

where $\mathbf{s}_0$ is independent of $\mathbf{r}$,

$$\mathbf{H}_k = \mathbf{J}_k^{-1}, \quad \mathbf{J}_k = \int_{\mathcal{S}} \underbrace{\mathbf{r}\otimes\mathbf{r}\cdots\otimes\mathbf{r}}_{k \text{ times}},$$

and in rectangular Cartesian coordinates with summation implied on repeated indices,

$$\big[\mathbf{H}_3(\mathbf{r}\otimes\mathbf{r})\big]_i = H_{3ijk}r_jr_k, \quad H_{ijkl}J_{jklm} = \delta_{im}. \tag{A.2}$$

Substitution from (A1) into $(24)_{14}$ and $(28)_2$ gives

$$\begin{aligned} &a_1 + a_2 + a_3 + a_4 + a_5 = 1, \\ &a_1 \operatorname{tr}(\mathbf{CH}_2) + 3a_3 \operatorname{tr}(\mathbf{CH}_4\mathbf{J}_2) + 4a_4 \operatorname{tr}(\mathbf{CH}_5\mathbf{J}_3) + 5a_5 \operatorname{tr}(\mathbf{CH}_6\mathbf{J}_4) = 0. \end{aligned} \tag{A.3}$$

Equation $(\text{A}.3)_2$ yields four equations for four different choices of the 2×2 matrix $\mathbf{C}$ which together with $(\text{A}.3)_1$ determine $a_1, a_2, \ldots, a_5$. Setting $\mathbf{s}_0 = \mathbf{T}(\xi)/A$ in (A.1), substituting this expression for $\mathbf{s}$ in $\mathcal{H}$, and taking the variations with respect to $\mathbf{C}$, we obtain

$$\begin{aligned} &\widehat{\mathbf{E}}_0 = \widehat{\mathbf{E}}_{op} + \frac{1}{A}\mathbb{F}_{Es}\mathbf{Ck}, \qquad \boldsymbol{\gamma}_0 = \boldsymbol{\gamma}_{op} + \frac{1}{A}\mathbb{F}_{\gamma s}\mathbf{Ck}, \quad \varepsilon_0 = \varepsilon_{op} + \frac{\mathbb{F}_{\gamma\sigma}^{\mathrm{T}}}{A}\mathbf{Ck}, \\ &\omega_0 = \omega_{op} + \frac{1}{A}\mathbb{M}_{\gamma\delta}^{\mathrm{T}}\mathbf{Ck}, \\ &\varepsilon_1 = \mathbf{J}^{-1}\big(\mathbb{F}_{\varepsilon\sigma}\mathbf{M} + \mathbf{C}^{\mathrm{T}}\mathbb{F}_{\gamma\sigma}\big), \\ &\boldsymbol{\Gamma} = \frac{1}{A}\big(\mathbb{F}_{Es}^{\mathrm{T}}\boldsymbol{\Sigma} + \mathbb{F}_{\gamma s}\mathbf{T} + \mathbb{F}_{\gamma\sigma}N + \mathbb{M}_{\gamma\delta}\Delta\big)\otimes\mathbf{k} \\ &\qquad + (\mathbb{F}_{\gamma\sigma}\otimes\mathbf{M})\mathbf{J}^{-1} + \mathbb{F}_{\gamma s}\mathbf{CK}, \end{aligned} \tag{A.4}$$

where

$$\mathbf{k} = a_2\mathbf{H}_3\mathbf{J}_2 + a_3\mathbf{H}_4\mathbf{J}_3 + a_4\mathbf{H}_5\mathbf{J}_4 + a_5\mathbf{H}_6\mathbf{J}_5, \tag{A.5}$$

and $\mathbf{K}$ is the following 2×2 matrix:

$$\begin{aligned} \mathbf{K}_{ij} &= a_1^2\mathbf{H}_{2ij} + 2a_1(1-a_1)\mathbf{H}_{2ij} \\ &+ a_2(a_2\mathbf{H}_{3ihk}\mathbf{J}_{4hklm}\mathbf{H}_{3lmj} + \cdots + a_5\mathbf{H}_{6ihklmn}\mathbf{J}_{7hklmnpq}\mathbf{H}_{3pqj}) \\ &+ a_3(a_2\mathbf{H}_{3ihk}\mathbf{J}_{5hklmn}\mathbf{H}_{4lmnj} + \cdots + a_5\mathbf{H}_{6ihklmn}\mathbf{J}_{8hklmnpqr}\mathbf{H}_{4pqrj}) \\ &+ a_4(a_2\mathbf{H}_{3ihk}\mathbf{J}_{6hklmnp}\mathbf{H}_{5lmnpj} + \cdots + a_5\mathbf{H}_{6ihklmn}\mathbf{J}_{9hklmnpqrs}\mathbf{H}_{5pqrsj}) \\ &+ a_5(a_2\mathbf{H}_{3ihk}\mathbf{J}_{7hklmnpq}\mathbf{H}_{6lmnpqj} + \cdots + a_5\mathbf{H}_{6ihklmn}\mathbf{J}_{10hklmnpqrst}\mathbf{H}_{6pqrstj}). \end{aligned} \tag{A.6}$$

Expressions for $\widehat{\mathbf{E}}_{op}$, $\boldsymbol{\gamma}_{op}$, etc. are identical to those of $\widehat{\mathbf{E}}_0$, $\boldsymbol{\gamma}_0$, etc. and are given as equations (31).

We note that for a rod made of an isotropic material, equation $(A.4)_3$ becomes

$$2\mu A\boldsymbol{\gamma}_0 = \mathbf{T} + \mathbf{C}\mathbf{k},$$

where we have used $\boldsymbol{\gamma}_{op} = \mathbf{T}/2\mu A$ for an isotropic body. If $\mathbf{V}(\xi)$ in equation (21) is a skew-symmetric matrix, then the cross-section of the rod undergoes rigid rotations and hence is not deformed, and $\mathbf{s}$ in (A.1) could be an affine function of $\mathbf{r}$ and still satisfy $(24)_{14}$ and $(28)_2$.

References

1. R.D. Mindlin, *An Introduction to the Mathematical Theory of Vibrations of Elastic Plates*. U.S. Army Signal Corps Engineering Laboratories, Fort Monmouth, NJ (1955).
2. R.D. Mindlin, Forced thickness-shear and flexural vibrations of piezoelectric crystal plates. *J. Appl. Phys.* **23** (1952) 83–88.
3. E. Reissner, On the theory of elastic plates. *J. Math. Phys.* **23** (1944) 184–191.
4. P.G. Ciarlet and P. Destuynder, A justification of the 2-dimensional linear plate model. *J. de Mécanique* **18** (1979) 315–344.
5. W.T. Koiter and J.G. Simmonds, Foundations of shell theory. In: *Proc. IUTAM Congress*, Moscow (1972) pp. 150–176.
6. P.M. Naghdi, The theory of shells and plates. In: C. Truesdell (ed.), *Handbuch der Physik*, Vol. VIa/2. Springer, Berlin (1972).
7. S.S. Antman, The theory of rods. In: C. Truesdell (ed.), *Handbuch der Physik*, Vol. VIa/2. Springer, Berlin (1972).
8. A.W. Leissa, Recent research in plate vibrations, 1981-85, Part I: Classical theory; Part II: Complicating effects. *The Shock and Vibration Digest* **19**(2) (1987) 11–18; **19**(3) (1987) 10–24.
9. E. Cosserat and F. Cosserat, Sur la statique de la ligne deformable. *Compt. Rend.* **145** (1907) 1409–1412.
10. J.L. Ericksen, Plane infinitesimal waves in homogeneous elastic plates. *J. Elasticity* **3** (1973) 161–167.
11. P. Podio-Guidugli, An exact derivation of the thin plate equations. *J. Elasticity* **22** (1989) 121–133.
12. V. Nicotra and P. Podio-Guidugli, Piezoelectric plates with changing thickness. *J. Structural Control* **5** (1998).
13. L. Teresi and A. Tiero, On variational approaches to plate models. *Meccanica* **32** (1997) 143–156.
14. J.S. Yang and R.C. Batra, Mixed variational principles in nonlinear piezoelectricity. *Internat. J. Non-Linear Mech.* **30** (1995) 719–726.
15. S.S. Vel and R.C. Batra, Analytical solution for rectangular thick laminated plates subjected to arbitrary boundary conditions. *AIAA J.* **37** (1999) 1464–1473.
16. N.N. Rogacheva, *The Theory of Piezoelectric Shells and Plates*. CRC Press, Boca Raton, FL (1994).

Journal of Elasticity **59:** 51–65, 2000.

Geometric Effects in an Elastic Tensegrity Structure

I.J. OPPENHEIM[1] and W.O. WILLIAMS[2]
[1]*Depts. of Civil and Environmental Engineering and Architecture, Carnegie Mellon University. E-mail: ijo@andrew.cmu.edu*
[2]*Dept. of Mathematical Sciences, Carnegie Mellon University. E-mail: wow@cmu.edu*

Received 14 January 1999; in revised form 26 January 2000

Dedicated to Roger Fosdick on the occasion of his 60th birthday.

Abstract. Tensegrity structures are under-constrained, 3-dimensional, self-stressing structural systems. They demonstrate an infinitesimal flex and when loaded they display a nonlinear geometric stiffening. In earlier work many examples of the resulting force–displacement relationship have been demonstrated numerically, and some aspects of the force–displacement relationship have been derived analytically. In this article an energy formulation is presented for the case of a simple but representative tensegrity structure, yielding an exact solution for the force–displacement relationship. The solution makes understandable the different appearance of the force–displacement relationship when comparing a system at zero prestress to one at high prestress, or when comparing a system with almost-inextensible members to one with highly extensible members. The exact solution also is offered as a benchmark against which numerical solutions should be tested. Furthermore, the formulation and the solution reveal conditions of asymmetry of response that have not been noted previously.

Mathematics Subject Classifications (2000): 52C25, 70B10, 70B15, 73K05.

Key words: tensegrity, structural stability, elastic structures.

1. Introduction

1.1. A BASIC TENSEGRITY STRUCTURE

Following the pattern of sculptures by Kenneth Snelson in 1948, in 1961 Buckminster Fuller patented a class of cable-bar assemblages which he called tensegrity structures. Figure 1 pictures a well-known example of such a structure constructed with three bars and six cables. Tensegrity structures have attracted interest recently because of their potential advantages as lightweight structures, as variable geometry trusses, as shape adaptive structures, and as deployable structures. The structure pictured represents fully the distinctive features of tensegrity structures.

It is a form-finding structure, an under-constrained structural system, prestressible while displaying an infinitesimal flexure even when the constituent elements are undeformable. It forms only at nodal geometries in which the statics matrix

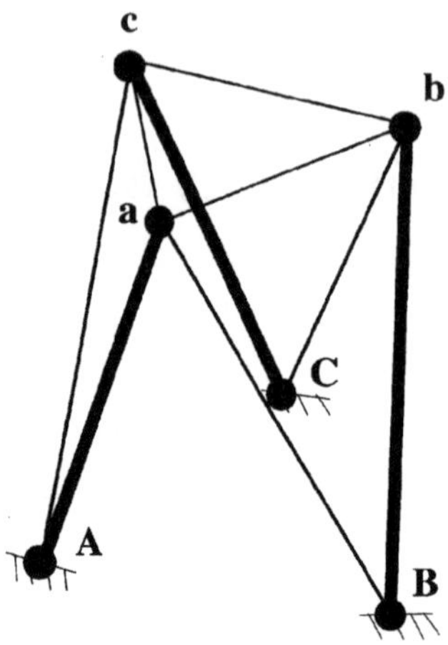

Figure 1. Simple tensegrity structure.

becomes rank deficient. The mechanics of the structure at that equilibrium geometry, in the absence of external loads, is treated in various sources and will not be repeated here. The general principles of the equilibrium mechanics [1, 9, 10, 14] are based on the properties of the statics matrix and apply only in an infinitesimal neighborhood of the equilibrium geometry. Those principles have been discussed in application to this particular tensegrity structure [7], an equilibrium analysis of this same structure can be found within a comprehensive text on under-constrained structures [5], and this same example has been treated in a recent article [4] that addresses questions of nonlinear response and control.

If constructed with 9 rigid bars, at a generic nodal geometry, a truss with the topology of Figure 1 would be a statically determinate rigid body. It would neither admit a prestress nor admit any nodal movement whatsoever. As a consequence, if any bar were replaced by a cable, the position would become unstable. However, at a tensegrity geometry such a structure admits a prestress and also displays an infinitesimal flexure, that is, a infinitesimal nodal displacement which does not require lengths of the structural elements to change. The prestress places six of the bars in tension, allowing them to be replaced by cables, as shown in Figure 1, without affecting either the infinitesimal flexibility or the stability. Greater (finite) nodal movement then is observed to require some cable elongation but no cable shortening, thereby explaining physically why the equilibrium position is a stable one.

In an actual tensegrity structure constructed with almost-inextensible members (such as steel cables) the infinitesimal flex is plainly evident (and consists in a simultaneous rotation and elevation change of the upper triangle). Following the nomenclature of [5], nodal loadings in the direction of the infinitesimal flex are termed perturbation loads, and nodal loadings orthogonal thereto are termed equilibrium loads. While equilibrium loads can be resolved directly by statics, resolution of perturbation loads requires nodal displacements, mobilizing the (nonlinear) geometric stiffening. The main contribution of this article is the derivation and examination of an exact solution for the geometric stiffening, the nonlinear force–displacement relationship displayed by the tensegrity structure when subject to perturbational loadings.

In the case just described the geometric stiffening effect is pronounced. In the close vicinity of the equilibrium geometry the structure is quite soft, but the stiffness increases dramatically with finite, but relatively small, displacement. However, we note that many laboratory examples are purposely constructed using highly extensible members such as rubber cords, because such assemblies allow both graceful form-finding and easy demonstration of shape changing. In such cases it may be difficult, from casual observation, to distinguish the infinitesimal flex and its nonlinear geometric stiffening from the large nodal displacements resulting from elastic effects. Nonetheless, analyzing any response to nodal forces requires the inclusion of the nonlinear geometric stiffening effect.

Characterization of tensegrity positions of an arbitrary assemblage of bars and cables is a difficult problem (cf. [14]). Here we shall be concerned with the behavior of our structure from an established tensegrity position.

1.2. THE TENSEGRITY POSITION AND MOTIONS

The problem will be examined for a tensegrity structure as pictured in Figure 1, further simplified by assumptions of symmetry and uniformity. The structure is taken as a right regular prism, in which the end faces are equilateral triangles, perpendicular to the axis joining their centroids. The tensegrity position, shown in [2] to be globally unique, then is known to require a $5\pi/6$ relative rotation between the triangles.

The three bars are assumed to be rigid, and the three cables comprising the edges of the end face inextensible, while the three diagonal cables are linearly elastic with equal moduli. Any applied force will be required to be symmetric about the axis of the prism. The result of these simplifications is that the triangular faces remain parallel and centered on the z-axis, so that all nodal displacements can be described by the angle of relative rotation or by the concomitant vertical displacement. The infinitesimal flexure of the system is an infinitesimal rotation of the upper triangle about its axis.

1.3. FORCE–DISPLACEMENT RELATIONS: PREVIOUS WORK

It has been observed that the initial stiffness in the direction of the infinitesimal flex depends upon the prestress; that if the system is assembled without prestress then the initial stiffness would be zero, whereas prestress creates an initial stiffness proportionate in magnitude. It is also understood that the stiffness increases as the structure is displaced along the direction of flex. This effect is of practical interest. In some instances the variation in stiffness with displacement is a feature that may prove useful in controlling the compliance of a structure and in controlling vibration (cf. [8]). Therefore, several investigators have addressed the question of generating the nonlinear force–displacement relationship resulting when such a structure is subjected to perturbation loads.

In principle, performing such analyses numerically should be fairly straightforward. Systematic methods have been outlined [5, 9] for doing so in general. Skelton and coworkers in [12, 13] present extensive results for structures with multistage topology. Within that work they arrive at equations of motion relating nodal forces and nodal displacements by using a Lagrangian formulation and show, in [11], numerically generated plots, recording such force–displacement relationships. Similar numerical analysis is contained in work recently described by Murakami [6] and by Djouadi et al. [4] describing multistage variants of structures based upon the topology depicted in Figure 1. The authors [7] have shown results obtained using a commercial finite element analysis package, applied to a structure with almost-inextensible members. Those numerical results show the initial stiffening effect of prestress and then the dramatic geometric stiffening effect with displacement.

2. Kinematics, Elasticity and Equilibrium

We consider the structure shown in Figure 1. The nodes A, B, and C are pinned to ground, the legs Aa, Bb, and Cc are inextensible bars, the elements ab, bc, and ca are inextensible cables, while the cross-cables Ac, Ba, and Cb are taken to be linearly elastic. Considering only the cross-cables as elastic simplifies the calculations, but captures the essence of the elastic response of such systems.

Because of the inextensibility, the upper triangle moves as a rigid body, and hence its position is easily described by the location of one point and a rotation matrix. We assume here that

- The triangles ABC and abc are congruent equilateral triangles.
- The legs have the same length, L.
- The cross-cables are identical in length and in elastic modulus.

As we will see, this complete symmetry leads to simple calculations.

We set the origin of coordinates at the centroid of the lower triangle, and locate its three vertices by the vectors $\mathbf{p}_A$, $\mathbf{p}_B$, and $\mathbf{p}_C$ in the x–y plane; for definiteness, we take $\mathbf{p}_A = a\,\mathbf{i}$, to be directed in the x-coordinate direction. We take $\mathbf{x}$ to be the vector locating the centroid of the upper triangle, so that the locations of the upper points are given as

$$\begin{aligned} \mathbf{p}_a &= \mathbf{x} + Q\mathbf{p}_A, \\ \mathbf{p}_b &= \mathbf{x} + Q\mathbf{p}_B, \\ \mathbf{p}_c &= \mathbf{x} + Q\mathbf{p}_C, \end{aligned} \tag{2.1}$$

where Q is the rotation matrix. Because of the symmetry of our system we have that Q is a rotation of angle ϕ (counterclockwise) about the z-axis, and $\mathbf{x} = h\mathbf{k}$. Thus, for example,

$$\mathbf{p}_a = h\,\mathbf{k} + a\cos\phi\,\mathbf{i} - a\sin\phi\,\mathbf{j}. \tag{2.2}$$

The feasible range of ϕ is from $-\pi/3$, where the cross-cables interfere, to π, where the legs interfere.

Now the constraint that the legs are of fixed length L becomes

$$(\mathbf{p}_a - \mathbf{p}_A)^2 = \left(h\mathbf{k} + a(\cos\phi - 1)\mathbf{i} - a\sin\phi\mathbf{j}\right)^2 = L^2 \tag{2.3}$$

or

$$h^2 = L^2 - 2a^2(1 - \cos\phi). \tag{2.4}$$

Next, we calculate the length λ of each cross-cable from

$$\lambda^2 = (\mathbf{p}_c - \mathbf{p}_A)^2 = (h\mathbf{k} + Q\mathbf{p}_C - \mathbf{p}_A)^2, \tag{2.5}$$

and since $\mathbf{p}_C = a(-(1/2)\mathbf{i} + (\sqrt{3}/2)\mathbf{j})$, we arrive at

$$\begin{aligned} \lambda^2 &= h^2 + 2a^2 + a^2\left(\cos\phi - \sqrt{3}\sin\phi\right) \\ &= L^2 + a^2\left(3\cos\phi - \sqrt{3}\sin\phi\right), \end{aligned} \tag{2.6}$$

where we have used (2.4). For later use, we calculate the derivatives:

$$2\lambda\lambda' = a^2\left(-3\sin\phi - \sqrt{3}\cos\phi\right), \tag{2.7}$$

$$2\lambda\lambda'' = a^2\left(\sqrt{3}\sin\phi - 3\cos\phi\right) - 2\left(\lambda'\right)^2. \tag{2.8}$$

If k is the spring constant, and λ_N is the **natural length** of the cross-cables, then the total elastic energy is

$$U(\phi) = \frac{3}{2}k\left[(\lambda - \lambda_N)^2\right] \tag{2.9}$$

(when $\lambda \geqslant \lambda_N$; otherwise it is zero). The applied moment or torque associated to a rotation ϕ is then the derivative of the energy with respect to ϕ:

$$M(\phi) = U'(\phi) = 3k(\lambda - \lambda_N)\lambda'(\phi), \tag{2.10}$$

and the equilibria of the unloaded system occur when this is equal to zero. This occurs when $\lambda = \lambda_N$ or when $\lambda' = 0$. The latter occurs at

$$\phi = -\frac{\pi}{6} \quad \text{or} \quad \phi = \frac{5\pi}{6}. \tag{2.11}$$

We note that an increment in ϕ from either position results in an *infinitesimal flexure*, which neither lengthens nor shortens any bar or cable.

Continuing,

$$U''(\phi) = 3k\lambda(\phi)^2 - 3k(\lambda - \lambda_N)\lambda''(\phi), \tag{2.12}$$

and it is easy to see that the point $\phi = -\pi/6$ always yields a maximum of energy, so that we can disregard it.*

On the other hand $\phi = 5\pi/6$, the classical *tensegrity placement* for the symmetric structure, yields

$$U''\left(\frac{5\pi}{6}\right) = 3k\lambda'\left(\frac{5\pi}{6}\right)^2 = 3\sqrt{3}ka^2\left(1 - \frac{\lambda_N}{\lambda_0}\right). \tag{2.13}$$

Here we have introduced the **equilibrium length** of the cross-cables:

$$\lambda_0 = \lambda\left(\frac{5\pi}{6}\right) = \sqrt{L^2 - 2\sqrt{3}a^2}. \tag{2.14}$$

If $\lambda_0 < \lambda_N$, then the cables would be fully slack in this placement, and it follows that the position is unstable [3]; henceforth, we assume that $\lambda_0 \geqslant \lambda_N$. If $\lambda_0 > \lambda_N$, the second derivative of the energy is positive, and we confirm that the tensegrity placement is a local minimum of the energy (and a minimum of the cross-cable length λ).

The point where $\lambda = \lambda_N$, if it is attainable and if it is not the tensegrity placement, corresponds to a position in which the structure has zero prestress and the cables may be further shortened by rotation. This is unstable [3]. On the other hand, if $\lambda = \lambda_N$ *coincides* with the tensegrity placement, the cable cannot be further shortened by rotation. Both second and third derivatives of the energy are zero in this case, but the fourth derivative is positive, and hence this again is a local minimum of energy, confirming its stability. This corresponds to the case of zero pre-stress in the tensegrity placement.

3. Response to Applied Loads

To continue in a symmetric geometry, we can consider only symmetric loadings. For convenience, we re-express all functions in terms of the angular deflection θ away from the equilibrium placement:

$$\phi = \theta + \frac{5}{6}. \tag{3.1}$$

We find

$$\lambda(\theta) = \sqrt{L^2 - 2\sqrt{3}a^2 \cos\theta}, \tag{3.2}$$

* In fact, one can show that this position would require compression in the edges of the upper triangle, and hence it is not a tensegrity position.

and

$$M(\theta) = 3\sqrt{3}ka^2\left[1 - \frac{\lambda_N}{\lambda(\theta)}\right]\sin\theta. \tag{3.3}$$

Note that $M(\theta)$ is *anti-symmetric* about $\theta = 0$.

3.1. APPLIED TORQUE

Consider first an applied torque, generated, for example, by equal tangential loads applied at each vertex. The derivative of the torque-angle relation (3.3) is

$$M'(\theta) = 3\sqrt{3}\,ka^2\left(1 - \frac{\lambda_N}{\lambda}\right)\cos\theta + 9ka^4\left(\frac{\lambda_N}{\lambda^3}\right)\sin^2\theta, \tag{3.4}$$

and thus the initial modulus of the elastic response is

$$M'(0) = 3\sqrt{3}ka^2\left(1 - \frac{\lambda_N}{\lambda_0}\right) = 3\sqrt{3}a^2\left(\frac{k(\lambda_0 - \lambda_N)}{\lambda_0}\right). \tag{3.5}$$

Note that the modulus is a linear function of the prestress

$$k(\lambda_0 - \lambda_N), \tag{3.6}$$

so that, in particular, the initial modulus is zero if there is no prestress.

The second derivative of the relation is

$$\begin{aligned} M''(\theta) &= -\frac{9\lambda_N a^4(\sin(\theta))^3}{2\lambda(\theta)^{5/2}} + \frac{3\lambda_N a^2\sqrt{3}\sin(\theta)\cos(\theta)}{2\lambda(\theta)^{3/2}} \\ &\quad -\frac{1}{2}\left(1 - \frac{\lambda_N}{\lambda(\theta)}\right)\sin(\theta), \end{aligned} \tag{3.7}$$

which is zero at $\theta = 0$ regardless of the value of the prestress.

Figures 2, 3 show plots of $M(\theta)$ for a structure with $L/a = 18/7$ for various choices of prestress (prestrains, $(\lambda_0 - \lambda_N)/\lambda_N$, from 0% to 300%). Figure 2 shows the response over a small range of rotations for zero prestress and for a small prestress (prestrain of 0.2%). Figure 3 shows the response under several prestresses over the feasible range $-7\pi/6 < \theta < \pi/6$. Note, in particular, that the curvature for small positive θ is negative for small prestresses, but positive for large prestresses, and for the latter the response appears much more nearly linear. The curvature sign-change occurs at a prestrain of

$$\frac{L + 6a^4 + 3\sqrt{3}L^2a^2}{L^4 - 12a^4}, \tag{3.8}$$

which in this geometry is 165%.

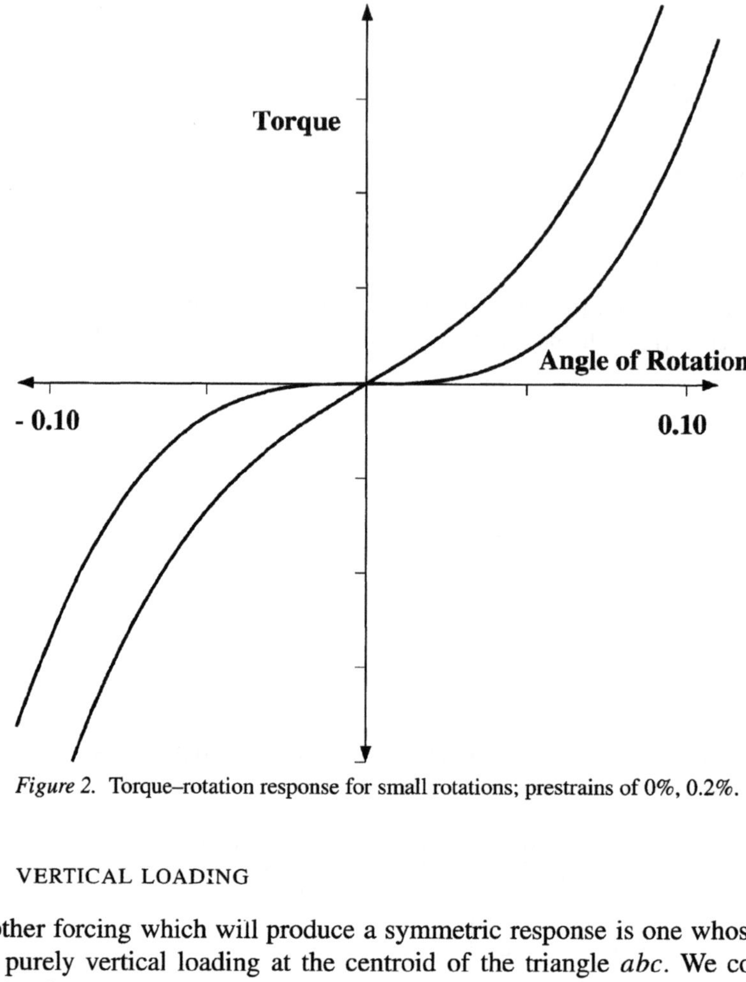

Figure 2. Torque–rotation response for small rotations; prestrains of 0%, 0.2%.

3.2. VERTICAL LOADING

Another forcing which will produce a symmetric response is one whose resultant is a purely vertical loading at the centroid of the triangle *abc*. We consider the relation between the applied force and the height h of the triangle. Since h and ϕ are related by (2.4), we can find h as a function of θ as

$$h(\theta) = \sqrt{L^2 - a^2(2 + \sqrt{3}\cos\theta + \sin\theta)}. \tag{3.9}$$

This function is invertible in the feasible domain. Using (2.4) and (2.6), we can solve to find the elastic energy as a function of h; the derivative of this function with respect to h is the force F required to displace the triangle vertically. The resulting equations are not particularly enlightening. It is easier to express this force (taking it positive if directed downwards) in terms of θ as

$$F(\theta) = -U'(\theta)\frac{\mathrm{d}\theta}{\mathrm{d}h} = M(\theta)/h'(\theta) = M(\theta)\left\{\frac{2h(\theta)}{a^2(\cos\theta - \sqrt{3}\sin\theta)}\right\}. \tag{3.10}$$

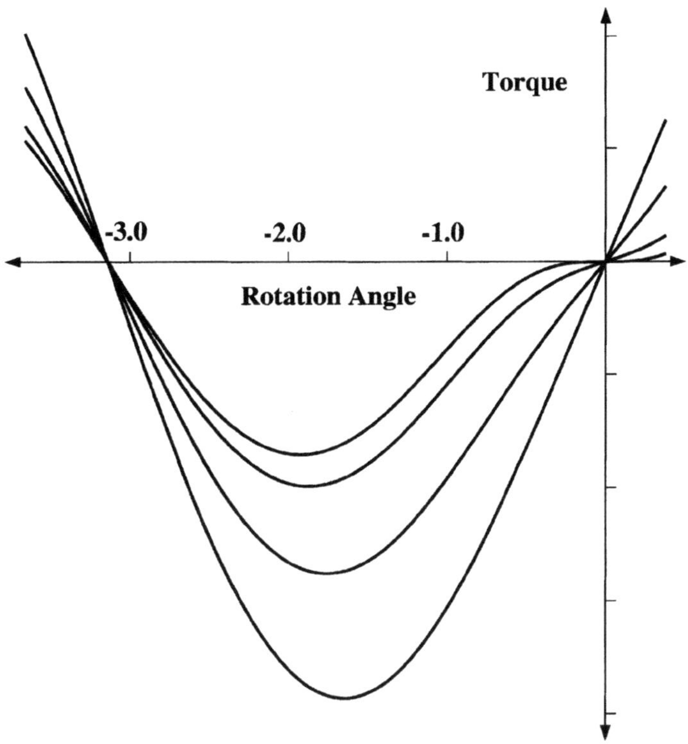

Figure 3. Torque–rotation response over feasible range; prestrains of 0%, 10%, 50%, and 300%.

This force must approach infinity as the derivative $h'(\theta)$ approaches zero, which occurs at $\theta = \pi/6$, the point where the legs interfere, and at $\theta = -5\pi/6$, the point where the legs are vertical.

Next we define the (downwards) displacement from the equilibrium position as

$$\delta(\theta) = h(0) - h(\theta) \tag{3.11}$$

and regard (3.10) and (3.11), using (3.9), as parametric equations describing the response function.

The derivatives of the force–displacement relation are calculated in these terms as

$$F'(h) = \frac{M'(\theta)}{h'(\theta)^2} - \frac{M(\theta)}{h'(\theta)^3}\big(h''(\theta)\big), \tag{3.12}$$

$$F''(h) = \frac{M''(\theta)}{h'(\theta)^3} - \frac{3M'(\theta)}{h'(\theta)^4} + M(\theta)\left[\frac{3h''(\theta)^2}{h'(\theta)^5} - \frac{h'''(\theta)}{h'(\theta)^4}\right]. \tag{3.13}$$

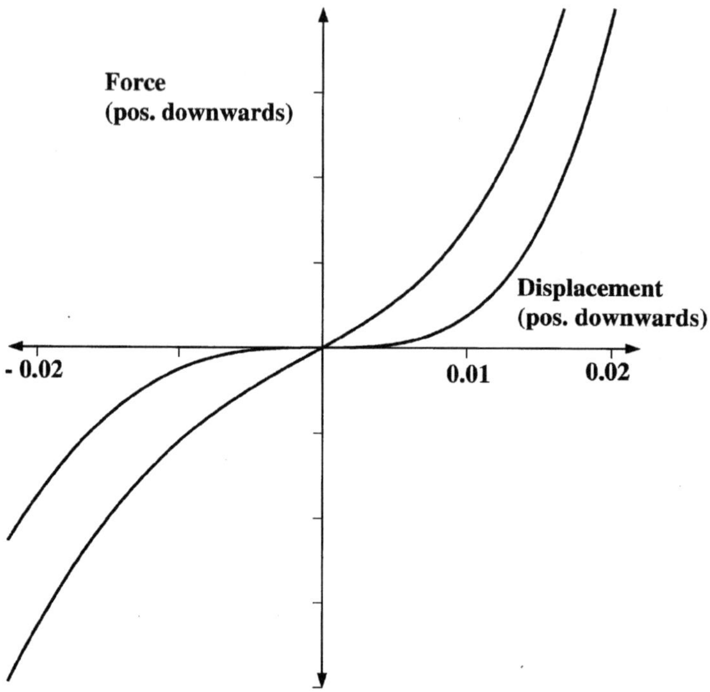

Figure 4. Force–displacement relations for a downwards force; no prestress and small prestress.

Evaluation at $\theta = 0$ gives the initial modulus and the initial curvature of the response as

$$F'(h_0) = -\frac{M'(0)}{h'(0)^2} = \frac{3\sqrt{3}}{h_0\lambda_0}k(\lambda_0 - \lambda_N), \tag{3.14}$$

$$F''(h_0) = -\frac{3M'(0)}{h'(0)^4} = -\frac{144\sqrt{3}h_0{}^4k(\lambda_0 - \lambda_N)}{a^6\lambda_0}. \tag{3.15}$$

The initial modulus is zero if there is no prestress, and is linear in the prestress, as for the previous loading. The initial curvature of the response, however, now is zero only for null prestress, and otherwise is negative and proportional to the prestress.

Figures 4–6 are graphs of F versus δ for the same values of L and a and the same assortment of prestresses as in the previous calculations. Figure 4 shows the response over a small range of displacement for the cases of zero prestress and small prestress (prestrain of 0.2%). The next two figures show more distinctly the discrepancy between this response and the "perturbation" loading previously analysed. Figure 5 shows the response over a range of θ values from -0.8 to 0.3 for various prestresses. Finally, Figure 6 shows the response over the range of θ between the infinities of F, for the zero prestress and the highest prestress (300%

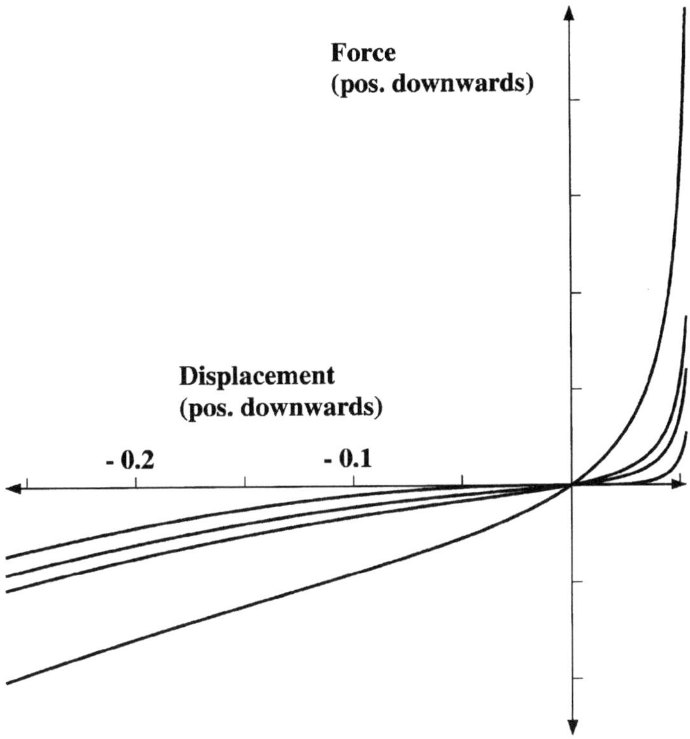

Figure 5. Force–displacement relations for a downwards force; various prestresses.

prestrain). The responses for the other prestress values are trapped between these two.

3.3. NATURAL MODE RESPONSE

Finally, to relate the current calculations to those obtained using the structural-matrix approach, we may look at the direction of infinitesimal flexure of the structure. The direction of infinitesimal flexure is that in which the structure may move with no initial extension of any of its edges; this vector is the null vector of the transpose of the structural matrix. Of course, once the structure is forced out of the tensegrity placement by an imposed force, this direction is no longer defined. To model what might be meant by a finite "displacement in the direction of the natural mode", we shall consider the path of a node of our structure in the continued symmetric deformation. We will calculate *the relation between arc-length along this path and the corresponding force directed tangential to that path.* The tangent vector to that path at the equilibrium position is the direction of infinitesimal flexure, and the corresponding force the initial forcing needed in that direction to drive the deformation.

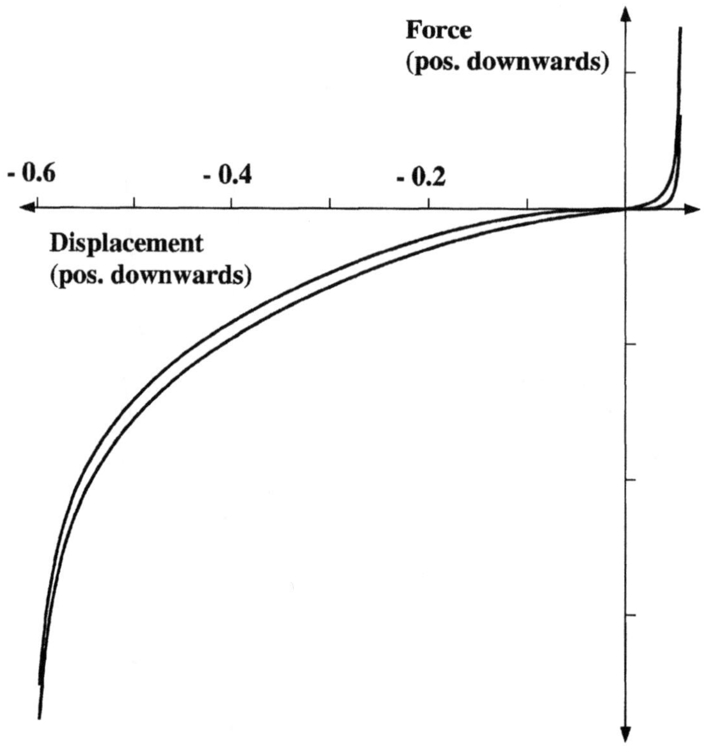

Figure 6. Force–displacement relations for a downwards force; zero and large prestress.

As a function of the angle ϕ of rotation, the position of node a is given by

$$\mathbf{p}_a(\phi) = h(\phi)\mathbf{k} + a\cos\phi\mathbf{i} - a\sin\phi\mathbf{j}, \tag{3.16}$$

where $h(\phi)$ is given by (2.4). Its derivative is

$$\mathbf{p}_a'(\phi) = a\left[-\frac{a\sin\phi}{h(\phi)}\mathbf{k} - \sin\phi\mathbf{i} - \cos\phi\mathbf{j}\right], \tag{3.17}$$

which has squared length

$$a^2 + \frac{a^4}{h(\phi)^2}\sin^2\phi. \tag{3.18}$$

Thus the arc-length along the curve is

$$s(\phi) = a\int_{5\pi/6}^{\phi}\sqrt{1 + \frac{a^4}{h(\phi)^2}\sin^2\phi\,\mathrm{d}\phi}, \tag{3.19}$$

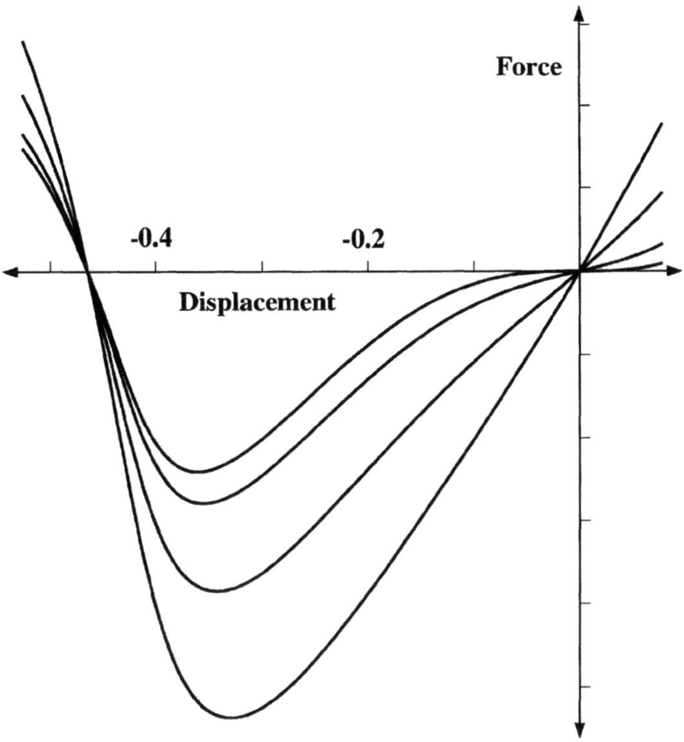

Figure 7. Force–displacement relations for a natural-mode force under various prestresses.

and we can, in principle, invert this to determine ϕ in terms of s. Then we could express the elastic energy in terms of s; its derivative with respect to s is the magnitude T of the tangential force required to drive the deformation. In this way we find

$$T(\phi) = \frac{M(\phi)}{a\sqrt{1 + (a^4/h(\phi)^2)\sin^2\phi}}. \tag{3.20}$$

Finally, we can express both T and s in term of the rotation-past-equilibrium θ. As before, we use this parameterization to plot the relation of T and s.

Figure 7 shows the response for the geometry already used, for various prestresses, verifying that the response in this mode is essentially the same as the torque-rotation response.

4. Remarks

The response of tensegrity structures has been observed to show a geometric stiffening, and we have found an analytic demonstration verifying this behavior, despite our assumption that the elastic elements are linear in response for all extensions. We hope that this solution may be useful, in particular, as a benchmark for development of numerical analysis of tensegrity structures.

The torque-rotation relation describes the fundamental response of the system. As we have noted, the initial modulus is linear in the prestress, and, since the initial value of the second derivative is zero regardless of the prestress, the graph shows an initially cubic behavior. The geometric non-linearization of response clearly is more pronounced for lower values of the prestress; for higher values the linear term dominates within the small-deformation range of motion. There is even a change in values of curvature near equilibrium (always zero *at* equilibrium) from negative to positive as the prestress becomes large. The departure from the upwards curvature form of response occurs at values of prestress which would be obtainable only in materials which could tolerate very large extensions (for the geometry we use for the plots, 165% strain); with metal cross-cables, for example, it would not be observable in practice.

As we have noted, the system may allow large rotations in the "unwinding" direction (θ negative), and the torque reaches a maximum, which can be calculated from (3.4). It is interesting that this occurs *before* the point where the cross cables are at maximum length (the point of maximum energy). As the unwinding continues, the geometry of the truss begins to allow the torque to be exercised more efficiently in counteracting the elastic forces. Continuing, one arrives at the point where the cross-cables are at their maximum length and the torque is zero. This is a point of unstable equilibrium.

It can be seen that the torque-angle loading and the natural mode loading yield essentially the same response. The natural-mode loading effects the torquing of the structure with no extraneous "equilibrium" loading, in contrast to the vertical loading or an x–y plane loading.

The vertical loading response presumably is more typical of the response of an tensegrity structure to an arbitrary applied force. The force–displacement relation is clearly distinguished from that of the natural mode loading. To begin, the response is not symmetric about zero, as in the other cases. And, although the same sort of prestress-initial modulus relation and nonlinear response occur as in the other cases, the stiffening is much more pronounced under this loading, due to the geometric relation prescribed by the constraints. There is also a distinct difference in the curvature of the response for small values of displacement. Finally, we note the approach to infinity of the response shown in Figure 6 is related to, but not directly caused by, the approach to physical interference of the structural elements.

This analysis is interestingly similar to that of a well-known model problem (cf. [1]). The system shown in Figure 8 is an under-constrained prestressible system.

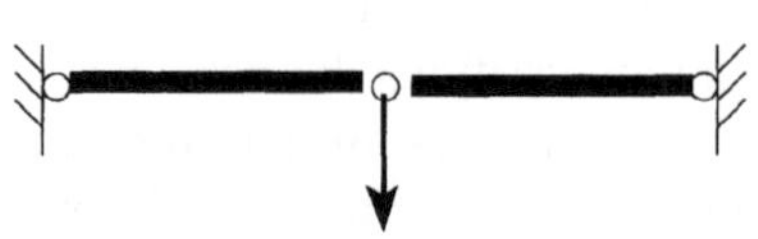

Figure 8. Two-dimensional model.

When the nodes are colinear a prestress state is possible and an infinitesimal flex is present. If the distance between the supports is $2L$, the natural length of each bar is L_N, and each bar has a spring constant of k, then it is easy to see that the force required to displace the central node by a distance x is

$$2k\left(1 - \frac{L_N}{\sqrt{L^2 + x^2}}\right)x. \tag{4.1}$$

The response shows a geometric stiffening effect, asymptotic to a linear graph. The small displacement approximation to (4.1) is

$$2k\left(1 - \frac{L_N}{L}\right)x + \frac{kL_N}{L^3}x^3 + \mathrm{o}\left(x^5\right), \tag{4.2}$$

verifying that the initial modulus is zero if the system is not prestressed, but proportional to the prestress if it exists. Our results for natural-mode loading for our tensegrity structure are remarkably similar to (4.1).

References

1. C.R. Calladine, Buckminster Fuller's "Tensegrity" strutures and Clerk Maxwell's rules for the construction of stiff frames. *Internat. J. Solids Struct.* **14** (1978) 161–172.
2. R. Connelly and M. Terrell, Globally rigid symmetric tensegrities. *Structural Topology* **21** (1995) 59–78.
3. R. Connelly and W. Whiteley, Second-order rigidity and prestress stability for tensegrity frameworks. *SIAM J. Discrete Math.* **9** (1996) 453–491.
4. S. Djouadi, R. Motro, J.C. Pons and B. Crosnier, Active control of tensegrity systems. *J. Aerospace Engrg.* **11** (1998) 37–44.
5. E.N. Kuznetsov, *Underconstrained Structural Systems*. Springer, New York (1991).
6. H. Murakami, Y. Nishimura, T.J. Impelluso and R.E. Skelton, A virtual reality-based CAD system for tensegrity structures. In: *Proc. of the 12th Engineering Mechanics Conf.*, Vol. 9 (1998) pp. 197–200.
7. I.J. Oppenheim and W.O. Williams, Tensegrity prisms as on adaptive structures. In: *Proc. of Symp. on Adaptive Structures and Material Systems* (1997) pp. 113–120.
8. I.J. Oppenheim and W.O. Williams, Vibration effects in a tensegrity prism. In: *Proc. of the 12th Engineering Mechanics Conf.* (1998) pp. 28–31.
9. S. Pellegrino, Analysis of prestressed mechanisms. *Internat. J. Solids Structures* **26** (1990) 1329–1350.
10. S. Pellegrino and C.R. Calladine, Matrix analysis of statically and kinematically indeterminate frameworks. *Internat. J. Solids Structures* **22** (1986) 409–428.
11. R.E. Skelton and R. Adhikari, An Introduction to smart tensegrity structures. In: *Proc. of the 12th Engineering Mechanics Conf.* (1998) pp. 24–27.
12. R.E. Skelton and C. Sultan, Controllable tensegrity, a new class of smart structures. In: V.V. Varadan and J. Chandra (eds), *Smart Materials and Structures*, Vol. 3039 (1997) pp. 166–177.
13. C. Sultan and R.E. Skelton, Integrated design of controllable tensegrity structures. In: *Adaptive Structures and Material Systems*, Vol. 54 (1997) pp. 27–36.
14. W. Whiteley, Tensegrity frameworks. In: H. Crapo and W. Whiteley (eds), *The Geometry of Rigid Structures* (1997).

Journal of Elasticity **59:** 67–81, 2000.

Theory for Atomic Diffusion on Fixed and Deformable Crystal Lattices

ELIOT FRIED[1] and SHAUN SELLERS[2]
[1]*Department of Theoretical and Applied Mechanics, University of Illinois at Urbana-Champaign, Urbana, IL 61801-2935, USA*
[2]*School of Mathematics, University of East Anglia, Norwich NR4 7TJ, England*

Received 15 January 1999; in revised form 1 June 1999

Dedicated to Roger Fosdick, whose fundamental and broad-reaching work in continuum thermomechanics has provided us with abundant inspiration.

Abstract. We develop a theoretical framework, for the diffusion of a single unconstrained species of atoms on a crystal lattice, that provides a generalization of the classical theories of atomic diffusion and diffusion-induced phase separation to account for constitutive nonlinearities, external forces, and the deformation of the lattice. In this framework, we regard atomic diffusion as a microscopic process described by two independent kinematic variables: (i) the atomic flux, which reckons the local motion of atoms relative to the motion of the underlying lattice, and (ii) the time-rate of the atomic density, which encompasses nonlocal interactions between migrating atoms and characterizes the kinematics of phase separation. We introduce generalized forces power-conjugate to each of these rates and require that these forces satisfy ancillary microbalances distinct from the conventional balance involving the forces that expend power over the rate at which the lattice deforms. A mechanical version of the second law, which takes the form of an energy imbalance accounting for all power expenditures (including those due to atomic diffusion and phase separation), is used to derive restrictions on constitutive equations. With these restrictions, the microbalance involving the forces conjugate to the atomic flux provides a generalization of the usual constitutive relation between the atomic flux and the gradient of the diffusion potential, a relation that in conjunction with the atomic balance yields a generalized Cahn–Hilliard equation.

1. Introduction

The diffusion of a single unconstrained atomic species on a fixed crystal lattice is commonly modeled by the Cahn–Hilliard equation (see, for example, Cahn [1–3] and Cahn and Hilliard [4–6])

$$\dot{\nu} = Div\left(D_1(\nu)Grad\,\nu - D_2(\nu)Grad\,(Div\,Grad\,\nu)\right), \tag{1.1}$$

with ν the *atomic density*, D_1 the nonnegative density-dependent *diffusivity*, and D_2 a small, nonnegative density-dependent coefficient. The first term on the right side is due to simple diffusion, whereas the second one is a higher order term that takes into account spatial inhomogeneities on a fine scale, such as those that may arise during processes involving phase separation. This equation can be derived by adjoining to the *atomic balance*

$$\dot{\nu} = -Div\, \jmath, \tag{1.2}$$

the constitutive relation

$$\jmath = -\kappa(\nu) Grad\, \mu, \tag{1.3}$$

determining the *atomic flux* as proportional to the gradient $Grad\, \mu$ of the *diffusion potential* μ through a nonnegative scalar *mobility* κ that, in general, depends on the atomic density. In *simple atomic diffusion*, the diffusion potential is given by the derivative of a constitutive response function $\hat{\psi}$ that determines the *energy density* ψ in terms of the atomic density, viz.,

$$\mu = \hat{\psi}'(\nu), \tag{1.4}$$

in which case the diffusive atomic flux takes the form

$$\jmath = -D(\nu) Grad\, \nu, \tag{1.5}$$

with diffusivity D determined by

$$D(\nu) = \kappa(\nu)\hat{\psi}''(\nu). \tag{1.6}$$

However, when strong spatial inhomogeneities in the density may be present, the response function determining the energy density is taken to depend on both the density ν and its gradient $Grad\, \nu$, so that the diffusion potential is given by the *variational derivative*

$$\mu = \frac{\partial \hat{\psi}}{\partial \nu}(\nu, Grad\, \nu) - Div\left(\frac{\partial \hat{\psi}}{\partial (Grad\, \nu)}(\nu, Grad\, \nu)\right). \tag{1.7}$$

Often, in this context, it is presumed that $\hat{\psi}$ has the particular form

$$\hat{\psi}(\nu, Grad\, \nu) = f(\nu) + \frac{\alpha}{2}|Grad\, \nu|^2, \tag{1.8}$$

with α a positive constant that is usually interpreted as proportional to a squared characteristic length of interaction and, hence, taken to be small, so that the atomic flux becomes

$$\jmath = -D_1(\nu) Grad\, \nu + D_2(\nu) Grad\,(Div\, Grad\, \nu), \tag{1.9}$$

with coefficients

$$D_1(\nu) = \kappa(\nu) f''(\nu) \quad \text{and} \quad D_2(\nu) = \alpha\kappa(\nu). \tag{1.10}$$

Recently, Gurtin [7] presented a thermodynamically consistent framework that provides a broad generalization of (1.1) that accounts for constitutive nonlinearities, kinetics, and deformation of the underlying lattice. This framework is based on the belief that to each independent kinematic process there should correspond a system of power-conjugate forces subject to a generalized force balance.* Specifically, Gurtin [7] identifies atomic diffusion and lattice deformation as distinct kinematical processes, with the density-rate $\dot{\nu}$ being a generalized velocity associated with atomic diffusion and the deformation-rate $\dot{y}$ being the velocity of the underlying lattice. Associated with $\dot{\nu}$ is a system of generalized forces that satisfy a scalar *microforce balance*. This balance is a postulate distinct from and ancillary to the conventional force balance involving forces that act power conjugate to the velocity of the underlying lattice. With appropriate thermodynamically consistent constitutive equations, the microforce balance yields an expression for the diffusion potential that generalizes (1.7). Finally, the generalized Cahn–Hilliard equation arises on combining with the atomic balance the thermodynamically derived expression determining the atomic flux as proportional to the gradient of the diffusion potential.

Although this approach provides a powerful and appealing means of deriving (1.1) and various generalizations thereof, we believe that it is lacking in the sense that it accounts only for one aspect of the kinematics associated with atomic diffusion. In particular, we take the view that the diffusive flux of atoms, as described by $\jmath$, should be considered as a kinematic process distinct from that described by the density-rate $\dot{\nu}$. Specifically, we conceive of atomic diffusion as a microscopic process described by

(i) the atomic flux $\jmath$, which measures the local motion of atoms relative to macroscopic motion of the lattice as characterized by the velocity $\dot{y}$, and by

(ii) the rate $\dot{\nu}$ of the atomic density, which encompasses nonlocal interactions between diffusing atoms and characterizes the kinematics of phase separation.

Adopting the view promoted by Fried and Gurtin [16, 17], Fried [18], and Gurtin [7], we arrive at an alternative derivation of the Cahn–Hilliard equation (1.1), where, to account for power expenditures associated with the microscopic motion of atoms, we introduce two separate systems of forces – one being power-conjugate to the atomic flux $\jmath$ and the other being power-conjugate to the rate $\dot{\nu}$ of the atomic density. Further, we require that these forces be consistent with separate microforce balances, one vectorial and involving the forces power-conjugate to $\jmath$ and the other scalar and involving the forces power-conjugate to $\dot{\nu}$.

We first focus on the development of a simple theory, where, to highlight the roles of the vectorial and scalar microforces, the lattice is fixed. Next, we extend the theory to include deformation of the lattice. For simplicity, we restrict our atten-

* A foundation for this viewpoint can be devised using the *principle of virtual power*, see, for example, [8, 9, 10, 11]. We follow here an alternative, but essentially equivalent, approach taken by Ericksen [12], Goodman and Cowin [13], Capriz and Podio-Guidugli [14], Capriz [15], Fried and Gurtin [16, 17], Fried [18], and Gurtin [7].

tion to processes involving only atomic diffusion and lattice deformation, ignoring thermal and other effects, so that the first and second laws of thermodynamics are replaced by an energy imbalance that accounts for all relevant power expenditures. As a further simplification, we ignore all forms of inertia. We employ a purely referential description and consistently use direct notation (see, for example, Gurtin [19]).

2. Simple Theory for Atomic Diffusion on Fixed Lattice

2.1. BALANCE AND IMBALANCE LAWS

The theory is based upon the following laws:

- atomic balance;
- flux-conjugate microforce balance;
- density-rate-conjugate microforce balance;
- energy imbalance.

In contrast to standard formulations of diffusion in rigid bodies, we include in this list of balances *two* microforce balances, which will be seen to provide appropriate generalizations of commonly assumed constitutive relations. In particular, the flux-conjugate microforce balance provides the generalization of (1.3), whereas the density-rate-conjugate microforce balance provides the generalization of (1.7).

We articulate our basic laws of balance and imbalance in global form over a generic subregion $\mathcal{P}$ of the region $\mathcal{B}$ occupied by the body in a fixed reference configuration. The outward unit-normal field to $\partial\mathcal{P}$, directed outward from $\mathcal{P}$, is designated by $\boldsymbol{n}$.

2.1.1. *Atomic Balance*

We introduce the fields

ν atomic density,

$\boldsymbol{\jmath}$ atomic flux,

m external atomic supply,

in which case the integrals

$$\int_{\mathcal{P}} \nu \, \mathrm{d}v, \qquad \int_{\partial\mathcal{P}} \boldsymbol{\jmath}\cdot\boldsymbol{n} \, \mathrm{d}a, \quad \text{and} \quad \int_{\mathcal{P}} m \, \mathrm{d}v \tag{2.1}$$

represent, respectively, the number of atoms in $\mathcal{P}$, the number of atoms, per unit time, added to $\mathcal{P}$ by diffusion across $\partial\mathcal{P}$, and the number of atoms added, per unit time, to $\mathcal{P}$ by agencies external to $\mathcal{P}$.*

* The supply m should not be confused with a reaction term. Rather, we view m as an external supply that accounts for the insertion of material by an external observer. Its role is analogous to that of the external heat supply that is routinely included in modern continuum-thermodynamical statements of energy balance (see, for example, Truesdell [20]). By introducing the external mass supply m, we follow Gurtin [7].

Atomic balance is the postulate that, for each region $\mathcal{P}$ and each instant, the rate at which the number of atoms within $\mathcal{P}$ changes with respect to time be equal to the rate at which atoms are added to $\mathcal{P}$ by the diffusion across $\partial\mathcal{P}$ and by external supplies to $\mathcal{P}$:

$$\overline{\int_{\mathcal{P}} \nu \,\mathrm{d}v}^{\,\boldsymbol{\cdot}} = -\int_{\partial\mathcal{P}} \boldsymbol{\jmath}\cdot\boldsymbol{n}\,\mathrm{d}a + \int_{\mathcal{P}} m\,\mathrm{d}v. \tag{2.2}$$

Using the transport and divergence theorems, we obtain the equivalent local field equation

$$\dot{\nu} = -Div\,\boldsymbol{\jmath} + m \tag{2.3}$$

enforcing atomic balance.

2.1.2. *Flux-Conjugate Microforce Balance*

Associated with the evolution of the atomic flux $\boldsymbol{\jmath}$, we introduce a system of power-conjugate *microforces*, consisting of

- $\boldsymbol{\Sigma}$ flux-conjugate microstress,
- $\boldsymbol{h}$ flux-conjugate internal body microforce,
- $\boldsymbol{\ell}$ flux-conjugate external body microforce,

in which case the integrals

$$\int_{\mathcal{P}} \boldsymbol{\Sigma}\boldsymbol{n}\,\mathrm{d}a, \qquad \int_{\mathcal{P}} \boldsymbol{h}\,\mathrm{d}v, \quad \text{and} \quad \int_{\mathcal{P}} \boldsymbol{\ell}\,\mathrm{d}v \tag{2.4}$$

represent the flux-conjugate microforces exerted on the region $\mathcal{P}$ by the flux-conjugate microtraction distributed over $\partial\mathcal{P}$, by agencies within $\mathcal{P}$, and by flux-conjugate agencies external to $\mathcal{P}$, respectively.

Flux-conjugate microforce balance is the postulate that, for each region $\mathcal{P}$ and each instant, the resultant, taking both internal and external sources into account, of the flux-conjugate microforces acting on $\mathcal{P}$ vanishes:

$$\int_{\partial\mathcal{P}} \boldsymbol{\Sigma}\boldsymbol{n}\,\mathrm{d}a + \int_{\mathcal{P}} (\boldsymbol{h} + \boldsymbol{\ell})\,\mathrm{d}v = \boldsymbol{0}. \tag{2.5}$$

Using the divergence theorem, we arrive at the equivalent local field equation

$$Div\,\boldsymbol{\Sigma} + \boldsymbol{h} + \boldsymbol{\ell} = \boldsymbol{0} \tag{2.6}$$

enforcing the flux-conjugate microforce balance.

2.1.3. *Density-Rate-Conjugate Microforce Balance*

Associated with the evolution of the atomic density ν, we introduce a system of power-conjugate *microforces*, consisting of

- $\boldsymbol{\xi}$ density-rate-conjugate microstress vector,
- π density-rate-conjugate internal body microforce,
- γ density-rate-conjugate external body microforce,

in which case the integrals

$$\int_{\mathcal{P}} \boldsymbol{\xi}\cdot\boldsymbol{n}\,\mathrm{d}a, \qquad \int_{\mathcal{P}} \pi\,\mathrm{d}v, \quad \text{and} \quad \int_{\mathcal{P}} \gamma\,\mathrm{d}v \tag{2.7}$$

represent the density-rate-conjugate microforces exerted on the region $\mathcal{P}$ by the density-rate-conjugate microtraction distributed over $\partial\mathcal{P}$, by agencies within $\mathcal{P}$, and by density-rate-conjugate agencies external to $\mathcal{P}$, respectively.

Density-rate-conjugate microforce balance is the postulate that, for each region $\mathcal{P}$ and each instant, the resultant, taking both internal and external sources into account, of the density-rate-conjugate microforces acting on $\mathcal{P}$ vanishes:

$$\int_{\partial\mathcal{P}} \boldsymbol{\xi}\cdot\boldsymbol{n}\,\mathrm{d}a + \int_{\mathcal{P}} (\pi + \gamma)\,\mathrm{d}v = 0. \tag{2.8}$$

Using the divergence theorem, we arrive at the equivalent local field equation

$$Div\,\boldsymbol{\xi} + \pi + \gamma = 0 \tag{2.9}$$

enforcing the density-rate-conjugate microforce balance.

2.1.4. *Energy Imbalance*

To formulate this imbalance, we introduce fields

- ψ free energy density,
- μ diffusion potential,

in which case the integrals

$$\int_{\mathcal{P}} \psi\,\mathrm{d}v \quad \text{and} \quad \int_{\mathcal{P}} \mu m\,\mathrm{d}v \tag{2.10}$$

represent the free energy in $\mathcal{P}$ and rate at which energy is added to $\mathcal{P}$ through the external supply of atoms to $\mathcal{P}$.

Further, the integrals

$$\int_{\partial\mathcal{P}} (\boldsymbol{\Sigma}^{\mathrm{T}}\boldsymbol{\jmath} + \boldsymbol{\xi}\dot{\nu})\cdot\boldsymbol{n}\,\mathrm{d}a \quad \text{and} \quad \int_{\mathcal{P}} (\boldsymbol{\ell}\cdot\boldsymbol{\jmath} + \gamma\dot{\nu})\,\mathrm{d}v \tag{2.11}$$

provide an accounting of the power expended on $\mathcal{P}$ by the tractions acting on $\partial\mathcal{P}$ and by the agencies acting external to that region.

Energy imbalance is the postulate that, for each region $\mathcal{P}$ and each instant:*

$$\overline{\int_{\mathcal{P}} \psi \, \mathrm{d}v} \leqslant \int_{\partial\mathcal{P}} (\boldsymbol{\Sigma}^{\mathrm{T}} \boldsymbol{\jmath} + \boldsymbol{\xi}\dot{\nu}) \cdot \boldsymbol{n} \, \mathrm{d}a + \int_{\mathcal{P}} (\boldsymbol{\ell} \cdot \boldsymbol{\jmath} + \gamma \dot{\nu} + \mu m) \, \mathrm{d}v. \tag{2.12}$$

Using the transport and divergence theorems as before, we arrive at the local inequality

$$-\dot{\psi} + (\mu - \pi)\dot{\nu} + \boldsymbol{\xi} \cdot \mathit{Grad}\,\dot{\nu} - \boldsymbol{h} \cdot \boldsymbol{\jmath} + (\boldsymbol{\Sigma} + \mu \mathbf{1}) \cdot \mathit{Grad}\,\boldsymbol{\jmath} \geqslant 0 \tag{2.13}$$

enforcing energy imbalance.

2.2. CONSTITUTIVE EQUATIONS

We assume that the free energy density ψ, the flux-congugate microstress $\boldsymbol{\Sigma}$, the flux-congugate internal body microforce $\boldsymbol{h}$, the density-rate-conjugate microstress $\boldsymbol{\xi}$, and the density-rate-conjugate internal scalar body microforce π are determined constitutively in terms of the atomic density ν, the gradient $\mathit{Grad}\,\nu$ of the atomic density, the diffusion potential μ, the gradient $\mathit{Grad}\,\mu$ of the diffusion potential, and the atomic flux $\boldsymbol{\jmath}$:

$$\left.\begin{aligned} \psi &= \hat{\psi}(\nu, \mathit{Grad}\,\nu, \mu, \mathit{Grad}\,\mu, \boldsymbol{\jmath}), \\ \boldsymbol{\Sigma} &= \widehat{\boldsymbol{\Sigma}}(\nu, \mathit{Grad}\,\nu, \mu, \mathit{Grad}\,\mu, \boldsymbol{\jmath}), \\ \boldsymbol{h} &= \hat{\boldsymbol{h}}(\nu, \mathit{Grad}\,\nu, \mu, \mathit{Grad}\,\mu, \boldsymbol{\jmath}), \\ \boldsymbol{\xi} &= \hat{\boldsymbol{\xi}}(\nu, \mathit{Grad}\,\nu, \mu, \mathit{Grad}\,\mu, \boldsymbol{\jmath}), \\ \pi &= \hat{\pi}(\nu, \mathit{Grad}\,\nu, \mu, \mathit{Grad}\,\mu, \boldsymbol{\jmath}). \end{aligned}\right\} \tag{2.14}$$

We emphasize that, in contrast to the standard approach to the theory of atomic diffusion, neither the diffusion potential μ nor the atomic flux $\boldsymbol{\jmath}$ is given by a constitutive relation.

Inserting the constitutive relations (2.14) into the local energy imbalance (2.13) and writing, for brevity,

$$z = (\nu, \mathit{Grad}\,\nu, \mu, \mathit{Grad}\,\mu, \boldsymbol{\jmath}), \tag{2.15}$$

we arrive at the functional inequality

$$\left(\frac{\partial\hat{\psi}}{\partial\nu}(z) - \mu + \hat{\pi}(z)\right)\dot{\nu} + \left(\frac{\partial\hat{\psi}}{\partial(\mathit{Grad}\,\nu)}(z) - \hat{\boldsymbol{\xi}}(z)\right) \cdot \mathit{Grad}\,\dot{\nu} + \frac{\partial\hat{\psi}}{\partial\mu}(z)\dot{\mu}$$

---

* To account for the diffusive flux of energy, Gurtin [7] included in his statement of energy imbalance the term

$$-\int_{\partial\mathcal{P}} \mu \boldsymbol{\jmath} \cdot \boldsymbol{n} \, \mathrm{d}a.$$

Since, in our context, the diffusive flux of energy can be interpreted as the surface power of the atomic diffusion, our theory accounts for this effect through the term with integrand $\boldsymbol{\Sigma}^{\mathrm{T}} \boldsymbol{\jmath} \cdot \boldsymbol{n}$.

$$+\frac{\partial\hat{\psi}}{\partial(\mathit{Grad}\,\mu)}(z)\cdot\mathit{Grad}\,\dot{\mu}+\frac{\partial\hat{\psi}}{\partial\boldsymbol{\jmath}}(z)\cdot\dot{\boldsymbol{\jmath}}+\hat{\boldsymbol{h}}(z)\cdot\boldsymbol{\jmath}$$
$$-\big(\widehat{\boldsymbol{\Sigma}}(z)+\mu\mathbf{1}\big)\cdot\mathit{Grad}\,\boldsymbol{\jmath}\leqslant 0. \tag{2.16}$$

Hence, following the procedure founded by Coleman and Noll [21] in their incorporation of the second law into continuum thermomechanics, we find that:

(i) the constitutive response functions $\hat{\psi}$ and $\hat{\boldsymbol{\xi}}$ delivering the free energy density ψ and the density-rate-conjugate microstress $\boldsymbol{\xi}$ must be independent of the diffusion potential μ, the gradient $\mathit{Grad}\,\mu$ of the diffusion potential, and the atomic flux $\boldsymbol{\jmath}$, and obey

$$\hat{\boldsymbol{\xi}}(\nu,\mathit{Grad}\,\nu)=\frac{\partial\hat{\psi}}{\partial(\mathit{Grad}\,\nu)}(\nu,\mathit{Grad}\,\nu); \tag{2.17}$$

(ii) the constitutive response function $\widehat{\boldsymbol{\Sigma}}$ delivering the flux-conjugate microstress $\boldsymbol{\Sigma}$ must be independent of the atomic density ν, the gradient $\mathit{Grad}\,\nu$ of the atomic density, the gradient $\mathit{Grad}\,\mu$ of the diffusion potential, and the atomic flux $\boldsymbol{\jmath}$, and obey

$$\widehat{\boldsymbol{\Sigma}}=-\mu\mathbf{1}; \tag{2.18}$$

(iii) the constitutive response function $\hat{\pi}$ delivering the density-rate-conjugate internal body microforce π must be independent of the gradient $\mathit{Grad}\,\mu$ of the diffusion potential and the atomic flux $\boldsymbol{\jmath}$, and obey

$$\hat{\pi}(\nu,\mathit{Grad}\,\nu,\mu)=\mu-\frac{\partial\hat{\psi}}{\partial\nu}(\nu,\mathit{Grad}\,\nu); \tag{2.19}$$

(iv) the constitutive response function $\hat{\boldsymbol{h}}$ for the flux-conjugate internal body microforce $\boldsymbol{h}$ must be consistent with the *residual inequality*

$$\hat{\boldsymbol{h}}(\nu,\mathit{Grad}\,\nu,\mu,\mathit{Grad}\,\mu,\boldsymbol{\jmath})\cdot\boldsymbol{\jmath}\leqslant 0. \tag{2.20}$$

Further, granted smoothness of the response function $\hat{\boldsymbol{h}}$, a result due to Gurtin and Voorhees [22] yields a general solution

$$\boldsymbol{h}=-\boldsymbol{Z}(\nu,\mathit{Grad}\,\nu,\mu,\mathit{Grad}\,\mu,\boldsymbol{\jmath})\boldsymbol{\jmath} \tag{2.21}$$

of (2.20), where the *reciprocal mobility* tensor $\boldsymbol{Z}$ must obey

$$\boldsymbol{\jmath}\cdot\boldsymbol{Z}(\nu,\mathit{Grad}\,\nu,\mu,\mathit{Grad}\,\mu,\boldsymbol{\jmath})\boldsymbol{\jmath}\geqslant 0. \tag{2.22}$$

The governing equations that arise on substituting the foregoing thermodynamically consistent constitutive equations in the local field equations (2.9) and (2.6) expressing flux-conjugate microforce balance and density-rate-conjugate microforce balance read

$$\left.\begin{aligned}\boldsymbol{Z}(\nu,\mathit{Grad}\,\nu,\mu,\mathit{Grad}\,\mu,\boldsymbol{\jmath})\,\boldsymbol{\jmath}+\mathit{Grad}\,\mu&=\boldsymbol{\ell},\\ \frac{\partial\hat{\psi}}{\partial\nu}(\nu,\mathit{Grad}\,\nu)-\mathit{Div}\left(\frac{\partial\hat{\psi}}{\partial(\mathit{Grad}\,\nu)}(\nu,\mathit{Grad}\,\nu)\right)-\mu&=\gamma.\end{aligned}\right\} \tag{2.23}$$

Together with the atomic balance (2.3), (2.23) form the final governing equations of our theory for the diffusion of a single unconstrained atomic species on a fixed lattice.

The foregoing results show that the behavior of a medium of the sort considered here is completely determined by the provision of two constitutive response functions:

- $\hat{\psi}$ determining the free energy density as a function of the density ν and its gradient $\mathit{Grad}\,\nu$; and
- the reciprocal mobility tensor $\boldsymbol{Z}$, which, in general, may depend on ν, $\mathit{Grad}\,\nu$, μ, $\mathit{Grad}\,\mu$, and $\boldsymbol{\jmath}$.

Provided that the external supplies $\boldsymbol{m}$, $\boldsymbol{\ell}$, and γ all vanish, that the constitutive function $\hat{\psi}$ is quadratic in the gradient of the atomic density and twice differentiable with respect to the atomic density, and that $\boldsymbol{Z} = \zeta\mathbf{1}$ with ζ depending at most on the atomic density, the system formed by combining (2.23) with the atomic balance (2.3) yields precisely the Cahn–Hilliard equation (1.1). If, furthermore, $\hat{\psi}$ is independent of the gradient of the atomic density, we obtain

$$\boldsymbol{\jmath} = -\frac{\hat{\psi}''(\nu)}{\zeta(\nu)}\mathit{Grad}\,\nu, \tag{2.24}$$

for the atomic flux $\boldsymbol{\jmath}$ in terms of the density ν and its gradient $\mathit{Grad}\,\nu$. Comparing (2.24) with the conventional relation (1.5), we obtain an expression

$$D(\nu) = \frac{\hat{\psi}''(\nu)}{\zeta(\nu)} \tag{2.25}$$

for the diffusivity. Hence, within our framework, a constitutive relation of the form (1.5) can, therefore, be interpreted as an expression of flux-conjugate microforce balance, granted that the flux-conjugate internal body microforce is isotropic and linear in the atomic flux and that the flux-conjugate external body microforce vanishes. If, moreover, we assume that $\hat{\psi}$ has the particular form

$$\hat{\psi}(\nu) = k_{\mathrm{B}}\theta\nu\log\nu \tag{2.26}$$

corresponding to the classical entropic contribution to the free energy of a dilute mixture, with k_{B} *Boltzmann's constant* and θ the *absolute temperature*, then (2.25) reduces to

$$D(\nu) = \frac{k_{\mathrm{B}}\theta}{\zeta(\nu)}, \tag{2.27}$$

the standard relation between diffusivity and mobility (see, for example, [23, 24]). More generally, granted that the reciprocal mobility tensor $\boldsymbol{Z}$ is invertible, our theory in conjunction with the assumption (2.26) yields the diffusivity as a tensor

$$\boldsymbol{D}(\nu, \mathit{Grad}\,\nu, \mu, \mathit{Grad}\,\mu, \boldsymbol{\jmath}) = k_{\mathrm{B}}\theta\boldsymbol{Z}^{-1}(\nu, \mathit{Grad}\,\nu, \mu, \mathit{Grad}\,\mu, \boldsymbol{\jmath}). \tag{2.28}$$

3. Theory Accounting for Both Atomic Diffusion and Lattice Deformation

3.1. BALANCE AND IMBALANCE LAWS

In addition to the mass balance (2.2), flux-conjugate microforce balance (2.5), and density-rate-conjugate microforce balance (2.8), the theory is based upon the following laws:

- lattice-velocity-conjugate force balance;
- lattice-velocity-conjugate torque balance;
- energy imbalance.

Again, the standard formulation of diffusion in a deformable body is supplemented with two microforce balances, which will provide appropriate generalizations of (1.3) and (1.7).

3.1.1. *Lattice-Velocity-Conjugate Force Balance and Lattice-Velocity-Conjugate Torque Balance*

The motion of the lattice is described by a mapping $\boldsymbol{y}$ of points $\boldsymbol{x}$ in $\mathcal{B}$ and times t into points $\boldsymbol{y}(\boldsymbol{x}, t)$ of space. We require that $Det(Grad\,\boldsymbol{y}) > 0$. Associated with this motion, we introduce a system of power-conjugate forces consisting of

$\boldsymbol{S}$ lattice-velocity-conjugate (Piola) stress,
$\boldsymbol{b}$ lattice-velocity-conjugate body force,

in which case the integrals

$$\int_{\mathcal{P}} \boldsymbol{S}\boldsymbol{n}\,\mathrm{d}a \quad \text{and} \quad \int_{\mathcal{P}} \boldsymbol{b}\,\mathrm{d}v \tag{3.1}$$

represent the lattice-velocity-conjugate forces exerted on the region $\mathcal{P}$ by the lattice-velocity-conjugate traction distributed over $\partial\mathcal{P}$ and by lattice-velocity-conjugate agencies external to $\mathcal{P}$, respectively. Further, the integrals

$$\int_{\mathcal{P}} \boldsymbol{y} \times \boldsymbol{S}\boldsymbol{n}\,\mathrm{d}a \quad \text{and} \quad \int_{\mathcal{P}} \boldsymbol{y} \times \boldsymbol{b}\,\mathrm{d}v \tag{3.2}$$

represent the lattice-velocity-conjugate torques exerted on the region $\mathcal{P}$ by the lattice-velocity-conjugate traction distributed over $\partial\mathcal{P}$ and by lattice-velocity-conjugate agencies external to $\mathcal{P}$, respectively.

Lattice-velocity-conjugate force balance is the postulate that, for each region $\mathcal{P}$ and each instant, the resultant of the lattice-velocity-conjugate forces acting on $\mathcal{P}$ vanishes:

$$\int_{\partial\mathcal{P}} \boldsymbol{S}\boldsymbol{n}\,\mathrm{d}a + \int_{\mathcal{P}} \boldsymbol{b}\,\mathrm{d}v = \boldsymbol{0}. \tag{3.3}$$

Further, *lattice-velocity-conjugate torque balance* is the postulate that, for each region $\mathcal{P}$ and each instant, the resultant of the torques associated with the lattice-velocity-conjugate forces acting on $\mathcal{P}$ vanishes:

$$\int_{\partial\mathcal{P}} \boldsymbol{y} \times \boldsymbol{S}\boldsymbol{n}\,\mathrm{d}a + \int_{\mathcal{P}} \boldsymbol{y} \times \boldsymbol{b}\,\mathrm{d}v = \mathbf{0}. \tag{3.4}$$

In writing (3.4), we have assumed that the diffusing atoms do not induce torques, as would be expected at least for approximately spherical atoms. Using the divergence theorem, we arrive at the equivalent local field equations

$$\mathit{Div}\,\boldsymbol{S} + \boldsymbol{b} = \mathbf{0} \quad \text{and} \quad (\mathit{Grad}\,\boldsymbol{y})\boldsymbol{S}^{\mathrm{T}} = \boldsymbol{S}(\mathit{Grad}\,\boldsymbol{y})^{\mathrm{T}}. \tag{3.5}$$

3.1.2. *Energy Imbalance*

To account for power expenditures associated with the motion of the lattice, we modify the energy imbalance (2.12) to read

$$\overline{\int_{\mathcal{P}} \psi\,\mathrm{d}v}^{\,\boldsymbol{\cdot}} \leqslant \int_{\partial\mathcal{P}} (\boldsymbol{\Sigma}^{\mathrm{T}}\boldsymbol{\jmath} + \boldsymbol{\xi}\dot{\nu} + \boldsymbol{S}^{\mathrm{T}}\dot{\boldsymbol{y}})\cdot\boldsymbol{n}\,\mathrm{d}a + \int_{\mathcal{P}} (\boldsymbol{\ell}\cdot\boldsymbol{\jmath} + \pi\dot{\nu} + \mu m + \boldsymbol{b}\cdot\dot{\boldsymbol{y}})\,\mathrm{d}v. \tag{3.6}$$

Proceeding as before, we arrive at the local inequality

$$-\dot{\psi} + (\mu - \pi)\dot{\nu} + \boldsymbol{\xi}\cdot\mathit{Grad}\,\dot{\nu} - \boldsymbol{h}\cdot\boldsymbol{\jmath} + \boldsymbol{S}\cdot\mathit{Grad}\,\dot{\boldsymbol{y}} + (\boldsymbol{\Sigma} + \mu\mathbf{1})\cdot\mathit{Grad}\,\boldsymbol{\jmath} \geqslant 0. \tag{3.7}$$

3.2. CONSTITUTIVE EQUATIONS

We add the lattice-velocity-conjugate stress and the lattice-motion gradient *Grad* $\boldsymbol{y}$ to the lists of dependent and independent constitutive variables, respectively, so that

$$\left.\begin{aligned}
\psi &= \hat{\psi}(\nu, \mathit{Grad}\,\nu, \mu, \mathit{Grad}\,\mu, \boldsymbol{\jmath}, \mathit{Grad}\,\boldsymbol{y}),\\
\boldsymbol{\Sigma} &= \widehat{\boldsymbol{\Sigma}}(\nu, \mathit{Grad}\,\nu, \mu, \mathit{Grad}\,\mu, \boldsymbol{\jmath}, \mathit{Grad}\,\boldsymbol{y}),\\
\boldsymbol{h} &= \hat{\boldsymbol{h}}(\nu, \mathit{Grad}\,\nu, \mu, \mathit{Grad}\,\mu, \boldsymbol{\jmath}, \mathit{Grad}\,\boldsymbol{y}),\\
\boldsymbol{\xi} &= \hat{\boldsymbol{\xi}}(\nu, \mathit{Grad}\,\nu, \mu, \mathit{Grad}\,\mu, \boldsymbol{\jmath}, \mathit{Grad}\,\boldsymbol{y}),\\
\pi &= \hat{\pi}(\nu, \mathit{Grad}\,\nu, \mu, \mathit{Grad}\,\mu, \boldsymbol{\jmath}, \mathit{Grad}\,\boldsymbol{y}),\\
\boldsymbol{S} &= \widehat{\boldsymbol{S}}(\nu, \mathit{Grad}\,\nu, \mu, \mathit{Grad}\,\mu, \boldsymbol{\jmath}, \mathit{Grad}\,\boldsymbol{y}).
\end{aligned}\right\} \tag{3.8}$$

Inserting the constitutive relations (3.8) into the local energy imbalance (3.7) and writing

$$z = (\nu, \mathit{Grad}\,\nu, \mu, \mathit{Grad}\,\mu, \boldsymbol{\jmath}, \mathit{Grad}\,\boldsymbol{y}), \tag{3.9}$$

we arrive at the functional inequality

$$\begin{aligned}&\Big(\frac{\partial}{\partial \nu}\hat{\psi}(z) - \mu + \hat{\pi}(z)\Big)\dot{\nu} + \Big(\frac{\partial \hat{\psi}}{\partial(\mathit{Grad}\, \nu)}(z) - \hat{\boldsymbol{\xi}}(z)\Big)\cdot \mathit{Grad}\,\dot{\nu} + \frac{\partial \hat{\psi}}{\partial \mu}(z)\dot{\mu}\\ &+\frac{\partial \hat{\psi}}{\partial(\mathit{Grad}\, \mu)}(z)\cdot \mathit{Grad}\,\dot{\mu} + \Big(\frac{\partial \hat{\psi}}{\partial(\mathit{Grad}\, \mathbf{y})})(z) - \widehat{\mathbf{S}}(z)\Big)\cdot \mathit{Grad}\,\dot{\mathbf{y}}\\ &+\frac{\partial \hat{\psi}}{\partial \boldsymbol{\jmath}}(z)\cdot \dot{\boldsymbol{\jmath}} + \hat{\boldsymbol{h}}(z)\cdot \boldsymbol{\jmath} - \big(\widehat{\boldsymbol{\Sigma}}(z) + \mu \mathbf{1}\big)\cdot \mathit{Grad}\,\boldsymbol{\jmath} \leqslant 0. \end{aligned} \tag{3.10}$$

Hence, proceeding as before, we obtain the results

(i) the constitutive response functions $\hat{\psi}$, $\widehat{\mathbf{S}}$, and $\hat{\boldsymbol{\xi}}$ delivering the energy density ψ, the lattice-velocity-conjugate stress $\mathbf{S}$, and the density-rate-conjugate microstress $\boldsymbol{\xi}$ must be independent of the diffusion potential μ, the gradient $\mathit{Grad}\,\mu$ of the diffusion potential, and the atomic flux $\boldsymbol{\jmath}$, and obey

$$\left.\begin{aligned}\widehat{\mathbf{S}}(\nu, \mathit{Grad}\,\nu, \mathit{Grad}\,\mathbf{y}) &= \frac{\partial \hat{\psi}}{\partial(\mathit{Grad}\,\mathbf{y})}(\nu, \mathit{Grad}\,\nu, \mathit{Grad}\,\mathbf{y}),\\ \hat{\boldsymbol{\xi}}(\nu, \mathit{Grad}\,\nu, \mathit{Grad}\,\mathbf{y}) &= \frac{\partial \hat{\psi}}{\partial(\mathit{Grad}\,\nu)}(\nu, \mathit{Grad}\,\nu, \mathit{Grad}\,\mathbf{y});\end{aligned}\right\} \tag{3.11}$$

(ii) the constitutive response function $\widehat{\boldsymbol{\Sigma}}$ delivering the flux-conjugate microstress $\boldsymbol{\Sigma}$ must be independent of the atomic density ν, the gradient $\mathit{Grad}\,\nu$ of the atomic density, the gradient $\mathit{Grad}\,\mu$ of the diffusion potential, and the atomic flux $\boldsymbol{\jmath}$, and obey

$$\widehat{\boldsymbol{\Sigma}} = -\mu \mathbf{1}; \tag{3.12}$$

(iii) the constitutive response function $\hat{\pi}$ delivering the density-rate-conjugate internal body microforce π must be independent of the gradient $\mathit{Grad}\,\mu$ of the diffusion potential and the atomic flux $\boldsymbol{\jmath}$, and obey

$$\hat{\pi}(\nu, \mathit{Grad}\,\nu, \mu) = \mu - \frac{\partial \hat{\psi}}{\partial \nu}(\nu, \mathit{Grad}\,\nu); \tag{3.13}$$

(iv) the constitutive response function $\hat{\boldsymbol{h}}$ for the flux-conjugate internal body microforce $\boldsymbol{h}$ must be consistent with the *residual inequality*

$$\hat{\boldsymbol{h}}(\nu, \mathit{Grad}\,\nu, \mu, \mathit{Grad}\,\mu, \boldsymbol{\jmath}, \mathit{Grad}\,\mathbf{y})\cdot \boldsymbol{\jmath} \leqslant 0. \tag{3.14}$$

Analogous to the result (2.21), we have the general solution

$$\boldsymbol{h} = -\mathbf{Z}(\nu, \mathit{Grad}\,\nu, \mu, \mathit{Grad}\,\mu, \boldsymbol{\jmath}, \mathit{Grad}\,\mathbf{y})\, \boldsymbol{\jmath} \tag{3.15}$$

of (3.14), where the *reciprocal mobility* tensor $\mathbf{Z}$ must obey

$$\boldsymbol{\jmath}\cdot \mathbf{Z}(\nu, \mathit{Grad}\,\nu, \mu, \mathit{Grad}\,\mu, \boldsymbol{\jmath})\boldsymbol{\jmath} \geqslant 0. \tag{3.16}$$

Further, the constitutive relations are restricted by objectivity; that is, under the observer transformation

$$\left.\begin{array}{llll} \nu \mapsto \nu, & \mathit{Grad}\,\nu \mapsto \mathit{Grad}\,\nu, & \jmath \mapsto \jmath, & \\ \mu \mapsto \mu, & \mathit{Grad}\,\mu \mapsto \mathit{Grad}\,\mu, & \mathit{Grad}\,\boldsymbol{y} \mapsto \boldsymbol{Q}(\mathit{Grad}\,\boldsymbol{y}), & \end{array}\right\} \tag{3.17}$$

with $\boldsymbol{Q}$ orthogonal, we must have

$$\left.\begin{array}{lll} \boldsymbol{\Sigma} \mapsto \boldsymbol{Q}\boldsymbol{\Sigma}, & \boldsymbol{h} \mapsto \boldsymbol{h}, & \boldsymbol{\ell} \mapsto \boldsymbol{\ell}, \\ \boldsymbol{\xi} \mapsto \boldsymbol{\xi}, & \pi \mapsto \pi, & \gamma \mapsto \gamma, \\ \boldsymbol{S} \mapsto \boldsymbol{Q}\boldsymbol{S}, & \boldsymbol{b} \mapsto \boldsymbol{b}, & \psi \mapsto \psi. \end{array}\right\} \tag{3.18}$$

Granted the foregoing thermodynamic results, these transformations require, in particular, that

$$\left.\begin{array}{l} \psi = \tilde{\psi}(\nu, \mathit{Grad}\,\nu, \boldsymbol{C}), \\ \boldsymbol{Z} = \tilde{\boldsymbol{Z}}(\nu, \mathit{Grad}\,\nu, \mu, \mathit{Grad}\,\mu, \jmath, \boldsymbol{C}), \\ \boldsymbol{S} = (\mathit{Grad}\,\boldsymbol{y})\dfrac{\partial\tilde{\psi}}{\partial\boldsymbol{C}}(\nu, \mathit{Grad}\,\nu, \boldsymbol{C}), \end{array}\right\} \tag{3.19}$$

with $\boldsymbol{C} = (\mathit{Grad}\,\boldsymbol{y})^{\mathrm{T}}\mathit{Grad}\,\boldsymbol{y}$ the *right Cauchy–Green tensor*.

The governing equations that arise on substituting the foregoing thermodynamically consistent constitutive equations in the local field equations $(3.5)_1$, (2.9) and (2.6) expressing lattice-velocity-conjugate force balance, flux-conjugate microforce balance, and density-rate-conjugate microforce balance read

$$\left.\begin{array}{r} \mathit{Div}\left((\mathit{Grad}\,\boldsymbol{y})\dfrac{\partial\tilde{\psi}}{\partial\boldsymbol{C}}(\nu, \mathit{Grad}\,\nu, \boldsymbol{C})\right) + \boldsymbol{b} = \boldsymbol{0}, \\ \tilde{\boldsymbol{Z}}(\nu, \mathit{Grad}\,\nu, \mu, \mathit{Grad}\,\mu, \jmath, \boldsymbol{C})\,\jmath + \mathit{Grad}\,\mu = \boldsymbol{\ell}, \\ \dfrac{\partial\tilde{\psi}}{\partial\nu}(\nu, \mathit{Grad}\,\nu, \boldsymbol{C}) - \mathit{Div}\left(\dfrac{\partial\tilde{\psi}}{\partial(\mathit{Grad}\,\nu)}(\nu, \mathit{Grad}\,\nu, \boldsymbol{C})\right) - \mu = \gamma. \end{array}\right\} \tag{3.20}$$

Together with the atomic balance (2.3), (3.20) form the final governing equations of our theory for the diffusion of a single unconstrained atomic species on a deformable lattice lattice.

We find that the behavior of a medium of the sort considered here is still completely determined by the provision of two constitutive response functions:

- $\tilde{\psi}$ determining the energy density as a function of the density ν, the gradient $\mathit{Grad}\,\nu$ of the atomic density, and right Cauchy–Green tensor $\boldsymbol{C}$ associated with the lattice deformation; and
- the reciprocal mobility tensor $\tilde{\boldsymbol{Z}}$, which, in general, may depend on the atomic density ν, the gradient $\mathit{Grad}\,\nu$ of the atomic density, the diffusion potential μ, the gradient $\mathit{Grad}\,\mu$ of the diffusion potential, the atomic flux $\jmath$, and the right Cauchy–Green tensor $\boldsymbol{C}$ associated with the lattice deformation.

4. Discussion

Our theory is based upon introducing generalized velocities for diffusion and corresponding power-conjugate forces, which we call microforces, into an otherwise conventional continuum-mechanical description. In particular, we identify two distinct generalized velocities – a vector-valued diffusive atomic flux that characterizes simple diffusion and a scalar-valued atomic density rate that characterizes phase separation. A systematic, thermodynamically consistent derivation generates a final system of governing equations – (2.3) and (2.23) for a fixed lattice, and (2.3) and (3.20) for a deformable lattice – that constitutes a generalization of common diffusion equations, a generalization that accounts for constitutive nonlinearities and external forces acting on the diffusing atoms.

We find that the constitutive response of the material is determined by the provision of two functions, one for the energy density and the other for the reciprocal mobility tensor. In particular, the Cahn–Hilliard equation (1.1) results on assuming that the energy response function is quadratic in $Grad\, \nu$ and that the reciprocal mobility tensor is proportional to the identity tensor by a scalar coefficient that is at most a function of ν. Importantly, we arrive at our theory without introducing a constitutive relation such as (1.3) for the diffusive flux.

Additionally, our proposed format also leads to a generalization, (2.25), of the standard relation between diffusivity and mobility. We find that the classical form for the relation (2.27) holds when the free energy is given by the special constitutive relation (2.26).

A distinction between the final governing equations of our theory and those arising in more conventional derivations is that, in our approach, a vector-valued microforce balance provides a generalization between diffusive atomic flux and gradient of the diffusion potential.

Acknowledgements

This work was performed with financial support from the U.S. Department of Energy and the EPSRC (U.K.).

References

1. J.W. Cahn, Free energy of a nonuniform system. II. Thermodynamic basis. *J. Chem. Phys.* **30** (1959) 1121–1124.
2. J.W. Cahn, On spinodal decomposition. *Acta Metallurgica* **9** (1961) 795–801.
3. J.W. Cahn, On spinodal decomposition in cubic crystals. *Acta Metallurgica* **10** (1962) 179–183.
4. J.W. Cahn and J.E. Hilliard, Free energy of a nonuniform system. I. Interfacial free energy. *J. Chem. Phys.* **28** (1958) 258–267.
5. J.W. Cahn and J.E. Hilliard, Free energy of a nonuniform system. III. Nucleation in a two-component incompressible fluid. *J. Chem. Phys.* **31** (1959) 688–699.
6. J.W. Cahn and J.E. Hilliard, Spinodal decomposition: A reprise. *Acta Metallurgica* **19** (1971) 151–161.

7. M.E. Gurtin, Generalized Ginzburg–Landau and Cahn–Hilliard equations based on a microforce balance. *Phys. D* **92** (1996) 178–192.
8. P. Germain, Sur l'application de la méthode des puissances virtuelles en mécanique des milieux continus. *C. R. Acad. Sci. Paris Sér. A* **274** (1973) 1051–1055.
9. S.S. Antman and J.E. Osborn, The principle of virtual work and integral laws of motion. *Arch. Rational Mech. Anal.* **69** (1979) 231–262.
10. G.A. Maugin, The principle of virtual power in continuum mechanics – application to coupled fields. *Acta Mechanica* **35** (1980) 1–70.
11. M. Frémond, Endommagement et principe des puissances virtuelles. *C. R. Acad. Sci. Paris Sér. II* **317** (1993) 857–863.
12. J.L. Ericksen, Conservation laws for liquid crystals. *Trans. Soc. Rheology* **5** (1961) 23–34.
13. M.A. Goodman and S.C. Cowin, A continuum theory for granular materials. *Arch. Rational Mech. Anal.* **44** (1972) 249–266.
14. G. Capriz and P. Podio-Guidugli, Structured continua from a Lagrangian point of view. *Annali Mat. Pura Appl.* **135** (1983) 1–25.
15. G. Capriz, *Continua with Microstructure*. Springer, New York (1989).
16. E. Fried and M.E. Gurtin, Continuum theory of thermally induced phase transitions based on an order parameter. *Phys. D* **68** (1993) 326–343.
17. E. Fried and M.E. Gurtin, Dynamic solid-solid transitions with phase characterized by an order parameter. *Phys. D* **72** (1994) 287–308.
18. E. Fried, Continua described by a microstructural field. *Zeitschrift für angewandte Mathematik und Physik* **47** (1996) 168–175.
19. M.E. Gurtin, *An Introduction to Continuum Mechanics*. Academic Press, New York/London (1981).
20. C. Truesdell, *Rational Thermodynamics*, 2nd edn. Springer, Berlin (1984).
21. B.D. Coleman and W. Noll, The thermodynamics of elastic materials with heat conduction and viscosity. *Arch. Rational Mech. Anal.* **13** (1963) 167–178.
22. M.E. Gurtin and P.W. Voorhees, The continuum mechanics of coherent two-phase elastic solids with mass transport. *Proc. Roy. Soc. London A* **444** (1993) 323–343.
23. W. Nernst, Zur Kinetik der in Lösung befindlichen Körper: I. Theorie der Diffusion. *Zeitschrift für physikalische Chemie* **2** (1888) 613–637.
24. A. Einstein, Über die von der molekularkinetischen Theorie der Wärme geforderten Bewegung von in ruhenden Flüssigkeiten suspendierten Teilchen. *Ann. Phys.* **17** (1905) 549–560.

Journal of Elasticity **59:** 83–113, 2000.

On the Existence of Minimizers with Prescribed Singular Points in Nonlinear Elasticity

JEYABAL SIVALOGANATHAN[1] and SCOTT J. SPECTOR[2]
[1] *Department of Mathematical Sciences, University of Bath, Bath BA2 7AY, U.K.*
E-mail: js@maths.bath.ac.uk
[2] *Department of Mathematics, Southern Illinois University, Carbondale, IL 62901-4408, U.S.A.*
E-mail: sspector@math.siu.edu

Received 15 January 1999; in revised form 13 March 2000

For Roger Fosdick, colleague, friend, and mentor.

Abstract. Experiments on elastomers have shown that sufficiently large triaxial tensions induce the material to exhibit holes that were not previously evident. In this paper, conditions are presented that allow one to use the direct method of the calculus of variations to deduce the existence of hole creating deformations that are global minimizers of a nonlinear, purely elastic energy. The crucial physical assumption used is that there are a finite (possibly large) number of material points in the undeformed body that constitute the only points at which cavities can form. Each such point may be viewed as a preexisting flaw or an infinitesimal microvoid in the material.

Mathematics Subject Classifications (2000): 74G65, 74G70, 74B20, 49J40, 35J50.

Key words: cavitation, degree, distributional Jacobian, elastic, elastomer, elastostatics, equilibrium, finite elasticity, fracture, injective, microvoid, one-to-one, polyconvexity, singular minimizer.

1. Introduction

Experiments on elastomers have shown that holes, which were not previously evident to optical observation, appear in portions of the elastomer that experience a sufficiently large triaxial tension. A reasonable expectation, especially in view of the analysis of Gent and Lindley [13] (see [18] for a brief description), is that such a cavitating deformation, i.e., an injective mapping that exhibits holes, should be a global minimizer of the elastic energy

$$E(\mathbf{u}) := \int_\Omega W\big(\mathbf{x}, \nabla\mathbf{u}(\mathbf{x})\big)\mathrm{d}\mathbf{x},$$

although it is possible that such a deformation is only a relative minimizer or that nonelastic effects, such as surface energy or plasticity, are needed to explain cavitation.

In this paper we present conditions on the stored energy density W that allow us to deduce the existence of a cavitating deformation that is a global minimizer of E among those deformations that are injective (almost everywhere) and satisfy sufficiently large displacement boundary conditions on the boundary of a bounded region $\Omega \subset \mathbb{R}^3$.

The main physical assumption that we use to obtain our existence result is that there are a finite (possibly large) number of material points in the undeformed body that constitute the only points at which cavities can form. We view each such point as a preexisting flaw or weakness in the material (see [12]), however, one could just as well view it as an infinitesimal precursor microvoid (see [16, 26]). Our theory assumes that such points are not detectable when the material is subjected to small (finite) deformations or purely compressive stresses. However, when any such point is subjected to a sufficiently large, triaxial tension, a hole, which was not previously in the material, will be created at that point.

Our purely elastic model may not explain the refined experiments of Cho and Gent [7]. These experiments indicate that a triaxial tension that induces a hole to appear in a large sample may not be sufficient to open a cavity when the same triaxial tension is only experienced in a small portion of the sample. In Section 4 we therefore also briefly consider a model that might better explain such results. In addition to the elastic energy E this model includes the energy of creation of each new hole. This creation energy may be different at each cavitation point and therefore model differing flaw strengths or preexisting microvoid sizes.

The main mathematical tools we use to prove our results are contained in [24]. These tools are based, in large part, on results on fine properties of deformations previously obtained by Šverák [28] and results on the distributional Jacobian previously obtained by Müller [23]. Our proof makes use of the notion of polyconvexity and the direct method of the calculus of variations, as first presented by Ball [2] in the context of nonlinear elasticity, to deduce the existence of minimizers. Our approach to cavitation follows that of Ball [3] (see also [17] and the reference therein) who views cavitation as the creation of a new hole in the body rather than the growth of a preexisting finite hole, as had previously been analyzed in [13].

In addition to the above mentioned existence results, we show, in Section 5, that energy minimizers satisfy various weak forms of the equilibrium equations. Therefore, since the cavitation points can be prescribed arbitrarily, our results imply that, for the stored-energy functions considered in this paper, there exist infinitely many singular weak solutions to the equations of nonlinear elastostatics. The proof of the results in this section follows the suggestion of Ball [4] that has been carried out in detail by Bauman et al. [6] (see also [14]). We also show that if the minimizer is smooth away from the finite number of prescribed cavitation points then, in the absence of body forces, the divergence of the Cauchy stress is zero in the deformed configuration and the divergence of the both the Piola–Kirchhoff stress and the the energy-momentum tensor are zero in the reference configuration.

An existence theorem that allows for cavitation has previously been obtained by Müller and Spector [24]. However, their model is not a purely elastic one since it includes an additional energy that is proportional to the area of newly created cavity surface. Our model needs no such additional energy to obtain existence, but must instead predetermine the finite set of points at which cavitation may occur. The model in [24] makes no such prespecification. It also allows for a countable number of new holes to be created.

Finally, we mention that our paper has a second purpose. Suppose that the body, in its reference configuration, is a ball centered at the origin that is composed of a homogeneous, isotropic, nonlinearly elastic material and is subject to radial boundary displacements. Then our intuition leads us to believe that, for many stored energy densities W, a global minimizer of the energy should be a radial deformation of the form

$$\mathbf{u}(\mathbf{x}) = \frac{r(|\mathbf{x}|)}{|\mathbf{x}|}\mathbf{x},$$

that minimizes the energy among radial, injective mappings and creates a single hole at the center of the ball. Unfortunately, except for an elastic fluid it has not been determined whether this conjecture is correct.* Thus an interesting, possibly simpler, open problem is to determine whether the radial minimizer is a global minimizer among injective deformations that open a single hole at the center. This paper establishes that the latter class contains a function that minimizes the energy within that class.

2. Preliminaries

In the following, Ω will denote a nonempty, bounded, open subset of $\mathbb{R}^n$, $n \geqslant 2$, whose boundary, $\partial\Omega$, is (strongly) Lipschitz (see [10, 22]). By $L^p(\Omega)$ and $W^{1,p}(\Omega)$ we denote the usual spaces of p-summable and Sobolev functions, respectively. We use the notation $L^p(\Omega;\mathbb{R}^m)$, etc., for vector-valued maps. A function is in $L^p_{\mathrm{loc}}(\Omega)$ if $\varphi \in L^p(U)$ for all open sets $U \subset\subset \Omega$, i.e., $U \subset K_U \subset \Omega$ for some compact set K_U. We point out that we *do not identify functions that agree almost everywhere.* We use the short-hand notation $\varphi \in W^{1,1}(\Omega)$, etc., to indicate that φ is a representative of an equivalence class that is contained in $W^{1,1}(\Omega)$. Weak convergence in these spaces will be indicated by the half arrow $\rightharpoonup$. Weak* (weak-star) convergence in $L^\infty(\Omega)$ will be denoted by $\overset{*}{\rightharpoonup}$ and the norm on this space will be denoted by $\|\cdot\|_\infty$. Composition of functions will be denoted by $\circ$.

The n-dimensional Lebesgue measure will be denoted by $\mathcal{L}^n$ and k-dimensional Hausdorff measure by $\mathcal{H}^k$. We write

$$B(\mathbf{b}, r) := \{\mathbf{x} \in \mathbb{R}^n\colon |\mathbf{x} - \mathbf{b}| < r\}$$

* Counterexamples have been obtained in [18]. However, these examples require W to grow so slowly at infinity that they are incompatible with the theory in this paper. See also [27] for a model problem in which the conjecture is true.

for the ball of radius r centered at $\mathbf{b} \in \mathbb{R}^n$ with volume $\omega_n r^n = \mathcal{L}^n(B(\mathbf{b}, r))$. For $\mathbf{b} \in \Omega$ we let

$$r_{\mathbf{b}} := \operatorname{dist}(\mathbf{b}, \partial\Omega),$$

i.e., the distance from $\mathbf{b}$ to the boundary of Ω.

We write Lin for the set of all linear maps from $\mathbb{R}^n$ into $\mathbb{R}^n$ with inner product and norm

$$\mathbf{K} \cdot \mathbf{L} = \operatorname{tr}(\mathbf{K}^{\mathrm{T}}\mathbf{L}), \qquad |\mathbf{L}| = (\mathbf{L} \cdot \mathbf{L})^{1/2},$$

respectively, where tr(**M**) denotes the trace of $\mathbf{M} \in \mathrm{Lin}$. We denote by $\mathrm{Lin}^{>}$ those $\mathbf{L} \in \mathrm{Lin}$ with positive determinant. The mapping adj : Lin $\to$ Lin will be the unique continuous function that satisfies

$$\mathbf{L}(\operatorname{adj}\mathbf{L}) = (\det\mathbf{L})\mathbf{I}$$

for all $\mathbf{L} \in \mathrm{Lin}$, where $\det\mathbf{L}$ is the determinant of $\mathbf{L}$ and $\mathbf{I} \in \mathrm{Lin}$ is the identity mapping. Thus, with respect to any orthonormal basis, the matrix corresponding to $\operatorname{adj}\mathbf{L}$ is the transpose of the cofactor matrix corresponding to $\mathbf{L}$. We denote by $\mathbf{id}$: $\mathbb{R}^n \to \mathbb{R}^n$ the identity mapping on $\mathbb{R}^n$, i.e., $\mathbf{id}(\mathbf{x}) \equiv \mathbf{x}$. We write div for the divergence operator in $\mathbb{R}^n$: for a vector field $\mathbf{u}$, $\operatorname{div}\mathbf{u} = \operatorname{tr}\nabla\mathbf{u}$; for a tensor field $\mathbf{S}$, $\operatorname{div}\mathbf{S}$ is the vector field with components $\Sigma_j \partial S_{ij}/\partial x_j$.

We briefly recall some facts about the Brouwer degree (see, e.g., [11, 25] for more details). Let $\mathbf{u}$: $\overline{\Omega} \to \mathbb{R}^n$ be a C^1 map. If $\mathbf{y}_0 \in \mathbb{R}^n \backslash \mathbf{u}(\partial\Omega)$ is such that $\det\nabla\mathbf{u}(\mathbf{x}) \neq 0$ for all $\mathbf{x} \in \mathbf{u}^{-1}(\mathbf{y}_0)$, one defines

$$\deg(\mathbf{u}, \Omega, \mathbf{y}_0) := \sum_{\mathbf{x}\in\mathbf{u}^{-1}(\mathbf{y}_0)} \operatorname{sgn}\big[\det\nabla\mathbf{u}(\mathbf{x})\big]. \tag{2.1}$$

Thus, in particular, if $\mathbf{g}$: $\overline{\Omega} \to \mathbb{R}^n$ is a diffeomorphism with $\det\nabla\mathbf{g} > 0$ on Ω, then from (2.1) we conclude that

$$\deg(\mathbf{g}, \Omega, \mathbf{y}_0) = \begin{cases} 1, & \text{if } \mathbf{y}_0 \in \mathbf{g}(\Omega), \\ 0, & \text{if } \mathbf{y}_0 \in \mathbb{R}^n \backslash \mathbf{g}(\overline{\Omega}). \end{cases} \tag{2.2}$$

If φ is a C^∞ function supported in the connected component of $\mathbb{R}^n \backslash \mathbf{u}(\partial\Omega)$ that contains $\mathbf{y}_0$, one can show that

$$\int_\Omega \varphi(\mathbf{u}(\mathbf{x})) \det\nabla\mathbf{u}(\mathbf{x})\, d\mathbf{x} = \deg(\mathbf{u}, \Omega, \mathbf{y}_0) \int_{\mathbb{R}^n} \varphi(\mathbf{y})\, d\mathbf{y}.$$

Using this formula and approximating by C^∞ functions, one can define $\deg(\mathbf{u}, \Omega, \mathbf{y})$ for any continuous function $\mathbf{u}$: $\overline{\Omega} \to \mathbb{R}^n$ and any $\mathbf{y} \in \mathbb{R}^n \backslash \mathbf{u}(\partial\Omega)$. Moreover, the degree only depends on $\mathbf{u}|_{\partial\Omega}$. Accordingly we will henceforth write $\deg(\mathbf{u}, \partial\Omega, \mathbf{y})$ instead of $\deg(\mathbf{u}, \Omega, \mathbf{y})$. More generally, since a continuous function on a compact set can be extended to a continuous function on all of $\mathbb{R}^n$ one has the following result.

PROPOSITION 2.1. *Suppose that* $\bar{\mathbf{u}} \in C(\partial B(\mathbf{b}, r); \mathbb{R}^n)$ *for some* $\mathbf{b} \in \mathbb{R}^n$ *and real number* $r > 0$. *Then* $\deg(\bar{\mathbf{u}}, \partial B(\mathbf{b}, r), \mathbf{y})$ *is well defined for every* $\mathbf{y} \in \mathbb{R}^n \setminus \bar{\mathbf{u}}(\partial B(\mathbf{b}, r))$. *Moreover,*

(i) *if* $\bar{\mathbf{u}}_j \in C(\partial B(\mathbf{b}, r); \mathbb{R}^n)$ *is such that* $\bar{\mathbf{u}}_j \to \bar{\mathbf{u}}$ *uniformly then*

$$\lim_{j\to\infty} \deg(\bar{\mathbf{u}}_j, \partial B(\mathbf{b}, r), \mathbf{y}) = \deg(\bar{\mathbf{u}}, \partial B(\mathbf{b}, r), \mathbf{y})$$

for every $\mathbf{y} \in \mathbb{R}^n \setminus \bar{\mathbf{u}}(\partial B(\mathbf{b}, r))$;

(ii) *if* $\mathbf{g}\colon \mathbb{R}^n \to \mathbb{R}^n$ *is a diffeomorphism that satisfies* $\det \nabla \mathbf{g} > 0$ *on* $\mathbb{R}^n$ *then*

$$\deg(\mathbf{g} \circ \bar{\mathbf{u}}, \partial B(\mathbf{b}, r), \mathbf{g}(\mathbf{y})) = \deg(\bar{\mathbf{u}}, \partial B(\mathbf{b}, r), \mathbf{y})$$

for every $\mathbf{y} \in \mathbb{R}^n \setminus \bar{\mathbf{u}}(\partial B(\mathbf{b}, r))$.

REMARKS. 1. In (i) one might have $\mathbf{y} \in \bar{\mathbf{u}}_j(\partial B(\mathbf{b}, r))$ and so $\deg(\bar{\mathbf{u}}_j, \partial B(\mathbf{b}, r), \mathbf{y})$ would not be defined. However, the uniform convergence $\bar{\mathbf{u}}_j \to \bar{\mathbf{u}}$ implies that the degree is well defined for all sufficiently large j.

2. For a ball one can find an explicit formula for the degree by using the extension $\mathbf{u} \in C^2(B(\mathbf{b}, r); \mathbb{R}^n) \cap C^0(\overline{B(\mathbf{b}, r)}; \mathbb{R}^n)$ that solves the vector Laplacian in $B(\mathbf{b}, r)$ with boundary values $\bar{\mathbf{u}}$. This function is defined by the Poisson integral formula (see, e.g., [15]).

3. Property (ii) is a consequence of (2.2) and the multiplicative property of the degree under compositions (see, e.g., [11, Theorem 2.10]; [25, Theorem 3.20]; [30, p. 578]).

Let $\mathbf{u} \in W^{1,p}(\Omega; \mathbb{R}^n)$, with $1 \leqslant p < n$. We will be interested in pointwise properties of $\mathbf{u}$ as well as restrictions of $\mathbf{u}$ to lower dimensional sets. In these cases, it is useful to consider an alternative representative in the same equivalence class. We define the **precise representative** $\mathbf{u}^*\colon \Omega \to \mathbb{R}^n$ by

$$\mathbf{u}^*(\mathbf{x}) = \begin{cases} \lim\limits_{\rho\to 0^+} ⨍_{B(\mathbf{x},\rho)} \mathbf{u}(\mathbf{z})\, d\mathbf{z}, & \text{if the limit exists,} \\ \mathbf{0}, & \text{otherwise.} \end{cases}$$

where $⨍_A$ denotes the integral average over A, i.e., the integral of the function over A divided by the n-dimensional Lebesgue measure of A.

We shall make use of the fact that, if $\mathbf{u} \in W^{1,p}(\Omega; \mathbb{R}^n)$ with $1 \leqslant p < n$, the above limit exists for every $\mathbf{x} \in \Omega \setminus P$, where $\mathcal{H}^{n-1}(P) = 0$. Thus, in particular, one can use the precise representative as a representative of the trace on $(n-1)$-dimensional surfaces. Moreover, if $p > n - 1$ then $\mathcal{H}^1(P) = 0$ and consequently for each $\mathbf{b} \in \Omega$ the above limit is defined at *every* point on $\partial B(\mathbf{b}, r)$ for almost every $r \in (0, r_{\mathbf{b}})$. For a thorough discussion of precise representatives we refer to [10].

The following observation will be useful in our later development in this paper.

PROPOSITION 2.2 (see, e.g., [24, Lemma 2.9]). *Assume that*

$$\mathbf{u}_j \rightharpoonup \mathbf{u} \quad \text{in } W^{1,p}(\Omega; \mathbb{R}^m).$$

Let $\mathbf{b} \in \Omega$ *and* $r_{\mathbf{b}} := \operatorname{dist}(\mathbf{b}, \partial\Omega)$. *Then there is an* $N_{\mathbf{b}} \subset \mathbb{R}$ *with* $\mathcal{L}^1(N_{\mathbf{b}}) = 0$ *such that for any* $r \in (0, r_{\mathbf{b}})\backslash N_{\mathbf{b}}$ *there exists a subsequence* $\mathbf{u}_j$ *(not relabeled), which will in general depend on r, such that*

$$\mathbf{u}_j^* \rightharpoonup \mathbf{u}^* \quad \text{in } W^{1,p}(\partial B(\mathbf{b}, r); \mathbb{R}^m).$$

Furthermore, if $p > n - 1$, *then* $\mathbf{u}_j^*|_{\partial B(\mathbf{b},r)}$ *and* $\mathbf{u}^*|_{\partial B(\mathbf{b},r)}$ *are continuous and*

$$\mathbf{u}_j^* \to \mathbf{u}^* \quad \text{uniformly on } \partial B(\mathbf{b}, r). \tag{2.3}$$

In nonlinear elasticity one is interested in globally invertible maps since, in general, matter cannot interpenetrate itself. We say that $\mathbf{u} \in W^{1,1}(\Omega; \mathbb{R}^n)$ is **invertible almost everywhere** (or equivalently, **one-to-one almost everywhere**) if there is a Lebesgue null set $N \subset \Omega$ such that $\mathbf{u}|_{\Omega\backslash N}$ is injective. We note that invertibility almost everywhere is a property of the equivalence class and not merely of the representative. However, the notion of invertibility almost everywhere is not sufficient for the analysis in function classes that allow for the formation of cavities. In fact the topological properties of such maps can differ drastically from everywhere invertible maps. The source of the difficulties is that a cavity formed at one point may be filled by material from elsewhere. In order to exclude such behavior the invertibility condition (INV) was introduced in [24].

DEFINITION 2.3. Let $B(\mathbf{b}, r) \subset \Omega$ and suppose that $\bar{\mathbf{u}} : \partial B(\mathbf{b}, r) \to \mathbb{R}^n$ is continuous. We define the **topological image** of $B(\mathbf{b}, r)$ under $\bar{\mathbf{u}}$ by

$$\operatorname{im}_T(\bar{\mathbf{u}}, B(\mathbf{b}, r)) := \{\mathbf{y} \in \mathbb{R}^n\backslash\bar{\mathbf{u}}(\partial B(\mathbf{b}, r)) : \deg(\bar{\mathbf{u}}, \partial B(\mathbf{b}, r), \mathbf{y}) \neq 0\}.$$

REMARK. If $\bar{\mathbf{u}}: \partial B(\mathbf{b}, r) \to \mathbb{R}^n$ happens to be the restriction of a homeomorphism $\mathbf{h} : \overline{B(\mathbf{b}, r)} \to \mathbb{R}^n$ to $\partial B(\mathbf{b}, r)$ then

$$\operatorname{im}_T(\bar{\mathbf{u}}, B(\mathbf{b}, r)) = \mathbf{h}(B(\mathbf{b}, r)).$$

DEFINITION 2.4. We say that $\mathbf{u}: \Omega \to \mathbb{R}^n$ satisfies condition (INV) provided that for every $\mathbf{b} \in \Omega$ there exists an $\mathcal{L}^1$ null set $N_{\mathbf{b}}$ such that, for all $r \in (0, r_{\mathbf{b}})\backslash N_{\mathbf{b}}$, $\mathbf{u}|_{\partial B(\mathbf{b},r)}$ is continuous:

(i) $\mathbf{u}(\mathbf{x}) \in \operatorname{im}_T(\mathbf{u}, B(\mathbf{b}, r)) \cup \mathbf{u}(\partial B(\mathbf{b}, r))$ for $\mathcal{L}^n$ a.e. $\mathbf{x} \in \overline{B(\mathbf{b}, r)}$,
(ii) $\mathbf{u}(\mathbf{x}) \in \mathbb{R}^n\backslash\operatorname{im}_T(\mathbf{u}, B(\mathbf{b}, r))$ for $\mathcal{L}^n$ a.e. $\mathbf{x} \in \Omega\backslash\overline{B(\mathbf{b}, r)}$.

REMARKS. 1. Fix $\mathbf{b} \in \Omega$. Then (i) and (ii) can be thought of as the requirement that (almost) every spherical shell centered at $\mathbf{b}$ is a *solid, impenetrable 2-dimensional body* that is subjected to a continuous deformation. Thus, all matter that was originally inside such a shell must remain inside and all matter that was originally outside such a shell must remain outside.

2. Another way one might attempt to eliminate physically inappropriate maps is to require that each deformation $\mathbf{u}$ lie in the same homotopy class as the identity map. For example, one might require that there exists $\mathbf{g} \in C^0([0,1]; W^{1,p}(\Omega;\mathbb{R}^n))$ with $\mathbf{g}(\cdot, t)$ one-to-one a.e. for every $t \in [0,1]$, $\mathbf{g}(\mathbf{x}, 0) = \mathbf{x}$, and $\mathbf{g}(\mathbf{x}, 1) = \mathbf{u}(\mathbf{x})$ for a.e. $\mathbf{x} \in \Omega$. However, there are mathematical difficulties with such conditions. In particular, it is not clear that the weak limit of an energy minimizing sequence of deformations that lies in such a homotopy class also lies in the same homotopy class. In addition, one would need to show that the weak limit of a sequence of deformations that are one-to-one a.e. is also one-to-one a.e.

Deformations that satisfy condition (INV) and have nonzero Jacobian are more regular than other elements of the Sobolev spaces $W^{1,p}$, $n-1 < p < n$. In particular, in [24] it is shown that such deformations have a representative that is continuous at every $\mathbf{x} \in \Omega \backslash \widehat{P}$, where $\mathcal{H}^{n-p}(\widehat{P}) = 0$ (and hence, $\mathcal{L}^n(\widehat{P}) = 0$) and, more significantly in the context of this work, are one-to-one a.e.

PROPOSITION 2.5 ([*24*, Lemma 3.4]). *Let* $\mathbf{u} \in W^{1,p}(\Omega;\mathbb{R}^n)$ *with* $p > n-1$. *Suppose that* $\det \nabla \mathbf{u} \neq 0$ *a.e. and that* $\mathbf{u}^*$ *satisfies condition (INV). Then* $\mathbf{u}$ *is one-to-one almost everywhere.*

REMARK. When $p < n$ the condition $\det\nabla\mathbf{u} > 0$ a.e. alone does not imply that $\mathbf{u}$ is one-to-one almost everywhere. For example, (cf. [19]) if $\Omega = B(\mathbf{0}, 2) \subset \mathbb{R}^3$ the radial mapping

$$\mathbf{u}(\mathbf{x}) := \big(|\mathbf{x}| - 1\big)\frac{\mathbf{x}}{|\mathbf{x}|},$$

which is contained in $W^{1,p}(B(\mathbf{0},2);\mathbb{R}^3)$ for $p \in [1,3)$, is *two-to-one a.e.*, yet has positive Jacobian at all points apart from the unit sphere and the origin.

For a diffeomorphism with nonnegative Jacobian the degree can only assume the values 1 and 0. The following is a counterpart of this result for the situation at hand.

PROPOSITION 2.6 (see [*24*, Lemma 3.5]). *Let* $\mathbf{u} \in W^{1,p}(\Omega;\mathbb{R}^n)$ *with* $p > n-1$. *Suppose that* $\mathbf{u}^*$ *satisfies condition (INV) and that* $\det\nabla\mathbf{u} \neq 0$ *a.e. Fix* $\mathbf{b} \in \Omega$.

(i) *Assume, in addition, that* $\det\nabla\mathbf{u} > 0$ *a.e. Then there exists an* $\mathcal{L}^1$ *null set* $N_\mathbf{b}$ *such that for every* $r \in (0, r_\mathbf{b})\backslash N_\mathbf{b}$

$$\deg\big(\mathbf{u}, \partial B(\mathbf{b}, r), \mathbf{y}\big) \in \{0, 1\} \quad \textit{for all } \mathbf{y} \in \mathbb{R}^n\backslash\mathbf{u}\big(\partial B(\mathbf{b}, r)\big). \tag{2.4}$$

(ii) *Conversely, if there is an* $r_0 \in (0, r_\mathbf{b})$ *such that (2.4) is satisfied for* $\mathcal{L}^1$ *a.e.* $r \in (0, r_0)$ *then*

$$\det\nabla\mathbf{u} > 0 \quad \textit{a.e. in } B(\mathbf{b}, r_0).$$

Another result we will make use of is the change of variables formula for Sobolev mappings. (For a proof see [24, Proposition 2.6], and the references therein.)

PROPOSITION 2.7. *Let* $\mathbf{u}$ *be (a representative of an equivalence class that is contained) in* $W^{1,1}(\Omega;\mathbb{R}^n)$. *Then there is a Lebesgue null set* $N_{\mathbf{u}} \subset \Omega$ *such that for any measurable function* $\varphi\colon \mathbb{R}^n \to \mathbb{R}$ *and any measurable set* $A \subset \Omega$

$$\int_A \varphi(\mathbf{u}(\mathbf{x}))\big|\det \nabla\mathbf{u}(\mathbf{x})\big|\,\mathrm{d}\mathbf{x} = \int_{\mathbf{u}(A\setminus N_{\mathbf{u}})} \mathbb{N}(\mathbf{u}, A, \mathbf{y})\varphi(\mathbf{y})\,\mathrm{d}\mathbf{y} \tag{2.5}$$

whenever either integral exists. Here

$$\mathbb{N}(\mathbf{u}, A, \mathbf{y}) := \text{cardinality}\big\{\mathbf{x} \in A\colon \mathbf{u}(\mathbf{x}) = \mathbf{y}\big\}.$$

REMARK. Marcus and Mizel [20] showed that if $p > n$ and if $\mathbf{u}$ is (the continuous representative of an equivalence class) in $W^{1,p}(\Omega;\mathbb{R}^n)$ then $\mathbf{u}$ maps null sets onto null sets and one can replace $A\setminus N_{\mathbf{u}}$ by A in (2.5).

We call a (nonnegative outer) measure μ on Ω a *Radon measure* provided that every Borel subset of Ω is μ-measurable, every subset of Ω is contained in a Borel set of the same μ-measure, and every compact subset of Ω has finite μ-measure. Thus, $\mathcal{L}^n$ is a Radon measure on Ω, but $\mathcal{H}^k$ is not when $k < n$. We say that a sequence of Radon measures μ_j on Ω converges *weak* in the sense of measures* to a Radon measure μ, denoted $\mu_j \overset{*}{\rightharpoonup} \mu$, provided

$$\int_\Omega \phi(\mathbf{x})\,\mathrm{d}\mu_j(\mathbf{x}) \to \int_\Omega \phi(\mathbf{x})\,\mathrm{d}\mu(\mathbf{x})$$

for every $\phi \in C_0(\Omega)$, i.e., every continuous function $\phi\colon \Omega \to \mathbb{R}$ that is supported in a compact subset of Ω. We will make use of the following result concerning Radon measures. For a proof, as well as a thorough discussion of Radon measures, see [10].

PROPOSITION 2.8 (Weak* compactness of Radon measures). *Let* μ_j *be a sequence of Radon measures on* Ω *that satisfy*

$$\sup_j \mu_j(K) < \infty \quad \textit{for each compact} \quad K \subset \Omega.$$

Then there exists a subsequence (not relabeled) and a Radon measure μ *such that*

$$\mu_j \overset{*}{\rightharpoonup} \mu \quad \textit{in the sense of measures.}$$

If $\mathbf{u} \in W^{1,p}_{\mathrm{loc}}(\Omega;\mathbb{R}^n)$, with $p > n^2/(n+1)$, then the linear functional $(\mathrm{Det}\nabla\mathbf{u})$: $C^1_0(\Omega) \to \mathbb{R}$ given by

$$(\mathrm{Det}\nabla\mathbf{u})(\phi) := -\frac{1}{n}\int_\Omega \nabla\phi \cdot (\mathrm{adj}\nabla\mathbf{u})\mathbf{u}\,\mathrm{d}\mathbf{x} \tag{2.6}$$

is a well-defined distribution, which is called the **distributional Jacobian**. If $\mathbf{u} \in W^{1,p}_{\mathrm{loc}}(\Omega; \mathbb{R}^n)$, with $p \geqslant n$ then the identity $\operatorname{div}(\operatorname{adj}\nabla\mathbf{u})^{\mathrm{T}} = \mathbf{0}$ can be used to show that $\mathrm{Det}\nabla\mathbf{u}$ is the distribution induced by the function $\det\nabla\mathbf{u}$. Indeed, for simplicity let $\mathbf{u}$ be C^2 on $\overline{\Omega}$ then, by the product rule and the above identity,

$$\begin{aligned}\operatorname{div}\big(\phi(\operatorname{adj}\nabla\mathbf{u})\mathbf{u}\big) &= \nabla\phi\cdot(\operatorname{adj}\nabla\mathbf{u})\mathbf{u} + \phi(\operatorname{adj}\nabla\mathbf{u})^{\mathrm{T}}\cdot\nabla\mathbf{u} + \phi\mathbf{u}\cdot\operatorname{div}(\operatorname{adj}\nabla\mathbf{u})^{\mathrm{T}}\\ &= \nabla\phi\cdot(\operatorname{adj}\nabla\mathbf{u})\mathbf{u} + n\phi(\det\nabla\mathbf{u})\end{aligned}$$

and hence, since ϕ has compact support, we conclude with the aid of the divergence theorem and (2.6) that

$$(\mathrm{Det}\nabla\mathbf{u})(\phi) = \int_\Omega \phi(\mathbf{x})\det\nabla\mathbf{u}(\mathbf{x})\,\mathrm{d}\mathbf{x}. \tag{2.7}$$

In general, this need not be the case and in fact (2.7) will not be satisfied when cavitation occurs. For example, the deformation

$$\mathbf{w}(\mathbf{x}) = \frac{r(|\mathbf{x}|)}{|\mathbf{x}|}\mathbf{x}$$

with $r \in C([0,1];\mathbb{R}) \cap C^1((0,1);\mathbb{R})$, $r(0) > 0$, and $r' > 0$, which creates a spherical hole at the center of the body $\Omega = B(\mathbf{0}, 1)$, has

$$(\mathrm{Det}\nabla\mathbf{w})(\phi) = \int_\Omega \phi(\mathbf{x})\det\nabla\mathbf{w}(\mathbf{x})\,\mathrm{d}\mathbf{x} + \omega_n r(0)^n\phi(\mathbf{0}) \quad \forall\phi\in C^1_0(\Omega), \tag{2.8}$$

where ω_n is the volume of the n-dimensional unit ball $B(\mathbf{0}, 1)$.

The above paragraph shows that, for radial cavitation, *the distributional Jacobian is a Radon measure*, that is, there exists a (nonnegative) Radon (outer) measure μ_J on Ω such that

$$(\mathrm{Det}\nabla\mathbf{u})(\phi) = \int_\Omega \phi(\mathbf{x})\,\mathrm{d}\mu_J(\mathbf{x}) \quad \text{for all } \phi\in C^1_0(\Omega). \tag{2.9}$$

In particular, if $\mathrm{Det}\nabla\mathbf{u} \geqslant 0$, i.e., $(\mathrm{Det}\nabla\mathbf{u})(\phi) \geqslant 0$ for all $\phi \in C^1_0(\Omega)$ that are nonnegative, then the Riesz representation theorem (see, e.g., [10, pp. 49–54]) can be used to show that there exists a Radon measure μ_J that satisfies (2.9). To simplify notation, whenever such a measure exists we will denote it by $\mathrm{Det}\nabla\mathbf{u}$ and thus not distinguish between the distribution $\mathrm{Det}\nabla\mathbf{u}$ and the measure μ_J. Thus, for any open set $U \subset \Omega$,

$$(\mathrm{Det}\nabla\mathbf{u})(U) := \mu_J(U) := \sup\big\{(\mathrm{Det}\nabla\mathbf{u})(\phi) \colon \phi\in C^1_0(U),\ \|\phi\|_\infty \leqslant 1\big\},$$

while, for an arbitrary set $A \subset \Omega$,

$$(\mathrm{Det}\nabla\mathbf{u})(A) := \mu_J(A) := \inf\big\{(\mathrm{Det}\nabla\mathbf{u})(U) \colon A\subset U\subset\Omega,\ U \text{ open}\big\}.$$

Hence, we will write

$$\mathrm{Det}\nabla\mathbf{w} = (\det\nabla\mathbf{w})\mathcal{L}^n + \omega_n r(0)^n\delta_0$$

instead of (2.8). Here, and in what follows, $\delta_{\mathbf{b}}$ denotes the Dirac measure centered at the point $\mathbf{b} \in \mathbb{R}^n$.

Now suppose $\mathbf{u} \in W^{1,p}_{\mathrm{loc}}(\Omega; \mathbb{R}^n)$, with $p > n - 1$ (rather than $p > n^2/(n+1)$). Then the precise representative $\mathbf{u}^*$ is continuous on the sphere $\partial B(\mathbf{b}, r)$ for almost every r and hence $\mathbf{u}^*(\partial B(\mathbf{b}, r))$ is compact for such r. If, in addition, $\mathbf{u}^*$ satisfies condition (INV) then it follows that $\mathbf{u}^* \in L^\infty_{\mathrm{loc}}(\Omega; \mathbb{R}^n)$ and hence that (2.6) is once again a well-defined distribution on Ω. The next result shows that in fact this distribution is a Radon measure.

PROPOSITION 2.9 (Müller [*23*] and [*24*, Lemma 8.1]). *Let* $\mathbf{u} \in W^{1,p}_{\mathrm{loc}}(\Omega, \mathbb{R}^n)$ *with* $p > n - 1$. *Suppose that* $\det\nabla\mathbf{u} > 0$ *a.e. and that* $\mathbf{u}^*$ *satisfies condition (INV). Then* $\mathrm{Det}\nabla\mathbf{u} \geqslant 0$ *and hence* $\mathrm{Det}\nabla\mathbf{u}$ *is a Radon measure. Furthermore,*

$$\mathrm{Det}\nabla\mathbf{u} = (\det\nabla\mathbf{u})\mathcal{L}^n + m,$$

where m is singular with respect to Lebesgue measure, and for $\mathcal{L}^1$ a.e. $r \in (0, r_{\mathbf{b}})$ one has

$$(\mathrm{Det}\nabla\mathbf{u})\big(\overline{B(\mathbf{b}, r)}\big) = \mathcal{L}^n\big(\mathrm{im}_T\big(\mathbf{u}, B(\mathbf{b}, r)\big)\big). \tag{2.10}$$

3. Main Convergence Result

We now fix a finite set of points in the material and restrict our attention to deformations that may only open new holes in the interior of the body at these points.

DEFINITION 3.1. Let $p > n - 1$ and suppose that $\mathbf{a}_i \in \Omega$, $i = 1, 2, 3, \ldots, M$ are given. Define $\mathrm{Def}^p(\Omega, \mathbf{a}_1, \ldots, \mathbf{a}_M)$ to be those maps $\mathbf{u} \in W^{1,p}(\Omega; \mathbb{R}^n)$ that satisfy:

(i) $\mathbf{u}^*$ satisfies condition (INV);
(ii) $\det\nabla\mathbf{u} > 0$ a.e.;
(iii) $\mathrm{Det}\nabla\mathbf{u} = (\det\nabla\mathbf{u})\mathcal{L}^n + \sum_{i=1}^M \alpha^i \delta_{\mathbf{a}_i}$,

where α^i are nonnegative real numbers that may depend on $\mathbf{u}$.

Thus, a *deformation* is a Sobolev mapping with positive Jacobian that can only open holes at a finite number of specified points in the body and which satisfies condition (INV) (and consequently is one-to-one a.e.).

REMARKS. 1. Results in [24] imply that each deformation $\mathbf{u} \in \mathrm{Def}^p$ has a representative that, in addition to being continuous $\mathcal{H}^{n-p}$ a.e., maps sets of measure zero onto such sets. This representative satisfies (i) and (ii) of condition (INV) for *every* $\mathbf{x} \in B(\mathbf{b}, r)$ and, when one uses this representative, $A \backslash N_{\mathbf{u}}$ can be replaced by A in equation (2.5) of the change of variables formula.

2. Note that deformations in $\mathrm{Def}^p(\Omega, \mathbf{a}_1, \ldots, \mathbf{a}_M)$ may not be smooth or even continuous on the entire complement of (potential) cavitation points $\mathbf{a}_i$. See [29] for an example of a deformation that is not continuous yet opens no new holes.

3. Examples in [24] show that a deformation in $\mathrm{Def}^p(\Omega, \mathbf{a}_1, \ldots, \mathbf{a}_M)$ may open additional holes at the boundary of the region.

We now state the main result of this section.

LEMMA 3.2. *Let $p > n-1$ and suppose that $\mathbf{u}_j \in \mathrm{Def}^p(\Omega, \mathbf{a}_1, \ldots, \mathbf{a}_M)$ satisfy*

$$\mathbf{u}_j \rightharpoonup \mathbf{u} \quad \text{in } W^{1,p}(\Omega; \mathbb{R}^n) \tag{3.1}$$

and

$$\det\nabla\mathbf{u}_j \rightharpoonup \theta \quad \text{in } L^1(\Omega) \tag{3.2}$$

for some $\mathbf{u} \in W^{1,p}(\Omega; \mathbb{R}^n)$ and $\theta \in L^1(\Omega)$. Assume, further, that $\theta > 0$ a.e. in Ω. Then

$$\mathbf{u} \in \mathrm{Def}^p(\Omega, \mathbf{a}_1, \ldots, \mathbf{a}_M) \quad \text{and} \quad \theta = \det\nabla\mathbf{u}. \tag{3.3}$$

In order to prove the above lemma we will need the following result. This result shows that the set of maps whose precise representatives satisfy (INV) is sequentially weakly closed in $W^{1,p}$. In addition, mappings that satisfy condition (INV) are in $L^\infty_{\mathrm{loc}}(\Omega)$ and consequently sequences of such deformations that converge weakly satisfy an additional convergence property.

LEMMA 3.3. *Let $p > n-1$ and suppose that $\mathbf{u}_j^*$ is a sequence in $W^{1,p}(\Omega; \mathbb{R}^n)$ that satisfies condition (INV). Assume that*

$$\mathbf{u}_j \rightharpoonup \mathbf{u} \quad \text{in } W^{1,p}(\Omega; \mathbb{R}^n).$$

Then $\mathbf{u}^*$ satisfies condition (INV). Moreover, there exists a subsequence (not relabeled) that satisfies

$$\mathbf{u}_j \to \mathbf{u} \quad \text{in } L^q_{\mathrm{loc}}(\Omega; \mathbb{R}^n) \tag{3.4}$$

for every $1 \leqslant q < \infty$.

Proof of Lemma 3.3. Without loss of generality we take $\mathbf{u}_j = \mathbf{u}_j^*$ and $\mathbf{u} = \mathbf{u}^*$. For a proof that $\mathbf{u}$ satisfies (INV) see Lemma 3.3 in [24]. To prove (3.4) we first note that, by the Rellich compactness theorem, there is a subsequence that converges strongly to $\mathbf{u}$ in $L^p(\Omega; \mathbb{R}^n)$ and hence a further subsequence (not relabeled) that satisfies $\mathbf{u}_j \to \mathbf{u}$ a.e. Let $1 \leqslant q < \infty$, $\mathbf{b} \in \Omega$, and suppose that $r \in (0, r_{\mathbf{b}})$ is such that $\mathbf{u}$ satisfies (i) and (ii) of condition (INV) on $B(\mathbf{b}, r)$ and (2.3) of Proposition 2.2. We will show that

$$\mathbf{u}_j \to \mathbf{u} \quad \text{in } L^q(B(\mathbf{b}, r); \mathbb{R}^n).$$

This will imply the desired result since any compact subset of Ω can be covered by a finite number of such balls.

Define

$$U := \{\mathbf{y} \in \mathbb{R}^n\colon \operatorname{dist}\big(\mathbf{y}, \mathbf{u}(\partial B(\mathbf{b}, r))\big) \leqslant 1\}.$$

Then by (2.3) of Proposition 2.2 one has that, for all $\mathbf{z} \in \partial B(\mathbf{b}, r)$ and for all j sufficiently large, $|\mathbf{u}_j(\mathbf{z}) - \mathbf{u}(\mathbf{z})| < 1$ and hence $\mathbf{u}_j(\mathbf{z}) \in U$. Thus for all such j the boundaries of each of the open sets $\operatorname{im}_T(\mathbf{u}_j, B(\mathbf{b}, r))$ are contained in the bounded set U. Therefore, the sets $\operatorname{im}_T(\mathbf{u}_j, B(\mathbf{b}, r))$ are bounded uniformly in j and hence, in view of (i) of (INV), there is a $K > 0$ such that

$$\|\mathbf{u}_j\|_{L^\infty(B(\mathbf{b},r))} \leqslant K$$

for all j sufficiently large. Since $\mathbf{u}_j \to \mathbf{u}$ a.e., the desired result now follows from the Lebesgue dominated convergence theorem. □

Proof of Lemma 3.2. Without loss of generality we take $\mathbf{u}_j = \mathbf{u}_j^*$ and $\mathbf{u} = \mathbf{u}^*$. Let $\mathbf{u}_j \in \operatorname{Def}^p(\Omega, \mathbf{a}_1, \ldots, \mathbf{a}_M)$ satisfy (3.1) and (3.2) where $\theta > 0$ a.e. Then, by (3.1) and Lemma 3.3, $\mathbf{u}^*$ satisfies condition (INV) and

$$\mathbf{u}_j \to \mathbf{u} \quad \text{in } L^q_{\text{loc}}\big(\Omega; \mathbb{R}^n\big) \quad \text{for every } q \in (1, \infty). \tag{3.5}$$

Moreover, since $p > n - 1$ (see, e.g., [1, Theorem 3.4] or [8, Theorem 7.5-1])

$$\operatorname{adj}\nabla\mathbf{u}_j \rightharpoonup \operatorname{adj}\nabla\mathbf{u} \quad \text{in } L^{p/(n-1)}\big(\Omega; \mathbb{R}^n\big). \tag{3.6}$$

We now show that $\operatorname{Det}\nabla\mathbf{u}$ is a Radon measure. By (3.5) and (3.6)

$$(\operatorname{adj}\nabla\mathbf{u}_j)\mathbf{u}_j \rightharpoonup (\operatorname{adj}\nabla\mathbf{u})\mathbf{u} \quad \text{in } L^1_{\text{loc}}\big(\Omega; \mathbb{R}^n\big) \tag{3.7}$$

and hence

$$(\operatorname{Det}\nabla\mathbf{u}_j)(\varphi) \to (\operatorname{Det}\nabla\mathbf{u})(\varphi) \quad \text{for every } \varphi \in C_0^\infty(\Omega). \tag{3.8}$$

Next, let $K \subset \Omega$ be compact. Then there exists $\psi = \psi_K \in C_0^1(\Omega; [0, 1])$ with $\psi = 1$ on K. Let $K' \subset \Omega$ be the closure of the set where ψ is strictly positive. If we write μ_j for the measure $\operatorname{Det}\nabla\mathbf{u}_j$ we find, with the aid of (2.9), that

$$(\operatorname{Det}\nabla\mathbf{u}_j)(K) = \int_K 1 \,\mathrm{d}\mu_j(\mathbf{x}) \leqslant \int_\Omega \psi(\mathbf{x}) \,\mathrm{d}\mu_j(\mathbf{x}) = (\operatorname{Det}\nabla\mathbf{u}_j)(\psi)$$

and, by definition of the distributional Jacobian,

$$\begin{aligned}(\operatorname{Det}\nabla\mathbf{u}_j)(\psi) &:= -\frac{1}{n}\int_\Omega \nabla\psi \cdot (\operatorname{adj}\nabla\mathbf{u}_j)\mathbf{u}_j \,\mathrm{d}\mathbf{x} \\ &\leqslant n^{-1} \sup\{|\nabla\psi(\mathbf{x})|\colon \mathbf{x} \in K'\} \big\|(\operatorname{adj}\nabla\mathbf{u}_j)\mathbf{u}_j\big\|_{L^1(K')}.\end{aligned}$$

However, in view of (3.7), $\|(\operatorname{adj}\nabla\mathbf{u}_j)\mathbf{u}_j\|_{L^1(K')}$ is bounded uniformly in j and hence

$$\sup_j\,(\mathrm{Det}\nabla\mathbf{u}_j)(K) < +\infty \quad \text{for every compact } K \subset \Omega. \tag{3.9}$$

Therefore, by the weak* compactness result for measures (Proposition 2.8), a subsequence of these measures converges to a Radon measure. This, together with (3.8), shows that $\mathrm{Det}\nabla\mathbf{u}$ is a Radon measure and that

$$\mathrm{Det}\nabla\mathbf{u}_j \overset{*}{\rightharpoonup} \mathrm{Det}\nabla\mathbf{u} \quad \text{in the sense of measures.} \tag{3.10}$$

Next, since $\mathbf{u}_j \in \mathrm{Def}^p(\Omega, \mathbf{a}_1, \ldots, \mathbf{a}_M)$

$$\mathrm{Det}\nabla\mathbf{u}_j = (\det\nabla\mathbf{u}_j)\mathcal{L}^n + \sum_{i=1}^{M} \alpha_j^i \delta_{\mathbf{a}_i}$$

with $\det\nabla\mathbf{u}_j > 0$ a.e. and $\alpha_j^i \geqslant 0$. In particular, it follows that for any $r \in (0, r_{\mathbf{a}_i})$

$$\alpha_j^i \leqslant (\mathrm{Det}\nabla\mathbf{u}_j)\big(\overline{B(\mathbf{a}_i, r)}\big).$$

Thus, by (3.9) with $K := \overline{B(\mathbf{a}_i, r)}$, for any k

$$0 \leqslant \alpha_k^i \leqslant \sup_j \alpha_j^i \leqslant \sup_j\,(\mathrm{Det}\nabla\mathbf{u}_j)(K) < +\infty.$$

Therefore, there exist $\alpha^i \in [0, \infty)$ such that (for a subsequence) $\alpha_j^i \to \alpha^i$ and consequently, as $j \to \infty$,

$$\alpha_j^i \delta_{\mathbf{a}_i} \overset{*}{\rightharpoonup} \alpha^i \delta_{\mathbf{a}_i} \quad \text{in the sense of measures.}$$

In view of (3.2) we therefore conclude that, as $j \to \infty$,

$$\mathrm{Det}\nabla\mathbf{u}_j \overset{*}{\rightharpoonup} \theta\mathcal{L}^n + \sum_{i=1}^{M} \alpha^i \delta_{\mathbf{a}_i} \quad \text{in the sense of measures,}$$

which together with (3.10) yields

$$\mathrm{Det}\nabla\mathbf{u} = \theta\mathcal{L}^n + \sum_{i=1}^{M} \alpha^i \delta_{\mathbf{a}_i}. \tag{3.11}$$

If $p > n^2/(n+1)$ we are done since a result of Müller [23] then yields $\theta = \det\nabla\mathbf{u}$, which together with (3.11) and the hypothesis $\theta > 0$ a.e. give (ii) and (iii) of Definition 3.1. In general, the desired result will follow if we show that $\det\nabla\mathbf{u} > 0$ a.e. since we can then apply Proposition 2.9 together with (3.11) to arrive at the same conclusion for all $p > n - 1$.

In order to get $\det\nabla\mathbf{u} > 0$ a.e. we first show that $\det\nabla\mathbf{u} \neq 0$ a.e. By the change of variables formula for Sobolev mappings (Proposition 2.7) and the fact that each $\mathbf{u}_j$ is one-to-one a.e. (Proposition 2.5) one has that for all $\varphi \in C_0(\mathbb{R}^n)$

$$\int_\Omega \varphi\big(\mathbf{u}_j(\mathbf{x})\big)\,\det\nabla\mathbf{u}_j(\mathbf{x})\,\mathrm{d}\mathbf{x} = \int_{\mathbb{R}^n} \varphi(\mathbf{y})\chi_{\mathbf{u}_j(\Omega\setminus N_j)}(\mathbf{y})\,\mathrm{d}\mathbf{y}, \tag{3.12}$$

for some Lebesgue null sets N_j.

In preparation to letting $j \to \infty$ in (3.12) we first note that, by the Rellich compactness theorem, we may assume that $\mathbf{u}_j \to \mathbf{u}$ a.e. by passing, if necessary, to a subsequence. Thus, since φ is continuous,

$$\varphi\big(\mathbf{u}_j(\mathbf{x})\big) \to \varphi\big(\mathbf{u}(\mathbf{x})\big) \quad \text{for a.e. } \mathbf{x} \in \Omega. \tag{3.13}$$

Moreover, each of the compositions, $\varphi \circ \mathbf{u}_j$, is contained in $L^\infty(\Omega)$ since φ is bounded. Next, the sequence of characteristic functions satisfies $\|\chi_{\mathbf{u}_j(\Omega\setminus N_j)}\|_\infty \leqslant 1$ and hence, since the unit ball is compact in the weak* topology, there is a $\zeta \in L^\infty(\mathbb{R}^n)$ with $\|\zeta\|_\infty \leqslant 1$ and a subsequence (not relabeled) such that

$$\chi_{\mathbf{u}_j(\Omega\setminus N_j)} \overset{*}{\rightharpoonup} \zeta \quad \text{in } L^\infty(\Omega). \tag{3.14}$$

We now take the limit of (3.12) as $j \to \infty$ to conclude, with the aid of (3.13), (3.14), and Lemma 6.7, that

$$\int_\Omega \varphi\big(\mathbf{u}(\mathbf{x})\big)\theta(\mathbf{x})\,\mathrm{d}\mathbf{x} = \int_{\mathbb{R}^n} \varphi(\mathbf{y})\zeta(\mathbf{y})\,\mathrm{d}\mathbf{y}. \tag{3.15}$$

Equation (3.15) is satisfied for all continuous φ with compact support and hence, by approximation (using the monotone convergence theorem) for φ the characteristic function of any open set.

We now show that $\det\nabla\mathbf{u} \neq 0$ a.e. Let $N_\mathbf{u}$ be the Lebesgue null set from the change of variables formula (Proposition 2.7) and define

$$M := \big\{\mathbf{x} \in \Omega\colon \det\nabla\mathbf{u}(\mathbf{x}) = 0\big\}\setminus N_\mathbf{u}.$$

Then Proposition 2.7 (with $A = M$) yields

$$0 = \int_{\mathbf{u}(M)} \mathbb{N}(\mathbf{u}, M, \mathbf{y})\psi(\mathbf{y})\,\mathrm{d}\mathbf{y}$$

for all measurable functions ψ and consequently $\mathbf{u}(M)$ is a null set. Let $\varepsilon > 0$ and suppose that $U \supset \mathbf{u}(M)$ is open with $\mathcal{L}^n(U) < \varepsilon$. Then $\chi_{\mathbf{u}(M)} \leqslant \chi_U$ and hence,

$$\chi_M(\mathbf{x}) \leqslant \chi_{\mathbf{u}(M)}\big(\mathbf{u}(\mathbf{x})\big) \leqslant \chi_U\big(\mathbf{u}(\mathbf{x})\big) \tag{3.16}$$

for a.e. $\mathbf{x} \in \Omega$. We note that $\|\zeta\|_\infty \leqslant 1$ and $\theta > 0$ a.e. and apply (3.15) with $\varphi = \chi_U$ to obtain, with the aid of (3.16),

$$\int_M \theta(\mathbf{x})\,\mathrm{d}\mathbf{x} \leqslant \int_\Omega \chi_U\big(\mathbf{u}(\mathbf{x})\big)\theta(\mathbf{x})\,\mathrm{d}\mathbf{x} = \int_U \zeta(\mathbf{y})\,\mathrm{d}\mathbf{y} \leqslant \mathcal{L}^n(U) < \varepsilon.$$

Thus, since $\theta > 0$ a.e. and since ε was arbitrary, we deduce that $\mathcal{L}^n(M) = 0$. Therefore, $\det \nabla \mathbf{u} \neq 0$ a.e. since $N_{\mathbf{u}}$ is itself a Lebesgue null set.

Finally, we show that $\det \nabla \mathbf{u} > 0$ a.e. Let $\mathbf{b} \in \Omega$ and let $N_{\mathbf{b}}$ be the $\mathcal{L}^1$ null set of Proposition 2.2. Then by Proposition 2.6(i) there exist $\mathcal{L}^1$ null sets N_j such that for every $r \in (0, r_{\mathbf{b}}) \backslash N_j$

$$\deg\big(\mathbf{u}_j, \partial B(\mathbf{b}, r), \mathbf{y}\big) \in \{0, 1\} \quad \text{for all } \mathbf{y} \in \mathbb{R}^n \backslash \mathbf{u}_j\big(\partial B(\mathbf{b}, r)\big). \tag{3.17}$$

Define $N := N_{\mathbf{b}} \cup (\bigcup_{j=1}^{\infty} N_j)$. Fix $r \in (0, r_{\mathbf{b}}) \backslash N$. Then by Proposition 2.2 there is a subsequence (not relabeled) such that $\mathbf{u}_j \to \mathbf{u}$ uniformly on $\partial B(\mathbf{b}, r)$. Therefore, in view of (3.17) and Proposition 2.1, we conclude that

$$\deg\big(\mathbf{u}, \partial B(\mathbf{b}, r), \mathbf{y}\big) \in \{0, 1\} \quad \text{for all } \mathbf{y} \in \mathbb{R}^n \backslash \mathbf{u}\big(\partial B(\mathbf{b}, r)\big). \tag{3.18}$$

Since (3.18) is satisfied for all $r \in (0, r_{\mathbf{b}}) \backslash N$ and since $\det \nabla \mathbf{u} \neq 0$ a.e. we can apply Proposition 2.6(ii) to conclude that $\det \nabla \mathbf{u} > 0$ a.e. in $B(\mathbf{b}, r_{\mathbf{b}})$, for every $\mathbf{b} \in \Omega$. The desired result, (3.3), now follows from Proposition 2.9 and equation (3.11). □

4. Existence of Minimizers

We consider an elastic body that, for convenience, we identify with the bounded, open, connected region $\mathcal{B} \subset \mathbb{R}^n$ with (strongly) Lipschitz boundary that it occupies in a fixed reference configuration.

4.1. THE DISPLACEMENT PROBLEM

Let $\mathcal{B} \subset\subset \Omega \subset \mathbb{R}^n$, where Ω is bounded, open, and connected and has (strongly) Lipschitz boundary. Suppose that a diffeomorphism $\mathbf{d}\colon \overline{\Omega} \to \mathbb{R}^n$ with strictly positive Jacobian is given. If $\mathbf{u} \in W^{1,p}(\mathcal{B}; \mathbb{R}^n)$ satisfies $\mathbf{u} - \mathbf{d} \in W_0^{1,p}(\mathcal{B}; \mathbb{R}^n)$ then we define its *extension* $\mathbf{u}^e\colon \Omega \to \mathbb{R}^n$ by

$$\mathbf{u}^e(\mathbf{x}) := \begin{cases} \mathbf{u}(\mathbf{x}), & \mathbf{x} \in \mathcal{B}, \\ \mathbf{d}(\mathbf{x}), & \mathbf{x} \notin \mathcal{B}, \end{cases}$$

and note that $\mathbf{u}^e \in W^{1,p}(\Omega; \mathbb{R}^n)$.

We first ignore any possible energy due to hole formation (see the Introduction), fix $\mathbf{a}_k \in \mathcal{B}$, $k = 1, 2, 3, \dots, M$, and seek a minimizer for the total elastic energy

$$E(\mathbf{u}) := \int_{\mathcal{B}} W\big(\mathbf{x}, \mathbf{u}(\mathbf{x}), \nabla \mathbf{u}(\mathbf{x})\big)\, d\mathbf{x} \tag{4.1}$$

in the class of admissible functions

$$\mathcal{A}^p := \big\{\mathbf{u} \in W^{1,p}\big(\mathcal{B}; \mathbb{R}^n\big) \colon \mathbf{u} - \mathbf{d} \in W_0^{1,p},\ \mathbf{u}^e \in \mathrm{Def}^p(\Omega, \mathbf{a}_1, \dots, \mathbf{a}_M)\big\}.$$

REMARKS. 1. The seemingly artificial requirement that admissible deformations possess local extensions that satisfy condition (INV) is necessitated by the problem of cavitation at the boundary (see [24]). In particular, given $\mathbf{a} \in \partial\mathcal{B}$ and (sufficiently small) $\alpha > 0$ one can construct deformations that satisfy

$$\mathrm{Det}\nabla\mathbf{u} = (\det\nabla\mathbf{u})\mathcal{L}^n, \qquad \mathrm{Det}\nabla\mathbf{u}^e = \big(\det\nabla\mathbf{u}^e\big)\mathcal{L}^n + \alpha\delta_{\mathbf{a}}.$$

Thus, the distributional Jacobian of a Sobolev mapping does not detect holes created at the boundary of the domain on which the mapping is defined. In the next subsection we will prove existence without such an extension requirement.

2. Note that an admissible mapping $\mathbf{u} \in \mathcal{A}^p$ must lie in $L^\infty(\mathcal{B}; \mathbb{R}^n)$ and satisfy $\|\mathbf{u}\|_\infty \leqslant \|\mathbf{d}\|_\infty$. To see this first suppose that Ω is sufficiently large so that $\mathcal{B} \subset B(\mathbf{b}, R) \subset\subset \Omega$ for some $\mathbf{b} \in \mathcal{B}$ and $R > 0$. Then for any sufficiently small $\varepsilon > 0$, $\mathcal{B} \subset B(\mathbf{b}, R+\varepsilon) \subset \Omega$. Since $\mathbf{u}^e = \mathbf{d}$ on $\partial B(\mathbf{b}, R+\varepsilon)$ and $\mathbf{d}$ is a homeomorphism it follows that (see the remark following the definition of the topological image) $\mathrm{im}_T(\mathbf{u}^e, B(\mathbf{b}, R+\varepsilon)) = \mathbf{d}(B(\mathbf{b}, R+\varepsilon))$. Consequently, since $\mathbf{u}^e$ satisfies (i) of (INV), for a.e. ε we have that $\mathbf{u}(\mathbf{x}) = \mathbf{u}^e(\mathbf{x}) \in \mathbf{d}(\overline{B(\mathbf{b}, R+\varepsilon)}) \subset \mathbf{d}(\overline{\Omega})$ for a.e. $\mathbf{x} \in \mathcal{B} \subset \overline{B(\mathbf{b}, R+\varepsilon)}$, which proves the claim for such Ω. For general Ω, one first replaces the set $B(\mathbf{b}, R)$ by an open set U, with C^1 boundary, that satisfies $\mathcal{B} \subset U \subset\subset \Omega$. The desired result then follows from the above argument with the balls $B(\mathbf{b}, R+\varepsilon)$ replaced by sets U_ε that are constructed in [24, Theorem 9.1]. (The sets U_ε are, *essentially*, defined by $\partial U_\varepsilon = \{\mathbf{x} + \varepsilon\mathbf{n}(\mathbf{x}) : \mathbf{x} \in \partial U\}$, where $\mathbf{n}$ is the outward unit normal to U. Theorem 9.1 shows that condition (INV) together with $\det\nabla\mathbf{u} > 0$ a.e. implies that (i) and (ii) of (INV) are also satisfied when balls are replaced by the sets U_ε.)

THEOREM 4.1. *Let $n = 3$, $p > 2$, $\mathcal{D} := \mathrm{Lin}^3 \times \mathrm{Lin}^3 \times (0, \infty)$, and suppose that the following conditions are satisfied:*

(i) *(Polyconvexity) There is a function $\Phi\colon \mathcal{B} \times \mathbb{R}^3 \times \mathcal{D} \to \mathbb{R}$ such that, for every $\mathbf{u} \in \mathbb{R}^3$ and a.e. $\mathbf{x} \in \mathcal{B}$,*

$$W(\mathbf{x}, \mathbf{u}, \mathbf{F}) = \Phi\big(\mathbf{x}, \mathbf{u}, (\mathbf{F}, \mathrm{adj}\,\mathbf{F}, \det\mathbf{F})\big) \quad \textit{whenever } \det\mathbf{F} > 0,$$

where $\Phi(\mathbf{x}, \mathbf{u}, \cdot)\colon \mathcal{D} \to \mathbb{R}$ is convex for the same $\mathbf{u}$ and $\mathbf{x}$.

(ii) *(Continuity) The function $\Phi(\mathbf{x}, \cdot, \cdot)\colon \mathbb{R}^3 \times \mathcal{D} \to \mathbb{R}$ is continuous for a.e. $\mathbf{x} \in \mathcal{B}$ and $\Phi(\cdot, \mathbf{u}, \mathcal{N})\colon \mathcal{B} \to \mathbb{R}$ is measurable for every $(\mathbf{u}, \mathcal{N}) \in \mathbb{R}^3 \times \mathcal{D}$.*

(iii) *(Coercivity) For each compact set $K \subset \mathbb{R}^3$*

$$W(\mathbf{x}, \mathbf{u}, \mathbf{F}) \geqslant \phi_K(\mathbf{x}) + c|\mathbf{F}|^p + \Gamma(\det\mathbf{F})$$

for a.e. $\mathbf{x} \in \mathcal{B}$ and every $(\mathbf{u}, \mathbf{F}) \in K \times \mathrm{Lin}^3$ with $\det\mathbf{F} > 0$, where $c > 0$, $\phi_K \in L^1(\mathcal{B})$ may depend on K, and $\Gamma\colon (0, \infty) \to \mathbb{R}$ is a convex function that satisfies $\Gamma(t)/t \to +\infty$ as $t \to +\infty$.

(iv) *$\Gamma(t) \to +\infty$ as $t \to 0^+$.*

Then E attains its infimum on $\mathcal{A}^p$.

REMARKS. 1. Body forces have been included in the function W. In particular, the dead-load body force $\mathbf{b}_0 \in L^1(\mathcal{B};\mathbb{R}^3)$ would contribute a term of the form $\beta(\mathbf{x},\mathbf{u}(\mathbf{x})) = \mathbf{b}_0(\mathbf{x})\cdot\mathbf{u}(\mathbf{x})$ to W. Such a term is usually not included in W due to difficulties in satisfying the coercivity hypothesis. However, in view of Remark 2 above, there is no such difficulty for the displacement problem since

$$\left|\beta(\mathbf{x},\mathbf{u}(\mathbf{x}))\right| \leqslant |\mathbf{b}_0(\mathbf{x})|\,\|\mathbf{d}\|_\infty.$$

2. Suppose that the boundary deformation is given by $\mathbf{d}(\mathbf{x}) = \lambda\mathbf{x}$. Then results of Ball [3] and many others (see [17]) imply that, for a large class of stored-energy functions of slow growth and for sufficiently large λ, any minimizer of E in the class $\mathcal{A}^p$ has strictly less energy than those deformations in $\mathcal{A}^p$ that satisfy $\mathrm{Det}\nabla\mathbf{u}^e = (\det\nabla\mathbf{u}^e)\mathcal{L}^3$. In other words, under these conditions the minimizer(s) of E will exhibit cavitation.

Proof of Theorem 4.1. Since $\mathbf{d} \in \mathcal{A}^p$ the set $\mathcal{A}^p$ is nonempty. Further, by (ii), $E(\mathbf{d})$ is finite and hence so is the infimum. We note that $\mathbf{u} \in \mathcal{A}^p$ satisfies $\|\mathbf{u}\|_\infty \leqslant \|\mathbf{d}\|_\infty$ and therefore choose $K := \overline{B(\mathbf{0},\|\mathbf{d}\|_\infty)}$. Thus, the coercivity of W implies that E is bounded below.

Let $\mathbf{u}_j \in \mathcal{A}^p$ be a minimizing sequence. Then by the coercivity of W and the Poincaré inequality (see, e.g., [22]) the sequence is bounded in $W^{1,p}(\mathcal{B};\mathbb{R}^3)$ and, hence, we may assume (for a subsequence) that $\mathbf{u}_j \rightharpoonup \mathbf{u}$ in $W^{1,p}(\mathcal{B};\mathbb{R}^3)$, $\mathbf{u}_j \to \mathbf{u}$ in $L^p(\mathcal{B};\mathbb{R}^3)$, and $\mathbf{u}_j \to \mathbf{u}$ a.e. We take $\mathbf{u}_j = \mathbf{u}_j^*$ and $\mathbf{u} = \mathbf{u}^*$.

In order to show that $\mathbf{u} \in \mathcal{A}^p$ we first note that $(\mathbf{u}_j - \mathbf{d}) \in W_0^{1,p}(\mathcal{B};\mathbb{R}^3)$ and consequently $(\mathbf{u} - \mathbf{d}) \in W_0^{1,p}(\mathcal{B};\mathbb{R}^3)$. In particular, $\mathbf{u}^e \in W^{1,p}(\Omega;\mathbb{R}^3)$ is well-defined. Next, In view of the superlinear growth of Γ at $+\infty$ we find, by the de la Vallée Poussin and Dunford–Pettis criteria (see, e.g., [21]) that there is a $\theta \in L^1(\mathcal{B})$ such that (for a subsequence)

$$\det\nabla\mathbf{u}_j \rightharpoonup \theta \quad \text{in } L^1(\mathcal{B}).$$

Define $\theta^e \in L^1(\Omega)$ by

$$\theta^e(\mathbf{x}) := \begin{cases} \theta(\mathbf{x}), & \mathbf{x} \in \mathcal{B}, \\ \det\nabla\mathbf{d}(\mathbf{x}), & \mathbf{x} \notin \mathcal{B}, \end{cases}$$

and note that

$$\det\nabla\mathbf{u}_j^e \rightharpoonup \theta^e \quad \text{in } L^1(\Omega).$$

Clearly $\theta \geqslant 0$ a.e. If θ was equal to zero on a set $A \subset \mathcal{B}$ of positive ($\mathcal{L}^3$) measure then we would have $\det\nabla\mathbf{u}_j \to 0$ in $L^1(A)$ and hence, by hypothesis (iv), (for a subsequence) $\Gamma(\det\nabla\mathbf{u}_j(\mathbf{x})) \to +\infty$ for a.e. $\mathbf{x} \in A$. In this case we would have $E(\mathbf{u}_j) \to +\infty$, by the coercivity of W and Fatou's lemma. Therefore $\theta > 0$ a.e. and hence $\theta^e > 0$ a.e. Consequently, we can apply Lemma 3.2 to conclude that

$$\det\nabla\mathbf{u}_j^e \rightharpoonup \det\nabla\mathbf{u}^e \quad \text{in } L^1(\Omega)$$

and $\mathbf{u}^e \in \mathrm{Def}^p(\Omega, \mathbf{a}_1, \ldots, \mathbf{a}_M)$. Therefore, $\mathbf{u} \in \mathcal{A}^p$ and

$$\det\nabla\mathbf{u}_j \rightharpoonup \det\nabla\mathbf{u} \quad \text{in } L^1(\mathcal{B}). \tag{4.2}$$

Finally, we show that $\mathbf{u}$ is a minimizer of E. We note that, since $p > 2$, $\mathrm{adj}\nabla\mathbf{u}_j \rightharpoonup \mathrm{adj}\nabla\mathbf{u}$ in $L^1(\mathcal{B}; \mathrm{Lin}^3)$ (see, e.g., [1, Theorem 3.4]; [8, Theorem 7.5-1]) and, hence, in view of (4.2) and hypotheses (i)–(iii), we can apply the lower-semicontinuity theorem of Ball et al. [5, Theorem 5.4] to conclude that

$$E(\mathbf{u}) \leqslant \liminf_{j\to\infty} E(\mathbf{u}_j) = \inf_{\mathcal{A}^p} E.$$

Therefore, $\mathbf{u} \in \mathcal{A}^p$ is the desired minimizer. □

As mentioned in the Introduction, our purely elastic model may not explain the refined experiments of [7]. We now examine the effect of including an energy associated with the formation of a new hole in order to address this difficulty. Let $\eta_i \geqslant 0$ for $i = 1, \ldots, M$ and let $\upsilon : [0, \infty)^M \to [0, \infty)$ be lower semicontinuous. Define

$$I(\mathbf{u}) := E(\mathbf{u}) + \upsilon(\alpha_\mathbf{u}^1, \ldots, \alpha_\mathbf{u}^M) + \sum_{i=1}^{M} H(\alpha_\mathbf{u}^i)\eta_i, \tag{4.3}$$

where

$$H(t) = \begin{cases} 1, & \text{if } t > 0, \\ 0, & \text{if } t \leqslant 0 \end{cases}$$

and $\mathbf{u} \in \mathcal{A}^p$ satisfies

$$\mathrm{Det}\nabla\mathbf{u} = (\det\nabla\mathbf{u})\mathcal{L}^3 + \sum_{i=1}^{M} \alpha_\mathbf{u}^i \delta_{\mathbf{a}_i}.$$

The term $H(\alpha_\mathbf{u}^i)\eta_i$ represents the energy associated with initiating a cavity at the point $\mathbf{a}_i$. Its value may be different at each cavitation point due to differing strengths of the preexisting flaws (sizes of the preexisting microvoids) in the material. The term $\upsilon(\alpha_\mathbf{u}^1, \ldots, \alpha_\mathbf{u}^M)$ represents an energy associated with cavity growth. It might, for example, only be positive when the holes are small so as to measure the relatively large surface energy needed to change a microvoid into a visible hole. Note that results on radial cavitation lead us to expect that the *cavitation load*, i.e., the minimal value of the principal stretches required to induce cavitation at a point, is an increasing function of the value of η at the point. A sufficiently small portion of the material might only have precursors with large η, which could explain the results of [7].

We note that the proofs of Lemma 3.2 and Theorem 4.1 show that $\mathbf{u} \mapsto \alpha_\mathbf{u}^k$ is sequentially weakly continuous on (an appropriately chosen subsequence of) each minimizing sequence in $\mathcal{A}^p$. Since both υ and H are lower semicontinuous it follows that each of the terms that constitute I is sequentially weakly lower semicontinuous on minimizing sequences in $\mathcal{A}^p$. The proof of Theorem 4.1, therefore, yields the following result.

THEOREM 4.2. *Let $n = 3$, $p > 2$ and suppose that W satisfies the hypotheses of Theorem 4.1. Assume further that I is given by (4.3), where $\eta_i \geqslant 0$ for $i = 1, \ldots, M$ and v: $[0, \infty)^M \to [0, \infty)$ are lower semicontinuous. Then I attains its infimum on $\mathcal{A}^p$.*

REMARK. The above result remains valid (see [24]) if one adds to I a surface energy that is proportional to the deformed surface area of the new cavities created in the material.

4.2. THE MIXED PROBLEM

In order to address the mixed problem let $\Omega = \mathcal{B}$ and $\partial\mathcal{B}_d \subset \partial\mathcal{B}$ with strictly positive $(n-1)$-dimensional measure ($\mathcal{H}^{n-1}(\partial\mathcal{B}_d) > 0$). We suppose that the deformation $\mathbf{d}$: $\partial\mathcal{B}_d \to \mathbb{R}^n$ is given on $\partial\mathcal{B}_d$, while dead-load tractions are prescribed on the remainder of the boundary, $\partial\mathcal{B}_t := \partial\mathcal{B}\backslash\partial\mathcal{B}_d$. We let E and I be given by (4.1) and (4.3), respectively, and seek a minimizer for each of the energies

$$\begin{aligned}\widehat{E}(\mathbf{u}) &:= E(\mathbf{u}) - L\mathbf{u},\\ \hat{I}(\mathbf{u}) &:= I(\mathbf{u}) - L\mathbf{u},\end{aligned}$$

where

$$L\mathbf{u} := \int_{\mathcal{B}} \mathbf{b}_0(\mathbf{x}) \cdot \mathbf{u}(\mathbf{x})\, \mathrm{d}\mathbf{x} + \int_{\partial\mathcal{B}_t} \mathbf{s}_0(\mathbf{x}) \cdot \mathbf{u}(\mathbf{x})\, \mathrm{d}\mathcal{H}^{n-1},$$

in the class of admissible functions

$$\mathcal{A}_m^p := \left\{\mathbf{u} \in W^{1,p}\left(\mathcal{B}; \mathbb{R}^n\right): \mathbf{u} = \mathbf{d} \text{ on } \partial\mathcal{B}_d,\ \mathbf{u} \in \mathrm{Def}^p(\mathcal{B}, \mathbf{a}_1, \ldots, \mathbf{a}_M)\right\}.$$

Here the equality on $\partial\mathcal{B}_d$ is taken to be in the sense of trace.

Due to the lack of an apriori L^∞-bound on such admissible deformations we must strengthen hypothesis (iii) to

(iii′) (coercivity)

$$W(\mathbf{x}, \mathbf{u}, \mathbf{F}) \geqslant \phi(\mathbf{x}) + c|\mathbf{F}|^p + \Gamma(\det \mathbf{F})$$

for a.e. $\mathbf{x} \in \mathcal{B}$ and every $(\mathbf{u}, \mathbf{F}) \in \mathbb{R}^3 \times \mathrm{Lin}^3$ with $\det \mathbf{F} > 0$, where $c > 0$, $\phi \in L^1(\mathcal{B})$, and Γ: $(0, \infty) \to \mathbb{R}$ is a convex function that satisfies $\Gamma(t)/t \to +\infty$ as $t \to +\infty$.

A slight modification of the proof of Theorem 4.1 then allows us to conclude

THEOREM 4.3. *Let* $n = 3$, $p > 2$, $\mathbf{d} \in L^p(\partial\mathcal{B}_d; \mathbb{R}^3)$, $\mathbf{b}_0 \in L^r(\mathcal{B}; \mathbb{R}^3)$, *and* $\mathbf{s}_0 \in L^q(\partial\mathcal{B}_t; \mathbb{R}^3)$ *be given, where* q *and* r *are chosen so that the linear mapping* L: $W^{1,p}(\mathcal{B}; \mathbb{R}^3) \to \mathbb{R}$ *is continuous. Suppose that* W *satisfies the hypotheses of Theorem 4.1 with (iii) replaced by (iii)′. Assume further that* I *is given by (4.3), where* $\eta_i \geqslant 0$ *for* $i = 1, \ldots, M$ *and* υ: $[0, \infty)^M \to [0, \infty)$ *are lower semicontinuous. Then, if* $\mathcal{A}_m^p$ *is nonempty,* $\widehat{E}$ *and* $\hat{I}$ *each attain their infimum on* $\mathcal{A}_m^p$.

REMARK. In particular, standard imbedding and trace theorems yield the continuity of the linear mapping L: $W^{1,p}(\mathcal{B}; \mathbb{R}^3) \to \mathbb{R}$ when $r \geqslant 3p/(4p-3)$ and $q \geqslant 2p/3(p-1)$. See [2, 8] for the appropriate modification to the proof when such dead-load body forces and surface tractions are included.

5. The Equilibrium Equations

In this section we take $\mathcal{B} \subset \mathbb{R}^n$ and consider equilibrium conditions that are satisfied by minimizers of the energy

$$I(\mathbf{u}) = \int_{\mathcal{B}} W(\nabla\mathbf{u}(\mathbf{x}))\,\mathrm{d}\mathbf{x} + \upsilon(\alpha^1, \ldots, \alpha^M) + \sum_{i=1}^{M} H(\alpha^i)\eta_i$$

when W: $\mathrm{Lin}^> \to \mathbb{R}$ is a smooth function. For simplicity we have restricted our attention to the case where W does not depend on $\mathbf{x}$ or $\mathbf{u}$. If such a minimizer is C^2 it must satisfy the Euler–Lagrange equations

$$\operatorname{div}\mathbf{S}(\nabla\mathbf{u}) \equiv \mathbf{0}, \tag{5.1}$$

where

$$\mathbf{S}(\mathbf{F}) := \frac{\mathrm{d}W}{\mathrm{d}\mathbf{F}}$$

is the **Piola–Kirchhoff stress.**

If $\mathbf{u}$ is only in $W^{1,p}(\mathcal{B}; \mathbb{R}^n)$ then it is not known whether (5.1) holds (in the sense of distributions), even if $p > n$. The difficulty is that to derive (5.1) one usually considers variations $\mathbf{u}+s\mathbf{v}$, $\mathbf{v} \in C_0^\infty(\mathcal{B}; \mathbb{R}^n)$. However, when $\mathbf{u}$ is in such a Sobolev space it is not clear that any such variation has finite energy since $\det(\nabla\mathbf{u} + s\nabla\mathbf{v})$ may be negative on a set of positive measure. Ball [4] observed, when $\upsilon = 0$ and $\eta_i = 0$, that one can still derive other equilibrium equations if one instead considers outer variations $\mathbf{g}_s \circ \mathbf{u}$ or inner variations $\mathbf{u} \circ \mathbf{g}_s$ where $\mathbf{g}_s$ is a one-parameter family of diffeomorphisms with $\mathbf{g}_0 = \mathbf{id}$. The former variations lead to conditions in the deformed configuration while the latter give the energy-momentum equations. We begin with outer variations and first fix some notation.

Let $\mathbf{u} \in W^{1,1}(\mathcal{B}; \mathbb{R}^n)$ be one-to-one a.e. and satisfy $\det\nabla\mathbf{u} > 0$ a.e. Suppose that $N_{\mathbf{u}}$ is the null set given by the change of variables formula (Proposition 2.7),

$M_{\mathbf{u}}$ is the null set where $\det\nabla\mathbf{u} = 0$, and $P_{\mathbf{u}}$ is a null set such that $\mathbf{u}|_{\mathcal{B}\backslash P_{\mathbf{u}}}$ is one-to-one. Define $Q_{\mathbf{u}} := M_{\mathbf{u}} \cup N_{\mathbf{u}} \cup P_{\mathbf{u}}$. Then we define the **Cauchy stress** $\mathbf{T} : \mathbf{u}(\mathcal{B}\backslash Q_{\mathbf{u}}) \to \mathrm{Lin}$ by

$$\mathbf{T}(\mathbf{y}) := \frac{\mathbf{S}(\nabla\mathbf{u}(\mathbf{x}))[\nabla\mathbf{u}(\mathbf{x})]^{\mathrm{T}}}{\det\nabla\mathbf{u}(\mathbf{x})}, \tag{5.2}$$

or, in components,

$$T^i_j = \sum_{\alpha=1}^{3} \frac{\partial W}{\partial F^i_\alpha}\frac{\partial u^j}{\partial x_\alpha}[\det\nabla\mathbf{u}]^{-1}.$$

Here $\mathbf{x}$ is the unique point in $\mathcal{B}\backslash Q_{\mathbf{u}}$ that satisfies $\mathbf{u}(\mathbf{x}) = \mathbf{y}$.

THEOREM 5.1. *Let $\upsilon \equiv 0$. Suppose that W is C^1 on* $\mathrm{Lin}^>$ *and that there are constants $C > 0$ and $\varepsilon > 0$ such that*

$$\left|\mathbf{S}(\mathbf{AF})\mathbf{F}^{\mathrm{T}}\right| \leqslant C(W(\mathbf{F}) + 1) \tag{5.3}$$

for all $\mathbf{F} \in \mathrm{Lin}^>$ and all $\mathbf{A} \in \mathrm{Lin}^>$ that satisfy $|\mathbf{A} - \mathbf{I}| < \varepsilon$. If $\mathbf{u} \in \mathcal{A}^p$ is any minimizer of I with finite energy then $\mathbf{T} \in L^1(\mathbf{u}(\mathcal{B}\backslash Q_{\mathbf{u}}); \mathrm{Lin})$ and

$$\int_{\mathbf{u}(\mathcal{B}\backslash Q_{\mathbf{u}})} \mathbf{T}\cdot\nabla\mathbf{v}\,\mathrm{d}\mathbf{y} = 0 \quad \forall\mathbf{v} \in C^1(\mathbb{R}^n;\mathbb{R}^n) \quad \textit{with } \mathbf{v} = \mathbf{0} \textit{ on } \mathbb{R}^n\backslash\mathbf{d}(\mathcal{B}). \tag{5.4}$$

Moreover, if $\mathbf{u} \in C^2(\mathcal{B}\backslash\{\mathbf{a}_1, \dots, \mathbf{a}_M\}; \mathbb{R}^n)$ satisfies $\det\nabla\mathbf{u} > 0$ on $\mathcal{B}\backslash\{\mathbf{a}_1, \dots, \mathbf{a}_M\}$ then

$$\mathrm{div}_{\mathbf{x}}\,\mathbf{S}(\nabla\mathbf{u}) = \mathbf{0} \quad \textit{in } \mathcal{B}\backslash\{\mathbf{a}_1, \dots, \mathbf{a}_M\} \tag{5.5}$$

and if, in addition, $\mathbf{u}$ is one-to-one on $\mathcal{B}\backslash\{\mathbf{a}_1, \dots, \mathbf{a}_M\}$ then

$$\mathrm{div}_{\mathbf{y}}\mathbf{T} = \mathbf{0} \quad \textit{in } \mathbf{u}(\mathcal{B}\backslash\{\mathbf{a}_1, \dots, \mathbf{a}_M\}). \tag{5.6}$$

REMARKS. 1. When $\mathbf{u} \in \mathcal{A}^p_m$ a slight modification of our proof will also yield (5.4)–(5.6).

2. If the energy term that depends on hole volume, υ, is not zero then the equilibrium equations will contain an additional term, see, e.g., [24].

Proof. Let $\mathbf{v} \in C^1(\mathbb{R}^n;\mathbb{R}^n)$ satisfy $\mathbf{v} = \mathbf{0}$ on $\mathbb{R}^n\backslash\mathbf{d}(\mathcal{B})$. Then $\mathbf{g}_s := \mathbf{id} + s\mathbf{v}$ is a diffeomorphism of $\mathbb{R}^n$ for small s. Define $\mathbf{u}_s := \mathbf{g}_s \circ \mathbf{u}$ so that, by Corollary 6.4, $\mathbf{u}_s \in \mathcal{A}^p$ for sufficiently small s. With the help of (5.3) one can use the dominated convergence theorem to show that (see [4], [6], and [14]) $\mathbf{T} \in L^1(\mathbf{u}(\mathcal{B}\backslash Q_{\mathbf{u}}); \mathrm{Lin})$ and

$$\begin{aligned}\left.\frac{\mathrm{d}}{\mathrm{d}s}\right|_{s=0}\int_{\mathcal{B}} W(\nabla\mathbf{u}_s)\,\mathrm{d}\mathbf{x} &= \int_{\mathcal{B}}\left.\frac{\mathrm{d}}{\mathrm{d}s}\right|_{s=0} W(\nabla\mathbf{u}_s)\,\mathrm{d}\mathbf{x}\\ &= \int_{\mathcal{B}} \mathbf{S}(\nabla\mathbf{u})\cdot\left[\big((\nabla\mathbf{v})\circ\mathbf{u}\big)\nabla\mathbf{u}\right]\mathrm{d}\mathbf{x}\\ &= \int_{\mathcal{B}} \mathbf{T}\big(\mathbf{u}(\mathbf{x})\big)\cdot\left[(\nabla\mathbf{v})\big(\mathbf{u}(\mathbf{x})\big)\right]\det\nabla\mathbf{u}(\mathbf{x})\,\mathrm{d}\mathbf{x}\end{aligned}$$

and hence, by the change of variables formula (Proposition 2.7),

$$\frac{\mathrm{d}}{\mathrm{d}s}\bigg|_{s=0}\int_{\mathscr{B}} W(\nabla\mathbf{u}_s)\,\mathrm{d}\mathbf{x} = \int_{\mathbf{u}(\mathscr{B}\setminus Q_{\mathbf{u}})} \mathbf{T}(\mathbf{y})\cdot\nabla\mathbf{v}(\mathbf{y})\,\mathrm{d}\mathbf{y}.$$

We note that $\mathbf{g}_s$ is uniformly close to **id** and hence that $\mathbf{g}_s$ cannot close (or open) any holes when s is sufficiently small. It follows that the additional cavitation energy does not change. Thus, since $\mathbf{u}$ minimizes I in $\mathcal{A}^p$ we conclude that (5.4) is satisfied.

Now suppose $\mathbf{u} \in C^2(\mathscr{B}\setminus\{\mathbf{a}_1,\dots,\mathbf{a}_M\};\mathbb{R}^n)$ satisfies $\det\nabla\mathbf{u} > 0$ on $\mathscr{B}\setminus\{\mathbf{a}_1,\dots,\mathbf{a}_M\}$ and let $\mathbf{b} \in \mathscr{B}\setminus\{\mathbf{a}_1,\dots,\mathbf{a}_M\}$. Then since $\det\nabla\mathbf{u}(\mathbf{b}) > 0$ we can apply the inverse function theorem to conclude that, for r sufficiently small, $\mathbf{u}|_{B(\mathbf{b},r)}$ is a diffeomorphism and $\mathbf{u}(B(\mathbf{b},r))$ is open. Therefore we can define the Cauchy stress $\mathbf{T}$ by (5.2) *on the entire set* $\mathbf{u}(B(\mathbf{b},r))$.

Next let $\mathbf{v} \in C^1(\mathbb{R}^n;\mathbb{R}^n)$ be supported in $\mathbf{u}(B(\mathbf{b},r))$. Then by the previous argument we find that

$$0 = \int_{\mathbf{u}(B(\mathbf{b},r))} \mathbf{T}\cdot\nabla\mathbf{v}\,\mathrm{d}\mathbf{y}$$

and hence, since $\mathbf{v} \in C^1$ is arbitrary and $\mathbf{T}$ is C^1, a standard result yields

$$\mathrm{div}_{\mathbf{y}}\mathbf{T} = \mathbf{0} \quad \text{in } \mathbf{u}\big(B(\mathbf{b},r)\big). \tag{5.7}$$

Since $\mathbf{u}$ is a diffeomorphism on $B(\mathbf{b},r)$ we can now apply the identity (see, e.g., [8, Theorem 1.7-1])

$$\mathrm{div}_{\mathbf{x}}\,\mathbf{S}\big(\nabla\mathbf{u}(\mathbf{x})\big) = \big[\det\nabla\mathbf{u}(\mathbf{x})\big]\,\mathrm{div}_{\mathbf{y}}\mathbf{T}\big(\mathbf{u}(\mathbf{x})\big)$$

to conclude

$$\mathrm{div}_{\mathbf{x}}\mathbf{S}(\nabla\mathbf{u}) = \mathbf{0} \quad \text{in } B(\mathbf{b},r).$$

Therefore (5.5) is satisfied since $\mathbf{b} \in \mathscr{B}\setminus\{\mathbf{a}_1,\dots,\mathbf{a}_M\}$ is arbitrary. Finally, if $\mathbf{u}$ is one-to-one the Cauchy stress $\mathbf{T}$ can be defined uniquely on $\mathbf{u}(\mathscr{B}\setminus\{\mathbf{a}_1,\dots,\mathbf{a}_M\})$ by (5.2) and (5.6) follows immediately from (5.7). □

We next consider inner variations.

THEOREM 5.2. *Let W be C^1 on* Lin$^>$ *and suppose that there are constants $C > 0$ and $\varepsilon > 0$ such that*

$$\big|\mathbf{F}^{\mathrm{T}}\mathbf{S}(\mathbf{F}\mathbf{A})\big| \leqslant C\big(W(\mathbf{F}) + 1\big) \tag{5.8}$$

for all $\mathbf{F} \in$ Lin$^>$ and all $\mathbf{A} \in$ Lin$^>$ that satisfy $|\mathbf{A} - \mathbf{I}| < \varepsilon$. If $\mathbf{u} \in \mathcal{A}^p$ (or $\mathcal{A}^p_m$) is any minimizer of I with finite energy then

$$\int_{\mathscr{B}} \big[W(\nabla\mathbf{u})\mathbf{I} - (\nabla\mathbf{u})^{\mathrm{T}}\mathbf{S}(\nabla\mathbf{u})\big]\cdot\nabla\mathbf{v}\,\mathrm{d}\mathbf{x} = 0 \tag{5.9}$$

for every $\mathbf{v} \in C^1(\overline{\Omega}; \mathbb{R}^n)$ *that satisfies* $\mathbf{v} = \mathbf{0}$ *on* $\{\mathbf{a}_1, \ldots, \mathbf{a}_M\} \cup (\overline{\Omega} \setminus \mathcal{B})$. *Moreover, if* $\mathbf{u} \in C^2(\mathcal{B} \setminus \{\mathbf{a}_1, \ldots, \mathbf{a}_M\}; \mathbb{R}^n)$ *then*

$$\operatorname{div}\big[W(\nabla\mathbf{u})\mathbf{I} - (\nabla\mathbf{u})^{\mathrm{T}}\mathbf{S}(\nabla\mathbf{u})\big] = \mathbf{0} \quad \text{in } \mathcal{B}\setminus\{\mathbf{a}_1, \ldots, \mathbf{a}_M\} \tag{5.10}$$

or, in components,

$$\sum_{\alpha,j} \frac{\partial}{\partial x^\alpha}\left\{W(\nabla\mathbf{u})\delta^\alpha_\beta - \frac{\partial W}{\partial F^j_\alpha}(\nabla\mathbf{u})\frac{\partial u_j}{\partial x_\beta}\right\} = 0.$$

REMARK. The above result is *not* valid when W depends on $\mathbf{x}$.

Proof. Let $\mathbf{v} \in C^1(\overline{\Omega}; \mathbb{R}^n)$ satisfy $\mathbf{v} = \mathbf{0}$ on $\{\mathbf{a}_1, \ldots, \mathbf{a}_M\} \cup (\overline{\Omega} \setminus \mathcal{B})$. Then $\mathbf{h}_s := \mathbf{id} + s\mathbf{v}$ is a diffeomorphism of $\mathcal{B}$ for small s. Let $\mathbf{u} \in \mathcal{A}^p$ ($\mathbf{u} \in \mathcal{A}^p_m$) have finite energy and minimize I. Define $\mathbf{u}_s := \mathbf{u} \circ \mathbf{g}_s$, where $\mathbf{g}_s := (\mathbf{h}_s)^{-1}$. Then, by Corollary 6.6, $\mathbf{u}_s \in \mathcal{A}^p$ ($\mathbf{u}_s \in \mathcal{A}^p_m$) for sufficiently small s.

Define $\mathbf{F}(\mathbf{x}) := \nabla\mathbf{u}(\mathbf{x})$, $\mathbf{G}_s(\mathbf{z}) := \nabla\mathbf{g}_s(\mathbf{z})$, and $\mathbf{H}_s(\mathbf{x}) := \nabla\mathbf{h}_s(\mathbf{x})$. Then if one takes the gradient, with respect to $\mathbf{x}$, of the identity $\mathbf{g}_s(\mathbf{h}_s(\mathbf{x})) = \mathbf{x}$ one concludes that $\mathbf{G}_s(\mathbf{h}_s(\mathbf{x}))\mathbf{H}_s(\mathbf{x}) = \mathbf{I}$ and, hence, that

$$\mathbf{G}_s\big(\mathbf{h}_s(\mathbf{x})\big) = \mathbf{H}_s(\mathbf{x})^{-1}. \tag{5.11}$$

For future use we note that upon differentiating the identity $\mathbf{H}_s^{-1}\mathbf{H}_s = \mathbf{I}$ with respect to s one deduces that

$$\frac{\mathrm{d}}{\mathrm{d}s}\mathbf{H}_s(\mathbf{x})^{-1} = -\mathbf{H}_s(\mathbf{x})^{-1}\left[\frac{\mathrm{d}}{\mathrm{d}s}\mathbf{H}_s(\mathbf{x})\right]\mathbf{H}_s(\mathbf{x})^{-1},$$

and therefore, since $\mathbf{h}_s = \mathbf{id} + s\mathbf{v}$,

$$\left.\frac{\mathrm{d}}{\mathrm{d}s}\mathbf{H}_s(\mathbf{x})^{-1}\right|_{s=0} = -\nabla\mathbf{v}. \tag{5.12}$$

Now, by (5.11) and the change of variables $Z = h_s(x)$,

$$\begin{aligned}\int_{\mathcal{B}} W\big(\nabla\mathbf{u}_s(\mathbf{z})\big)\mathrm{d}\mathbf{z} &= \int_{\mathcal{B}} W\big(\mathbf{F}(\mathbf{g}_s(\mathbf{z}))\mathbf{G}_s(\mathbf{z})\big)\det\mathbf{G}_s(\mathbf{z})\det\mathbf{H}_s\big(\mathbf{g}_s(\mathbf{z})\big)\,\mathrm{d}\mathbf{z}\\ &= \int_{\mathcal{B}} W\big(\mathbf{F}(\mathbf{x})\mathbf{H}_s(\mathbf{x})^{-1}\big)\det\mathbf{H}_s(\mathbf{x})\,\mathrm{d}\mathbf{x}\end{aligned}$$

and consequently, with the aid of (5.8), one can use the Lebesgue dominated convergence theorem to show that (see [4, 6, 14])

$$\begin{aligned}\left.\frac{\mathrm{d}}{\mathrm{d}s}\right|_{s=0}\int_{\mathcal{B}} W\big(\nabla\mathbf{u}_s(\mathbf{z})\big)\mathrm{d}\mathbf{z} &= \left.\frac{\mathrm{d}}{\mathrm{d}s}\right|_{s=0}\int_{\mathcal{B}} W(\mathbf{F}(\mathbf{x})\mathbf{H}_s(\mathbf{x})^{-1})\det\mathbf{H}_s(\mathbf{x})\,\mathrm{d}\mathbf{x}\\ &= \int_{\mathcal{B}}\left.\frac{\mathrm{d}}{\mathrm{d}s}\right|_{s=0}\big[W\big(\mathbf{F}\mathbf{H}_s^{-1}\big)\det\mathbf{H}_s\big]\mathrm{d}\mathbf{x}.\end{aligned} \tag{5.13}$$

We note that

$$\frac{\mathrm{d}}{\mathrm{d}s} \det \mathbf{H}_s(\mathbf{x}) = \big(\operatorname{adj} \mathbf{H}_s(\mathbf{x})\big) \cdot \frac{\mathrm{d}}{\mathrm{d}s}\mathbf{H}_s(\mathbf{x})$$

and hence, since $\mathbf{h}_s = \mathbf{id} + s\mathbf{v}$,

$$\left.\frac{\mathrm{d}}{\mathrm{d}s}\right|_{s=0} \det \mathbf{H}_s(\mathbf{x}) = \mathbf{I} \cdot \nabla\mathbf{v}. \tag{5.14}$$

Thus, by the product rule, (5.12), and (5.14),

$$\left.\frac{\mathrm{d}}{\mathrm{d}s}\right|_{s=0} \big[W\big(\mathbf{F}\mathbf{H}_s^{-1}\big) \det \mathbf{H}_s\big] = \big[W(\nabla\mathbf{u})\mathbf{I} - (\nabla\mathbf{u})^{\mathrm{T}}\mathbf{S}(\nabla\mathbf{u})\big] \cdot \nabla\mathbf{v},$$

which together with (5.13) yields

$$\left.\frac{\mathrm{d}}{\mathrm{d}s}\right|_{s=0} \int_{\mathcal{B}} W\big(\nabla\mathbf{u}_s(\mathbf{z})\big)\mathrm{d}\mathbf{z} = \int_{\mathcal{B}} \big[W(\nabla\mathbf{u})\mathbf{I} - (\nabla\mathbf{u})^{\mathrm{T}}\mathbf{S}(\nabla\mathbf{u})\big] \cdot \nabla\mathbf{v}\, \mathrm{d}\mathbf{x}.$$

In view of Theorem 6.5 the additional energy due to cavitation is invariant under composition on the right with a diffeomorphism. Thus, since $\mathbf{u}$ minimizes I in $\mathcal{A}^p$ (or $\mathcal{A}^p_m$) we conclude that (5.9) is satisfied. Finally, if $\mathbf{u}$ is C^2 away from the cavitation points a standard argument yields (5.10). □

6. Invariance under Composition by Diffeomorphisms

Let $\mathcal{B} \subset \mathbb{R}^n$ and suppose that $\mathbf{g}\colon \overline{\mathcal{B}} \to \mathbb{R}^n$ is an orientation preserving diffeomorphism of $\mathcal{B}$. Then one can view either $\mathcal{B}$ or $\mathbf{g}(\mathcal{B})$ as the reference configuration for the body. Any restrictions that one imposes on deformations, such as $\mathbf{u} \in W^{1,p}$, $\det\nabla\mathbf{u} > 0$, a.e., or condition (INV); or any restrictions that one imposes on the constitutive relation, such as polyconvexity, should not depend on this arbitrary choice of reference configuration. In this section we will show that our class of admissible deformations is invariant under such a change in reference configuration. We will also show that if after deforming the body by an admissible deformation $\mathbf{u}$ one further deforms the body by using an orientation preserving diffeomorphism $\mathbf{g}$ of $\mathbb{R}^n$ then the composition $\mathbf{g} \circ \mathbf{u}$ is also admissible. We first consider the invariance of our class of deformations under composition on the left by a diffeomorphism.

LEMMA 6.1. *Let* $\mathbf{g}\colon \mathbb{R}^n \to \mathbb{R}^n$ *be a diffeomorphism that satisfies* $\det \nabla\mathbf{g} > 0$ *on* $\mathbb{R}^n$. *Suppose that* $\mathbf{u} \in W^{1,p}(\Omega; \mathbb{R}^n)$, *with* $p > n-1$, *satisfies* $\det\nabla\mathbf{u} > 0$ *a.e. If* $\mathbf{u}^*$ *satisfies condition (INV) then so does* $\mathbf{g} \circ \mathbf{u}^*$. *Moreover, for every* $\mathbf{b} \in \Omega$,

$$\mathbf{g}\big(\operatorname{im}_T\big(\mathbf{u}^*, B(\mathbf{b}, r)\big)\big) = \operatorname{im}_T\big(\mathbf{g} \circ \mathbf{u}^*, B(\mathbf{b}, r)\big) \tag{6.1}$$

for a.e. $r \in (0, r_{\mathbf{b}})$.

Proof. As usual we will write $\mathbf{u}$ for $\mathbf{u}^*$. Let $\mathbf{b} \in \Omega$ and $r \in (0, r_{\mathbf{b}})$ be such that $\mathbf{u}|_{\partial B(\mathbf{b},r)}$ is continuous and (i) and (ii) of condition (INV) are satisfied. We first consider (i) of (INV). Let $N \subset \overline{B(\mathbf{b}, r)}$ be a Lebesgue null set such that

$$\mathbf{u}(\mathbf{x}) \in \operatorname{im}_T\big(\mathbf{u}, B(\mathbf{b}, r)\big) \cup \mathbf{u}\big(\partial B(\mathbf{b}, r)\big) \quad \text{for every } \mathbf{x} \in \overline{B(\mathbf{b}, r)} \backslash N.$$

Fix $\mathbf{x} \in \overline{B(\mathbf{b}, r)} \backslash N$. If $\mathbf{u}(\mathbf{x}) \in \mathbf{u}(\partial B(\mathbf{b}, r))$ then

$$\mathbf{g}\big(\mathbf{u}(\mathbf{x})\big) \in \mathbf{g}\big(\mathbf{u}\big(\partial B(\mathbf{b}, r)\big)\big). \tag{6.2}$$

Otherwise, $\mathbf{u}(\mathbf{x}) \in \operatorname{im}_T(\mathbf{u}, B(\mathbf{b}, r))$ and hence by the definition of the topological image $\deg(\mathbf{u}, \partial B(\mathbf{b}, r), \mathbf{u}(\mathbf{x})) \neq 0$. Thus, by the multiplicative property of degree (Proposition 2.1(ii)) $\deg(\mathbf{g} \circ \mathbf{u}, \partial B(\mathbf{b}, r), \mathbf{g}(\mathbf{u}(\mathbf{x}))) \neq 0$ and consequently, by the definition of the topological image,

$$\mathbf{g}\big(\mathbf{u}(\mathbf{x})\big) \in \operatorname{im}_T\big(\mathbf{g} \circ \mathbf{u}, B(\mathbf{b}, r)\big). \tag{6.3}$$

Therefore, by (6.2) and (6.3),

$$\mathbf{g}\big(\mathbf{u}(\mathbf{x})\big) \in \operatorname{im}_T\big(\mathbf{g} \circ \mathbf{u}, B(\mathbf{b}, r)\big) \cup (\mathbf{g} \circ \mathbf{u})\big(\partial B(\mathbf{b}, r)\big) \quad \text{for every } \mathbf{x} \in \overline{B(\mathbf{b}, r)} \backslash N,$$

which shows that $\mathbf{g} \circ \mathbf{u}$ satisfies (i) of (INV). The proof of (ii) of (INV) is similar.

We now consider (6.1). By the multiplicative property of degree, Proposition 2.1(ii),

$$\deg\big(\mathbf{g} \circ \mathbf{u}, \partial B(\mathbf{b}, r), \mathbf{g}(\mathbf{y})\big) = \deg\big(\mathbf{u}, \partial B(\mathbf{b}, r), \mathbf{y}\big)$$

for every $\mathbf{y} \in \mathbb{R}^n \backslash \mathbf{u}(\partial B(\mathbf{b}, r))$. Thus the definition of the topological image yields

$$\mathbf{g}(\mathbf{y}) \in \operatorname{im}_T\big(\mathbf{g} \circ \mathbf{u}, B(\mathbf{b}, r)\big) \Longleftrightarrow \mathbf{y} \in \operatorname{im}_T\big(\mathbf{u}, B(\mathbf{b}, r)\big).$$

Equation (6.1) now follows. □

LEMMA 6.2. *Let* $\mathbf{g}$ *and* $\mathbf{u}$ *satisfy the hypotheses of Lemma 6.1. Suppose, in addition, that* $\det \nabla \mathbf{g} \in L^\infty(\mathbb{R}^n)$. *Then*

$$\big(\operatorname{Det}\nabla(\mathbf{g} \circ \mathbf{u})\big)(A) \leqslant \| \det \nabla \mathbf{g}\|_\infty (\operatorname{Det}\nabla \mathbf{u})(A), \tag{6.4}$$

for every $A \subset \Omega$ *and hence the Radon measure* $\operatorname{Det}\nabla(\mathbf{g} \circ \mathbf{u})$ *is absolutely continuous with respect to* $\operatorname{Det}\nabla\mathbf{u}$.

Proof. As usual we write $\mathbf{u}$ for $\mathbf{u}^*$. Fix $\mathbf{b} \in \Omega$. Then by Lemma 6.1 and Proposition 2.9 we have that for $\mathcal{L}^1$ a.e. $r \in (0, r_{\mathbf{b}})$

$$(\operatorname{Det}\nabla\mathbf{u})(\overline{B(\mathbf{b}, r)}) = \mathcal{L}^n\big(\operatorname{im}_T\big(\mathbf{u}, B(\mathbf{b}, r)\big)\big) \tag{6.5}$$

and

$$\big(\operatorname{Det}\nabla(\mathbf{g} \circ \mathbf{u})\big)(\overline{B(\mathbf{b}, r)}) = \mathcal{L}^n\big(\operatorname{im}_T\big(\mathbf{g} \circ \mathbf{u}, B(\mathbf{b}, r)\big)\big).$$

Thus, in view of (6.1),

$$\big(\mathrm{Det}\nabla(\mathbf{g}\circ\mathbf{u})\big)(\overline{B(\mathbf{b},r)}) = \mathcal{L}^n\big(\mathbf{g}\big(\mathrm{im}_T(\mathbf{u},B(\mathbf{b},r))\big)\big). \tag{6.6}$$

However, by the change of variables formula for diffeomorphisms (see Proposition 2.7 and the remark that follows it), for any Lebesgue measurable set $A \subset \Omega$,

$$\mathcal{L}^n\big(\mathbf{g}(A)\big) = \int_A \det\nabla\mathbf{g}(\mathbf{x})\,\mathrm{d}\mathbf{x} \leqslant \|\det\nabla\mathbf{g}\|_\infty \mathcal{L}^n(A). \tag{6.7}$$

Therefore, if we take $A = \overline{B(\mathbf{b},r)}$ in (6.7) we find, with the aid of (6.5) and (6.6), that

$$\big(\mathrm{Det}\nabla(\mathbf{g}\circ\mathbf{u})\big)(\overline{B(\mathbf{b},r)}) \leqslant \|\det\nabla\mathbf{g}\|_\infty(\mathrm{Det}\nabla\mathbf{u})(\overline{B(\mathbf{b},r)}).$$

A standard result on Radon measures (see, e.g., [10, Lemma 1(i), Section 1.6.1]) then yields (6.4). □

THEOREM 6.3. *Let* $\mathbf{g}\colon \mathbb{R}^n \to \mathbb{R}^n$ *be a diffeomorphism that satisfies* $\det\nabla\mathbf{g} \in L^\infty(\mathbb{R}^n)$ *and* $\det\nabla\mathbf{g} > 0$ *on* $\mathbb{R}^n$. *Suppose that*

$$\mathbf{u} \in \mathrm{Def}^p(\Omega,\mathbf{a}_1,\ldots,\mathbf{a}_M).$$

Then

$$\mathbf{g}\circ\mathbf{u} \in \mathrm{Def}^p(\Omega,\mathbf{a}_1,\ldots,\mathbf{a}_M).$$

Proof. Let $\mathbf{u} \in \mathrm{Def}^p(\Omega,\mathbf{a}_1,\ldots,\mathbf{a}_M)$; then $\mathbf{u}^*$ satisfies condition (INV), $\det\nabla\mathbf{u} > 0$ a.e., and

$$\mathrm{Det}\nabla\mathbf{u} = (\det\nabla\mathbf{u})\mathcal{L}^n + \sum_{i=1}^{M}\alpha^i\delta_{\mathbf{a}_i}. \tag{6.8}$$

Let $\mathbf{g}\colon \mathbb{R}^n \to \mathbb{R}^n$ be a diffeomorphism that satisfies $\det\nabla\mathbf{g} \in L^\infty(\mathbb{R}^n)$ and $\det\nabla\mathbf{g} > 0$ on $\mathbb{R}^n$. Then, by Lemma 6.1, $\mathbf{g}\circ\mathbf{u}^*$ satisfies condition (INV). Since $\det\nabla(\mathbf{g}\circ\mathbf{u}) = (\det\nabla\mathbf{g})(\det\nabla\mathbf{u})$ it is clear that $\det\nabla(\mathbf{g}\circ\mathbf{u}) > 0$ a.e. Thus the desired result will follow once we show that there exist $\beta^i \geqslant 0$ such that

$$\mathrm{Det}\nabla(\mathbf{g}\circ\mathbf{u}) = \big(\det\nabla(\mathbf{g}\circ\mathbf{u})\big)\mathcal{L}^n + \sum_{i=1}^{M}\beta^i\delta_{\mathbf{a}_i}. \tag{6.9}$$

By Proposition 2.9 there is a Radon measure m, which is singular with respect to Lebesgue measure, such that

$$\mathrm{Det}\nabla(\mathbf{g}\circ\mathbf{u}) = \big(\det\nabla(\mathbf{g}\circ\mathbf{u})\big)\mathcal{L}^n + m. \tag{6.10}$$

Let $N \subset \Omega$ be a Lebesgue null set. Define $N^- := N\backslash\{\mathbf{a}_1,\ldots,\mathbf{a}_M\}$. Then, by (6.8), N^- is a $\mathrm{Det}\nabla\mathbf{u}$ null set and hence, by Lemma 6.2, N^- is a $\mathrm{Det}\nabla(\mathbf{g}\circ\mathbf{u})$ null set.

Therefore, $m(N\backslash\{\mathbf{a}_1, \ldots, \mathbf{a}_M\}) = 0$ for every Lebesgue null set N. Since m is singular with respect to Lebesgue measure it follows that there exist $\beta^i \geqslant 0$ such that

$$m = \sum_{i=1}^{M} \beta^i \delta_{\mathbf{a}_i},$$

which together with (6.10) yields (6.9). □

COROLLARY 6.4. *Let* $\mathbf{u} \in \mathcal{A}^p$. *Suppose that* $\mathbf{v} \in C^1(\mathbb{R}^n; \mathbb{R}^n)$ *satisfies* $\mathbf{v} = \mathbf{0}$ *on* $\mathbb{R}^n\backslash\mathbf{d}(\mathcal{B})$. *Define*

$$\mathbf{g}_s := \mathbf{id} + s\mathbf{v} \quad \textit{and} \quad \mathbf{u}_s := \mathbf{g}_s \circ \mathbf{u}.$$

Then, for all sufficiently small s,

$$\mathbf{u}_s \in \mathcal{A}^p.$$

REMARK. A similar result is valid for $\mathbf{u} \in \mathcal{A}^p_m$.

Proof. Let $\mathbf{v} \in C^1(\mathbb{R}^n; \mathbb{R}^n)$ satisfy $\mathbf{v} = \mathbf{0}$ on $\mathbb{R}^n\backslash\mathbf{d}(\mathcal{B})$. Then it follows that (see, e.g., [8, Section 5.5]), for sufficiently small s, the mapping $\mathbf{g}_s := \mathbf{id} + s\mathbf{v}$ is a diffeomorphism of $\mathbb{R}^n$, $\det\nabla\mathbf{g}_s \in L^\infty(\mathbb{R}^n)$, and $\det\nabla\mathbf{g}_s > 0$ on $\mathbb{R}^n$.

Let $\mathbf{u} \in \mathcal{A}^p$ and define $\mathbf{u}_s := \mathbf{g}_s \circ \mathbf{u}$. Then $\mathbf{u}^e \in \mathrm{Def}^p(\Omega, \mathbf{a}_1, \ldots, \mathbf{a}_M)$ and we may therefore apply Theorem 6.3 to conclude

$$\left(\mathbf{u}^e\right)_s := \mathbf{g}_s \circ \mathbf{u}^e \in \mathrm{Def}^p(\Omega, \mathbf{a}_1, \ldots, \mathbf{a}_M).$$

Next, we note that $\mathbf{v} = \mathbf{0}$ on $\mathbf{d}(\overline{\Omega}\backslash\mathcal{B}) = \mathbf{d}(\overline{\Omega})\backslash\mathbf{d}(\mathcal{B})$ and consequently $(\mathbf{u}^e)_s(\mathbf{x}) = \mathbf{id}(\mathbf{u}^e(\mathbf{x})) = \mathbf{d}(\mathbf{x})$ for $\mathbf{x} \in \overline{\Omega}\backslash\mathcal{B}$. Therefore, $(\mathbf{u}_s)^e = (\mathbf{u}^e)_s \in \mathrm{Def}^p(\Omega, \mathbf{a}_1, \ldots, \mathbf{a}_M)$ and $\mathbf{u} - \mathbf{d} \in W_0^{1,p}(\mathcal{B}; \mathbb{R}^n)$, which completes the proof. □

We now consider the invariance of the class of deformations under composition on the right by a diffeomorphism.

THEOREM 6.5. *Let* $\mathbf{g}\colon \overline{\Omega} \to \overline{\Omega}$ *be a diffeomorphism that satisfies* $\det\nabla\mathbf{g} > 0$ *on* Ω. *Suppose that*

$$\mathbf{u} \in \mathrm{Def}^p(\Omega, \mathbf{a}_1, \ldots, \mathbf{a}_M)$$

with

$$\mathrm{Det}\nabla\mathbf{u} = (\det\nabla\mathbf{u})\mathcal{L}^n + \sum_{i=1}^{M} \alpha^i \delta_{\mathbf{a}_i}. \tag{6.11}$$

Then

$$\mathbf{u} \circ \mathbf{g} \in \mathrm{Def}^p\big(\Omega, \mathbf{g}^{-1}(\mathbf{a}_1), \ldots, \mathbf{g}^{-1}(\mathbf{a}_M)\big)$$

and

$$\mathrm{Det}\nabla(\mathbf{u}\circ\mathbf{g}) = \big(\det\nabla(\mathbf{u}\circ\mathbf{g})\big)\mathcal{L}^n + \sum_{i=1}^{M}\alpha^i\delta_{\mathbf{g}^{-1}(\mathbf{a}_i)}. \tag{6.12}$$

Proof. Let $\mathbf{u} \in \mathrm{Def}^p(\Omega, \mathbf{a}_1, \ldots, \mathbf{a}_M)$ so that $\mathbf{u}^*$ satisfies condition (INV), $\det\nabla\mathbf{u} > 0$ a.e., and (6.11) is satisfied. Let $\mathbf{g}\colon \overline{\Omega} \to \overline{\Omega}$ be a diffeomorphism that satisfies $\det\nabla\mathbf{g} > 0$ on Ω. Then, by [24, Theorem 9.1], $\mathbf{u}^*\circ\mathbf{g}$ satisfies condition (INV). Since $\det\nabla(\mathbf{u}\circ\mathbf{g}) = (\det\nabla\mathbf{u})(\det\nabla\mathbf{g})$ it is clear that $\det\nabla(\mathbf{u}\circ\mathbf{g}) > 0$ a.e. Thus, the desired result will follow once we show that (6.12) is satisfied.

Let $\phi \in C_0^1(\Omega)$ and set $x = g(z)$. Then, by the definition of the distributional Jacobian, the change of variables formula for diffeomorphisms (see Proposition 2.7 and the remark that follows it), and the fact that $\mathbf{g}(\Omega) = \Omega$,

$$\begin{aligned}
&\big(\mathrm{Det}\nabla(\mathbf{u}\circ\mathbf{g})\big)(\phi)\\
&= -\frac{1}{n}\int_\Omega \nabla_{\mathbf{z}}\phi\cdot\big[\operatorname{adj}\nabla_{\mathbf{z}}(\mathbf{u}\circ\mathbf{g})\big](\mathbf{u}\circ\mathbf{g})\,\mathrm{d}\mathbf{z}\\
&= -\frac{1}{n}\int_\Omega \mathbf{G}^{\mathrm{T}}\nabla_{\mathbf{x}}(\phi\circ\mathbf{g}^{-1})(\mathbf{g}(\mathbf{z}))\cdot[\operatorname{adj}\mathbf{G}]\big[\operatorname{adj}\nabla_{\mathbf{x}}\mathbf{u}(\mathbf{g}(\mathbf{z}))\big]\mathbf{u}(\mathbf{g}(\mathbf{z}))\,\mathrm{d}\mathbf{z}\\
&= -\frac{1}{n}\int_\Omega \nabla_{\mathbf{x}}(\phi\circ\mathbf{g}^{-1})(\mathbf{g}(\mathbf{z}))\cdot\big[\operatorname{adj}\nabla_{\mathbf{x}}\mathbf{u}(\mathbf{g}(\mathbf{z}))\big]\mathbf{u}(\mathbf{g}(\mathbf{z}))\det\mathbf{G}\,\mathrm{d}\mathbf{z}\\
&= -\frac{1}{n}\int_\Omega \nabla(\phi\circ\mathbf{g}^{-1})(\mathbf{x})\cdot\big[\operatorname{adj}\nabla\mathbf{u}(\mathbf{x})\big]\mathbf{u}(\mathbf{x})\,\mathrm{d}\mathbf{x}\\
&= (\mathrm{Det}\nabla\mathbf{u})(\phi\circ\mathbf{g}^{-1}),
\end{aligned}$$

where $\mathbf{G} := \nabla\mathbf{g}(\mathbf{z})$ and we have made use of the identity $\mathbf{G}[\operatorname{adj}\mathbf{G}] = (\det\mathbf{G})\mathbf{I}$.

Next, let $U \subset \Omega$ be open and note that, since $\mathbf{g}$ is a diffeomorphism, to each $\phi \in C_0^1(U)$ there corresponds a unique $\psi := \phi\circ\mathbf{g}^{-1} \in C_0^1(\mathbf{g}(U))$ and, moreover, $\|\phi\|_\infty = \|\psi\|_\infty$. Consequently, the preceding computation yields

$$\begin{aligned}
\big(\mathrm{Det}\nabla(\mathbf{u}\circ\mathbf{g})\big)(U) :&= \sup\big\{\big(\mathrm{Det}\nabla(\mathbf{u}\circ\mathbf{g})\big)(\phi)\colon \phi\in C_0^1(U),\ \|\phi\|_\infty \leqslant 1\big\}\\
&= \sup\big\{(\mathrm{Det}\nabla\mathbf{u})(\phi\circ\mathbf{g}^{-1})\colon \phi\in C_0^1(U),\ \|\phi\|_\infty \leqslant 1\big\}\\
&= \sup\big\{(\mathrm{Det}\nabla\mathbf{u})(\psi)\colon \psi\in C_0^1(\mathbf{g}(U)),\ \|\psi\|_\infty \leqslant 1\big\}\\
&=: (\mathrm{Det}\nabla\mathbf{u})(\mathbf{g}(U)).
\end{aligned}$$

Finally, let $A \subset \Omega$ be arbitrary and note that, since $\mathbf{g}$ is a diffeomorphism, to each open set U with $A \subset U \subset \Omega$ there corresponds a unique open set $V = \mathbf{g}(U)$ with $\mathbf{g}(A) \subset V \subset \Omega = \mathbf{g}(\Omega)$. Thus, in view of the preceding computation,

$$\begin{aligned}
\big(\mathrm{Det}\nabla(\mathbf{u}\circ\mathbf{g})\big)(A) &:= \inf\big\{\big(\mathrm{Det}\nabla(\mathbf{u}\circ\mathbf{g})\big)(U)\colon A\subset U\subset\Omega,\ U \text{ open}\big\}\\
&= \inf\big\{(\mathrm{Det}\nabla\mathbf{u})(\mathbf{g}(U))\colon A\subset U\subset\Omega,\ U \text{ open}\big\}\\
&= \inf\big\{(\mathrm{Det}\nabla\mathbf{u})(V) : \mathbf{g}(A)\subset V\subset\Omega,\ V \text{ open}\big\}\\
&=: (\mathrm{Det}\nabla\mathbf{u})(\mathbf{g}(A)),
\end{aligned}$$

which together with (6.11) yields

$$\big(\mathrm{Det}\nabla(\mathbf{u}\circ\mathbf{g})\big)(A) = \int_{\mathbf{g}(A)} \det\nabla\mathbf{u}\,d\mathbf{x} + \left(\sum_{i=1}^{M} \alpha^i \delta_{\mathbf{g}^{-1}(\mathbf{a}_i)}\right)(A).$$

Equation (6.12) then follows from the change of variables formula. □

COROLLARY 6.6. *Let* $\mathbf{u} \in \mathcal{A}^p$ $(\mathbf{u} \in \mathcal{A}^p_m)$. *Suppose* $\mathbf{v} \in C^1(\overline{\Omega}; \mathbb{R}^n)$ *satisfies* $\mathbf{v} = \mathbf{0}$ *on* $\{\mathbf{a}_1, \dots, \mathbf{a}_M\} \cup (\overline{\Omega} \setminus \mathcal{B})$. *Define*

$$\mathbf{h}_s := \mathbf{id} + s\mathbf{v} \quad \textit{and} \quad \mathbf{u}_s := \mathbf{u} \circ (\mathbf{h}_s)^{-1}.$$

Then, for all sufficiently small s,

$$\mathbf{u}_s \in \mathcal{A}^p \quad (\mathbf{u}_s \in \mathcal{A}^p_m).$$

Proof. Let $\mathbf{v} \in C^1(\overline{\Omega}; \mathbb{R}^n)$ satisfy $\mathbf{v} = \mathbf{0}$ on $\{\mathbf{a}_1, \dots, \mathbf{a}_M\} \cup (\overline{\Omega} \setminus \mathcal{B})$. Then it follows that (see, e.g., [8, Section 5.5]), for sufficiently small s, the mapping $\mathbf{h}_s := \mathbf{id} + s\mathbf{v}$ is a diffeomorphism, $\det \nabla\mathbf{h}_s > 0$ on Ω, $\mathbf{h}_s(\Omega) = \Omega$, and $\mathbf{h}_s(\mathbf{x}) = \mathbf{x}$ for $\mathbf{x} \in \overline{\Omega}\setminus\mathcal{B}$. Clearly, $\mathbf{h}_s(\mathbf{a}_i) = \mathbf{a}_i$.

Let $\mathbf{u} \in \mathcal{A}^p$; define $\mathbf{u}_s := \mathbf{u} \circ (\mathbf{h}_s)^{-1}$. Then $\mathbf{u}^e \in \mathrm{Def}^p(\Omega, \mathbf{a}_1, \dots, \mathbf{a}_M)$ and we may therefore apply Theorem 6.5 to conclude

$$\left(\mathbf{u}^e\right)_s := \mathbf{u}^e \circ (\mathbf{h}_s)^{-1} \in \mathrm{Def}^p(\Omega, \mathbf{a}_1, \dots, \mathbf{a}_M).$$

Next, we note that $\mathbf{v} = \mathbf{0}$ on $\overline{\Omega}\setminus\mathcal{B}$ and hence $(\mathbf{u}^e)_s(\mathbf{x}) = \mathbf{u}^e(\mathbf{id}(\mathbf{x})) = \mathbf{d}(\mathbf{x})$ for $\mathbf{x} \in \overline{\Omega}\setminus\mathcal{B}$. Therefore, $(\mathbf{u}_s)^e = (\mathbf{u}^e)_s \in \mathrm{Def}^p(\Omega, \mathbf{a}_1, \dots, \mathbf{a}_M)$ and $\mathbf{u}-\mathbf{d} \in W_0^{1,p}(\mathcal{B}; \mathbb{R}^n)$, which completes the proof when $\mathbf{u} \in \mathcal{A}^p$. The proof for $\mathbf{u} \in \mathcal{A}^p_m$ is similar. □

Appendix

The following result concerns the weak convergence of a sequence of products. Since we could not find it in a standard reference, we include a proof for the convenience of the reader.

LEMMA 6.7. *Let* $\psi_j \in L^\infty(\Omega)$ *and* $\theta_j \in L^1(\Omega)$ *satisfy*

$$\psi_j \to \psi \quad \textit{pointwise a.e.},$$
$$\theta_j \rightharpoonup \theta \quad \textit{in } L^1(\Omega),$$

where $\|\psi_j\|_\infty \leqslant K$ *for some* $K > 0$, $\theta \in L^1(\Omega)$, *and* $\psi \in L^\infty(\Omega)$. *Then*

$$\theta_j\psi_j \rightharpoonup \theta\psi \quad \textit{in } L^1(\Omega).$$

Proof. Fix $\varphi \in L^\infty(\Omega)$. Without loss of generality let $K = \|\psi\|_\infty = \|\varphi\|_\infty$. With a view toward applying Egoroff's theorem let $E \subset \Omega$ be any measurable set that satisfies $\psi_j \to \psi$ uniformly on $\Omega \backslash E$. Then

$$\int_\Omega \varphi(\theta_j \psi_j - \theta\psi)\,\mathbf{dx} = \int_\Omega \varphi(\theta_j - \theta)\psi\,\mathbf{dx} + \int_{\Omega\backslash E} \varphi(\psi_j - \psi)\theta_j\,\mathbf{dx} + \int_E \varphi(\psi_j - \psi)\theta_j\,\mathbf{dx}. \tag{A.1}$$

The first integral in the right-hand side of (A.1) goes to zero, as $j \to \infty$, by the weak convergence hypothesis. The absolute value of the second integral in the right-hand side of (A.1) is bounded above by

$$K \sup_{\mathbf{x}\in\Omega\backslash E} \left|\psi_j(\mathbf{x}) - \psi(\mathbf{x})\right| \int_\Omega |\theta_j|\,\mathbf{dx}.$$

We note that the sequence θ_j converges weakly in $L^1(\Omega)$ and, hence, the L^1 norm of θ_j is bounded uniformly in j. Thus, this term goes to zero, as $j \to \infty$, since $\psi_j \to \psi$ uniformly on $\Omega \backslash E$.

The absolute value of the last integral in the right-hand side of (A.1) is bounded above by

$$\left(\|\psi_j\|_\infty + \|\psi\|_\infty\right)\|\varphi\|_\infty \int_E |\theta_j|\,\mathbf{dx} \leqslant 2K^2 \int_E |\theta_j|\,\mathbf{dx}. \tag{A.2}$$

We note that the Dunford–Pettis compactness criterion (see, e.g., [9, Corollary 4.8.11]) implies that the sequence θ_j is uniformly integrable, that is, the integrals in the right-hand side of (A.2) will go to zero, uniformly in j, as the measure of E is made arbitrarily small. Finally, Egoroff's theorem yields E with arbitrarily small measure. □

Acknowledgements

This research was initiated while SJS was an EPSRC Visiting Fellow at the University of Bath. It was completed while the authors were visiting the IMA in Minneapolis. The research of SJS was also supported by the National Science Foundation.

References

1. J.M. Ball, Constitutive inequalities and existence theorems in nonlinear elastostatics. In: R.J. Knops (ed.), *Nonlinear Analysis and Mechanics*, Vol. I. Pitman (1977) pp. 187–241.
2. J.M. Ball, Convexity conditions and existence theorems in nonlinear elasticity. *Arch. Rational Mech. Anal.* **63** (1977) 337–403.
3. J.M. Ball, Discontinuous equilibrium solutions and cavitation in nonlinear elasticity. *Phil. Trans. Roy. Soc. London A* **306** (1982) 557–611.

4. J.M. Ball, Minimizers and the Euler-Lagrange equations. In: P.G. Ciarlet and M. Roseau (eds), *Trends and Applications of Pure Mathematics to Mechanics*. Springer, Berlin (1984) pp. 1–4.
5. J.M. Ball, J.C. Currie and P.J. Olver, Null Lagrangians, weak continuity, and variational problems of arbitrary order. *J. Functional Anal.* **41** (1981) 135–174.
6. P. Bauman, N.C. Owen and D. Phillips, Maximum principles and a priori estimates for a class of problems from nonlinear elasticity. *Analyse non linéaire* **8** (1991) 119–157.
7. K. Cho and A.N. Gent, Cavitation in model elastomeric composites. *J. Materials Sci.* **235** (1988) 141–144.
8. P.G. Ciarlet, *Mathematical Elasticity*, Vol. 1. Elsevier, Amsterdam (1988).
9. N. Dunford and J.T. Schwartz, *Linear Operators*, part. 1. Interscience, New York (1958).
10. L.C. Evans and R.F. Gariepy, *Measure Theory and Fine Properties of Functions*. CRC Press, Boca Raton (1992).
11. I. Fonseca and W. Gangbo, Degree Theory in Analysis and Applications. Oxford Univ. Press, Oxford (1995).
12. A.N. Gent, Cavitation in rubber: A cautionary tale. *Rubber Chem. Technology* **63** (1990) G49–G53.
13. A.N. Gent and P.B. Lindley, Internal rupture of bonded rubber cylinders in tension. *Proc. Roy. Soc. London A* **249** (1985) 195–205.
14. M. Giaquinta, G. Modica and J. Souček, Cartesian currents, weak diffeomorphisms and existence theorems in nonlinear elasticity. *Arch. Rational Mech. Anal.* **106** (1989) 97–159.
15. D. Gilbarg and N.S. Trudinger, *Elliptic Partial Differential Equations of Second Order*, 2nd ed. Springer, Berlin (1983).
16. C.O. Horgan and R. Abeyaratne, A bifurcation problem for a compressible nonlinearly elastic medium: Growth of a microvoid. *J. Elasticity* **16** (1986) 189–200.
17. C.O. Horgan and D.A. Polignone, Cavitation in nonlinearly elastic solids: A review. *J. Appl. Mech.* **48** (1995) 471–485.
18. R.D. James and S.J. Spector, The formation of filamentary voids in solids. *J. Mech. Phys. Solids* **39** (1991) 783–813.
19. R.D. James and S.J. Spector, Remarks on $W^{1,p}$- quasiconvexity, interpenetration of matter, and function spaces for elasticity. *Anal. Non-Linéaire* **9** (1992) 263–280.
20. M. Marcus and V.J. Mizel, Transformations by functions in Sobolev spaces and lower semicontinuity for parametric variational problems. *Bull. Am. Math. Soc.* **79** (1973) 790–795.
21. P.A. Meyer, *Probability and Potentials*. Blaisdell (1966).
22. C.B. Morrey, *Multiple Integrals in the Calculus of Variations*. Springer, Berlin (1966).
23. S. Müller, Det=det – A remark on the distributional determinant. *C. R. Acad. Sci. Paris Ser. I* **311** (1990) 13–17.
24. S. Müller and S.J. Spector, An existence theory for nonlinear elasticity that allows for cavitation. *Arch. Rational Mech. Anal.* **131** (1995) 1–66.
25. J.T. Schwartz, *Nonlinear Functional Analysis*. Gordon and Breach (1969).
26. J. Sivaloganathan, Uniqueness of regular and singular equilibria for spherically symmetric problems of nonlinear elasticity. *Arch. Rational Mech. Anal.* **96** (1986) 97–136.
27. J. Sivaloganathan, Singular minimizers in the calculus of variations: A degenerate form of cavitation. *Anal. Non-Linéaire* **9** (1992) 657–681.
28. V. Šverák, Regularity properties of deformations with finite energy. *Arch. Rational Mech. Anal.* **100** (1988) 105–127.
29. Q. Tang, Almost-everywhere injectivity in nonlinear elasticity. *Proc. Roy. Soc. Edinburgh A* **109** (1988) 79–95.
30. E. Zeidler, *Nonlinear Functional Analysis and its Applications*. Springer, Berlin (1986).

Journal of Elasticity **59:** 115–130, 2000.

Feedback Linearization and Semilinearization for Smart Elastic Structures

STUART S. ANTMAN
Department of Mathematics, Institute for Physical Science and Technology, and Institute for Systems Research, University of Maryland, College Park, MD 20742-4015, U.S.A.
E-mail: ssa@math.umd.edu

Received 19 January 1999; in revised form 11 February 2000

This paper is dedicated to Roger Fosdick on the occasion of his sixtieth birthday.

Abstract. This paper shows how the geometrically exact quasilinear equations of motion of nonlinearly elastic and viscoelastic rods and shells whose response is sensitive to ambient magnetic, electric, or thermal fields can be converted to semilinear or linear equations by suitable feedback controls of the ambient fields. Indeed, in certain cases, the feedback can make the response of a nonhomogeneous structure be like that of a homogenous structure, enlarge or diminish the isotropy group of the structure, increase or decrease the internal dissipation in the structure, and cause naturally different wave speeds to be the same. The availability of such controls indicates that the shocks to which quasilinear hyperbolic partial differential equations for nonlinear elastic structures are susceptible need cause no difficulty in control problems. In particular, if the structure is subject to additional controls that cause it to perform specific tasks, then these additional controls are treated by the theory for (semi)linear partial differential equations, for which there is an extensive development.

Mathematics Subject Classifications (2000): 73D99, 73Gxx, 73Kxx, 93B18, 93B52.

Key words: smart materials, magnetoelasticity, rods, shells, feedback linearization.

1. Introduction

For the purposes of this paper, a *smart material* is one in which there is a strong coupling between mechanical effects and effects of electromagnetic or thermal origin. To be specific, however, we ignore electrical and thermal effects; they can be treated by the same mathematical methods used to treat magnetic effects. Of course, the actual material responses of each of these cases restrict the circumstances under which our theory is applicable, but we encounter the same difficulties within magneto(visco)elasticity.

There are several recently developed alloys of common metals, such as iron and nickel, and rare earths, such as terbium and dysprosium, whose mechanical responses are very sensitive to mild changes in magnetic field. These include Terfenol and MetGlas (a trademark of the Allied Chemical Corp). The available

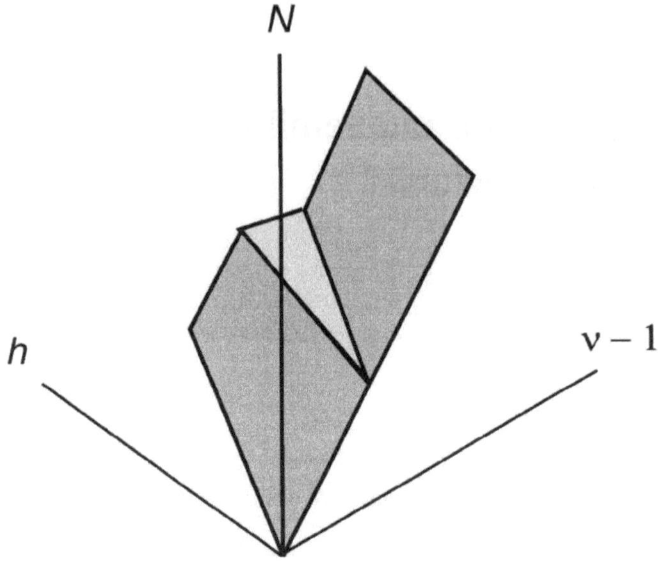

Figure 1. A piecewise linear approximation of the graph of the tension N as a function $\widehat{N}$ of the elongation $\nu - 1$ and the magnetic field h for MetGlas. The actual graph looks like a smoothed out version of this. The range of the ν shown here is smaller than that in which the purely mechanical response exhibits nonlinearity. The nonlinearity of N comes into play at small strain; it cannot be avoided. The smallest slope of $\widehat{N}(\cdot, h)$ is about $1/10$ the largest.

experimental results for the response of these materials is essentially limited to tension tests [26] giving the tension N as a function $\widehat{N}$ of the stretch ν and the magnetic field h. The graph of this function for metglas, which is fabricated in thin strips, is given in Figure 1. (The graphs for terfenol are like those for metglas with N representing the pressure. Terfenol is brittle in tension, so its use is slightly different from that for metglas. See [9, 23, 29] for information about these materials.) The function $\widehat{N}$ is even in h. There are critical values $h_c < h_s$ of h such that as h increases in the interval $[h_c, h_s]$, the effects of magnetization diminish to zero, so that $\widehat{N}(\nu, h_s) = \widehat{N}(\nu, 0)$. When h exceeds the saturation field h_s, $\widehat{N}(\nu, h)$ remains at its saturation value $\widehat{N}(\nu, 0)$. For our purpose of using the magnetic field to control material properties, we may accordingly limit our attention to h's in the interval $[0, h_c]$.

2. Longitudinal Motion of a Magneto(visco)elastic Rod

Let s identify a typical material section of a naturally straight rod and let $w(s, t)$ denote the longitudinal position of s at time t. Let $(\rho A)(s)$ denote the mass density per unit reference length of the rod at s. Let $N(s, t)$ denote the tension at material section s at time t. Let $h(s, t)$ be (the material form of) the magnetic field along the rod at (s, t). It will serve as the control variable for our problem. We suppose that it is generated by many independent coils attached to the rod along its length, so that its magnitude can essentially be varied at each section s by varying the current in the coil encircling that section. We are presuming that the rod is so thin that the magnetic field is independent of the thickness. (We comment on this point in Section 6.)

For a magnetoelastic rod, N has the form

$$N(s,t) = \widehat{N}\big(w_s(s,t), h(s,t), s\big) \tag{1}$$

and for a magnetoviscoelastic rod of strain-rate type,

$$N(s,t) = \widehat{N}\big(w_s(s,t), w_{st}(s,t), h(s,t), s\big). \tag{2}$$

Here the subscripts denote partial derivatives. We assume that $\widehat{N}$ of (2) is increasing in its second argument, and has the form of Figure 1 for any fixed value of its second argument.

The unforced longitudinal motion is governed by

$$(\rho A)(s) w_{tt}(s,t) = \partial_s N(s,t), \tag{3}$$

where N is given by (1) or (2). We are assuming that the magnetic field h can be imposed, so we do not have to concern ourselves with its evolution as governed by a version of Maxwell's equations. Thus h serves to change the constitutive function, acting as a body control. It therefore is especially effective because it controls the principal part of the partial differential operator.

This system (1), (3) with a given h typically exhibits shocks. As shown in [2], h can be chosen to prevent shocks: If there is a positive-valued function a and a constant c such that in (ν, h, x)-space, the plane $z = a\nu + c$ intersects the surface $z = \widehat{N}(\nu, h, x)$ for each fixed s, then the solution(s) $\tilde{h}(\nu, x)$ of

$$a(x)\nu + c = \widehat{N}(\nu, h, x), \tag{4}$$

which correspond to the intersection of the plane with the surface, are feedbacks that linearize (1), (3), producing the linear wave equation

$$(\rho A)(s) w_{tt}(s,t) = \partial_s\big[a(s) w_s(s,t)\big]. \tag{5}$$

(To effect this feedback, the stretch w_s at each section must be monitored with a strain gauge, not necessarily of the traditional sort; lasers have recently been used for this purpose.) Note that a can be taken to be constant. It can be chosen over a range to produce different wave speeds. In this case, if ρA is constant, then (5) is a linear wave equation with constant coefficients. Since c can be chosen in any convenient way, it is very easy to ensure that (4) holds. (We denote the last arguments of $\widehat{N}$ by x to distinguish between the derivative $\widehat{N}_x$ of $\widehat{N}$ with respect to its last argument and the total derivative of $\widehat{N}$ with respect to s, which is $\partial_s \widehat{N} = \widehat{N}_\nu w_{ss}(s,t) + \widehat{N}_h h_s(s,t) + \widehat{N}_x$, where the arguments of $\widehat{N}$ and its derivatives are $(w_s(s,t), h(s,t), s)$. We use the same convention for $\tilde{h}$.) In Section 6 we discuss more general choices for a and c, in particular, when they each depend on s and t, in which case they can be used for the control of the structure. Because of this important purpose, we avoid the facile linearization for materials like that of Figure 1 effected by turning off the magnetic field: $h = 0$.

Suppose that the magnetoelastic rod is non-uniform, so that ρA or $\widehat{N}$ depend explicitly on s. Are there controls $\tilde{h}(\nu, x)$ that lead to wave equations for w with constant coefficients? If so, there would be a positive constant α^2 such that

$$\widehat{N}\big(w_s, \tilde{h}(w_s, s), s\big)_s = \alpha^2(\rho A)(s)w_{ss}, \tag{6}$$

so that

$$\widehat{N}_\nu + \widehat{N}_h \tilde{h}_\nu = \alpha^2 \rho A, \qquad \widehat{N}_h \tilde{h}_x + \widehat{N}_x = 0. \tag{7}$$

We solve (7) for $\tilde{h}_\nu$ and $\tilde{h}_x$, and impose the compatibility condition that the mixed partial derivatives of $\tilde{h}$ must be equal:

$$\left[\frac{\alpha^2\rho A - \widehat{N}_\nu}{\widehat{N}_h}\right]_x = -\left[\frac{\widehat{N}_x}{\widehat{N}_h}\right]_\nu, \tag{8}$$

where the arguments of the derivatives of $\widehat{N}$ are $\nu, \tilde{h}(\nu, x), x$ and the argument of ρA is x. We carry out the indicated derivations and substitute into the resulting equation the representations for $\tilde{h}_\nu$ and $\tilde{h}_x$ in terms of the derivatives of $\widehat{N}$ coming from (7) to get a nasty nonlinear partial differential equation, parametrized by α, for $\widehat{N}$, which means that such a feedback is possible only for very special materials (not necessarily physically realistic), so this method lacks any usefulness.

It is possible, however, to find a feedback that causes $z := (\rho A)w$ to satisfy the wave equation with constant coefficients: Clearly

$$z_s = \rho A w_s + \frac{(\rho A)_s}{\rho A} z, \tag{9}$$

so that (3) becomes

$$z_{tt} = \partial_s \widehat{N}\left(\frac{z_s}{\rho A} - \frac{(\rho A)_s}{(\rho A)^2} z, h, s\right). \tag{10}$$

This becomes the wave equation with constant coefficients if h can be chosen so that

$$\widehat{N}\left(\frac{z_s}{\rho A} - \frac{(\rho A)_x}{(\rho A)^2} z, h, x\right) = \alpha^2 z_x + c \tag{11}$$

or, equivalently,

$$\widehat{N}(\nu, h, x) = \alpha^2\left(\rho A \nu + \frac{(\rho A)_x}{\rho A} z\right) + c. \tag{12}$$

(Note that z_t is the linear momentum.) A feedback $\tilde{h}(\nu, z, x)$ satisfying (12) depends not only on the stretch but also the position of the section. For each fixed z, (12) is described by the intersection of a plane with the graph of $\widehat{N}$. Again the freedom in picking c makes such feedbacks feasible.

Now we look at magnetoviscoelastic rods of strain-rate type, governed by (2), (3). If a feedback $\tilde{h}(\nu, \dot{\nu}, x)$ can be found to satisfy

$$N(\nu, \dot{\nu}, h, x) = a(x)\nu + b(x)\dot{\nu} + c, \tag{13}$$

then the original quasilinear parabolic-hyperbolic partial differential equation is replaced by a linear equation of the same type:

$$\rho A w_{tt} = \partial_s\big[a(s)w_s + b(s)w_{st}\big]. \tag{14}$$

Note that the feedback requires the monitoring of not only the strain but also the strain rate. The graphical interpretation of (13) is not as simple as that of (4). Again the availability of the additive constant c gives (12) great feasibility. Note that if there are feedback solutions for large b, they increase damping in the rod, and if there are solutions for small or zero b, they reduce or annihilate damping in the rod. We could even contemplate taking b negative to destabilize the rod. (Of course, the mechanical power saved in the rod by a reduction of b is spent in the generator supplying electrical power to the coils.)

Equation (3) with N having a different form also describes the planar shearing of an incompressible medium. Our same methods apply to it.

3. Planar Motions of Rods

We now study the far richer problems for rods, not necessarily naturally straight, undergoing flexure as well as extension and shear, for which we use a geometrically exact theory [1]: Let $\{\boldsymbol{i}, \boldsymbol{j}, \boldsymbol{k}\}$ be a fixed right-handed orthonormal basis for Euclidean space. We interpret s as the arc-length parameter for the base curve of mass centroids of a rod in its natural reference state. We interpret $\boldsymbol{r}(s,t)$ as the position of the material point on the base curve with coordinate s at time t. Here we restrict our attention to deformations for which $\boldsymbol{r}$ lies in the $\{\boldsymbol{i}, \boldsymbol{j}\}$-plane. We interpret $\theta(s,t)$ as characterizing the orientation of the section at s, and set

$$\boldsymbol{a}(\theta) := \cos\theta\boldsymbol{i} + \sin\theta\boldsymbol{j}, \qquad \boldsymbol{b}(\theta) = -\sin\theta\boldsymbol{i} + \cos\theta\boldsymbol{j}. \tag{15}$$

We decompose

$$\boldsymbol{r}_s =: \nu\boldsymbol{a} + \eta\boldsymbol{b}. \tag{16}$$

ν measures volume change, but may be regarded as essentially measuring length change, η measures shear, and θ_s measures flexure. Let $N\boldsymbol{a}+H\boldsymbol{b}$ denote the contact force at s and $M\boldsymbol{k}$ the contact (bending) couple at s. The unforced equations of motion are

$$\rho A \boldsymbol{r}_{tt} = \partial_s (N\boldsymbol{a} + H\boldsymbol{b}), \tag{17}$$

$$\rho J \theta_{tt} = M_s + \boldsymbol{k} \cdot \big[\boldsymbol{r}_s \times (N\boldsymbol{a} + H\boldsymbol{b})\big]. \tag{18}$$

Here $(\rho A)(s)$ is the mass density per unit length at s in the reference configuration and $(\rho J)(s)$ is interpreted as the mass moment of inertia of the section s at about an axis parallel to $\boldsymbol{k}$ through the line of centroids at s. In components, these equations of motion are less attractive:

$$\rho A \boldsymbol{r}_{tt} \cdot \boldsymbol{a} = N_s - \theta_s H, \qquad \rho A \boldsymbol{r}_{tt} \cdot \boldsymbol{b} = H_s + \theta_s H, \tag{19}$$

$$\rho J \theta_{tt} = M_s + \nu H - \eta N. \tag{20}$$

The most important feature of these equations is the nonlinear term $\boldsymbol{k} \cdot (\boldsymbol{r}_s \times (N\boldsymbol{a} + H\boldsymbol{b})) = \nu H - \eta N$ appearing in (18) and (20). The nonlinearities on the right-hand sides of (19) are slightly artificial and will not cause difficulties.

Properly invariant constitutive equations for magnetoelastic rods have the form

$$N = \widehat{N}(\nu, \eta, \theta_s, \mathsf{h}, , s), \quad H = \widehat{H}(\nu, \eta, \theta_s, \mathsf{h}, s), \quad M = \widehat{M}(\nu, \eta, \theta_s, \mathsf{h}, s). \tag{21}$$

Here h represents a triple of magnetic controls corresponding to the strains ν, η, θ_s. There seem to be practical ways of designing composite rods consisting of assemblages of rods or strips of magnetoelastic materials surrounded by coils capable of influencing these strains. For example, if the composite rod consists of two magnetoelastic rods (forming a bimetallic strip) with each of the constituent rods surrounded by its own set of coils, then sending a current through the coils of one rod weakens it in extension. Weakening just one rod affects the flexural strain, while weakening both simultaneously affects the extensional strain. The shear could be influenced by disposing magnetoelastic elements surrounded by coils at a 45° angle to the axis $\boldsymbol{r}$. (See [3] for a description of how to construct $\widehat{M}$ from Figure 1.)

If $(\widehat{N}, \widehat{H}, \widehat{M})$ is strictly monotone in (ν, η, θ_s) (a sufficient condition for which is that the matrix of partial derivatives of $\widehat{N}$, $\widehat{H}$, $\widehat{M}$ with respect to ν, η, θ_s be positive-definite) and if h is regarded as given, then the substitution of these constitutive equations into the equations of motion yields a quasilinear hyperbolic system.

If we can find a feedback h so that

$$\begin{bmatrix} \widehat{N}(\nu, \eta, \mu, \mathsf{h}, x) \\ \widehat{H}(\nu, \eta, \mu, \mathsf{h}, x) \\ \widehat{M}(\nu, \eta, \mu, \mathsf{h}, x) \end{bmatrix} = \mathsf{C}(x) \begin{bmatrix} \nu \\ \eta \\ \mu \end{bmatrix} + \mathsf{c}, \tag{22}$$

where $\mathsf{C}(x)$ with components $C_{ij}(x)$ is a symmetric positive-definite matrix and c is a constant vector, then the governing equations reduce to a curious quasilinear hyperbolic system for $(\boldsymbol{r}, \theta)$:

$$\begin{aligned} \rho A \boldsymbol{r}_{tt} = \partial_s \big[&(C_{11} \boldsymbol{r}_s \cdot \boldsymbol{a} + C_{12} \boldsymbol{r}_s \cdot \boldsymbol{b} + C_{13} \theta_s) \boldsymbol{a} \\ &+ (C_{21} \boldsymbol{r}_s \cdot \boldsymbol{a} + C_{22} \boldsymbol{r}_s \cdot \boldsymbol{b} + C_{23} \theta_s) \boldsymbol{b} \big], \end{aligned} \tag{23}$$

$$\begin{aligned}\rho J\theta_{tt} &= \partial_s[C_{31}\boldsymbol{r}_s\cdot\boldsymbol{a}+C_{32}\boldsymbol{r}_s\cdot\boldsymbol{b}+C_{33}\theta_s]\\ &\quad+(C_{21}\boldsymbol{r}_s\cdot\boldsymbol{a}+C_{22}\boldsymbol{r}_s\cdot\boldsymbol{b}+C_{23}\theta_s)\boldsymbol{r}_s\cdot\boldsymbol{a}\\ &\quad-(C_{11}\boldsymbol{r}_s\cdot\boldsymbol{a}+C_{12}\boldsymbol{r}_s\cdot\boldsymbol{b}+C_{13}\theta_s)\boldsymbol{r}_s\cdot\boldsymbol{b}.\end{aligned}\tag{24}$$

The quasilinearity, which is of a very mild sort, is due to the presence of the base vectors $\boldsymbol{a}(\theta)$ and $\boldsymbol{b}(\theta)$ in (23). If we introduce the new variables $\omega = \rho J\theta_t$ and $(u\boldsymbol{i}+v\boldsymbol{j}) = \rho A\boldsymbol{r}_t$, then the system (23), (24) is equivalent to the following seventh-order system of first-order equations:

$$\begin{aligned}\theta_t &= (\rho J)^{-1}\omega,\\ \nu_t &= (\rho A)^{-1}[u_s\cos\theta+v_s\sin\theta]+\eta\omega,\\ \eta_t &= (\rho A)^{-1}[-u_s\sin\theta+v_s\cos\theta]-\nu\omega,\\ \mu_t &= \omega_s,\\ u_t &= N_s\cos\theta-H_s\sin\theta-\mu(N\sin\theta+H\cos\theta),\\ v_t &= N_s\sin\theta+H_s\cos\theta+\mu(N\cos\theta-H\sin\theta),\\ \rho J\omega_t &= M_s+\nu H-\eta N.\end{aligned}\tag{25}$$

Though this system is likewise quasilinear because of the disposition of the θ's, a straightforward calculation shows that the characteristic speeds are 0 and $\pm\sqrt{C_i}$, where the C_i are the (positive) eigenvalues of C. This fact suggests that this system lacks shocks. (A formal proof could be based on a demonstration that the characteristic equations subject to suitable initial and boundary conditions have solutions defined for all time. It does not appear that a phenomenon like this is mentioned in books on partial differential equations.)

We can, however, show that the system (23), (24) is equivalent to a semilinear system, known to lack shocks, by introducing $\alpha\boldsymbol{a}+\beta\boldsymbol{b} = \rho A\boldsymbol{r}_t$. The resulting system for $\nu, \eta, \mu, \alpha, \beta, \omega$, which is independent of θ (which, like $\boldsymbol{r}$, can be found afterwards), is immediately seen to be semilinear.

While (22) does not admit a simple graphical interpretation, the availability of the additive constant $\mathbf{c}$ greatly enhances the possibility of finding feedbacks.

If we choose

$$\mathsf{C}(x) = \operatorname{diag}\big(C(x), C(x), D(x)\big),\tag{26}$$

then the governing equations uncouple into linear wave equations:

$$\rho A\boldsymbol{r}_{tt} = \partial_s[C\boldsymbol{r}_s], \qquad \rho J\theta_{tt} = \partial_s[D\theta_s].\tag{27}$$

If ρA and ρJ are constant, and if we further take C and D to be constant with

$$C/\rho A = D/\rho J,\tag{28}$$

then these two equations become linear wave equations with the same constant wave speeds.

In the theory of shearable linearly elastic rods, C is usually taken to be diagonal with the extensional stiffness C_{11} denoted EA, with the shear stiffness C_{22}

denoted GA, and with the bending stiffness C_{33} denoted EJ. Here E and G are the (effective) elastic and shear moduli, A is the cross-sectional area, and J is the areal moment of inertia of the cross section. Since $G < E$ for natural materials, the equality of C_{11} with C_{22} presumed in (26) does not occur for such materials (while (28) is standard). Thus the feedback enables us to synthesize a completely artificial material.

The conversion of the quasilinear hyperbolic equations of motion to semilinear equations removes the possibility of shocks. But equilibrium problems for the semilinear equations can still have multiple solutions. In particular, the rods can buckle as in Euler's problem. The presence of multiple equilibrium solutions, many of which are unstable, organizes the (infinite-dimensional) phase space for dynamical problems and gives the solutions a very complicated structure (even in the presence of damping). The artificial materials leading to (27) do not have multiple equilibrium solutions under common boundary conditions, e.g., those that usually give rise to buckling. We can easily find a class of feedbacks that prevent buckling for rods with nonlinear constitutive equations: We choose h for which there is a function $\widehat{G}$ such that

$$\widehat{N}(\nu, \eta, \mu, \mathsf{h}, s) = \nu\widehat{G}(\nu, \eta, \mu, s), \quad \widehat{H}(\nu, \eta, \mu, \mathsf{h}, s) = \eta\widehat{G}(\nu, \eta, \mu, s) \quad (29)$$

so that $\nu\widehat{H} - \eta\widehat{N} = 0$ with the new constitutive functions for (N, H, M) being strictly monotone functions of (ν, η, μ). Then the equilibrium equations for the rod under dead loads have unique solutions. (In general, the strict monotonicity of $(\widehat{N}, \widehat{H}, \widehat{M})$ makes the differential operator in the equilibrium equations semi-monotone. This operator becomes strictly monotone when the lower-order term $\nu\widehat{H} - \eta\widehat{N}$ vanishes.)

Much analysis is carried out for inextensible, unshearable rods, which are generated by the constraints $\nu = 1$ and $\eta = 0$. In this case, the equations of motion are the same, but N and H are now Lagrange multipliers maintaining these constraints, and the only constitutive equation is $M = \widehat{M}(\theta_s, h, s)$. The governing equations are not so easily classified. The choice of a feedback h to satisfy $\widehat{M}(\mu, h, s) = a(s)\mu + c$ replaces the quasilinear operator $\partial_s\widehat{M}(\theta_s, h, s)$ with a linear one, and produces the linear equation

$$\rho J\theta_{tt} = \partial_s[a\theta_s] + H, \quad (30)$$

which is supplemented with the nonlinear (17) or (19) and with the constraint in the form $\boldsymbol{r}_s = \boldsymbol{a}(\theta)$. The resulting system is well-behaved (see [8]). It does not seem possible to simplify the system further, as for the unconstrained rod.

The treatment of feedbacks for magnetoviscoelastic rods follows the pattern established in this and the preceding section; the details are left to the reader. Here the properly invariant constitutive equations for magnetoelastic rods have the form

$$\begin{aligned} N &= \widehat{N}(\eta, \eta, \theta_s, \mathsf{h}, \nu_t, \eta_t, \theta_{st}, s), \\ H &= \widehat{H}(\nu, \eta, \theta_s, \mathsf{h}, \nu_t, \eta_t, \theta_{st}, s), \\ M &= \widehat{M}(\nu, \eta, \theta_s, \mathsf{h}, \nu_t, \eta_t, \theta_{st}, s). \end{aligned} \quad (31)$$

4. Spatial Motions of Rods

Similar methods handle the partial differential equations for the motion of rods (not necessarily naturally straight) in space, which we briefly describe, with the aim of showing those features that differ from those for the planar motion of rods. We study the geometrically exact equations of motion of rods that can suffer flexure, extension, torsion, and shear [1]. We take s and $\boldsymbol{r}(s,t)$ to be defined as in the preceding section. We characterize the orientation at time t of the section at s by an orthonormal pair $\boldsymbol{d}_1(s,t)$, $\boldsymbol{d}_2(s,t)$ of vectors. We set $\boldsymbol{d}_3 := \boldsymbol{d}_1 \times \boldsymbol{d}_2$. The orthonormality of $\{\boldsymbol{d}_k(s,t)\}$ for each (s,t) implies that there are vector-valued functions $\boldsymbol{u}$ and $\boldsymbol{w}$ such that

$$\partial_s \boldsymbol{d}_k = \boldsymbol{u} \times \boldsymbol{d}_k, \qquad \partial_t \boldsymbol{d}_k = \boldsymbol{w} \times \boldsymbol{d}_k. \tag{32}$$

By using the equality of mixed partial derivatives of the $\boldsymbol{d}_k$, we can prove that

$$\boldsymbol{u}_t = \boldsymbol{w}_s - \boldsymbol{u} \times \boldsymbol{w}. \tag{33}$$

Since the basis $\{\boldsymbol{d}_k\}$ is natural for the intrinsic description of deformation, we decompose relevant vector-valued functions with respect to it:

$$\boldsymbol{u} = u_k \boldsymbol{d}_k, \qquad \boldsymbol{r}_s = v_k \boldsymbol{d}_k, \qquad \boldsymbol{w} = w_k \boldsymbol{d}_k. \tag{34}$$

The triples

$$\mathsf{u} := (u_1, u_2, u_3), \qquad \mathsf{v} := (v_1, v_2, v_3) \tag{35}$$

are the *strain* variables corresponding to the motion of $\boldsymbol{r}, \boldsymbol{d}_1, \boldsymbol{d}_2$ with u_1 and u_2 measuring flexure, u_3 measuring torsion, v_1 and v_2 measuring shear, and v_3 measuring dilatation.

In the configuration at time t, the resultant contact force and contact couple at the section s are respectively denoted $\boldsymbol{n}(s,t)$ and $\boldsymbol{m}(s,t)$. The unforced equations of motion are

$$\rho A \boldsymbol{r}_{tt} = \boldsymbol{n}_s, \tag{36}$$

$$\rho J_{kl}[(\partial_t w_l)\boldsymbol{d}_k + w_l \boldsymbol{w} \times \boldsymbol{d}_k] \equiv [\rho J_{kl} w_l \boldsymbol{d}_k]_t = \boldsymbol{m}_s + \boldsymbol{r}_s \times \boldsymbol{n}, \tag{37}$$

where ρA is the mass density per unit reference length and where $(\rho J_{\gamma\delta})(s)$, with Greek indices ranging over 1, 2, are the prescribed components of a positive-definite symmetric 2×2 matrix mass-moments of inertia of the section s, $\rho J_{\gamma 3} = \rho J_{3\gamma} = 0$, $\rho J_{33} = \rho J_{\gamma\gamma}$. Here we have used the summation convention. Note that these equations not only have the nonlinearity $\boldsymbol{r}_s \times \boldsymbol{n}$ possessed by the equations for planar motion, but also have the quadratic nonlinearity in the acceleration terms, which is just like that which appears in Euler's equations of motion for a rigid body.

Let

$$\mathsf{m} := (m_1, m_2, m_3), \quad \mathsf{n} := (n_1, n_2, n_3), \quad m_k := \boldsymbol{m} \cdot \boldsymbol{d}_k,\ n_k := \boldsymbol{n} \cdot \boldsymbol{d}_k. \tag{38}$$

m_1 and m_2 are the *bending couples*, m_3 is the *twisting couple*, n_1 and n_2 are the *shear forces*, and $\boldsymbol{n} \cdot \boldsymbol{r}_s/|\boldsymbol{r}_s|$ is the *tension*. (These terms are not strictly analogous to those used in structural mechanics, in which it is usually assumed that $\boldsymbol{d}_3 = \boldsymbol{r}_s/|\boldsymbol{r}_s|$.)

The rod is magnetoelastic if it has constitutive equations of the form

$$\begin{aligned} \mathbf{m}(s,t) &= \widehat{\mathbf{m}}\big(\mathbf{u}(s,t), \mathbf{v}(s,t), \mathbf{h}(s,t), s\big), \\ \mathbf{n}(s,t) &= \hat{\mathbf{n}}\big(\mathbf{u}(s,t), \mathbf{v}(s,t), \mathbf{h}(s,t), s\big), \end{aligned} \tag{39}$$

and it is magnetoviscoelastic of strain-rate type if

$$\begin{aligned} \mathbf{m}(s,t) &= \widehat{\mathbf{m}}\big(\mathbf{u}(s,t), \mathbf{v}(s,t), \mathbf{u}_t(s,t), \mathbf{v}_t(s,t), \mathbf{h}(s,t), s\big), \\ \mathbf{n}(s,t) &= \hat{\mathbf{n}}\big(\mathbf{u}(s,t), \mathbf{v}(s,t), \mathbf{u}_t(s,t), \mathbf{v}_t(s,t), \mathbf{h}(s,t), s\big). \end{aligned} \tag{40}$$

Here $\mathbf{h}$ is a sextuple of magnetic controls presumed capable of influencing the six strains. As in the preceding section, one can contrive devices for which such controls are effective. These constitutive equations are invariant under rigid motion.

If $(\widehat{\mathbf{m}}, \hat{\mathbf{n}})$ is uniformly monotone in $(\mathbf{u}, \mathbf{v})$, then the equations of motion for magnetoelastic rods form a twelfth-order quasilinear hyperbolic system, As in the preceding section, we can make these equations of motion equivalent to those of a semilinear system by finding a feedback $\mathbf{h}$ for which $(\widehat{\mathbf{m}}, \hat{\mathbf{n}})$ is linear in $(\mathbf{u}, \mathbf{v})$:

$$\begin{bmatrix} \widehat{\mathbf{m}}(\mathbf{u}, \mathbf{v}, \mathbf{h}, x) \\ \hat{\mathbf{n}}(\mathbf{u}, \mathbf{v}, \mathbf{h}, x) \end{bmatrix} = \mathbf{C}(x) \begin{bmatrix} \mathbf{u} \\ \mathbf{v} \end{bmatrix} + \mathbf{c}, \tag{41}$$

where $\mathbf{C}$ is a (6×6)-matrix and $\mathbf{c}$ is a 6-vector. We can take $\mathbf{C}$ diagonal: $\mathbf{C} = \operatorname{diag}(C_1, C_2, C_3, C_4, C_5, C_6)$. (This corresponds to simple, reasonable models of linearly elastic rods.) In particular, if $\mathbf{h}$ can be chosen so that $C_1 = C_2$ and $C_4 = C_5$, then the feedback system corresponds to that for a transversely isotropic body [1]. Alternatively, if the original material is transversely isotropic, then by violating these equations, we destroy the transverse isotropy of the feedback system. The isotropy of the feedback system might strongly effect the motion of the rod: It is known that travelling waves in transversely isotropic hyperelastic rods are governed by a totally integrable Hamiltonian system [4], whereas those for a system without such isotropy are not so governed, and have solutions that behave chaotically [22]. If we can find a feedback for which (41) holds with the artificial $C_1 = C_2 = C_3 = C$ and $C_4 = C_5 = C_6 = D$, then (36), (37), (33) reduce to the system

$$\rho A \boldsymbol{r}_{tt} = \partial_s[C\boldsymbol{r}_s], \tag{42}$$

$$\rho J_{kl}\big[(\partial_t w_l)\boldsymbol{d}_k + w_l \boldsymbol{w} \times \boldsymbol{d}_k\big] = \partial_s[D\boldsymbol{u}], \tag{43}$$

$$\boldsymbol{u}_t = \boldsymbol{w}_s - \boldsymbol{u} \times \boldsymbol{w}. \tag{44}$$

(Note that the availability of the additive constant $\mathbf{c}$ as before greatly enhances the possibility of synthesizing such an artificial material.) Here the equation for $\boldsymbol{r}$ is a linear wave equation uncoupled from the others, whose semilinearity is irreducible.

If the rod is constrained to be inextensible and unshearable (as in the Kirchhoff theory [21]) so that $\boldsymbol{r}_s = \boldsymbol{d}_3$, then $\boldsymbol{n}$ is a Lagrange multiplier whose presence in the term $\boldsymbol{r}_s \times \boldsymbol{n}$ prevents its removal by the class of feedbacks we are discussing.

5. Motion of Shells

The ideas of the last two sections apply to shells. We emphasize only those points of novelty.

We locate a typical material point on the midsurface of a shell in its reference configuration by a pair (s^1, s^2) of curvilinear coordinates. Let $\boldsymbol{r}(s^1, s^2, t)$ denote the position at time t of material point (s^1, s^2). Let $\boldsymbol{d}(s^1, s^2, t)$ be a unit vector that characterizes the orientation of the material line normal to the midsurface in the reference configuration. Let the differential surface area on the midsurface in its reference configuration be denoted by $\gamma(s^1, s^2)\, ds^1\, ds^2$. We set $\boldsymbol{r}_{,\beta} := \partial \boldsymbol{r}/\partial s^\beta$, etc. We assume that $\{\boldsymbol{r}_{,1}, \boldsymbol{r}_{,2}, \boldsymbol{d}\}$ are independent. (We can impose constitutive restrictions that would ensure that solutions of the equations of motion have this independence, at least almost everywhere.)

The unforced equations of motion [1] for a shell that can suffer flexure, in-plane extension, and shear are

$$\rho a \gamma \boldsymbol{r}_{tt} = \left(\gamma \boldsymbol{n}^\beta\right)_{,\beta}, \tag{45}$$

$$\rho J \gamma \boldsymbol{d} \times \boldsymbol{d}_{tt} = \left(\gamma \boldsymbol{m}^\beta\right)_{,\beta} + \gamma \boldsymbol{r}_{,\beta} \times \boldsymbol{n}^\beta, \tag{46}$$

where Greek indices range over 1, 2, and diagonally repeated Greek indices are summed over this range. Here $(\rho a \gamma)(s^1, s^2)$ is the mass per unit reference area of the midsurface at (s^1, s^2), $(\rho J \gamma)(s^1, s^2)$ is the mass-moment of inertia about the midsurface of the material line normal to the midsurface in the reference configuration, and $\boldsymbol{n}^\beta$ and $\boldsymbol{m}^\beta$ are the scaled resultant contact force and couple acting across the material curve $s^\beta = \text{constant}$.

The material of the shell is magnetoelastic if $\boldsymbol{n}^\beta$ and $\boldsymbol{m}^\beta$ are functions of $\boldsymbol{r}_{,\beta}$, $\boldsymbol{d}$, $\boldsymbol{d}_\beta$, h, $\beta = 1, 2$. Here h is a sextuple of magnetic controls presumed capable of influencing each mode of strain of the shell. Let $\boldsymbol{a}^3 := (\boldsymbol{r}_{,1} \times \boldsymbol{r}_{,2})/\,|\boldsymbol{r}_{,1} \times \boldsymbol{r}_{,2}|$. Then the constitutive equations are invariant under rigid motions [1] if they specify the components of $\boldsymbol{n}^\alpha$ and $\boldsymbol{m}^\alpha$ with respect to either of the bases $\{\boldsymbol{r}_{,1}, \boldsymbol{r}_{,2}, \boldsymbol{d}\}$ or $\{\boldsymbol{r}_{,1}, \boldsymbol{r}_{,2}, \boldsymbol{a}^3\}$ as functions of $\boldsymbol{r}_{,\alpha} \cdot \boldsymbol{r}_{,\beta}$, $\boldsymbol{r}_{,\alpha} \cdot \boldsymbol{d}$, $\boldsymbol{r}_{,\beta} \cdot \boldsymbol{d}_{,\beta}$, $\boldsymbol{d} \cdot \boldsymbol{a}^3$, $\alpha, \beta = 1, 2$.

The resemblance of (45), (46) to (36), (37) means that exactly the same methods used there can be applied here to replace (45), (46) with a system equivalent to a semilinear system that is generally irreducible to a linear system. One phenomenon possessed by these shell equations that is lacking in the rod equations of Section 4 is the Poisson-ratio effect: A tension in one direction produces an elongation in that direction and produces contractions in transverse directions. Suitable feedbacks could lead to artificial materials in which the transverse contraction is very small or zero or even negative (i.e., an expansion), or alternatively, to materials in which the transverse contraction is large, say the order of the longitudinal extension. In a similar vein, we could use a feedback to reduce the reaction across certain material curves. An extreme manifestation of this process would be to synthesize a material that behaves like an array of fibers. Such a model was used in [28] (cf. [24]) to describe parachutes. To examine some related ideas in a concrete setting, we consider the axisymmetric motion of shells.

We again take $\{\boldsymbol{i}, \boldsymbol{j}, \boldsymbol{k}\}$ to be a fixed right-handed orthonormal basis, and take $\boldsymbol{k}$ to be the axis of symmetry of the shell. We set $\boldsymbol{e}_1(\phi) := \cos\phi\boldsymbol{i} + \sin\phi\boldsymbol{j}$, $\boldsymbol{e}_2(\phi) := \boldsymbol{k} \times \boldsymbol{e}_1(\phi)$. We identify s^1 with the arc-length parameter s along meridians of the axisymmetric midsurface in its reference configuration, and identify s^2 with ϕ. In keeping with the axisymmetry, we require that $\boldsymbol{r}$ and $\boldsymbol{d}$, which we rename as $\boldsymbol{b}$, have the forms

$$\begin{aligned} \boldsymbol{r}(s,\phi,t) &= r(s,t)\boldsymbol{e}_1(\phi) + z(s,t)\boldsymbol{k}, \\ \boldsymbol{b}(s,\phi,t) &= -\sin\theta(s,t)\boldsymbol{e}_1(\phi) + \cos\theta(s,t)\boldsymbol{k}, \end{aligned} \tag{47}$$

and define $\boldsymbol{a}(s,\phi,t) := \boldsymbol{e}_2(\phi) \times \boldsymbol{b}(s,\phi,t)$. Let r° denote the value of r in the reference configuration. Then $\gamma = r^\circ$. The deformation is measured by

$$\begin{aligned} &\tau := \frac{r}{r^\circ}, \qquad \nu := \boldsymbol{r}_s \cdot \boldsymbol{a}, \qquad \eta := \boldsymbol{r}_s \cdot \boldsymbol{b}, \qquad \sigma := \frac{\sin\theta}{r^\circ}, \\ &\mu := \theta_s. \end{aligned} \tag{48}$$

Also in accord with the axisymmetry we set

$$\boldsymbol{n}^1 = N\boldsymbol{a} + H\boldsymbol{b}, \qquad \gamma\boldsymbol{n}^2 = T\boldsymbol{e}_2, \qquad \boldsymbol{m}^1 = -M\boldsymbol{e}_2, \qquad \gamma\boldsymbol{m}^2 = \Sigma\boldsymbol{a}. \tag{49}$$

Then (45), (46) reduce to

$$\rho a r^\circ \boldsymbol{r}_{tt} = \partial_s\big[r^\circ(N\boldsymbol{a} + H\boldsymbol{b})\big] - T\boldsymbol{e}_1, \tag{50}$$

$$\rho J r^\circ \theta_{tt} = \partial_s\big[r^\circ M\big] - \Sigma\cos\theta + r^\circ(\nu H - \eta N), \tag{51}$$

which closely resemble (17), (18). The material of the shell is magnetoelastic if T, N, H, Σ, M have constitutive equations of the form

$$M = \widehat{M}(\tau, \nu, \eta, \sigma, \mu, \mathsf{h}, s), \quad \text{etc.} \tag{52}$$

and the material is magnetoviscoelastic if

$$M = \widehat{M}(\tau, \nu, \eta, \sigma, \mu, \tau_t, \nu_t, \eta_t, \sigma_t, \mu_t, \mathsf{h}, s), \quad \text{etc.} \tag{53}$$

Here h is a quintuple corresponding to the five deformations (48).

We can clearly use the methods of Section 3 to find feedbacks that semilinearize (50), (52). We can even annihilate the term $\nu H - \eta N$ by synthesizing an artificial material. But there seems to be no promising way to use our magnetoelastic feedbacks to get rid of the product $\Sigma\cos\theta$.

There are some attractive tasks that can be accomplished by suitable feedbacks. Consider the results of [5] for aeolotropic circular plates subject to compression or tension on the edge. (These results can doubtless be extended to shells that have a pole, say $r^\circ(0) = 0$.) Depending on the nature of the aeolotropy (near the center) the stress at the center can be zero for all compressive thrusts at the edge, or can be infinite for all nonzero compressive thrusts, or in general, can stay at a fixed state until the edge thrust is increased past one of a discrete collection of thresholds, at which the stress at the center jumps to another fixed state. When the edge is subject to tension, both isotropic and aeolotropic plates can suffer a cavitation in which a hole forms at the center. but in many cases the process of cavitation for the aeolotropic shell has a sudden jump, as in snap buckling. It is feasible to use the magnetic field to vary the isotropy near the pole (by making $\widehat{T}$ behave differently from $\widehat{N}$ and making $\widehat{\Sigma}$ behave differently from $\widehat{M}$). In particular, when the edge is under compression and there is a small hole at the center, it might be possible to make that hole squeeze tightly on a thin shaft placed in it or make the hole release the pressure on the shaft. When the edge is under tension, a feedback might be used to prevent cavitation, which has been adduced as a mechanism for fracture, or alternatively as a mechanism to promote a more violent cavitation.

6. Conclusion

The primary virtue of feedback linearization and semilinearization of the equations of motion for structures is that the simplified equations, when subject to additional controls with a bona fide objective, are much more amenable to the available methods of control theory. For example, suppose that our structure is further subject to boundary controls, which are independent of the magnetic controls that effect the (semi)linearization of the equations of motion. Then the control-theoretic problem of finding boundary controls to cause the structure to perform some task is one for the control of (semi)linear partial differential equations, for which there is an extensive body of work (cf. [19, 20], e.g.).

We can also exploit this (semi)linearization even when the additional controls are of the same sort that effect the (semi)linearization. To be specific, consider (3) subject to (1) or (2). Note that all our development goes through even if we replace (4) with

$$a(s,t)\nu + c(s,t) = \widehat{N}(\nu, h, s). \tag{54}$$

The function c adds a forcing term to the linear equation, while the dependence of a on t makes the linear system nonautonomous. Given $\widehat{N}$ there is a set $\mathcal{S}(x)$ of pairs (a, c) of numbers for which the graphs of $z = a\nu + c$ and $z = \widehat{N}(\nu, h, x)$ intersect. We can use the functions a and c as controls for additional tasks. The admissible class of such controls consists of all functions for which $(a(s,t), c(s,t)) \in \mathcal{S}(s)$ for each s. As before, we are considering the problem of a controlling a system governed by (semi)linear partial differential equations.

The quadratic acceleration terms in (37), (43) are exactly the same as those in Euler's equations of motion for a rigid body, where they play a crucial role in the stability of motion, in particular, in the stability of steady rotations. The methods of using feedback torques to stabilize and control these equations while preserving their Hamiltonian structure (cf. [6] and references therein) can certainly be applied to the quadratic acceleration terms in our partial differential equations.

The equations of motion of hyperelastic rods and shells subject to conservative loads are Hamiltonian [27]. Although there is an extensive theory on the stability and control of discrete Hamiltonian systems and on the Hamiltonian structure of partial differential equations, these methods have scarcely been applied to unsteady nonlinear problems of elasticity (cf. [18]) because they cannot handle the shocks to which the equations are subject. (One can argue that these equations are not truly Hamiltonian because shocks represent a concentration of dissipative effects.) Our ability to (semi)linearize these equations suggests that Hamiltonian methods could be profitably applied to them.

Clearly, our methods can be used to synthesize material responses other than linear. (We have mentioned a couple of instances of such syntheses in Sections 4 and 5.) E.g., in [3] it is shown how to synthesize the material of a planar elastic rod that has a prescribed first bifurcating branch for its buckling under terminal thrust. One could even contemplate replacing nonmonotone constitutive functions (like those used in [11, 15]) with monotone ones, or vice versa.

Some of the artificial materials whose synthesis we have described have constitutive functions lacking the nice mathematical property of monotonicity (or of convexity of energy). There have been some studies [16] of problems for such linearly elastic materials that can handle some of our artificial materials.

Rather straightforward modifications of the methods described here handle richer models for structures and handle larger classes of materials.

There is an extensive theory of feedback linearization of systems of ordinary differential equations of the form $(\mathrm{d}/\mathrm{d}t)\mathbf{x} = \mathbf{f}(\mathbf{x}) + \mathbf{F}(\mathbf{x})\mathbf{h}$ where the control $\mathbf{h}$ appears affinely [17]. (For simplicity of exposition, we limit our attention to autonomous equations.) The objective of this theory is to find a change of variables of $(\mathbf{x}, \mathbf{h}) \mapsto (\mathbf{y}, \mathbf{k})$ affine in $\mathbf{h}$ that converts the original system to a system linear in $\mathbf{y}$ and affine in $\mathbf{k}$. The affine dependence on the control $\mathbf{h}$ makes this problem rich and difficult. We never encounter comparable difficulties in our work, because the underlying physics ensures that the magnetic field that is our control affects the material response, which is the repository of the principal part of the nonlinearity. The analog of our equations for ordinary differential equations would be a system of the form $(\mathrm{d}/\mathrm{d}t)\mathbf{x} = \mathbf{f}(\mathbf{x}, \mathbf{h})$. (Spatial discretizations of our equations lead to ordinary differential equations of this and slightly more complicated form.) The feedback linearization of this system is effected by finding an $\mathbf{h}$ such that $\mathbf{f}(\mathbf{x}, \mathbf{h})$ has the form $\mathbf{f}(\mathbf{x}, \mathbf{h}) = \mathbf{A}\mathbf{x} + \mathbf{B}\mathbf{k}$, where $\mathbf{A}$ and $\mathbf{B}$ are constant matrices and $\mathbf{k}$ is a vector of controls. We have seen that it is easy to effect such controls.

For structures more complicated than rods and shells, namely, 2- and 3-dimensional bodies, one might only be able to prescribe the magnetic field on the boundary. In this case the generalization of the equations of motion must be supplemented with appropriate versions of Maxwell's equations (cf. [7, 10, 12, 13, 14, 25]).

Acknowledgements

The research reported here was supported in part by ARO-MURI97 Grant No. DAAG55-97-1-0114 to the Center for Dynamics and Control of Smart Structures and by the NSF Grant DMS 99 71823.

References

1. S.S. Antman, *Nonlinear Problems of Elasticity*. Springer, Berlin (1995).
2. S.S. Antman, Dynamical problems for geometrically exact theories of nonlinearly viscoelastic rods. *J. Nonlinear Sci.* **6** (1996) 1–18.
3. S.S. Antman, Synthesis of nonlinear constitutive functions. Applications to the electromagnetic control of snapping. *J. Appl. Mech.* **66** (1999) 280–283.
4. S.S. Antman and T.-P. Liu, Travelling waves in hyperelastic rods. *Quart. Appl. Math.* **36** (1979) 377–399.
5. S.S. Antman and P.V. Negrón-Marrero, The remarkable nature of radially symmetric equilibrium states of aeolotropic nonlinearly elastic bodies. *J. Elasticity* **18** (1987) 131–164.
6. A.M. Bloch, P.S. Krishnaprasad, J.E. Marsden and G. Sánchez de Alvarez, Stabilization of rigid body dynamics by internal and external torques. *Automatica* **28** (1992) 745–756.
7. W.F. Brown, *Magnetoelastic Interactions*. Springer, Berlin (1966).
8. R.E. Caflisch and J.H. Maddocks, Nonlinear dynamical theory of the elastica. *Proc. Roy. Soc. Edinburgh* A **99** (1984) 1–23.
9. D.M. Dozor, M.J. Gerver and R. Swenbeck, Nonlinear modeling for control of terfenol-D based actuators. In: *Proc. of the 4th SPIE Conf. on Smart Materials and Structures*, Vol. 3039 (1997) pp. 644–654.
10. A.C. Eringen and G.A. Maugin, *Electrodynamics of Continua*. Springer, Berlin (1990).
11. R.L. Fosdick and R.D. James, The elastica and the problem of pure bending for a non-convex stored energy function. *J. Elasticity* **11** (1981) 165–186.
12. A.E. Green and P.M. Naghdi, Electromagnetic effects in the theory of shells and plates. *Phil. Trans. Roy. Soc. London* A **309** (1983) 559–610.
13. A.E. Green and P.M. Naghdi, Electromagnetic effects in the theory of rods. *Phil. Trans. Roy. Soc. London* A **314** (1983) 311–352.
14. K. Hutter and A.A.F. van de Ven, *Field-Matter Interactions in Thermoelastic Solids*. Springer, Berlin (1978).
15. R.D. James, The equilibrium and post-buckling behavior of an elastic curve governed by a non-convex energy. *J. Elasticity* **11** (1981) 239–269.
16. R.J. Knops and L.E. Payne, *Uniqueness Theorems in Linear Elasticity*. Springer, Berlin (1971).
17. A.J. Krener, Feedback linearization. In: J. Baillieul and J.C. Willems (eds.), *Mathematical Control Theory*. Springer, Berlin (1998) pp. 66–98.
18. P.S. Krishnaprasad and J.E. Marsden, Hamiltonian structures and stability for rigid bodies with flexible attachments. *Arch. Rational Mech. Anal.* **98**, 71–93.
19. J. Lagnese and J.-L. Lions, *Modelling, Analysis, and Control of Thin Plates*. Masson/Springer (1989).
20. J.-L. Lions, *Optimal Control of Systems Governed by Partial Differential Equations*. Springer, Berlin (1971).
21. A.E.H. Love, *A Treatise on the Mathematical Theory of Elasticity*. 4th edn. Cambridge Univ. Press, Cambridge (1927).
22. A. Mielke and P.J. Holmes, Spatially complex equilibria of buckled rods. *Arch. Rational Mech. Anal.* **101** (1988) 319–348.

23. M.B. Moffett, A.E. Clark, M. Wun-Fogle, J.F. Lindberg, J.P. Teter and E.A. McLaughlin, Characterization of terfenol-D for magnetostrictive transducers. *J. Acoust. Soc. Amer.* **89** (1991) 1448–1455.
24. P.V. Negrón-Marrero and S.S. Antman, Singular global bifurcation problems for buckling of anisotropic plates. *Proc. Roy. Soc. London A* **427** (1990) 95–137.
25. R.C. Rogers and S.S. Antman, Steady-state problems of nonlinear electro-magneto-thermo-elasticity. *Arch. Rational Mech. Anal.* **95** (1986) 279–323.
26. H.T. Savage and M.L. Spano, Theory and application of highly magnetoelastic metglas 2605SC. *J. Appl. Phys.* (1982) 8092–8097.
27. J.C. Simo, J.E. Marsden and P.S. Krishnaprasad, The Hamiltonian structure of nonlinear elasticity: The material and convective representations of solids, rods, and plates. *Arch. Rational Mech. Anal.* **104** (1988) 125–183.
28. G.I. Taylor, On the shape of parachutes (1919). In: G.K. Batchelor (ed.), *The Scientific Papers of Sir Geoffrey Ingram Taylor*, Vol. III. Cambridge Univ. Press, Cambridge (1963) pp. 26–37.
29. M. Wun-Fogle, H.T. Savage, S.S. Antman and L.T. Kabacoff, The effect of longitudinal stress and torsional strain on the magnetization of amorphous magnetoelastic wires. *Anales de Fisica Serie B* **86** (1990) 277–283.

Journal of Elasticity **59:** 131–143, 2000.

Slip Bands and Stress Oscillations in Bars

GIANNI F. ROYER-CARFAGNI
Department of Civil Engineering, University of Parma, Parco Area delle Scienze 181/A, I43100 Parma, Italy

Received 19 January 1999; in revised form 14 February 2000

Dedicated to Professor Roger Fosdick.

Abstract. A model is proposed that is able to describe the phenomena of strain localization and serrated deformation, peculiar to many materials, but particularly evident in mild steel. A nonlinear elastic bar is conceived like the assemblage of thin filaments, whose energy is supposed to be a nonconvex function of the relative displacement of each filament's extremities. The model explains the onset of elastic, plastic and strain-hardening phases in the bar, interpreting the formation of slip bands and stress oscillations.

1. Introduction

Careful experimentation reveals that materials may exhibit highly irregular deformations, even when stressed far below their ultimate strength. To illustrate, Figure 1 shows a deformed metal crystal, from which it is evident that plasticity is due to slips concentrated in bands a few microns wide, spaced with surprising regularity throughout the specimen [7]. Structured deformations of this type also influence the material's gross response. A significant case is that of serrated plastic deformation in metals [1, Chapter 6], a phenomenon revealed by the stress oscillations at yielding in metallic bars stretched in a hard device (the Portevin–Le Chatelier effect). These are clearly recognizable in Figure 2, which reports average-stress as a function of average-strain for a tensile test on a mild-steel bar, in correspondence to the plastic *plateau*. A reasonable explanation of this effect consists in correlating each stress drop to the release occurring when one slip band forms. Stress recovery starts when the amount of slip allowed in the band is completed and ends when a new slip layer is nucleated. Consequently, yielding appears as a pseudo-equilibrium state, characterized by the ordered and successive nucleation of almost-equidistant slip bands, gradually invading the whole bar. This is the major difference with respect to elastic and strain hardening periods, where strain increments are *uniformly* distributed throughout the specimen. Experimental results have been discussed at length in [4] and interpreted in [5] through a simple mechanical model consisting in a particular arrangement of spring-dashpot units.

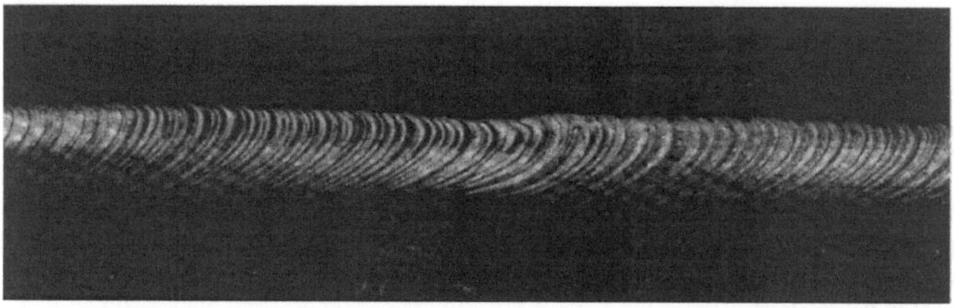

Figure 1. Extended copper–aluminum single crystal (from an experiment by Elam [7]).

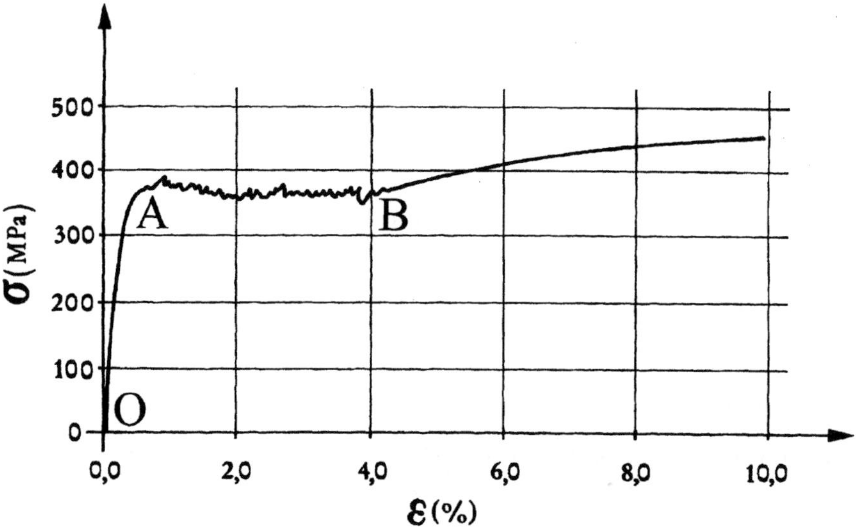

Figure 2. Mean-stress vs. average-strain diagram for a mild steel bar stretched in a hard device.

The literature contains numerous reports of attempts to model localized deformations in a continuum, though, in our opinion, two main difficulties exist. The first [8] is that if we start from a discrete model comprised of an assemblage of a finite number of grains (likewise in [5]), we do not ascertain in the limit the behavior of a continuous bar by letting the diameter of each grain tend towards zero and the number of grains towards infinity. For example, both width and wavelength of oscillations at yielding that the model in [5] was able to reproduce tend to zero as the length of the units becomes very small and their number consequently increased. The second obstacle is that if we consider "structured deformations" as the weak limit of discontinuous deformations, we must prescribe the exact location of the strain discontinuities *a priori*.

A simple model which describes the creation of slip bands is presented here and is based on three main ideas: the material is imagined to be composed of numerous small but finite elements, i.e., thin filaments (whiskers), staggered along their lengths; the energy of each fiber is a nonconvex function, not of the point-wise, but of the fiber's average strain value; the fibers are glued together on their lateral surface, so that deformation in any one will affect the behavior of its nearest neighbors. Assuming that the bar's ends are gradually displaced apart in a hard device, we discuss the structure of strain fields which minimize total energy. This is represented by the integral of a nonconvex function of the strain field's spatial average and, in this respect, differs from other recently proposed, non-local functionals (for example, [3]), which contain nonconvex functions of the point-wise strain and consider weighted spatial averages of the strain field as additional terms.

2. The Model

We consider a prismatic bar of total length $L + l_0$ and uniform cross-sectional area A, stretched in a testing-machine which responds to load increases with negligible movement (hard device). The axis of the bar is vertical and we choose a reference coordinate ζ which is zero at the bar's midpoint and increases upward, so that the interval $\mathcal{B} = [-(L+l_0)/2, +(L+l_0)/2]$ denotes the undistorted natural reference configuration of the body. The bar is conceived of as an assemblage of whiskers, all having the same length l_0 and cross section A_0, for which it is assumed that

$$l_0 \ll L, \qquad A_0 \ll A. \tag{1}$$

Imagining these constituent fibers to be perfectly adherent, we examine the deformations for which horizontal cross sections remain horizontal. Thus, $u(t)\colon \mathcal{B} \to \mathbb{R}$ denotes the upward displacement of all whisker points belonging to the bar cross section which was originally at t. In general, $u(\cdot)$ will be considered to be a function of bounded variation, so that its generalized derivative $\varepsilon(\cdot) = Du(\cdot)$ is a measure in $\mathcal{B}$. Our aim is to find the displacement field in the bar when its ends are displaced relatively by a quantity βL, with $\beta \geqslant 0$.

For any given fiber in position

$$z - \frac{l_0}{2} \leqslant \zeta \leqslant z + \frac{l_0}{2}, \tag{2}$$

with $-L/2 \leqslant z \leqslant L/2$, the corresponding average strain $e(z)$ can be represented in the form

$$e(z) = \int_{-\infty}^{+\infty} \rho(\zeta - z)\, \mathrm{d}\varepsilon(\zeta), \quad \forall z \in (-L/2, +L/2), \tag{3}$$

where the weight function $\rho(\cdot)$ is given by

$$\rho(\zeta - z) = \begin{cases} 1/l_0 & \text{for } (\zeta - z) \in (-l_0/2, l_0/2), \\ 0 & \text{for } (\zeta - z) \notin (-l_0/2, l_0/2). \end{cases} \tag{4}$$

Under the hypothesis that the energy of each fiber depends on its average elongation alone, for the fiber identified by (2) we have

$$\mathcal{E}_{\text{fiber}} = U\big(e(z)\big), \tag{5}$$

where the energy density $U(\cdot)$ is assumed, after a classical theory by Dunn and Fosdick [2], to be a continuous nonconvex function with superlinear growth at infinity and $U(0) = U'(0) = 0$.

Now, let the function $f(\cdot)$ indicate the distribution of fibers throughout the bar, so that $\mathrm{d}n = f(z)\,\mathrm{d}z$ is the number of fibers between, and not intersecting, the two cross sections at $\zeta = z - (l_0 + \mathrm{d}z)/2$ and at $\zeta = z + (l_0 + \mathrm{d}z)/2$. Recalling (3) and (5), the amount of energy associated to these fibers is

$$\mathrm{d}\mathcal{E} = f(z)\,U\big(e(z)\big)\mathrm{d}z.$$

By summing the contributions from all fibers, the total potential energy results:

$$\mathcal{E} = \int_{-L/2}^{L/2} f(z)\,U\big(e(z)\big)\mathrm{d}z. \tag{6}$$

We may assume the fiber distribution to be statistically uniform throughout the bar, so that $f(\cdot)$ takes the simple form

$$f(\cdot) = \text{constant} = \frac{N}{L/l_0}\frac{1}{l_0} = \frac{N}{L}, \tag{7}$$

where $N/(L/l_0)$ denotes the (average) number of fibers having nonzero intersection with any given plane orthogonal to the bar's axis, i.e. the (average) number of fibers making up the bar's section. Consequently, N is representative of the total number of fibers present in the bar. As a result of (6) and (7), the expression for the total potential energy stored in the bar now reads

$$\mathcal{E}_{\text{tot}}[\varepsilon] = \frac{N}{L}\int_{-L/2}^{L/2} U\big(e(z)\big)\mathrm{d}z = \frac{N}{L}\int_{-L/2}^{L/2} U\left(\int_{-\infty}^{+\infty} \rho(\zeta - z)\,\mathrm{d}\varepsilon(\zeta)\right)\mathrm{d}z. \tag{8}$$

Let us observe that, by reversing the integration order and using (4), we can write

$$\begin{aligned}\int_{-L/2}^{L/2}\left(\int_{-\infty}^{+\infty} \rho(\zeta - z)\,\mathrm{d}\varepsilon(\zeta)\right)\mathrm{d}z &= \int_{-\infty}^{+\infty}\left(\int_{-L/2}^{L/2} \rho(\zeta - z)\mathrm{d}z\right)\mathrm{d}\varepsilon(\zeta) \\ &\cong \int_{-L/2}^{L/2} \mathrm{d}\varepsilon(\zeta) = \beta L,\end{aligned} \tag{9}$$

provided that we accept an approximation in proximity of the bar's ends. The approximation is justified by hypothesis $(1)_1$ and, moreover, (9) is exact when $\varepsilon(\cdot) = Du(\cdot)$ is constant.

The minimum problem that will be here considered is therefore

$$\min_{\varepsilon \in \mathcal{A}_\beta} \mathcal{E}_{\mathrm{tot}}[\varepsilon], \tag{10}$$

where $\mathcal{E}_{\mathrm{tot}}[\varepsilon]$ has been defined in (8) and $\mathcal{A}_\beta$ denotes the class of admissible functions (measures) $\varepsilon(\cdot) = Du(\cdot)$ satisfying the constraint

$$\int_{-L/2}^{L/2} \left(\int_{-\infty}^{+\infty} \rho(\zeta - z)\, \mathrm{d}\varepsilon(\zeta) \right) \mathrm{d}z = \beta L. \tag{11}$$

In general, it could be shown that problem (10) admits a solution provided that ε is bounded from below. This condition is naturally satisfied by the usual assumption $\varepsilon = Du > -1$, equivalent to ruling out any interpenetration of material. However, in this paper we will consider only the tensile response of bars ($\beta \geqslant 0$) in the slightly stronger hypothesis

$$\varepsilon = Du \geqslant 0, \tag{12}$$

which, without loosing generality, allows great simplifications in the calculations that follow. Moreover, we will study the particular case in which $U(\cdot)$ is of the form

$$U\big(e(z)\big) = -\frac{1}{2} E e_1^2 + E e_1 e(z) + \inf \begin{cases} \frac{1}{2} E\big(e(z) - e_1\big)^2, \\ \frac{1}{2} E\big(e(z) - e_2\big)^2, \end{cases} \tag{13}$$

where e_1 and e_2 are two positive numbers, with $e_2 > e_1 > 0$, and E is a material constant having the dimensions of energy per fiber.

3. The Structure of Minimizers

In this introductory paper we present some explicit calculations for that particular energy described above as the average strain β, just as in a standard tensile test on steel bars, is gradually increased from $\beta = 1$.

Minimizers will be studied directly by constructing functions that attain lower bounds in energy. To this end, observe that because of (11), the minimizers of (8) are invariant if any linear expression is added to the energy density $U(\cdot)$. It is then useful to replace $U\big(e(z)\big)$ by

$$\begin{aligned} U^*\big(e(z)\big) &= U\big(e(z)\big) - \left[\frac{E}{2} e_1^2 + E e_1 (e(z) - e_1)\right] \\ &= \inf \begin{cases} \frac{1}{2} E\big(e(z) - e_1\big)^2, \\ \frac{1}{2} E\big(e(z) - e_2\big)^2, \end{cases} \end{aligned} \tag{14}$$

and consider the *auxiliary* problem

$$\min_{\varepsilon\in\mathcal{A}_\beta} \mathcal{E}^*_{\text{tot}}[\varepsilon] = \min_{\varepsilon\in\mathcal{A}_\beta} \frac{N}{L}\int_{-L/2}^{L/2} U^*\big(e(z)\big)\mathrm{d}z, \tag{15}$$

whose minimum points, for any given β, clearly coincide with those of (10), even if the infimum changes by the additive constant

$$\frac{N}{L}\int_{-L/2}^{L/2} U\big(e(z)\big)\mathrm{d}z - \frac{N}{L}\int_{-L/2}^{L/2} U^*\big(e(z)\big)\mathrm{d}z = -\frac{E}{2}Ne_1^2 + ENe_1\beta. \tag{16}$$

The necessary conditions for minimizers are obtained through the Euler–Lagrange equations. Considering variations $\delta\varepsilon_a(\zeta)$ of $\varepsilon(\zeta)$ which are absolutely continuous with respect to the Lebesgue's measure and recalling the expression (3) of $e(z)$, the first variation of (15) reads

$$\delta\mathcal{E}^*_{\text{tot}} = \frac{N}{L}\int_{-L/2}^{L/2} U^{*\prime}\big(e(z)\big)\cdot\left(\int_{-\infty}^{+\infty}\rho(t-z)\delta\varepsilon_a(t)\,\mathrm{d}t\right)\mathrm{d}z, \tag{17}$$

where, in order to respect (11), $\delta\varepsilon_a(\cdot)$ must satisfy

$$\int_{-L/2}^{L/2}\left(\int_{-\infty}^{+\infty}\rho(t-z)\delta\varepsilon_a(t)\,\mathrm{d}t\right)\mathrm{d}z = 0. \tag{18}$$

With standard arguments, assuming that minimizers are attained for $\varepsilon > 0$ so that two sided variations $\delta\varepsilon_a(\cdot)$ are possible, we obtain the necessary condition

$$\int_{-L/2}^{L/2} U^{*\prime}\big(e(z)\big)\rho(t-z)\,\mathrm{d}z = \lambda\int_{-L/2}^{L/2}\rho(t-z)\,\mathrm{d}z, \tag{19}$$

where λ is a constant and $t\in(-L/2-l_0/2,\ L/2+l_0/2)$ is variable. Differentiating with respect to t both sides of expression (19), provided $t\in(-L/2+l_0/2,\ L/2-l_0/2)$, we obtain

$$U^{*\prime}\big(e(t+l_0/2)\big) - U^{*\prime}\big(e(t-l_0/2)\big) = 0, \quad \forall t\in\left(\frac{-L+l_0}{2},\frac{L-l_0}{2}\right), \tag{20}$$

which implies, in general, that $U^{*\prime}(e(z))$ is a periodic function of period l_0. On the other hand, when $t\in(-L/2-l_0/2,\ -L/2+l_0/2)$, we find

$$U^{*\prime}\big(e(t+l_0/2)\big) = \lambda, \quad \forall t\in\left(\frac{-L-l_0}{2},\frac{-L+l_0}{2}\right), \tag{21}$$

and, analogously, when $t\in(L/2-l_0/2,\ L/2+l_0/2)$, we obtain

$$U^{*\prime}\big(e(t-l_0/2)\big) = \lambda, \quad \forall t\in\left(\frac{L-l_0}{2},\frac{L+l_0}{2}\right). \tag{22}$$

Combining (20) with (21) and (22), we achieve the final condition

$$U^{*\prime}\big(e(z)\big) = U^{*\prime}\bigg(\int_{-L/2}^{L/2} \rho(\zeta - z)\,\mathrm{d}\varepsilon(\zeta)\bigg) = \text{constant}. \tag{23}$$

Another noteworthy consideration is that, since $\varepsilon(z) = Du(z)$, equation (3) can be integrated so that $e(z)$, on recalling (12), takes the equivalent form

$$e(z) = \frac{u(z + l_0/2) - u(z - l_0/2)}{l_0}. \tag{24}$$

From this expression, it is clear that any two displacement fields $u(\cdot)$ differing by a function of period l_0, correspond to the same function $e(\cdot)$, that is, adding any l_0-periodic function *with zero average* to $\varepsilon(\cdot)$ does not modify the value taken by the functionals in (8) and (15). This implies that the minimizing fields presented in the following can be considered to be defined only up to zero-average, l_0-periodic functions. This inconvenience might be overcome if more elaborate expressions for the kernel $\rho(\cdot)$, other than the particular (4), were considered.

In order to discuss the course of equilibrium states in gradually extended bars more in detail, it is useful to distinguish three cases, referred to as the elastic, plastic and strain hardening periods.

Case 1: $0 < \beta \leqslant e_1$ (elastic period). If $U^{**}(\cdot)$ denotes the lower convex envelope of $U^*(\cdot)$, in light of $U^*(\cdot) \geqslant U^{**}(\cdot)$, Jensen's inequality and (11), we have

$$\begin{aligned}
&\frac{1}{L}\int_{-L/2}^{L/2} U^*\bigg(\int_{-\infty}^{+\infty} \rho(\zeta - z)\,\mathrm{d}\varepsilon(\zeta)\bigg)\,\mathrm{d}z \\
&\quad \geqslant \frac{1}{L}\int_{-L/2}^{L/2} U^{**}\bigg(\int_{-\infty}^{+\infty} \rho(\zeta - z)\,\mathrm{d}\varepsilon(\zeta)\bigg)\mathrm{d}z \\
&\quad \geqslant U^{**}\bigg(\frac{1}{L}\int_{-L/2}^{L/2}\bigg(\int_{-\infty}^{+\infty} \rho(\zeta - z)\,\mathrm{d}\varepsilon(\zeta)\bigg)\mathrm{d}z\bigg) = U^{**}(\beta).
\end{aligned} \tag{25}$$

But, since $U^{**}(\beta) = U^*(\beta)$ if $0 < \beta \leqslant e_1$, we obtain the lower bound

$$\frac{N}{L}\int_{-L/2}^{L/2} U^*\bigg(\int_{-\infty}^{+\infty} \rho(\zeta - z)\,\mathrm{d}\varepsilon(\zeta)\bigg)\mathrm{d}z \geqslant N \cdot U^*(\beta). \tag{26}$$

Minimizers of (15) are thus obtained from the integral equation

$$e(z) = \int_{-\infty}^{+\infty} \rho(\zeta - z)\,\mathrm{d}\varepsilon(\zeta) = \text{constant} = \beta, \tag{27}$$

among whose solutions, apart from pathological cases*, certainly one is represented by

$$Du(\zeta) = \varepsilon(\zeta) = \text{constant} = \beta. \tag{28}$$

Case 2: $e_1 < \beta \leqslant e_2$ (yielding period). Since $U^*(\cdot) \geqslant 0$, then $\mathcal{E}^*_{\text{tot}}[\varepsilon] \geqslant 0$, and from the structure of $U^*(\cdot)$ we find that this lower bound can be achieved if and only if $e(z)$ takes the sole values e_1 or e_2. Recalling (24), let us then discuss the solutions to the equation

$$e(z) = \frac{u(z + l_0/2) - u(z - l_0/2)}{l_0} = \begin{cases} e_1, \\ e_2, \end{cases} \tag{29}$$

which clearly satisfies the Euler–Lagrange's conditions (23). From (12), $u(\cdot)$ is a monotone, nondecreasing function, possibly discontinuous on a set of points Γ. As it is clear that (29) cannot be satisfied by continuous displacement fields $u(\cdot)$, apart from pathological cases (see the previous footnote), the simplest way to construct minimizers is to consider piece-wise continuous functions $u(\cdot)$, presenting constant slope equal to e_1 at the continuity points and a certain number of jumps, all equal to $(e_2 - e_1)l_0$, spaced at intervals *no less than* l_0. The required number of jumps can be obtained by imposing that the integral of $e(z)$ in the bar is equal to βL. The corresponding minimizing strain fields thus result to be of the form

$$\varepsilon(\zeta) = e_1 + \sum_{x\in\Gamma} \delta_x (e_2 - e_1) l_0, \tag{30}$$

where δ_x denotes the Dirac's unit mass centered at $\zeta = x$ and Γ represents the set of points where the jumps occur. On referring to Figure 1, it is natural to correlate the first contribution in (30) to the elastic state of the material comprised between any two consecutive slip bands, whereas the second contribution is representative of the amount of deformation concentrated in the slip bands themselves. In particular, the theory prescribes that the amount of slip is fixed and the same in each band.

It should be noticed, however, that the aforementioned construction does not cover every value $\beta \in (e_1, e_2)$. In fact, solution (30) does not allow to have $e(z) = e_2$ on a set whose measure is not a multiple of l_0, so that, from (11), β could only be of the form $\beta = e_1 + n\,(e_2 - e_1)l_0/L$, with n a nonnegative integer. The characterization of the minimum points in the transitory stage between the formation of two consecutive slip bands, i.e. when

$$\beta = e_1 + \frac{l_0}{L}(e_2 - e_1)(n + \alpha), \quad \alpha \in]0, 1[, \quad \beta \leqslant e_2, \tag{31}$$

is rather complicated. Heuristically, we try to consider a succession of equilibrium states that provides a smooth transition from $\beta = e_1 + n(e_2 - e_1)l_0/L$ to

* Denoting with u' the absolutely continuous part with respect to the Lesbegue's measure of Du, we may consider for example $u' = 0$ a.e. and jumps equal to β equally spaced at inteval l_0 throughout the bar.

$\beta = e_1 + (n+1)(e_2 - e_1)l_0/L$. The underlying idea is that causes of various nature (inertia forces, second order effects, etc.), which are neglected for simplicity in this introductory model, might instead play a decisive role in the transitory stage, regularizing the structure of minimizers. In other words, we now investigate local and not necessarily global minima, empirically trying to penalize the distance between two consecutive minimizers as β is varied. Therefore, we tentatively consider equilibrium displacement fields which satisfy the Euler–Lagrange's equation (23). From (14), it can be seen that in general $U^{*\prime}(e_1+\gamma) = U^{*\prime}(e_2+\gamma)$ for sufficiently small γ; we thus look for solutions such that $e(z)$ takes on the sole two values $e_1+\gamma$ or $e_2+\gamma$, for some γ to be determined. Now two cases are of interest.

(a) If the number of slip bands is n, $e(z)$ equals $e_2+\gamma$ on a set Ω_2 with $meas(\Omega_2) = nl_0$ and $e_1+\gamma$ on the set $\Omega_1 = (0, L)\backslash\Omega_2$. The quantity γ, from (11), must satisfy

$$\beta = \frac{(L - nl_0)(e_1+\gamma) + nl_0(e_2+\gamma)}{L}. \tag{32}$$

Equating (31) and (32), for any n we obtain the value $\gamma = \alpha l_0(e_2 - e_1)/L$, to which corresponds the energy

$$\begin{aligned}\mathcal{E}^{*(a)}_{\text{tot}} &= \frac{N}{L}\frac{E}{2}\left\{(L - nl_0)\left[\alpha\frac{l_0}{L}(e_2 - e_1)\right]^2 + nl_0\left[\alpha\frac{l_0}{L}(e_2 - e_1)\right]^2\right\} \\ &= \frac{1}{2}N\,E\left[\alpha\frac{l_0}{L}(e_2 - e_1)\right]^2.\end{aligned} \tag{33}$$

(b) If a new slip band is created, so that their total number results to be $n+1$, $e(z)$ is equal to $e_2+\theta$ on a set Θ_2 with $meas(\Theta_2) = (n+1)l_0$, and to $e_1+\theta$ in the remaining part $\Theta_1 = (0, L)\backslash\Theta_2$. Equating the expression for β obtained from (11), i.e.,

$$\beta = \frac{[L - (n+1)l_0](e_1+\theta) + (n+1)l_0(e_2+\theta)}{L}, \tag{34}$$

to (31), we find $\theta = -(1-\alpha)l_0(e_2 - e_1)/L$, and the value of the energy

$$\begin{aligned}\mathcal{E}^{*(b)}_{\text{tot}} &= \frac{N}{L}\frac{E}{2}\left\{[L - (n+1)l_0]\left[(1-\alpha)\frac{l_0}{L}(e_2 - e_1)\right]^2 + (n+1)l_0\left[(1-\alpha)\frac{l_0}{L}(e_2 - e_1)\right]^2\right\} \\ &= N\frac{E}{2}\left[(1-\alpha)\frac{l_0}{L}(e_2 - e_1)\right]^2.\end{aligned} \tag{35}$$

It is easily verified that the minimum between (33) and (35) is obtained in case (a) when $\alpha \in (0, \frac{1}{2})$, and in case (b) when $\alpha \in (\frac{1}{2}, 1)$. Supposing, in the aforementioned approximations, that the material responds each time by tending towards the lower energy, this analysis shows that the displacement jump associated to formation of one slip band does not increase gradually, but instantly reaches the value $(e_2 - e_1)l_0$ when $\alpha = \frac{1}{2}$. The yielding phase ends when the whole length of the bar has been affected by slip bands, that is, recalling that the minimum distance between consecutive bands is l_0, when their total number is of the order

$$n^* = \frac{L}{l_0}. \tag{36}$$

Case 3: $\beta > e_2$ (strain hardening period). An argument similar to that used in Case 1 to obtain equation (26) yields again the lower bound

$$\mathcal{E}^*_{\text{tot}}[\varepsilon] \geqslant N \cdot U^*(\beta). \tag{37}$$

Looking for solutions which connect with those obtained in Case 2 for $\beta \to e_2^-$, we consider the minimizing strain field

$$\varepsilon(\zeta) = e_1 + \beta - e_2 + \sum_{x \in \Lambda} \delta_x (e_2 - e_1) l_0, \tag{38}$$

where δ_x denotes the Dirac's unit mass centered at $\zeta = x$ and Λ represents the location of slip bands at the end of the yielding period, i.e. n^* points l_0-equidistant one from the other. In this solution, deformation increases uniformly in the bar portions lying between any two consecutive slip bands, whereas the amount of slip is maintained constant.

4. Average Stress-Strain Diagrams

For any given β, let us define

$$W(\beta) = \min_{\varepsilon \in \mathcal{A}_\beta} \mathcal{E}_{\text{tot}}[\varepsilon]. \tag{39}$$

More in particular, for the reasons discussed in Section 2, in the transitory stage when β is of the form (31), we consider for $W(\beta)$ the values corresponding to the equilibrium points (not-necessarily global minimizers) defined by the formulas (33) and (35). Therefore, the following expressions are considered.

When $0 < \beta \leqslant e_1$ (elastic period), from (26) and (16), we find

$$W(\beta) = \frac{1}{2} N E \beta^2. \tag{40}$$

In case $e_1 < \beta \leqslant e_2$ (yielding), using (33) and (16), we see that when (Case 2(a))

$$e_1 + n\frac{(e_2 - e_1)l_0}{L} < \beta < e_1 + \left(n + \frac{1}{2}\right)\frac{(e_2 - e_1)l_0}{L} \leqslant e_2,$$

we have

$$W(\beta) = N\left[-\frac{1}{2}E\,e_1^2 + E e_1 \beta + \frac{1}{2}E(\beta - e_1 - nq)^2\right]. \tag{41}$$

When instead (Case 2(b))

$$e_1 + \left(n + \frac{1}{2}\right)\frac{(e_2 - e_1)l_0}{L} \leqslant \beta < e_1 + (n+1)\frac{(e_2 - e_1)l_0}{L} \leqslant e_2,$$

from (35) and (16), we find

$$W(\beta) = N\left[-\frac{1}{2}Ee_1^2 + Ee_1\beta + \frac{1}{2}E\big(\beta - e_1 - (n+1)\,q\big)^2\right]. \tag{42}$$

And finally, when $\beta > e_2$, from (37), we get

$$W(\beta) = N\left[-\frac{1}{2}Ee_1^2 + Ee_1\beta + \frac{1}{2}E(\beta - e_2)^2\right]. \tag{43}$$

In (41) and (42) we have denoted

$$q = \frac{(e_2 - e_1)l_0}{L}$$

as the increment in β due to the formation of one slip band.

The axial load in the bar $P = \widehat{P}(\beta)$ can be obtained through a relation of the type

$$\widehat{P}(\beta) = \frac{1}{L}\frac{\partial W(\beta)}{\partial \beta}. \tag{44}$$

Consequently, on denoting A as the specimen's cross-sectional area and s as the number of fibers per unit volume, i.e., $s = N/AL$, the mean stress $\sigma = \widehat{\sigma}(\beta)$ is

$$\widehat{\sigma}(\beta) = \begin{cases} sE\beta & \text{if } 0 \leqslant \beta \leqslant e_1, \\ sE(\beta - nq) & \text{if } e_1 + nq \leqslant \beta \leqslant e_1 + \left(n + \frac{1}{2}\right)q \leqslant e_2, \\ sE\big[\beta - (n+1)q\big] & \text{if } e_1 + \left(n + \frac{1}{2}\right)q \leqslant \beta \leqslant e_1 + (n+1)q \leqslant e_2, \\ sE\big[\beta - (e_2 - e_1)\big] & \text{if } \beta > e_2. \end{cases} \tag{45}$$

The qualitative graph of $\sigma = \widehat{\sigma}(\beta)$ has been plotted in Figure 3 for $L/l_0 = 10$. Stress oscillations are clearly recognizable. The corresponding graph for $W(\beta)$ is instead represented in Figure 4. Here the stress oscillations of Figure 3 have their counterparts in the wiggly graph in the proximity of the tangent line that supports the nonconvex energy from below. Similar forms of energy were recently obtained, by Truskinovsky and Zanzotto [9], among others, through an apparently completely different theory. Models founded on the definition of such wiggly energies play a fundamental role in modern research on materials [6].

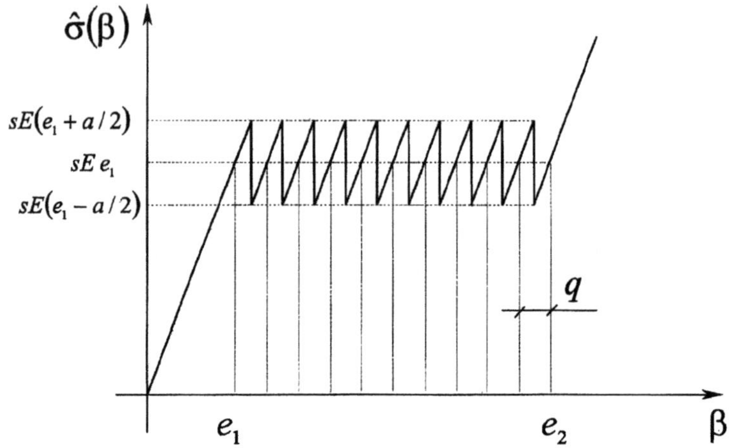

Figure 3. Mean-stress vs. average-strain relationship.

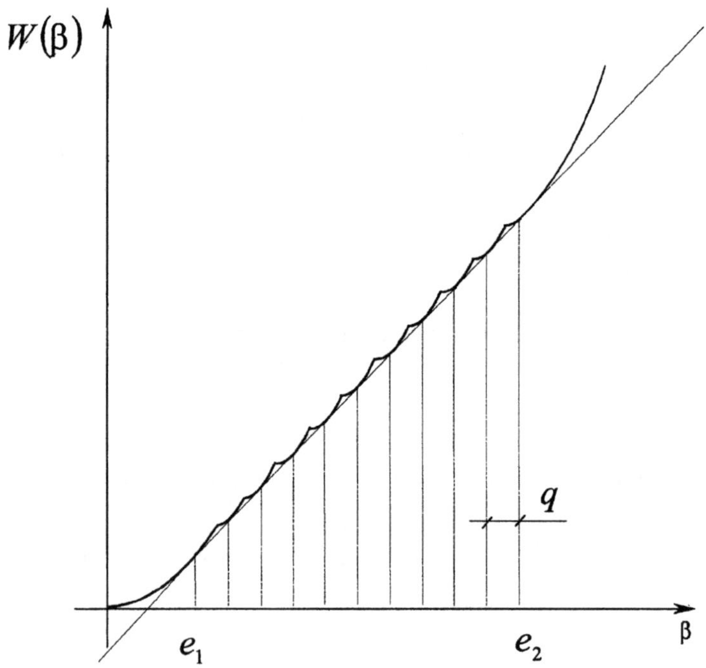

Figure 4. Energy stored in the bar as a function of average elongation.

5. Comparison with Experiments

The model considered in this paper appears able to accurately reproduce the three stages of metallic bars' tensile behavior, recognizable in the Figure 2 graph. In the first branch OA, the response of the bar is linear and elastic. Careful measurements [4] have shown that local strain develops *uniformly* throughout the specimen, as suggested in expression (28). The yielding phase begins in the vicinity of point A. This is characterized at the microscopic level by the development of almost equidistant slip bands: yielding strain does not proceed uniformly, but when one element or layer yields, it almost instantaneously undergoes a large strain whose growth then almost stops while the yielding process is transferred to another portion of the bar. At the macroscopic level, this process has its counterpart in the stress oscillations around an almost constant mean value, until the strain hardening period is reached (point B in Figure 2). The model accurately reproduces the yielding process through the solution discussed in Section 2 (Case 2), and the similarities between the graphs in Figures 2 and 3, with regard to the material's gross response, is also immediate. Moreover, the theory prescribes a fixed amount of slip for each band, a fact which is confirmed with good approximation by experimental evidence. The third, strain hardening stage, starts at point B and is characterized by a strain field well approximated by (38): strain increases uniformly in the bar portions between consecutive slip bands, whereas the amount of slip remains the same as that produced in the yielding period.

Acknowledgements

The author would like to express his grateful appreciation to Professor Giovanni Alberti of the University of Pisa, for helpful discussion during the preparation of this work, and to an unknown reviewer, for very valuable comments about the first version of the manuscript. The partial support of the European Community under contract SMT4-CT96-2130 is grateful acknowledged.

References

1. J.F. Bell, The experimental foundations of solid mechanics. *Handbuch der Physik*, Vol. VIa/1. Springer, Berlin (1973).
2. J.E. Dunn and R.L. Fosdick, The morphology and stability of material phases. *Arch. Rational. Mech. Anal.* **74** (1980) 1–99.
3. R.L. Fosdick and D.E. Mason, Single phase energy minimizers for materials with nonlocal spatial dependence. *Quart. Appl. Math.* **54** (1996) 161–195.
4. M. Froli and G.F. Royer-Carfagni, On discontinuous deformation of tensile steel bars: Experimental results. *ASCE J. Engrg. Mech.* **125** (11) (1999) 1243–1250.
5. M. Froli and G.F. Royer-Carfagni, A mechanical model for the elastic-plastic behavior of metallic bars. *Internat. J. Solids Struct.* **37** (29) (2000) 3901–3918.
6. R.D. James, Wiggly energies. In: R.C. Batra and M.F. Beatty (eds), *Contemporary Research in the Mechanics and Mathematics of Materials*. Cimne, Barcelona (1996) 275–287.
7. A. Nadai, *Theory of flow and Fracture of Solids*. McGraw-Hill, New York (1950).
8. L. Truskinovsky, Fracture as a phase transition. In: R.C. Batra and M.F. Beatty (eds), *Contemporary Research in the Mechanics and Mathematics of Materials*. Cimne, Barcelona (1996) 322–334.
9. L. Truskinovsky and G. Zanzotto, L. Ericksen's bar revisited: Energy wiggles. *J. Mech. Phys. Solids* **44** (1996) 1371–1408.

Journal of Elasticity **59:** 145–173, 2000.

Two Micromechanical Models in Acoustoelasticity: a Comparative Study

ROBERTO PARONI[1] and CHI-SING MAN[2]
[1]*Mathematical Institute, University of Oxford, 24–29 St. Giles', Oxford OX1 3LB, U.K.*
[2]*Department of Mathematics, University of Kentucky, Lexington, KY 40506-0027, U.S.A.*

Received 21 January 1999; in revised form 7 January 2000

To Roger Fosdick, natural philosopher par excellence.

Abstract. Herein we derive, under the micromechanical model we proposed earlier, Man and Paroni [14], a complete set of formulae for the twelve material constants in the acoustoelastic constitutive equation for orthorhombic aggregates of cubic crystallites. We present also a second model and compare its predictions on the material constants with those of the first model. Both these models lead to constitutive equations which are indifferent to rotation of reference placement. This allows us to appeal to a new representation theorem (Paroni and Man [15]), which greatly facilitates our derivation of the formulae for the material constants. The second model introduced in this paper is intimately related to some previous averaging theories in the literature. We explain why and in what sense our second model could be taken as a generalization of its predecessors.

Mathematics Subject Classifications (2000): 74B10, 74E25, 74M25.

Key words: residual stress, crystallographic texture, acoustoelasticity, micromechanics.

1. Introduction

The presence of stress in a body changes the speeds of elastic waves propagating in it. This phenomenon is called the acoustoelastic effect. The possibility of using the acoustoelastic effect for nondestructive evaluation of stress has been the impetus for much of the research in acoustoelasticity during the last four decades.

Early discussions bearing on a general theory of linear elasticity with initial stress can be traced back to Cauchy, Rayleigh, Love, Biot, and others (for references and historical comments, see [1, Section 55; 2, Section 68; 3]). Acoustoelasticity, however, has taken a different route at least in its main line of development until the eighties. The first major paper in this line of development is that of Hughes and Kelly [4], who use the theory of nonlinear elasticity as formulated by Murnaghan [5] and show, by making wave-speed measurements on three materials, how the three third-order elastic constants of an isotropic material can be determined. Hughes and Kelly's theory was developed to full generality for materials with arbitrary symmetry by Toupin and Bernstein [6] and Thurston and Brugger [7] in the sixties. In this approach the acoustoelastic effect is studied in the framework of small displacements superimposed on a finite deformation of a perfectly elastic material. Three configurations of the body play crucial roles in the theory: a natural or stress-free configuration, the initially deformed and stressed configuration, and the current configuration. The body is assumed to be hyperelastic, and it is the elastic deformation from the unstressed "natural state" to the initial configuration which gives rise to the stress in the latter. In order to capture the effect of the inital stress on the speeds of waves superimposed on the initial configuration, nonlinearity of the constitutive equation is taken into account to the extent that the second-order effects of the initial strain are considered. Besides the usual fourth-order elasticity tensor in classical linear elasticity, a sixth-order tensor with third-order elastic constants appears in the theory. This theory has been applied since for determination of applied stress in deformed elastic bodies and, despite the hyperelastic assumption, has also been used for evaluation of residual stress. As residual stresses often arise as a result of inhomogeneous plastic deformations, using a purely hyperelastic theory as the theoretical basis for acoustoelastic measurement of residual stress is objectionable. Pao et al. [8, p. 84] raised a further objection to such use: "For a genuine problem of residual stresses, the natural state is unknown. The specimen we receive for ultrasonic testing is already at an initial state ...." Indeed, we can go further and say that a natural state might not even exist for some materials. Despite these objections, the aforementioned approach dominated the literature in acoustoelasticity through two decades of intense activites from the early sixties till the early eighties. This approach remains a mainstay in acoustoelasticity to this day.

Most structural metals are polycrystalline aggregates. Manufacturing processes such as rolling, forging, extrusion, annealing and aging usually leave the grains in certain preferred orientations so that the metal acquires a slight anisotropy. This anisotropy causes shifts in speeds of ultrasonic waves, which are often of the same order of or even an order higher than those due to the presence of stress. From his experimental investigations Crecraft [9] concluded "that elastic anisotropy of small orders can be reliably measured, and that in cases where anisotropy due to preferred grain orientation is of a regular or predictable nature..., then residual stresses can be estimated from the remaining anisotropy not accounted for by grain alignment." In materials science preferred orientation of crystallites, or crystallographic texture, is mathematically described by the orientation distribution function (ODF) [10, 11]. It was Sayers [12] and Johnson [13] who introduced the ODF into acoustoelasticity. They considered an orthorhombic aggregate of cubic crystallites given in an unstressed natural state. Sayers [12] computed the second-order elastic constants of the textured polycrystal and expressed them in terms of the ODF and the elastic constants of the the single crystal. Johnson [13] attempted to do the same for the third-order elastic constants, but his work was not totally successful (see Remark 6.1 below). In their computations both of them employed the Voigt assumption or "the Voigt approximation of the first kind" (i.e., the strain is taken to be homogeneous across the crystallites and equal to that of the aggregate), and they obtained the second-order and third-order elastic constants of the polycrystal through orientational averaging (i.e., each physical property of the polycrystal is estimated by integrating the corresponding orientation-dependent property of the single crystal over the orientations with the ODF as weight). While the orientational averaging of the second-order elastic constants is quite simple, the averaging of the third-order elastic constants weighted by the ODF involves for each elastic constant many symbolic integrations, which make the computation very tedious and time-consuming. Today, with the latest generation of computers, the computation is still time expensive. It must have taken Johnson a long time to complete all the integrations, and it is understandable if in that effort some mistake was made (as we shall point out in Remark 6.1 below). Once the second-order and third-order elastic constants of the polycrystal in its natural state are estimated, the theories of Toupin and Bernstein [6] and Thurston and Brugger [7] can be applied. Thus, in this approach the theory of acoustoelasticity is divided into two parts. First, an estimate is made on the elastic property of the material in the given natural configuration; the effect of crystallographic texture is taken into account in this estimate. Then the phenomenological theory based on hyperelasticity is used to relate wave speeds to the initial deformation and to the elastic constants and texture of the polycrystal at its natural state. In both steps it is essential to make use of the natural configuration, and in both steps it is assumed that the material be hyperelastic. Clearly, if we want to develop a theory of acoustoelasticity for measurement of residual stress, these assumptions should not be made.

In 1987 Man and Lu [3] pointed out that linear elasticity with initial stress should provide a suitable theoretical framework for acoustoelasticity. There the initially stressed configuration serves as the only reference configuration for description of deformations. Moreover, no assumption is made on the existence of a natural configuration for the polycrystal, and the origin of the initial stresses is of no concern (e.g., they may arise from prior inhomogeneous plastic deformations of the polycrystal). In this formulation the elastic properties of the initially stressed polycrystal are embodied in the incremental elasticity tensor. Recently we introduced (cf. [14]) a simple micromechanical model by which we can in principle express the incremental elasticity tensor $\mathcal{L}$ explicitly in terms of the initial stress $\mathring{\mathbf{T}}$, the ODF w which describes the texture at the initial configuration, and the second-order and third-order elastic constants of the single crystal. We write $\mathcal{L}(\mathring{\mathbf{T}}, w)[\cdot] = \mathbb{C}(w)[\cdot] + \mathbb{D}(w)[\mathring{\mathbf{T}}, \cdot]$; here $\mathbb{C}(w)$ is the elasticity tensor, and $\mathbb{D}(w)$ is a sixth-order tensor which we call the acoustoelastic tensor. Hereafter we refer to this model as our first model. In this model the initial stresses in the crystallites are assumed to be the same and equal to that of the aggregate, the Voigt assumption is adopted for the incremental motion, and the incremental elasticity tensor of the polycrystal is obtained through orientational averaging. For an initially stressed orthorhombic aggregate of cubic crystallites, if the integrations required for orientational averaging in our first model are carried out by brute force, the task proves to be even more daunting than that faced by Johnson [13]. Handicapped by the time-consuming integrations, we could not give a complete characterization of $\mathcal{L}$ when we wrote our 1996 paper, but we managed to evaluate a sufficient number of integrals to arrrive at a birefringence formula which sheds much light on a central problem in acoustoelasticity, namely, that concerning the separation of stress-induced and texture-induced birefringence.

In the present paper we shall revisit our first model and, for orthorhombic aggregates of cubic crystallites, derive a complete set of formulae for all the components of the incremental elasticity tensor $\mathcal{L}$. The problem of time-expensive integrations is circumvented by appealing to a new representation theorem (see [15, Theorem 3.3.1]. Under our first model, the sixth-order acoustoelastic tensor $\mathbb{D}(w)$ is linear in the ODF w and satisfies (see Lemma 6.1 below) the relation

$$\mathbb{D}(\mathcal{T}_{\mathbf{Q}}w)\big[\mathbf{QaQ}^{\mathrm{T}}, \mathbf{QbQ}^{\mathrm{T}}\big] = \mathbf{Q}\mathbb{D}(w)[\mathbf{a}, \mathbf{b}]\mathbf{Q}^{\mathrm{T}},$$

for all rotations $\mathbf{Q}$ and for all second-order symmetric tensors $\mathbf{a}$ and $\mathbf{b}$. In the preceding equation, $\mathcal{T}_{\mathbf{Q}}w$ is the ODF defined by $\mathcal{T}_{\mathbf{Q}}w(\mathbf{R}) = w(\mathbf{Q}^{\mathrm{T}}\mathbf{R})$ for all rotations $\mathbf{R}$, and $\mathbf{Q}^{\mathrm{T}}$ denotes the transpose of $\mathbf{Q}$. The constraint above on $\mathbb{D}(w)$ dictates that under our first model the acoustoelastic tensor can be written as

$$\mathbb{D}(w) = \mathbb{D}_{\mathrm{iso}} + \sum_{p=1}^{5} \gamma_p \mathbf{\Phi}^{(p)}(w),$$

where $\mathbb{D}_{\mathrm{iso}}$ is a constant sixth-order isotropic tensor (it is what $\mathbb{D}$ would be if the crystallites have no preferred orientations), the γ_p's are material constants, and the $\mathbf{\Phi}^{(p)}(w)$'s (for $p = 1, \ldots, 5$) are explicitly determined sixth-order orthotropic tensors that depend only on w. Hence, in order to completely determine the sixth-order acoustoelastic tensor $\mathbb{D}(w)$, it suffices to evaluate only four material constants in $\mathbb{D}_{\mathrm{iso}}$ and the five material constants γ_p in the representation above. In this way, the task of orientational averaging for the sixth-order tensor $\mathbb{D}(w)$ is considerably simplified; it is just sligthly more challenging than its counterpart for the fourth-order elasticity tensor $\mathbb{C}(w)$. Explicit formulae which, under our first model, express the nine material constants of $\mathbb{D}(w)$ in terms of the single-crystal elastic constants are displayed in the Appendix.

In Section 3 we present a second micromechanical model. In our second model we adopt a Voigt-type assumption also for the crystallites in their initial configuration, namely that all the crystallites pertaining to a macroscopic aggregate point carry the same homogeneous left stretch tensor. We show how all the components of the incremental elasticity tensor $\widehat{\mathcal{L}}(\mathring{\mathbf{T}}, w)$ under our second model can be computed up to order $o(\|w - w_{\mathrm{iso}}\|)$; here $w_{\mathrm{iso}} = 1/(8\pi^2)$ denotes the ODF when the crystallites have no preferred orientations. Again, the representation theorem mentioned above plays a crucial simplifying role in the calculations. This second model turns out to be intimately related to several previous theories in acoustoelasticity. We explain why and in what sense our second model could be taken as a generalization of its predecessors in Section 5.

In Section 7 we compare the predictions of the two models for several materials at room temperature. These examples reveal two problems:

(i) for some materials (e.g., copper) the predictions of the two models can be widely different;

(ii) the predicted elastic constants of the polycrystal may be sensitive to the single-crystal elastic constants used in the computation – experimental errors in single-crystal constants may render the predictions dubious.

These problems notwithstanding, it seems that our two models might still be useful for some materials (e.g., aluminum), as we shall see in more details in Section 7.

2. Preliminaries and the First Model

For convenience, we briefly summarize some definitions and findings of Man and Paroni [14], which serve as a prerequisite to the present paper. The reader is referred to the quoted paper for a more detailed exposition.

We consider a placement $\boldsymbol{\kappa}$ of a polycrystalline aggregate B in the 3-dimensional space E^3. The placement $\boldsymbol{\kappa}$ defines the initial configuration of B. We assume that each macroscopic material point X in B consists of numerous crystallites $\mathcal{B}_\alpha$, which occupy subregions $\boldsymbol{\kappa}(\mathcal{B}_\alpha)$ in the initial configuration. Further, we suppose that each subregion $\boldsymbol{\kappa}(\mathcal{B}_\alpha)$ may be identified as the image of a homogeneous transplacement $\boldsymbol{\Phi}_\alpha\colon \boldsymbol{\lambda}(\mathcal{B}) \to E^3$ as restricted to some appropriate subset of a fixed natural configuration $\boldsymbol{\lambda}(\mathcal{B})$ of a perfect crystal $\mathcal{B}$ of the material in question. Let $\mathbf{P}_\alpha = \nabla\boldsymbol{\Phi}_\alpha$, $\mathbf{I}$ be the identity tensor, and

$$\mathbf{P}_\alpha = \mathring{\mathbf{R}}_\alpha\mathring{\mathbf{U}}_\alpha = \mathring{\mathbf{V}}_\alpha\mathring{\mathbf{R}}_\alpha, \tag{1}$$

$$\mathring{\mathbf{E}}_\alpha = \frac{1}{2}\big(\mathbf{P}_\alpha^{\mathrm{T}}\mathbf{P}_\alpha - \mathbf{I}\big) \tag{2}$$

be the polar decomposition of $\mathbf{P}_\alpha$ and the Lagrangian prestrain of $\mathcal{B}_\alpha$ with the configuration $\boldsymbol{\Phi}_\alpha^{-1}(\boldsymbol{\kappa}(\mathcal{B}_\alpha)) \subset \boldsymbol{\lambda}(\mathcal{B})$ as reference, respectively. Supposing that $\boldsymbol{\Phi}_\alpha$ is a transplacement with possibly large rotation but small prestrain $\mathring{\mathbf{E}}_\alpha$, and denoting with $\boldsymbol{\sigma}_\alpha$ and $\mathbf{e}_\alpha$ the second Piola–Kirchhoff stress with $\boldsymbol{\kappa}(\mathcal{B}_\alpha)$ as reference and an infinitesimal strain superimposed on $\boldsymbol{\kappa}(\mathcal{B}_\alpha)$, respectively, we have (cf. [14, equation (21)]):

$$\begin{aligned}\boldsymbol{\sigma}_\alpha &= \boldsymbol{\sigma}_\alpha\big(\mathbf{e}_\alpha; \mathring{\mathbf{E}}_\alpha, \mathring{\mathbf{R}}_\alpha\big)\\ &= \mathring{\mathbf{T}}_\alpha + \mathring{\mathbf{R}}_\alpha\mathbb{C}^\circ\big[\mathring{\mathbf{R}}_\alpha^{\mathrm{T}}\mathbf{e}_\alpha\mathring{\mathbf{R}}_\alpha\big]\mathring{\mathbf{R}}_\alpha^{\mathrm{T}} + \mathfrak{D}\big(\mathring{\mathbf{E}}_\alpha, \mathring{\mathbf{R}}_\alpha\big)[\mathbf{e}_\alpha],\end{aligned} \tag{3}$$

with

$$\mathfrak{D}\big(\mathring{\mathbf{E}}, \mathring{\mathbf{R}}\big)[\mathbf{e}] = \mathring{\mathbf{R}}\,\mathbb{D}^\circ\big[\mathring{\mathbf{E}}, \mathring{\mathbf{R}}^{\mathrm{T}}\mathbf{e}\mathring{\mathbf{R}}\big]\mathring{\mathbf{R}}^{\mathrm{T}} - \big(\mathrm{tr}\mathring{\mathbf{E}}\big)\mathring{\mathbf{R}}\,\mathbb{C}^\circ\big[\mathring{\mathbf{R}}^{\mathrm{T}}\mathbf{e}\mathring{\mathbf{R}}\big]\mathring{\mathbf{R}}^{\mathrm{T}}$$

$$+\,\mathring{\mathbf{R}}(\mathring{\mathbf{E}}\mathbb{C}^\circ[\mathring{\mathbf{R}}^{\mathrm{T}}\mathbf{e}\mathring{\mathbf{R}}] + \mathbb{C}^\circ[\mathring{\mathbf{R}}^{\mathrm{T}}\mathbf{e}\mathring{\mathbf{R}}]\mathring{\mathbf{E}})\mathring{\mathbf{R}}^{\mathrm{T}}$$
$$+\,\mathring{\mathbf{R}}\,\mathbb{C}^\circ[\mathring{\mathbf{E}}\mathring{\mathbf{R}}^{\mathrm{T}}\mathbf{e}\mathring{\mathbf{R}} + \mathring{\mathbf{R}}^{\mathrm{T}}\mathbf{e}\mathring{\mathbf{R}}\,\mathring{\mathbf{E}}]\mathring{\mathbf{R}}^{\mathrm{T}}, \tag{4}$$

where $\mathbb{C}^\circ$ and $\mathbb{D}^\circ$ are the fourth- and sixth-order elasticity tensors of the crystal $\mathscr{B}$ in the configuration $\boldsymbol{\lambda}(\mathscr{B})$ respectively, and $\mathring{\mathbf{T}}_\alpha$ is given by (cf. [14, equation (23)]):

$$\mathring{\mathbf{T}}_\alpha = \mathring{\mathbf{R}}_\alpha\mathbb{C}^\circ[\mathring{\mathbf{E}}_\alpha]\mathring{\mathbf{R}}_\alpha^{\mathrm{T}}. \tag{5}$$

Let $\mathbb{K}^\circ = (\mathbb{C}^\circ)^{-1}$ be the compliance tensor. By inverting equation (5) we obtain

$$\mathring{\mathbf{E}}_\alpha = \mathbb{K}^\circ[\mathring{\mathbf{R}}_\alpha^{\mathrm{T}}\mathring{\mathbf{T}}_\alpha\mathring{\mathbf{R}}_\alpha]. \tag{6}$$

To deduce the constitutive equation for the polycrystalline aggregate, we make the following two assumptions:

1-T. Let $\mathring{\mathbf{T}}$ be the initial stress at the macroscopic material point X in the initial configuration. Each crystallite $\mathscr{B}_\alpha$ pertaining to X carries the same homogeneous initial stress $\mathring{\mathbf{T}}_\alpha = \mathring{\mathbf{T}}$ at the initial configuration $\boldsymbol{\kappa}(\mathscr{B}_\alpha)$.

2-e. When the polycrystal undergoes a deformation from the initial configuration and the resulting infinitesimal strain at X is $\mathbf{e}$, the homogeneous infinitesimal strains $\mathbf{e}_\alpha$ imposed on the configurations $\boldsymbol{\kappa}(\mathscr{B}_\alpha)$ by the deformation are all equal to the symmetric second-order tensor $\mathbf{e}$.

Let $\mathcal{G} = \mathrm{SO}(3)$ be the rotation group and $\wp$ be the probability Borel measure on $\mathcal{G}$ which describes the texture at the point $\boldsymbol{\kappa}(X)$ in the initial configuration. Hereafter we adopt the postulate of orientational averaging, namely that any property of the polycrystalline aggregate in the initial configuration is equal to the average of the corresponding property of the single crystal over the orientation measure $\wp$. Thus, using equation (6), we can write:

$$\begin{aligned}\mathscr{L}(\mathring{\mathbf{T}}, \wp)[\mathbf{e}] &= \int_{\mathcal{G}} (\mathring{\mathbf{R}}\,\mathbb{C}^\circ[\mathring{\mathbf{R}}^{\mathrm{T}}\mathbf{e}\mathring{\mathbf{R}}]\mathring{\mathbf{R}}^{\mathrm{T}} + \mathfrak{D}(\mathbb{K}^\circ[\mathring{\mathbf{R}}^{\mathrm{T}}\mathring{\mathbf{T}}\mathring{\mathbf{R}}], \mathring{\mathbf{R}})[\mathbf{e}])\mathrm{d}\wp(\mathring{\mathbf{R}}) \\ &= \mathbb{C}(\wp)[\mathbf{e}] + \mathbb{D}(\mathring{\mathbf{T}}, \wp)[\mathbf{e}],\end{aligned} \tag{7}$$

where

$$\mathbb{C}(\wp)[\mathbf{e}] = \int_{\mathcal{G}} \mathring{\mathbf{R}}\,\mathbb{C}^\circ[\mathring{\mathbf{R}}^{\mathrm{T}}\mathbf{e}\mathring{\mathbf{R}}]\mathring{\mathbf{R}}^{\mathrm{T}}\,\mathrm{d}\wp(\mathring{\mathbf{R}}), \tag{8}$$

and

$$\mathbb{D}(\mathring{\mathbf{T}}, \wp)[\mathbf{e}] = \int_{\mathcal{G}} \mathfrak{D}(\mathbb{K}^\circ[\mathring{\mathbf{R}}^{\mathrm{T}}\mathring{\mathbf{T}}\mathring{\mathbf{R}}], \mathring{\mathbf{R}})[\mathbf{e}]\,\mathrm{d}\wp(\mathring{\mathbf{R}}). \tag{9}$$

Finally, from equations (3) and (7) we find

$$\sigma = \int_{\mathcal{G}} \sigma(\mathbf{e}; \mathbb{K}^\circ[\mathring{\mathbf{R}}^{\mathrm{T}}\mathring{\mathbf{T}}\mathring{\mathbf{R}}], \mathring{\mathbf{R}})\mathrm{d}\wp(\mathring{\mathbf{R}}) = \mathring{\mathbf{T}} + \mathscr{L}(\mathring{\mathbf{T}}, \wp)[\mathbf{e}]. \tag{10}$$

When no confusion should arise, we shall suppress the dependence on the orientation measure $\wp$ in $\mathcal{L}(\mathring{\mathbf{T}}, \wp)$, $\mathbb{C}(\wp)$ and $\mathbb{D}(\mathring{\mathbf{T}}, \wp)$ and simply write $\mathcal{L}(\mathring{\mathbf{T}})$, $\mathbb{C}$ and $\mathbb{D}(\mathring{\mathbf{T}})$, respectively.

3. The Second Model

One of the basic assumptions of the first model was the uniformity of the prestress $\mathring{\mathbf{T}}$ across the crystallites pertaining to the same macroscopic material point. In this section we replace this assumption on the prestress with one on the predeformation, and we derive a new model accordingly.

As in the first model, we take equations (3)–(5) to be the constitutive equation of the crystallite $\mathcal{B}_\alpha$, but we now make the following two assumptions:

1-V. Every crystallite $\mathcal{B}_\alpha$ pertaining to the macroscopic material point X carries the same homogeneous left stretch tensor $\mathring{\mathbf{V}} = \mathring{\mathbf{V}}(X)$ at the initial configuration $\kappa(\mathcal{B}_\alpha)$ (i.e., we suppose there exists a positive-definite symmetric tensor $\mathring{\mathbf{V}}$ such that $\mathring{\mathbf{V}}_\alpha = \mathring{\mathbf{V}}(X)$ for every crystallite $\mathcal{B}_\alpha$ in X).

2-e. When the polycrystal undergoes a deformation from the initial configuration and the resulting infinitesimal strain at X is $\mathbf{e}$, the homogeneous infinitesimal strains $\mathbf{e}_\alpha$ imposed on the configurations $\kappa(\mathcal{B}_\alpha)$ by the deformation are all equal to the symmetric second-order tensor $\mathbf{e}$.

REMARK 3.1. In 1-V we assume the uniformity of the left stretch tensor inside a material point, but several other possibilities can be considered. For instance, in place of 1-V, one may make the assumption

1-U. Every crystallite $\mathcal{B}_\alpha$ pertaining to the particle X carries the same homogeneous right stretch tensor $\mathring{\mathbf{U}} = \mathring{\mathbf{U}}(X)$ at the initial configuration $\kappa(\mathcal{B}_\alpha)$.

The assumption 1-U is equivalent to the uniformity of the right Cauchy–Green tensor $\mathring{\mathbf{C}} = \mathbf{P}^{\mathrm{T}}\mathbf{P} = \mathring{\mathbf{U}}^2$ and therefore is equivalent to the uniformity of the Lagrangian prestrain $\mathring{\mathbf{E}} = \frac{1}{2}(\mathring{\mathbf{C}} - \mathbf{I})$. On the other hand 1-V is equivalent to the uniformity of the left Cauchy–Green tensor $\mathring{\mathbf{B}} = \mathbf{P}\mathbf{P}^{\mathrm{T}} = \mathring{\mathbf{V}}^2$ and also (cf. [16, 17]) to the uniformity of the Eulerian prestrain (also called the Almansi tensor)

$$\mathring{\mathbf{A}} = \frac{1}{2}\big(\mathbf{I} - \mathring{\mathbf{B}}^{-1}\big) = \frac{1}{2}\big(\mathbf{I} - \mathring{\mathbf{V}}^{-2}\big). \tag{11}$$

Even if $\mathring{\mathbf{U}}$ and $\mathring{\mathbf{V}}$ have the same eigenvalues and therefore they correspond to the same stretch, the assumptions 1-V and 1-U are not equivalent. With assumption 1-U we postulate that all the crystallites have the same stretches in the same directions, while with assumption 1-V it is assumed that the crystallites have the same stretches but in different directions, for $\mathring{\mathbf{V}}$ acts after the rotation $\mathring{\mathbf{R}}_\alpha$. Hence, assumption 1-U is more restrictive than assumption 1-V. In what follows, in order to keep the notation as simple as possible, we shall restrict our attention to the 1-V assumption. Nevertheless, with minor changes, all the calculations that follow can be adapted to the model with the 1-U assumption.

Another possible replacement of assumption 1-V is the uniformity of the deformation gradient **P**; but in this case it would follow, from the uniqueness of the polar decomposition, not only the uniformity of the left and right stretch tensors, but also the uniformity of the rotation, making the averaging meaningless.

From equation (1) we have

$$\mathring{\mathbf{U}}_\alpha = \mathring{\mathbf{R}}_\alpha^{\mathrm{T}} \mathring{\mathbf{V}} \mathring{\mathbf{R}}_\alpha. \tag{12}$$

Thus $\mathring{\mathbf{V}}$ and $\mathring{\mathbf{U}}_\alpha$ correspond to the same stretches but in different directions related by the orthogonal transformation $\mathring{\mathbf{R}}_\alpha$. From the identity $(\mathring{\mathbf{U}}_\alpha)^2 = \mathbf{I} + 2\mathring{\mathbf{E}}_\alpha$, we deduce

$$\mathring{\mathbf{U}}_\alpha = \mathbf{I} + \mathring{\mathbf{E}}_\alpha + \mathrm{o}\big(\|\mathring{\mathbf{E}}_\alpha\|\big), \tag{13}$$

and from equations (12) and (13) it easily follows that

$$\mathring{\mathbf{V}} = \mathbf{I} + \mathring{\mathbf{R}}_\alpha \mathring{\mathbf{E}}_\alpha \mathring{\mathbf{R}}_\alpha^{\mathrm{T}} + \mathrm{o}\big(\|\mathring{\mathbf{E}}_\alpha\|\big), \tag{14}$$

which says that the quantity $\mathring{\mathbf{E}}^*$, as defined by

$$\mathring{\mathbf{E}}^* = \mathring{\mathbf{R}}_\alpha \mathring{\mathbf{E}}_\alpha \mathring{\mathbf{R}}_\alpha^{\mathrm{T}}, \tag{15}$$

is uniform, up to terms of order $\mathrm{o}(\|\mathring{\mathbf{E}}_\alpha\|)$, across the crystallites pertaining to the same material point.

REMARK 3.2. Under the assumption of small Lagrangian strain $\mathring{\mathbf{E}}_\alpha$, the tensor $\mathring{\mathbf{E}}^*$ is equal to the Eulerian prestrain $\mathring{\mathbf{A}}$ defined by equation (11). In fact from equation (14) we have

$$\mathring{\mathbf{V}}^2 = \mathbf{I} + 2\mathring{\mathbf{R}}_\alpha \mathring{\mathbf{E}}_\alpha \mathring{\mathbf{R}}_\alpha^{\mathrm{T}} + \mathrm{o}\big(\|\mathring{\mathbf{E}}_\alpha\|\big) = \mathbf{I} + 2\mathring{\mathbf{E}}^*,$$

and by dropping the $\mathrm{o}(\|\mathring{\mathbf{E}}_\alpha\|)$ terms we find

$$\big(\mathring{\mathbf{V}}^2\big)^{-1} = \mathbf{I} - 2\mathring{\mathbf{E}}^*. \tag{16}$$

Hence, from equations (11) and (16) we obtain $\mathring{\mathbf{A}} = \mathring{\mathbf{E}}^*$.

As in the first model, we adopt the postulate of orientational averaging. Let

$$\mathbb{B}(\mathring{\mathbf{E}}^*, \wp)[\mathbf{e}] = \int_{\mathcal{G}} \mathfrak{D}\big(\mathring{\mathbf{R}}^{\mathrm{T}} \mathring{\mathbf{E}}^* \mathring{\mathbf{R}}, \mathring{\mathbf{R}}\big)[\mathbf{e}]\, \mathrm{d}\wp(\mathring{\mathbf{R}}). \tag{17}$$

Similar to our convention regarding the tensors $\mathbb{C}$, $\mathbb{D}$ and $\mathcal{L}$, we shall suppress the dependence of $\mathbb{B}$ on $\wp$ when no confusion should arise. We note that by defining $\mathbb{C}$ as in equation (8) we find

$$\begin{aligned}\mathbb{B}(\mathring{\mathbf{E}}^*)[\mathbf{e}] &= \int_{\mathcal{G}} \Big(\mathring{\mathbf{R}}\,\mathbb{D}^\circ[\mathring{\mathbf{R}}^{\mathrm{T}}\mathring{\mathbf{E}}^*\mathring{\mathbf{R}}^{\mathrm{T}}\mathbf{e}\mathring{\mathbf{R}}]\mathring{\mathbf{R}}^{\mathrm{T}} - (\operatorname{tr}\mathring{\mathbf{E}}^*)\mathring{\mathbf{R}}\,\mathbb{C}^\circ[\mathring{\mathbf{R}}^{\mathrm{T}}\mathbf{e}\mathring{\mathbf{R}}]\mathring{\mathbf{R}}^{\mathrm{T}} \\ &\quad + \mathring{\mathbf{E}}^*\mathring{\mathbf{R}}\,\mathbb{C}^\circ[\mathring{\mathbf{R}}^{\mathrm{T}}\mathbf{e}\mathring{\mathbf{R}}]\mathring{\mathbf{R}}^{\mathrm{T}} + \mathring{\mathbf{R}}\,\mathbb{C}^\circ[\mathring{\mathbf{R}}^{\mathrm{T}}\mathbf{e}\mathring{\mathbf{R}}]\mathring{\mathbf{R}}^{\mathrm{T}}\mathring{\mathbf{E}}^* \\ &\quad + \mathring{\mathbf{R}}\,\mathbb{C}^\circ[\mathring{\mathbf{R}}^{\mathrm{T}}\mathring{\mathbf{E}}^*\mathbf{e}\mathring{\mathbf{R}} + \mathring{\mathbf{R}}^{\mathrm{T}}\mathbf{e}\mathring{\mathbf{E}}^*\mathring{\mathbf{R}}]\mathring{\mathbf{R}}^{\mathrm{T}}\Big)\mathrm{d}\wp(\mathring{\mathbf{R}}) \\ &= \int_{\mathcal{G}} \mathring{\mathbf{R}}\,\mathbb{D}^\circ[\mathring{\mathbf{R}}^{\mathrm{T}}\mathring{\mathbf{E}}^*\mathring{\mathbf{R}},\ \mathring{\mathbf{R}}^{\mathrm{T}}\mathbf{e}\mathring{\mathbf{R}}]\mathring{\mathbf{R}}^{\mathrm{T}}\,\mathrm{d}\wp(\mathring{\mathbf{R}}) - (\operatorname{tr}\mathring{\mathbf{E}}^*)\mathbb{C}[\mathbf{e}] \\ &\quad + \mathring{\mathbf{E}}^*\mathbb{C}[\mathbf{e}] + \mathbb{C}[\mathbf{e}]\mathring{\mathbf{E}}^* + \mathbb{C}[\mathring{\mathbf{E}}^*\mathbf{e} + \mathbf{e}\mathring{\mathbf{E}}^*]. \end{aligned} \tag{18}$$

From equations (3) and (17) we deduce

$$\sigma = \int_{\mathcal{G}} \sigma(\mathbf{e};\ \mathring{\mathbf{R}}^{\mathrm{T}}\mathring{\mathbf{E}}^*\mathring{\mathbf{R}},\ \mathring{\mathbf{R}})\mathrm{d}\wp(\mathring{\mathbf{R}}) = \mathring{\mathbf{T}} + \mathbb{C}[\mathbf{e}] + \mathbb{B}(\mathring{\mathbf{E}}^*)[\mathbf{e}], \tag{19}$$

where $\mathring{\mathbf{T}}(X)$ should be interpreted as the average, over the orientation measure $\wp$, of the pre-stresses $\mathring{\mathbf{T}}_\alpha$ of the crystallites $\mathcal{B}_\alpha$ pertaining to the macroscopic material point X.

In order to recast the constitutive equation (19) in a form similar to equation (10), we need to find the relation between $\mathring{\mathbf{E}}^*$ and $\mathring{\mathbf{T}}$. This can be done easily by using equations (5) and (15); thus

$$\mathring{\mathbf{T}} = \int_{\mathcal{G}} \mathring{\mathbf{R}}\mathbb{C}^\circ[\mathring{\mathbf{R}}^{\mathrm{T}}\mathring{\mathbf{E}}^*\mathring{\mathbf{R}}]\mathring{\mathbf{R}}^{\mathrm{T}}\,\mathrm{d}\wp(\mathring{\mathbf{R}}), \tag{20}$$

which, by equation (8), reduces to

$$\mathring{\mathbf{T}} = \mathbb{C}[\mathring{\mathbf{E}}^*]. \tag{21}$$

Now setting $\mathbb{K} = \mathbb{C}^{-1}$ and

$$\widehat{\mathbb{D}}(\mathring{\mathbf{T}})[\mathbf{e}] = \mathbb{B}(\mathbb{K}[\mathring{\mathbf{T}}])[\mathbf{e}], \tag{22}$$

and defining the incremental elasticity tensor in our second model as

$$\widehat{\mathcal{L}}(\mathring{\mathbf{T}})[\mathbf{e}] = \mathbb{C}[\mathbf{e}] + \widehat{\mathbb{D}}(\mathring{\mathbf{T}})[\mathbf{e}], \tag{23}$$

we obtain

$$\sigma = \mathring{\mathbf{T}} + \widehat{\mathcal{L}}(\mathring{\mathbf{T}})[\mathbf{e}]. \tag{24}$$

In the preceding equations we have suppressed the dependence of $\mathbb{K}$, $\widehat{\mathbb{D}}$ and $\widehat{\mathcal{L}}$ on $\wp$; we shall write $\mathbb{K}(\wp)$, $\widehat{\mathbb{D}}(\mathring{\mathbf{T}}, \wp)$ and $\widehat{\mathcal{L}}(\mathring{\mathbf{T}}, \wp)$ when we want to emphasize this dependence.

4. Material Symmetry and Symmetries of the Incremental Elasticity Tensor

Throughout this section it is understood that a polycrystalline aggregate with a texture defined by a specific orientation measure $\wp$ is given. For simplicity we shall generally suppress the dependence of constitutive functions on $\wp$.

In [14] the group of texture symmetry $\mathcal{G}_{\text{tex}}$ is defined as:

$$\mathcal{G}_{\text{tex}} = \{\mathbf{Q} \in \mathcal{G}\colon \wp(\mathbf{Q}A) = \wp(A) \text{ for every Borel set } A \in \mathcal{G}\}. \tag{25}$$

Let Lin, Orth and Sym be the set of all second-order tensors, second-order orthogonal and second-order symmetric tensors, respectively.

LEMMA 4.1. *For each* $\mathbf{Q} \in \mathcal{G}_{\text{tex}}$ *and for each* $\mathbf{e}, \mathbf{f} \in \text{Sym}$, *we have*

$$\begin{aligned} \mathbb{C}[\mathbf{Q}\mathbf{e}\mathbf{Q}^{\mathrm{T}}] &= \mathbf{Q}\mathbb{C}[\mathbf{e}]\mathbf{Q}^{\mathrm{T}}, \\ \mathbb{B}(\mathbf{Q}\mathbf{f}\mathbf{Q}^{\mathrm{T}})[\mathbf{Q}\mathbf{e}\mathbf{Q}^{\mathrm{T}}] &= \mathbf{Q}\mathbb{B}(\mathbf{f})[\mathbf{e}]\mathbf{Q}^{\mathrm{T}}, \\ \widehat{\mathbb{D}}(\mathbf{Q}\mathbf{f}\mathbf{Q}^{\mathrm{T}})[\mathbf{Q}\mathbf{e}\mathbf{Q}^{\mathrm{T}}] &= \mathbf{Q}\widehat{\mathbb{D}}(\mathbf{f})[\mathbf{e}]\mathbf{Q}^{\mathrm{T}}. \end{aligned} \tag{26}$$

Proof. The proof of $(26)_1$ and of $(26)_2$ is similar to the proof of [14, Theorem 4.2]. The proof of $(26)_3$ follows directly from equations (22) and $(26)_2$. □

From Lemma 4.1 and equation (23) follows easily the next corollary.

COROLLARY 4.2. *For each* $\mathbf{Q} \in \mathcal{G}_{\text{tex}}$, *we have*

$$\widehat{\mathcal{L}}(\mathbf{Q}\mathring{\mathbf{T}}\mathbf{Q}^{\mathrm{T}})[\mathbf{Q}\mathbf{e}\mathbf{Q}^{\mathrm{T}}] = \mathbf{Q}\widehat{\mathcal{L}}(\mathring{\mathbf{T}})[\mathbf{e}]\mathbf{Q}^{\mathrm{T}}. \tag{27}$$

In order to study the symmetries of $\widehat{\mathcal{L}}(\mathring{\mathbf{T}})$ we need to define the group of stress symmetry

$$\mathcal{G}_{\text{str}} = \{\mathbf{Q} \in \mathcal{G}\colon \mathbf{Q}\mathring{\mathbf{T}}(X)\mathbf{Q}^{\mathrm{T}} = \mathring{\mathbf{T}}(X)\}, \tag{28}$$

and the group of aggregate symmetry

$$\mathcal{G}_{\text{agg}} = \mathcal{G}_{\text{tex}} \cap \mathcal{G}_{\text{str}}. \tag{29}$$

This done, from Lemma 4.1 and Corollary 4.2 we obtain:

COROLLARY 4.3. *For each* $\mathbf{Q} \in \mathcal{G}_{\text{agg}}$, *we have*

$$\begin{aligned} \widehat{\mathbb{D}}(\mathring{\mathbf{T}})[\mathbf{Q}\mathbf{e}\mathbf{Q}^{\mathrm{T}}] &= \mathbf{Q}\widehat{\mathbb{D}}(\mathring{\mathbf{T}})[\mathbf{e}]\mathbf{Q}^{\mathrm{T}}, \\ \widehat{\mathcal{L}}(\mathring{\mathbf{T}})[\mathbf{Q}\mathbf{e}\mathbf{Q}^{\mathrm{T}}] &= \mathbf{Q}\widehat{\mathcal{L}}(\mathring{\mathbf{T}})[\mathbf{e}]\mathbf{Q}^{\mathrm{T}}. \end{aligned} \tag{30}$$

REMARK 4.1. From equation (23) we see that the fourth-order tensor $\widehat{\mathcal{L}}(\mathring{\mathbf{T}})$ is the sum of two fourth-order tensors $\mathbb{C}$ and $\widehat{\mathbb{D}}(\mathring{\mathbf{T}})$. By Lemma 4.1, the symmetry group of $\mathbb{C}$ contains $\mathcal{G}_{\text{tex}}$ as a subgroup; by Corollary 4.3, we know only that the symmetry group of $\widehat{\mathbb{D}}(\mathring{\mathbf{T}})$ has $\mathcal{G}_{\text{agg}}$ as a subgroup.

Hereafter we assume that the crystallites $\mathcal{B}_\alpha$ be hyperelastic; more precisely, we postulate the existence of a smooth function $W\colon \text{Sym} \to \mathbb{R}$, called the strain-energy function, such that

$$\boldsymbol{\sigma} = DW(\mathbf{E}). \tag{31}$$

Using the Taylor formula for W, we find:

$$\begin{aligned} \mathbb{C}^\circ &= D(DW)(\mathbf{0}), \\ \mathbb{D}^\circ &= D(D(DW))(\mathbf{0}). \end{aligned} \tag{32}$$

It follows directly from equation (32) that $\mathbb{C}^\circ$ and $\mathbb{D}^\circ(\mathbf{f})$, for every $\mathbf{f} \in \text{Sym}$, have the major symmetry, i.e.,

$$\begin{aligned} \mathbb{C}^{\circ\mathrm{T}} &= \mathbb{C}^\circ, \\ \mathbb{D}^\circ(\mathbf{f})^{\mathrm{T}} &= \mathbb{D}^\circ(\mathbf{f}), \quad \text{for every } \mathbf{f} \in \text{Sym}, \end{aligned} \tag{33}$$

and

$$\mathbb{D}^\circ(\mathbf{f})[\mathbf{g}] = \mathbb{D}^\circ(\mathbf{g})[\mathbf{f}], \quad \text{for every } \mathbf{f}, \mathbf{g} \in \text{Sym}; \tag{34}$$

here by definition the transpose of a fourth-order tensor $\mathbb{A}$ is the fourth-order tensor $\mathbb{A}^{\mathrm{T}}$ which satisfies

$$\mathbb{A}^{\mathrm{T}}\mathbf{g} \cdot \mathbf{f} = \mathbb{A}\mathbf{f} \cdot \mathbf{g}, \quad \text{for every } \mathbf{g}, \mathbf{f} \in \text{Sym}. \tag{35}$$

To prove that also $\widehat{\mathcal{L}}(\mathring{\mathbf{T}})$ has the major symmetry, it is convenient to define the tensor product of linear transformations. Given $\mathbf{G}, \mathbf{K} \in \text{Lin}$, we define the fourth-order tensor $\mathbf{G} \boxtimes \mathbf{K}$ by

$$(\mathbf{G} \boxtimes \mathbf{K})\mathbf{S} = \mathbf{G}\mathbf{S}\mathbf{K}^{\mathrm{T}}, \quad \text{for every } \mathbf{S} \in \text{Lin}. \tag{36}$$

THEOREM 4.4. *For each* $\mathbf{K} \in \text{Sym}$ *and* $\mathring{\mathbf{R}} \in \text{Orth}$, *we have*

$$\mathfrak{D}(\mathbf{K}, \mathring{\mathbf{R}})^{\mathrm{T}} = \mathfrak{D}(\mathbf{K}, \mathring{\mathbf{R}}). \tag{37}$$

Proof. Using the tensor product defined in equation (36), we can rewrite equation (4) as

$$\begin{aligned} \mathfrak{D}(\mathbf{K}, \mathring{\mathbf{R}}) = {} & (\mathring{\mathbf{R}} \boxtimes \mathring{\mathbf{R}})\{\mathbb{D}^\circ(\mathbf{K}) - (\operatorname{tr}\mathbf{K})\mathbb{C}^\circ + (\mathbf{K} \boxtimes \mathbf{I})\mathbb{C}^\circ \\ & + (\mathbf{I} \boxtimes \mathbf{K}^{\mathrm{T}})\mathbb{C}^\circ + \mathbb{C}^\circ(\mathbf{K} \boxtimes \mathbf{I}) + \mathbb{C}^\circ(\mathbf{I} \boxtimes \mathbf{K}^{\mathrm{T}})\} \\ & \times (\mathring{\mathbf{R}}^{\mathrm{T}} \boxtimes \mathring{\mathbf{R}}^{\mathrm{T}}), \end{aligned} \tag{38}$$

and since $\mathbf{K} \in \mathrm{Sym}$ this is equal to

$$\mathfrak{D}(\mathbf{K}, \mathring{\mathbf{R}}) = (\mathring{\mathbf{R}} \boxtimes \mathring{\mathbf{R}})\{\mathbb{D}^{\circ}(\mathbf{K}) - (\operatorname{tr}\mathbf{K})\mathbb{C}^{\circ} + (\mathbf{K} \boxtimes \mathbf{I})\mathbb{C}^{\circ} + (\mathbf{I} \boxtimes \mathbf{K})\mathbb{C}^{\circ} + \mathbb{C}^{\circ}(\mathbf{K} \boxtimes \mathbf{I}) + \mathbb{C}^{\circ}(\mathbf{I} \boxtimes \mathbf{K})\}(\mathring{\mathbf{R}}^{\mathrm{T}} \boxtimes \mathring{\mathbf{R}}^{\mathrm{T}}). \tag{39}$$

Now using the identity $(\mathbf{A} \boxtimes \mathbf{K})^{\mathrm{T}} = (\mathbf{A}^{\mathrm{T}} \boxtimes \mathbf{K}^{\mathrm{T}})$, which holds for every $\mathbf{A}, \mathbf{K} \in \mathrm{Lin}$, the theorem follows. □

COROLLARY 4.5. *The fourth-order tensors* $\mathbb{C}$, $\widehat{\mathbb{D}}(\mathring{\mathbf{T}})$ *and* $\widehat{\mathcal{L}}(\mathring{\mathbf{T}})$ *have the major symmetry, i.e.,*

$$\mathbb{C}^{\mathrm{T}} = \mathbb{C}, \qquad \widehat{\mathbb{D}}(\mathring{\mathbf{T}})^{\mathrm{T}} = \widehat{\mathbb{D}}(\mathring{\mathbf{T}}), \qquad \widehat{\mathcal{L}}(\mathring{\mathbf{T}})^{\mathrm{T}} = \widehat{\mathcal{L}}(\mathring{\mathbf{T}}). \tag{40}$$

Proof. The major symmetry of $\mathbb{C}$ follows from equation (8), which can be rewritten as

$$\mathbb{C} = \int_{\mathcal{G}} (\mathring{\mathbf{R}} \boxtimes \mathring{\mathbf{R}})\mathbb{C}^{\circ}(\mathring{\mathbf{R}} \boxtimes \mathring{\mathbf{R}})^{\mathrm{T}} \,\mathrm{d}\wp(\mathring{\mathbf{R}}), \tag{41}$$

and from equation $(33)_1$. A glance at equation (18) reveals that the relation

$$\mathbb{B}(\mathbf{K})^{\mathrm{T}} = \mathbb{B}(\mathbf{K}), \quad \text{for every } \mathbf{K} \in \mathrm{Sym} \tag{42}$$

follows as a direct consequence of Theorem 4.4. Then from equations (22), (23) and (42) we get the rest of the corollary. □

Similarly we can prove the following:

COROLLARY 4.6. $\mathbb{D}(\mathring{\mathbf{T}})$ *and* $\mathcal{L}(\mathring{\mathbf{T}})$ *have the major symmetry, i.e.,*

$$\mathbb{D}(\mathring{\mathbf{T}})^{\mathrm{T}} = \mathbb{D}(\mathring{\mathbf{T}}), \qquad \mathcal{L}(\mathring{\mathbf{T}})^{\mathrm{T}} = \mathcal{L}(\mathring{\mathbf{T}}). \tag{43}$$

REMARK 4.2. While $\mathbb{D}^{\circ}$ satisfies equation (34), $\mathbb{D}$ and $\widehat{\mathbb{D}}$ do not. In fact, for $\mathbf{f}, \mathbf{g} \in \mathrm{Sym}$, in general $\mathfrak{D}(\mathbf{f}, \mathring{\mathbf{R}})[\mathbf{g}] \neq \mathfrak{D}(\mathbf{g}, \mathring{\mathbf{R}})[\mathbf{f}]$. This fact implies that for a fixed texture symmetry $\mathbb{D}$ and $\widehat{\mathbb{D}}$ will have more independent coefficients than $\mathbb{D}^{\circ}$. For instance, for orthorhombic texture symmetry $\mathbb{D}$ and $\widehat{\mathbb{D}}$ have 39 independent coefficients each, while $\mathbb{D}^{\circ}$ has only 20.

5. Previous Theories and the Second Model. A Comparison

Here we briefly review some previous averaging theories in acoustoelasticity and compare them with the second model discussed above. As we shall explain below, our second model is intimately related to these theories and may be taken as a generalization of them.

Before the work of Man and Paroni [14], the averaging theories which enjoy some popularity in acoustoelasticity all refer to a *natural configuration* (i.e., a stress-free configuration) of the body. Let us denote one such configuration by $\boldsymbol{\kappa}^N$. We assume that the natural configuration $\boldsymbol{\kappa}^N(\mathcal{B}_\alpha)$ of the crystallite $\mathcal{B}_\alpha$ may be identified with the image of a rigid rotation $\mathbf{R}_\alpha$ of an appropriate subset of $\boldsymbol{\lambda}(\mathcal{B})$. Consider a deformation, whose gradient is $\mathbf{F}_\alpha^N$, superimposed on $\boldsymbol{\kappa}^N(\mathcal{B}_\alpha)$. Let $\mathbf{E}_\alpha^N = \frac{1}{2}((\mathbf{F}_\alpha^N)^{\mathrm{T}}\mathbf{F}_\alpha^N - \mathbf{I})$ and $\mathbf{E}_{\boldsymbol{\lambda},\alpha} = \frac{1}{2}((\mathbf{F}_\alpha^N\mathbf{R}_\alpha)^{\mathrm{T}}(\mathbf{F}_\alpha^N\mathbf{R}_\alpha) - \mathbf{I})$ be the Lagrangian strains of the crystallite $\mathcal{B}_\alpha$ with respect to the reference configurations $\boldsymbol{\kappa}^N(\mathcal{B}_\alpha)$ and $\boldsymbol{\lambda}(\mathcal{B})$, respectively. We have

$$\mathbf{E}_{\boldsymbol{\lambda},\alpha} = \mathbf{R}_\alpha^{\mathrm{T}}\mathbf{E}_\alpha^N\mathbf{R}_\alpha. \tag{44}$$

Let $\boldsymbol{\sigma}_{\boldsymbol{\lambda},\alpha}$ and $\boldsymbol{\sigma}_\alpha^N$ be the second Piola–Kirchhoff stress tensors that result from the deformation $\mathbf{F}_\alpha^N$ with $\boldsymbol{\lambda}(\mathcal{B})$ and $\boldsymbol{\kappa}^N(\mathcal{B}_\alpha)$ as reference, respectively. It is easy to see (cf. [14, equation (15)]), that

$$\boldsymbol{\sigma}_\alpha^N = \mathbf{R}_\alpha\boldsymbol{\sigma}_{\boldsymbol{\lambda},\alpha}\mathbf{R}_\alpha^{\mathrm{T}}. \tag{45}$$

Assuming the constitutive equation of $\boldsymbol{\sigma}_{\boldsymbol{\lambda},\alpha}$ to be

$$\boldsymbol{\sigma}_{\boldsymbol{\lambda},\alpha} = \mathbb{C}^\circ[\mathbf{E}_{\boldsymbol{\lambda},\alpha}] + \frac{1}{2}\mathbb{D}^\circ[\mathbf{E}_{\boldsymbol{\lambda},\alpha}, \mathbf{E}_{\boldsymbol{\lambda},\alpha}], \tag{46}$$

we obtain from equations (44) and (45) the equation

$$\boldsymbol{\sigma}_\alpha^N = \mathbf{R}_\alpha\mathbb{C}^\circ\big[\mathbf{R}_\alpha^{\mathrm{T}}\mathbf{E}_\alpha^N\mathbf{R}_\alpha\big]\mathbf{R}_\alpha^{\mathrm{T}} + \frac{1}{2}\mathbf{R}_\alpha\mathbb{D}^\circ\big[\mathbf{R}_\alpha^{\mathrm{T}}\mathbf{E}_\alpha^N\mathbf{R}_\alpha, \mathbf{R}_\alpha^{\mathrm{T}}\mathbf{E}_\alpha^N\mathbf{R}_\alpha\big]\mathbf{R}_\alpha^{\mathrm{T}}. \tag{47}$$

Denoting by $\wp^N$ the orientation measure which describes the texture at the *natural* configuration $\boldsymbol{\kappa}^N$, and assuming that the Lagrangian strain $\mathbf{E}_\alpha^N$ be equal to $\mathbf{E}^N(X)$ for all the crystallites $\mathcal{B}_\alpha$ comprising the material point X, we apply orientational averaging and obtain

$$\boldsymbol{\sigma}^N = \mathbb{C}^N\big[\mathbf{E}^N\big] + \frac{1}{2}\mathbb{B}^N\big[\mathbf{E}^N, \mathbf{E}^N\big], \tag{48}$$

where $\mathbb{C}^N[\cdot]$ and $\mathbb{B}^N[\cdot\,,\cdot]$ are defined by

$$\mathbb{C}^N[\mathbf{E}] = \int_{\mathcal{G}} \mathbf{R}\mathbb{C}^\circ\big[\mathbf{R}^{\mathrm{T}}\mathbf{E}\mathbf{R}\big]\mathbf{R}^{\mathrm{T}}\,\mathrm{d}\wp^N(\mathbf{R}), \tag{49}$$

$$\mathbb{B}^N[\mathbf{E}_1, \mathbf{E}_2] = \int_{\mathcal{G}} \mathbf{R}\mathbb{D}^\circ\big[\mathbf{R}^{\mathrm{T}}\mathbf{E}_1\mathbf{R}, \mathbf{R}^{\mathrm{T}}\mathbf{E}_2\mathbf{R}\big]\mathbf{R}^{\mathrm{T}}\,\mathrm{d}\wp^N(\mathbf{R}), \tag{50}$$

and for simplicity we have suppressed the dependence of $\mathbb{C}^N$ and $\mathbb{B}^N$ on the orientation measure $\wp^N$. Integral (49) has been evaluated for polycrystals with various texture and crystal symmetries (cf. [12, 18]). In the case of $\mathbf{E}_1 = \mathbf{E}_2$, integral (50) was computed by Bross [19] and Chang [20] for isotropic aggregates of cubic crystallites, by Barsch [21] for isotropic aggregates of triclinic crystallites, and by Johnson [13, 22] for transversally-isotropic and for orthotropic aggregates of cubic crystallites. (Some of the calculated results reported by Johnson [13] are erroneous; see Remark 6.1 below.)

Until the early eighties the dominant approach in acoustoelasticity was based on the continuum theory of small elastic motions (waves) superimposed on a large deformation of a hyperelastic material body, where the large deformation in question takes the body from a natural state to the initial configuration (cf. [6, 7, 23, 24]). This same approach was adopted by all researchers who use (48) as the constitutive equation of the polycrystal B at its natural state. Let $\boldsymbol{\kappa}^N(B)$ and $\boldsymbol{\kappa}(B)$ be the natural state and the initial configuration in question, respectively. Let $\overset{\diamond}{\mathbf{F}} = \overset{\diamond}{\mathbf{V}}\overset{\diamond}{\mathbf{R}}$ be the gradient of the map that takes $\boldsymbol{\kappa}^N(B)$ to $\boldsymbol{\kappa}(B)$, and $\overset{\diamond}{\mathbf{E}} = \frac{1}{2}(\overset{\diamond}{\mathbf{F}}{}^{\mathrm{T}}\overset{\diamond}{\mathbf{F}} - \mathbf{I})$ and $\overset{\diamond}{\mathbf{H}} = \overset{\diamond}{\mathbf{F}} - \mathbf{I}$ be the Lagrangian strain and the displacement gradient associated with the deformation gradient $\overset{\diamond}{\mathbf{F}}$, respectively. If, as we have done throughout this paper, we let $\mathbf{e}$ be an infinitesimal strain superimposed on the initial configuration $\boldsymbol{\kappa}(B)$, and $\boldsymbol{\sigma}$ be the second Piola–Kirchhoff stress, we find, after doing an analysis similar to that of [14, Section 2, equations (17), (20)] (see also [25, equation (3.10)]),

$$\boldsymbol{\sigma} = \mathring{\mathbf{T}} + \big(\det\overset{\diamond}{\mathbf{F}}\big)^{-1}\overset{\diamond}{\mathbf{F}}\big(\mathbb{C}^N\big[\overset{\diamond}{\mathbf{F}}{}^{\mathrm{T}}\mathbf{e}\overset{\diamond}{\mathbf{F}}\big] + \mathbb{B}^N\big[\overset{\diamond}{\mathbf{E}}, \overset{\diamond}{\mathbf{F}}{}^{\mathrm{T}}\mathbf{e}\overset{\diamond}{\mathbf{F}}\big]\big)\overset{\diamond}{\mathbf{F}}{}^{\mathrm{T}}, \tag{51}$$

and (see [14, equation (16)])

$$\mathring{\mathbf{T}} = (\det\overset{\diamond}{\mathbf{F}})^{-1}\overset{\diamond}{\mathbf{F}}\Big(\mathbb{C}^N[\overset{\diamond}{\mathbf{E}}] + \frac{1}{2}\mathbb{B}^N[\overset{\diamond}{\mathbf{E}}, \overset{\diamond}{\mathbf{E}}]\Big)\overset{\diamond}{\mathbf{F}}{}^{\mathrm{T}}. \tag{52}$$

Putting $\overset{\diamond}{\mathbf{F}} = \mathbf{I} + \overset{\diamond}{\mathbf{H}}$ and $\overset{\diamond}{\mathbf{e}} = (\overset{\diamond}{\mathbf{H}} + \overset{\diamond}{\mathbf{H}}{}^{\mathrm{T}})/2$ in equations (51) and (52), we obtain after discarding the $o(\|\overset{\diamond}{\mathbf{H}}\|)$ terms the equations

$$\begin{aligned}\boldsymbol{\sigma} &= \mathring{\mathbf{T}} + \mathbb{C}^N[\mathbf{e}] - \big(\operatorname{tr}\overset{\diamond}{\mathbf{e}}\big)\mathbb{C}^N[\mathbf{e}] + \overset{\diamond}{\mathbf{H}}\mathbb{C}^N[\mathbf{e}] + \mathbb{C}^N[\mathbf{e}]\overset{\diamond}{\mathbf{H}}{}^{\mathrm{T}} \\ &\quad + \mathbb{C}^N\big[\mathbf{e}\overset{\diamond}{\mathbf{H}} + \overset{\diamond}{\mathbf{H}}{}^{\mathrm{T}}\mathbf{e}\big] + \mathbb{B}^N\big[\overset{\diamond}{\mathbf{e}}, \mathbf{e}\big],\end{aligned} \tag{53}$$

and

$$\mathring{\mathbf{T}} = \mathbb{C}^N\big[\overset{\diamond}{\mathbf{e}}\big]. \tag{54}$$

Equations (53) and (54) are used in the review of acoustoelasticity by Pao et al. [8]. In their review Pao et al., however, never appeal to any micromechanical modelling; there, $\mathbb{C}^N$ and $\mathbb{B}^N$ are phenomenological constitutive functions. If we substitute equations (49) and (50) into equations (53) and (54), we arrive at an approximate and specialized version of our second model. In our second model each grain in the initial configuration is related to its preimage in the reference perfect crystal by a small elastic Lagrangian strain and a finite rotation. In equations (53) and (54), the initial configuration of the polycrystal differs from its given natural configuration by an infinitesimal elastic displacement.

EXAMPLE 5.1. To give one illustration that our second model is more general than equations (53) and (54), let us consider the special instance where $\overset{\diamond}{\mathbf{F}} = \mathbf{Q}$ is a pure rotation. Then $\overset{\circ}{\mathbf{T}} = \mathbf{0}$, $\overset{\diamond}{\mathbf{E}} = \mathbf{0}$, $\overset{\diamond}{\mathbf{H}} = \overset{\diamond}{\mathbf{W}} \equiv (\overset{\diamond}{\mathbf{H}} - \overset{\diamond}{\mathbf{H}}^{\mathrm{T}})/2$, and equation (53) becomes

$$\boldsymbol{\sigma} = \mathbb{C}^N[\mathbf{e}] + \overset{\diamond}{\mathbf{W}}\mathbb{C}^N[\mathbf{e}] - \mathbb{C}^N[\mathbf{e}]\overset{\diamond}{\mathbf{W}} + \mathbb{C}^N\big[\mathbf{e}\overset{\diamond}{\mathbf{W}} - \overset{\diamond}{\mathbf{W}}\mathbf{e}\big], \tag{55}$$

where

$$\mathbb{C}^N[\mathbf{e}] = \int_{\mathcal{G}} \mathbf{R}\mathbb{C}^{\circ}\big[\mathbf{R}^{\mathrm{T}}\mathbf{e}\mathbf{R}\big]\mathbf{R}^{\mathrm{T}}\,\mathrm{d}\wp^N(\mathbf{R})$$

refers to the texture at the natural configuration. On the other hand, under the second model

$$\boldsymbol{\sigma} = \int_{\mathcal{G}} \mathbf{R}\mathbb{C}^{\circ}\big[\mathbf{R}^{\mathrm{T}}\mathbf{e}\mathbf{R}\big]\mathbf{R}^{\mathrm{T}}\,\mathrm{d}\wp(\mathbf{R}),$$

which refers to the texture at the initial configuration. As $\wp$ and $\wp^N$ are related by $\wp(A) = \wp^N(\mathbf{Q}^{\mathrm{T}}A)$ for each Borel set A in SO(3), we obtain after some easy manipulations that $\boldsymbol{\sigma} = \mathbf{Q}\mathbb{C}^N[\mathbf{Q}^{\mathrm{T}}\mathbf{e}\mathbf{Q}]\mathbf{Q}^{\mathrm{T}}$ in our second model. This expression reduces to equation (55) if we put $\mathbf{Q} = \mathbf{I} + \overset{\diamond}{\mathbf{W}} + o(\|\overset{\diamond}{\mathbf{W}}\|)$ and drop all terms of order $o(\|\overset{\diamond}{\mathbf{W}}\|)$. By restricting attention to the initial configuration our second model automatically takes full account of the rotation $\mathbf{Q}$, whereas equation (55) reflects the effect of $\mathbf{Q}$ on the texture at the natural state only up to terms of order $o(\|\overset{\diamond}{\mathbf{W}}\|)$.

In his phenomenological theory Wallace ([26, p. 339]) restricts attention to the case $\overset{\diamond}{\mathbf{R}} = \mathbf{I}$ so that $\overset{\diamond}{\mathbf{H}} = \overset{\diamond}{\mathbf{E}} + o(\|\overset{\diamond}{\mathbf{E}}\|)$. Substituting $\overset{\diamond}{\mathbf{F}} = \mathbf{I} + \overset{\diamond}{\mathbf{E}} + o(\|\overset{\diamond}{\mathbf{E}}\|)$ in equations (51) and (52) and discarding the $o(\|\overset{\diamond}{\mathbf{E}}|)$ terms, we obtain

$$\begin{aligned}\boldsymbol{\sigma} &= \overset{\circ}{\mathbf{T}} + \mathbb{C}^N[\mathbf{e}] - \big(\operatorname{tr}\overset{\diamond}{\mathbf{E}}\big)\mathbb{C}^N[\mathbf{e}] + \overset{\diamond}{\mathbf{E}}\mathbb{C}^N[\mathbf{e}] + \mathbb{C}^N[\mathbf{e}]\overset{\diamond}{\mathbf{E}} \\ &\quad + \mathbb{C}^N\big[\mathbf{e}\overset{\diamond}{\mathbf{E}} + \overset{\diamond}{\mathbf{E}}\mathbf{e}\big] + \mathbb{B}^N\big[\overset{\diamond}{\mathbf{E}}, \mathbf{e}\big],\end{aligned} \tag{56}$$

and

$$\overset{\diamond}{\mathbf{E}} = \mathbb{K}^N\big(\overset{\circ}{\mathbf{T}}\big), \tag{57}$$

where $\mathbb{K}^N = (\mathbb{C}^N)^{-1}$. Sayers and Allen [27] adopt equations (56) and (57) from [26]. They take equation (49) as the expression for $\mathbb{C}^N$, which has been evaluated by Sayers [12] earlier. For them, however, $\mathbb{B}^N$ remains a phenomenological constitutive function. Comparing equations (49), (50), (56) and (57) with equations (8), (18) and (19), we see that should Sayers and Allen adopt equation (50) as the expression for $\mathbb{B}^N$ and evaluate it, their formulae would differ from the corresponding formulae of our second model only in the orientation measure used. In fact, in our second model we use the orientation measure of the initial configuration while the formulae of Sayers and Allen refer to the orientation measure of the perhaps fictitious natural configuration. Supposing that there indeed exists a natural configuration, we can make the two models formally equivalent by assuming that the orientation measure of the natural and initial configurations to be identically equal, i.e., $\wp^N = \wp$, and by making the Voigt assumption for the deformation from the natural state $\boldsymbol{\kappa}^N$ to the initial state $\boldsymbol{\kappa}$, i.e., in this deformation each grain pertaining to the same material point X undergoes the deformation $\overset{\diamond}{\mathbf{F}}(X)$. From the Voigt assumption we deduce that $\mathbf{P}_\alpha = \overset{\diamond}{\mathbf{F}}\mathbf{R}_\alpha$ (recall that $\mathbf{P}_\alpha$ was defined in equation (1)), and since we have supposed $\mathring{\mathbf{R}} = \mathbf{I}$ we have $\mathbf{P}_\alpha = \mathring{\mathbf{V}}\mathbf{R}_\alpha$. On the other hand, in the second model, we have $\mathbf{P}_\alpha = \mathring{\mathbf{V}}\mathring{\mathbf{R}}_\alpha$. From the uniqueness of the polar decomposition we obtain $\mathring{\mathbf{V}} = \overset{\diamond}{\mathbf{V}}$ and $\mathring{\mathbf{R}}_\alpha = \mathbf{R}_\alpha$, which shows that under the assumptions above Sayers and Allen would have arrived at the same formulae delivered by our second model if they evaluated $\mathbb{B}^N$ in equation (50). Even so, there remain basic differences in physical interpretation though. In the approach of Sayers and Allen [27] à la Wallace [26], the initial configuration is postulated to be the result of a small elastic stretch imposed on some natural state of the polycrystalline aggregate. Our second model neither refers to a natural configuration nor postulates the origin of the initial configuration of the polycrystal. While it allows what postulated in the other approach as a distinct possibility, any assumption regarding the origin of the initial configuration is simply unnecessary for the model itself. Indeed the validity of our second model remains a logical possibility even if individual grains in the polycrystal have had undergone complex thermomechanical histories before they arrive at their initial configuration.

Equation (51) served as a theoretical building block in the studies of Johnson and Springer [28, 29] where they compared the measured and predicted second- and third-order elastic constants for a rolled plate of 7039-T64 aluminum exhibiting orthorhombic texture. Equations (49) and (50) were taken by Johnson and Springer as the expressions for $\mathbb{C}^N$ and $\mathbb{B}^N$ in equation (51), respectively. Unlike Sayers and Allen [27], Johnson [13] had already evaluated $\mathbb{B}^N$ before this work. Unfortunately there were errors (see Remark 6.1 below) in the results on $\mathbb{B}^N$ reported by Johnson [13], and it is unclear whether those errors adversely affected the studies of Johnson and Springer. In their experiments the initial deformations were pure stretches, and obviously they assumed that the crystallographic textures of their samples remained unchanged during the elastic loadings which created the initially stressed configurations. Thus their theory reduces exactly to the one described in the preceding paragraph and becomes formally the same as our second model. Our comments at the end of the preceding paragraph as to why our second model should be regarded as physically more general apply verbatim.

6. Indifference to Rotation of Reference Placement

Throughout the rest of this paper we suppose the orientation measure $\wp$ to be absolutely continuous (cf. [30, 31]) with respect to the Haar measure $\wp_H$ on $\mathcal{G}$, so that we can define the orientation distribution function (ODF)

$$w = \frac{1}{8\pi^2}\frac{\mathrm{d}\wp}{\mathrm{d}\wp_H}, \tag{58}$$

where $\mathrm{d}\wp/\mathrm{d}\wp_H$ denotes the Radon–Nikodym derivative of $\wp$ with respect to $\wp_H$. It is common practice (cf. [10, 11]) to expand the orientation distribution function w as an infinite series of the generalized spherical harmonics

$$w(\psi, \theta, \phi) = \frac{1}{8\pi^2} + \sum_{l=1}^{\infty}\sum_{m=-l}^{l}\sum_{n=-l}^{l} W_{lmn} Z_{lmn}(\cos\theta)\mathrm{e}^{-im\psi}\mathrm{e}^{-in\phi}, \tag{59}$$

where Z_{lmn} are the augmented Jacobi polynomials (cf. [32], where an explicit formula of Z_{lmn} is displayed). This representation of w is particularly useful for our present purpose, for it turns out that both $\mathcal{L}(\mathring{\mathbf{T}}, \wp)$ and $\widehat{\mathcal{L}}(\mathring{\mathbf{T}}, \wp)$, the incremental elasticity tensors under our two models, can depend only on those W_{lmn} coefficients with $l \leqslant 6$. If the crystallites have no preferred orientations, all the W_{lmn} with $l \geqslant 1$ are zero and $w = 1/(8\pi^2)$; hereafter we write $w_{\mathrm{iso}} = 1/(8\pi^2)$. We call W_{lmn} $(l \geqslant 1)$ the texture coefficients.

In the next lemmas we prove that both of our models deliver constitutive equations which are indifferent to rotation of reference placement, a condition exploited earlier in a special context by Man [32] and defined in a more general context by Paroni and Man [15]. In what follows, for a given $\mathbf{Q} \in \mathcal{G}$, we denote by $\mathcal{T}_{\mathbf{Q}}w$ the ODF defined by

$$\mathcal{T}_{\mathbf{Q}}w(\mathbf{R}) = w(\mathbf{Q}^{\mathrm{T}}\mathbf{R}) \tag{60}$$

for each $\mathbf{R} \in \mathcal{G}$. To proceed, first we replace the orientation measure $\wp$ by the ODF w in the expressions for the sixth-order tensors $\mathbb{D}$ and $\mathbb{B}$: (see equations (9), (17) and (58))

$$\mathbb{D}(w)[\mathring{\mathbf{T}}, \mathbf{e}] = 8\pi^2 \int_{\mathcal{G}} \mathfrak{D}(\mathbb{K}^{\circ}[\mathbf{R}^{\mathrm{T}}\mathring{\mathbf{T}}\mathbf{R}], \mathbf{R})[\mathbf{e}]w(\mathbf{R})\,\mathrm{d}\wp_H(\mathbf{R}), \tag{61}$$

$$\mathbb{B}(w)[\mathring{\mathbf{E}}^{*}, \mathbf{e}] = 8\pi^2 \int_{\mathcal{G}} \mathfrak{D}(\mathbf{R}^{\mathrm{T}}\mathring{\mathbf{E}}^{*}\mathbf{R}, \mathbf{R})[\mathbf{e}]w(\mathbf{R})\,\mathrm{d}\wp_H(\mathbf{R}). \tag{62}$$

LEMMA 6.1. *For every* $\mathbf{a}, \mathbf{b} \in \mathrm{Sym}$ *and* $\mathbf{Q} \in \mathcal{G}$, *the identities*

$$\mathbb{D}(\mathcal{T}_{\mathbf{Q}}w)[\mathbf{Q}\mathbf{a}\mathbf{Q}^{\mathrm{T}}, \mathbf{Q}\mathbf{b}\mathbf{Q}^{\mathrm{T}}] = \mathbf{Q}\mathbb{D}(w)[\mathbf{a}, \mathbf{b}]\mathbf{Q}^{\mathrm{T}}, \tag{63}$$

$$\mathbb{B}(\mathcal{T}_{\mathbf{Q}}w)[\mathbf{Q}\mathbf{a}\mathbf{Q}^{\mathrm{T}}, \mathbf{Q}\mathbf{b}\mathbf{Q}^{\mathrm{T}}] = \mathbf{Q}\mathbb{B}(w)[\mathbf{a}, \mathbf{b}]\mathbf{Q}^{\mathrm{T}}, \tag{64}$$

hold.

Proof. Fix $\mathbf{a}, \mathbf{b} \in \mathrm{Sym}$ and $\mathbf{Q} \in \mathcal{G}$. From equation (4) we can easily see that

$$\mathfrak{D}(\mathbf{c}, \mathbf{QR})[\mathbf{b}] = \mathbf{Q}\mathfrak{D}(\mathbf{c}, \mathbf{R})\big[\mathbf{Q}^{\mathrm{T}}\mathbf{bQ}\big]\mathbf{Q}^{\mathrm{T}} \tag{65}$$

for every $\mathbf{c} \in \mathrm{Sym}$. By appealing to the invariance of the Haar measure, we have

$$\begin{aligned}
&\mathbb{D}(\mathcal{T}_{\mathbf{Q}}w)\big[\mathbf{QaQ}^{\mathrm{T}}, \mathbf{QbQ}^{\mathrm{T}}\big] \\
&\quad = 8\pi^2 \int_{\mathcal{G}} \mathfrak{D}\big(\mathbb{K}^{\circ}\big[\mathbf{R}^{\mathrm{T}}\mathbf{QaQ}^{\mathrm{T}}\mathbf{R}\big], \mathbf{R}\big)\big[\mathbf{QbQ}^{\mathrm{T}}\big]w\big(\mathbf{Q}^{\mathrm{T}}\mathbf{R}\big)\, \mathrm{d}p_H(\mathbf{R}) \\
&\quad = 8\pi^2 \int_{\mathcal{G}} \mathfrak{D}\big(\mathbb{K}^{\circ}\big[\mathbf{R}^{\mathrm{T}}\mathbf{aR}\big], \mathbf{QR}\big)\big[\mathbf{QbQ}^{\mathrm{T}}\big]w(\mathbf{R})\, \mathrm{d}p_H(\mathbf{R}) \\
&\quad = 8\pi^2 \int_{\mathcal{G}} \mathbf{Q}\mathfrak{D}\big(\mathbb{K}^{\circ}\big[\mathbf{R}^{\mathrm{T}}\mathbf{aR}\big], \mathbf{R}\big)[\mathbf{b}]\mathbf{Q}^{\mathrm{T}}w(\mathbf{R})\, \mathrm{d}p_H(\mathbf{R}) \\
&\quad = \mathbf{Q}\mathbb{D}(w)[\mathbf{a}, \mathbf{b}]\mathbf{Q}^{\mathrm{T}}.
\end{aligned}$$

Equation (64) can be proved in a similar way. □

LEMMA 6.2. *Let us denote (see equations* (8) *and* (58)*), for each* $\mathbf{a} \in \mathrm{Sym}$,

$$\mathbb{C}(w)[\mathbf{a}] = 8\pi^2 \int_{\mathcal{G}} \mathbf{R}\mathbb{C}^{\circ}\big[\mathbf{R}^{\mathrm{T}}\mathbf{aR}\big]\mathbf{R}^{\mathrm{T}}w(\mathbf{R})\, \mathrm{d}p_H(\mathbf{R}), \tag{66}$$

and let $\mathbb{K}(w) = (\mathbb{C}(w))^{-1}$. *Then*

$$\mathbb{C}(\mathcal{T}_{\mathbf{Q}}w)\big[\mathbf{QaQ}^{\mathrm{T}}\big] = \mathbf{Q}\mathbb{C}(w)[\mathbf{a}]\mathbf{Q}^{\mathrm{T}}, \tag{67}$$

$$\mathbb{K}(\mathcal{T}_{\mathbf{Q}}w)\big[\mathbf{QaQ}^{\mathrm{T}}\big] = \mathbf{Q}\mathbb{K}(w)[\mathbf{a}]\mathbf{Q}^{\mathrm{T}}, \tag{68}$$

for every $\mathbf{Q} \in \mathcal{G}$.

Proof. The proof of equation (67), which is similar to the proof of Lemma 6.1, has been given by Man [32]. We now prove equation (68). Let $\mathbb{I}$ be the fourth-order identity tensor. Let $\mathbf{Q} \in \mathcal{G}$ and $\mathbb{Q} = \mathbf{Q} \boxtimes \mathbf{Q}$. Then, using equation (67), we have $\mathbb{I} = \mathbb{K}(w)\mathbb{C}(w) = \mathbb{K}(w)\mathbb{Q}^{\mathrm{T}}\mathbb{C}(\mathcal{T}_{\mathbf{Q}}w)\mathbb{Q}$, and hence $\mathbb{Q}^{\mathrm{T}} = \mathbb{K}(w)\mathbb{Q}^{\mathrm{T}}\mathbb{C}(\mathcal{T}_{\mathbf{Q}}w)$. Multiplying both sides of this identity on the right by $\mathbb{K}(\mathcal{T}_{\mathbf{Q}}w)$, we obtain equation (68). □

LEMMA 6.3. *Let us denote (see equation* (22)*)*

$$\widehat{\mathbb{D}}(w)[\mathbf{a}, \mathbf{b}] = \mathbb{B}(w)\big[\mathbb{K}(w)[\mathbf{a}], \mathbf{b}\big], \tag{69}$$

for each $\mathbf{a}, \mathbf{b} \in \mathrm{Sym}$. *Then*

$$\widehat{\mathbb{D}}(\mathcal{T}_{\mathbf{Q}}w)\big[\mathbf{QaQ}^{\mathrm{T}}, \mathbf{QbQ}^{\mathrm{T}}\big] = \mathbf{Q}\widehat{\mathbb{D}}(w)[\mathbf{a}, \mathbf{b}]\mathbf{Q}^{\mathrm{T}}, \tag{70}$$

for each $\mathbf{Q} \in \mathcal{G}$ *and each* $\mathbf{a}, \mathbf{b} \in \mathrm{Sym}$.

Proof. Fix $\mathbf{a}, \mathbf{b} \in \mathrm{Sym}$, and $\mathbf{Q} \in \mathcal{G}$. Then using equations (64) and (68), we find

$$\begin{aligned}\widehat{\mathbb{D}}(\mathcal{T}_{\mathbf{Q}}w)\big[\mathbf{QaQ}^{\mathrm{T}}, \mathbf{QbQ}^{\mathrm{T}}\big] &= \mathbb{B}(\mathcal{T}_{\mathbf{Q}}w)\big[\mathbb{K}(\mathcal{T}_{\mathbf{Q}}w)\big[\mathbf{QaQ}^{\mathrm{T}}\big], \mathbf{QbQ}^{\mathrm{T}}\big] \\ &= \mathbb{B}(\mathcal{T}_{\mathbf{Q}}w)\big[\mathbf{Q}\mathbb{K}(w)[\mathbf{a}]\mathbf{Q}^{\mathrm{T}}, \mathbf{QbQ}^{\mathrm{T}}\big] \\ &= \mathbf{Q}\mathbb{B}(w)\big[\mathbb{K}(w)[\mathbf{a}], \mathbf{b}\big]\mathbf{Q}^{\mathrm{T}} \\ &= \mathbf{Q}\widehat{\mathbb{D}}(w)[\mathbf{a}, \mathbf{b}]\mathbf{Q}^{\mathrm{T}}. \qquad \square\end{aligned}$$

REMARK 6.1. Let us suppose that the measure $\wp^N$ considered in Section 5 is absolutely continuous with respect to the Haar measure $\wp_H$, so that

$$w^N = \frac{1}{8\pi^2}\frac{\mathrm{d}\wp^N}{\mathrm{d}\wp_H}$$

is the ODF associated with the measure $\wp^N$. Let us also rewrite $\mathbb{B}^N$ (cf. equation (50)) as

$$\mathbb{B}^N(w^N)[\mathbf{a}, \mathbf{a}] = 8\pi^2 \int_{\mathcal{G}} \mathbf{R}\mathbb{D}^{\circ}\big[\mathbf{R}^{\mathrm{T}}\mathbf{aR}, \mathbf{R}^{\mathrm{T}}\mathbf{aR}\big]\mathbf{R}^{\mathrm{T}} w^N(\mathbf{R})\, \mathrm{d}\wp_H(\mathbf{R}). \tag{71}$$

Then, following the proof of Lemma 6.1, we can show that

$$\mathbb{B}^N(\mathcal{T}_{\mathbf{Q}}w^N)\big[\mathbf{QaQ}^{\mathrm{T}}, \mathbf{QaQ}^{\mathrm{T}}\big] = \mathbf{Q}\mathbb{B}^N(w^N)[\mathbf{a}, \mathbf{a}]\mathbf{Q}^{\mathrm{T}}, \tag{72}$$

for every $\mathbf{a} \in \mathrm{Sym}$. As mentioned in Section 5, the integrals in equation (71) were computed by Johnson [13] for the case of an orthorhombic aggregate of cubic crystallites. Some of the results reported by Johnson [13], however, do not satisfy equation (72). For instance, take $\mathbf{Q}$ to be the rotation about the 2-axis by $\pi/2$ and use equation (72) to put in relation the 112323 and the 331212 components of $\mathbb{B}^N$. The relation that follows is different from the one reported in [13, Table 4]. Hence, we must conclude that some of the calculations reported by Johnson [13] were erroneous.

7. Numerical Comparison of the Two Models

Henceforth we restrict our attention to the case of an aggregate of cubic crystallites with orthorhombic texture symmetry. We find it convenient to choose a coordinate system whose coordinate axes are the axes of orthorhombic symmetry of the texture, and we pick for the crystallites a reference orientation whose three four-fold axes of cubic symmetry fall on the axes of the coordinate system.

From equation (67) and the linearity of $\mathbb{C}(w)$ in w, it follows (see [32]) that $\mathbb{C}(w)$ can be written as

$$\mathbb{C}(w) = \mathbb{C}_{\mathrm{iso}} + \gamma_0 \mathbf{\Phi}^{(0)}(w), \tag{73}$$

where $\mathbb{C}_{\mathrm{iso}}$ is a constant isotropic fourth-order tensor, γ_0 a constant and $\mathbf{\Phi}^{(0)}(w)$ a fourth-order tensor which depends only on w (see [32] for the definition of $\mathbf{\Phi}^{(0)}(w)$, where $\mathbf{\Phi}^{(0)}$ and γ_0 are denoted there by $\mathbf{\Phi}$ and c, respectively). The isotropic tensor $\mathbb{C}_{\text{iso}}$ is determined by its Lamé constants λ and μ. Under both our models, expressions of λ, μ, and γ_0 in terms of single-crystal elastic constants (i.e., those pertaining to $\mathbb{C}^{\circ}$) are the same as those given by the Voigt model [12, 33]:

$$\lambda = \frac{c_{11} + 4c_{12} - 2c_{44}}{5}, \quad \mu = \frac{c_{11} - c_{12} + 3c_{44}}{5}, \quad \gamma_0 = c_{11} - c_{12} - 2c_{44}.$$

As $\mathbb{D}(w)$ is linear in w and satisfies equation (63), we may appeal to [15, Theorem 3.3.1] (cf. also [34] for another statement of this theorem) to write

$$\mathbb{D}(w) = \mathbb{D}_{\text{iso}} + \sum_{p=1}^{5} \gamma_p \mathbf{\Phi}^{(p)}(w). \tag{74}$$

Here $\mathbb{D}_{\text{iso}}$ is a constant sixth-order isotropic tensor; it is what the sixth-order elasticity tensor would be if the crystallites are randomly oriented. By a well-known representation theorem of isotropic tensors (see, for example, [2, 35]), we can write

$$\begin{aligned} \mathbb{D}_{\text{iso}}[\mathbf{a}, \mathbf{b}] &= \beta_1 \text{tr}(\mathbf{a})\text{tr}(\mathbf{b})\mathbf{I} + \beta_2 \text{tr}(\mathbf{a})\mathbf{b} + \beta_3\big(\text{tr}(\mathbf{b})\mathbf{a} + \text{tr}(\mathbf{ab})\mathbf{I}\big) \\ &\quad + \beta_4(\mathbf{ab} + \mathbf{ba}), \end{aligned} \tag{75}$$

where the β_i's are material constants. The $\mathbf{\Phi}^{(p)}(w)$'s are explicitly determined sixth-order orthotropic tensors. For $p = 1, \ldots, 4$, the components of $\mathbf{\Phi}^{(p)}(w)$ are linear combinations of the texture coefficients W_{400}, W_{420}, and W_{440}; the components of $\mathbf{\Phi}^{(5)}(w)$ are linear combinations of W_{600}, W_{620}, W_{640}, and W_{660}. The coefficients of all these linear combinations are tabulated in the paper of Paroni and Man [15]. The γ_p's in equation (74) are material constants; they are related to the coefficients a_1, e_1, e_2, d and a_2 of [15] by $\gamma_1 = a_1$, $\gamma_2 = e_1$, $\gamma_3 = e_2$, $\gamma_4 = d_1$ and $\gamma_5 = a_2$.

Representation formula (74) tells us that we do not need to carry out all the integrations appearing in equation (9) to compute $\mathbb{D}(w)$. It suffices to evaluate by integration only the material constants β_i ($i = 1, \ldots, 4$) and γ_p ($p = 1, \ldots, 5$). To this end we just need to compute four suitable components of $\mathbb{D}(w)$ with $w = w_{\text{iso}}$ to get the β_i's, and three suitable components of $\mathbb{D}(w)$ to obtain the γ_p's. The resulting expressions which relate the β_i's and the γ_p's to the single-crystal elastic constants are reported in the Appendix. Under the first model, the coefficients s_{11}, s_{12}, and s_{44} in those expressions are single-crystal elastic compliances; they are related to the elastic stiffnesses through the following formulae:

$$\begin{aligned} s_{11} &= \frac{c_{11} + c_{12}}{(c_{11} - c_{12})(c_{11} + 2c_{12})}, \\ s_{12} &= \frac{-c_{12}}{(c_{11} - c_{12})(c_{11} + 2c_{12})}, \\ s_{44} &= \frac{1}{c_{44}}. \end{aligned}$$

We now turn our attention to the tensors pertaining to the second model. From equation (62), we see that $\mathbb{B}(w)$ depends linearly on the ODF w. Since $\mathbb{B}(w)$ satisfies equation (64), we can again appeal to [15, Theorem 3.3.1] to write

$$\mathbb{B}(w) = \mathbb{B}_{\text{iso}} + \sum_{p=1}^{5} \widetilde{\gamma}_p \mathbf{\Phi}^{(p)}(w), \tag{76}$$

where $\mathbb{B}_{\text{iso}}$ is given by the same representation formula (75) as $\mathbb{D}_{\text{iso}}$ with the β_i's replaced by $\widetilde{\beta}_i$'s. By carrying out the integration on the right hand side of equation (62), we determine the $\widetilde{\beta}_i$'s and the $\widetilde{\gamma}_p$'s; they are formally given by the same expressions for the β_i's and γ_p's in the Appendix, but the coefficients s_{11}, s_{12}, and s_{44} are replaced by the constants

$$\tilde{s}_{11} = 1, \qquad \tilde{s}_{12} = 0, \quad \text{and} \quad \tilde{s}_{44} = 2,$$

respectively. Now that we know $\mathbb{B}(w)$, we can compute $\mathbb{K}(w)$ by inverting equation (73) and evaluating $\widehat{\mathbb{D}}(w)$ from equation (69). We note, though, that $\widehat{\mathbb{D}}(w)$ is not linear in w. To simplify the comparison of the two models, henceforth we assume the texture to be weak. Then from equation (70) and the discussions on linearization given in [15, 32], we can express $\widehat{\mathbb{D}}(w)$ in the form

$$\widehat{\mathbb{D}}(w) = \widehat{\mathbb{D}}_{\text{iso}} + \sum_{p=1}^{5} \widehat{\gamma}_p \mathbf{\Phi}^{(p)}(w) + o\big(\|w - w_{\text{iso}}\|\big), \tag{77}$$

where representation formula (75) can be applied to $\widehat{\mathbb{D}}_{\text{iso}}$, with the β_i's replaced by $\widehat{\beta}_i$'s. Hence, to compare the two models it suffices to compare the β_i's with the $\widehat{\beta}_i$'s, and the γ_p's with the $\widehat{\gamma}_p$'s.

We now linearize $\widehat{\mathbb{D}}(w)$. We start by evaluating $\mathbb{K}(w)$ up to $o(\|w - w_{\text{iso}}\|)$. From equation (73) we obtain

$$\begin{aligned}
\mathbb{K}(w) &= \big(\mathbb{C}(w)\big)^{-1} = \big(\mathbb{C}_{\text{iso}}\big(\mathbb{I} + \gamma_0 \mathbb{C}_{\text{iso}}^{-1} \mathbf{\Phi}^{(0)}(w)\big)\big)^{-1} \\
&= \big(\mathbb{I} - \gamma_0 \mathbb{C}_{\text{iso}}^{-1} \mathbf{\Phi}^{(0)}(w) + o\big(\|w - w_{\text{iso}}\|\big)\big)\mathbb{C}_{\text{iso}}^{-1} \\
&= \mathbb{C}_{\text{iso}}^{-1} - \mathbb{C}_{\text{iso}}^{-1} \mathbf{\Phi}^{(0)}(w) \mathbb{C}_{\text{iso}}^{-1} + o\big(\|w - w_{\text{iso}}\|\big).
\end{aligned} \tag{78}$$

Thus, from equations (69) and (76) we obtain

$$\begin{aligned}
\widehat{\mathbb{D}}(w)\big[\mathring{\mathbf{T}}, \cdot\big] &= \mathbb{B}_{\text{iso}}\big[\mathbb{C}_{\text{iso}}^{-1}\big[\mathring{\mathbf{T}}\big], \cdot\big] - \gamma_0 \mathbb{B}_{\text{iso}}\big[\mathbb{C}_{\text{iso}}^{-1} \mathbf{\Phi}^{(0)}(w) \mathbb{C}_{\text{iso}}^{-1}\big[\mathring{\mathbf{T}}\big], \cdot\big] \\
&\quad + \sum_{p=1}^{5} \widetilde{\gamma}_p \mathbf{\Phi}^{(p)}(w)\big[\mathbb{C}_{\text{iso}}^{-1}\big[\mathring{\mathbf{T}}\big], \cdot\big] + o\big(\|w - w_{\text{iso}}\|\big),
\end{aligned} \tag{79}$$

which is the linearized expression we have been looking for. The first term in equation (79) is the isotropic part, i.e., $\widehat{\mathbb{D}}_{\text{iso}}[\mathring{\mathbf{T}}, \cdot] = \mathbb{B}_{\text{iso}}[\mathbb{C}_{\text{iso}}^{-1}[\mathring{\mathbf{T}}], \cdot]$. The second and the third term together represent the linear orthotropic part of $\widehat{\mathbb{D}}(w)$; it is this latter part that determines the values of the $\widehat{\gamma}_p$'s. We refrain from displaying explicit expressions for the $\widehat{\beta}_i$'s and $\widehat{\gamma}_p$'s, which are too complicated to be reproduced here.

In Table I we list the single-crystal elastic constants of several materials at room temperature. The data on third-order elastic constants listed under Al (T) and Al (SR) for aluminum are taken from [36] and from [37], respectively; the second-order elastic constants are from [36]. The data for copper is taken from [21], the data for germanium from [38, pp. 9, 259], the data from sodium chloride from [38, pp. 30, 258], and the data for silicon from [38, pp. 11, 266]. All the second-order (SOEC) and third-order elastic constants (TOEC) in Table I are in units of GPa.

Table II displays the values of the β_i's and $\widehat{\beta}_i$'s for the materials in question as delivered by the first and the second model, respectively. In Table III we report the computed values of the γ_i's and $\widehat{\gamma}_i$'s for the two models. The values of β_i, $\widehat{\beta}_i$, γ_p, and $\widehat{\gamma}_p$ are pure numbers.

Tables II and III show that the polycrystal elastic constants predicted by the two models can be quite close (e.g., aluminum) or hugely different (e.g., copper). A glance at the single-crystal SOEC and TOEC data under Al (T) and Al (SR) in Table I and the corresponding predicted values of β_1 in Table II reveals yet another potential problem, namely that the predicted values of polycrystal material constants may be sensitive to the SOEC and TOEC of the crystallites used in the computations. In particular, as the TOEC are difficult to measure and the values reported in the literature for TOEC often carry relatively wide margins of error, the truthfulness of the predictions may be dubious.

With these questions in mind, let us compute for illustration the acoustoelastic birefringence for aluminum and for copper sheets. The sheets are assumed to be orthorhombic with the rolling direction (RD), the transverse direction (TD) and the normal direction (ND) defining the axes of symmetry. Following the convention set out at the beginning of this section, we choose a Cartesian coordinate system such that RD, TD, and ND agree with the 1-, 2-, and 3-coordinate axis, respectively. Suppose the sheets in question are prestressed with $\mathring{\mathbf{T}} = \mathrm{diag}(\mathring{T}_{11}, \mathring{T}_{22}, \mathring{T}_{33})$. Let V_{IJ} be the phase velocity of shear waves propagating in the I-direction with polarization in the J-direction, and let $\overline{V}_T = (V_{31}+V_{32})/2$. The quantity $(V_{31}-V_{32})/\overline{V}_T$

Table I.

GPa	c_{11}	c_{12}	c_{44}	d_{111}	d_{112}	d_{123}	d_{144}	d_{155}	d_{456}
Al (T)	106.75	60.41	28.34	−1076	−315	36	−23	−340	−30
Al (SR)	106.75	60.41	28.34	−1224	−373	25	−64	−368	−27
Cu	166.1	119.9	75.6	−1271	−814	−50	−3	−780	−95
Ge	129	48	67.1	−720	−380	−30	−10	−305	−45
NaCl	49.1	12.8	12.8	−843	−50	46	29	−60	26
Si	165	64	79.2	−795	−445	−75	15	−310	−86

Table II.

GPa	β_1	β_2	β_3	β_4	$\widehat{\beta}_1$	$\widehat{\beta}_2$	$\widehat{\beta}_3$	$\widehat{\beta}_4$
Al (T)	0.19	1.43	−1.34	−4.86	0.19	1.47	−1.34	−4.92
Al (SR)	0.89	0.96	−2.63	−4.54	0.89	1.00	−2.64	−4.60
Cu	−0.71	1.31	−0.41	−4.17	−0.53	1.54	−0.68	−5.09
Ge	−0.57	−1.01	−0.57	−0.50	−0.60	−0.85	−0.53	−0.73
NaCl	−0.17	−0.59	−0.95	−2.03	0.00	−0.30	−1.21	−2.48
Si	−0.97	−0.63	0.07	−0.56	−1.02	−0.47	0.14	−0.79

Table III.

GPa	γ_1	γ_2	γ_3	γ_4	γ_5	$\widehat{\gamma}_1$	$\widehat{\gamma}_2$	$\widehat{\gamma}_3$	$\widehat{\gamma}_4$	$\widehat{\gamma}_5$
Al (T)	64.08	−10.80	−9.50	−4.93	200.14	53.45	−6.51	−7.91	−5.63	199.94
Al (SR)	66.90	−10.87	−9.47	−4.43	193.62	56.44	−6.75	−7.94	−5.11	193.57
Cu	238.26	−80.51	−35.24	11.74	370.67	95.73	−25.24	−15.57	1.40	261.50
Ge	76.26	−10.62	−12.73	−8.65	97.14	57.66	−5.38	−10.02	−8.29	90.61
NaCl	−292.90	27.30	36.01	21.60	118.47	−298.93	25.44	36.16	23.22	95.08
Si	72.55	−7.93	−12.17	−10.94	76.05	57.77	−4.17	−9.94	−10.23	69.22

is called the acoustoelastic birefringence. We refer the reader to [14, Section 5], for the derivation of a formula which expresses the acoustoelastic birefringence in terms of the prestress and the incremental elasticity tensor in question. In what follows we shall use this formula to compute the acoustoelastic birefringence. (Note that the symbols β_j's and γ_k's in [14] should not be confused with the β_i's and γ_p's in the present paper; these two sets of symbols represent totally different physical quantities in the two papers.)

By using the polycrystal elastic constants deduced from the second model and from the data for aluminum given by Thomas [36], we find

$$\begin{aligned}\frac{V_{31}-V_{32}}{\overline{V}_T} &= -3.97W_{420}+\big(-0.047-0.30W_{400}+0.14W_{440}\\&\quad+0.95W_{600}+3.56W_{640}\big)\big(\mathring{T}_{11}-\mathring{T}_{22}\big)\\&\quad-(0.062W_{420}+2.60\,W_{620})\big(\mathring{T}_{11}+\mathring{T}_{22}\big)\\&\quad+(0.60W_{420}+5.20W_{620})\mathring{T}_{33}+\mathrm{o}\big(\|w-w_{\mathrm{iso}}\|\big),\end{aligned} \tag{80}$$

while using the data reported by Sarma and Reddy [37] we get

$$\frac{V_{31}-V_{32}}{\overline{V}_T} = -3.97W_{420}+(-0.044-0.28W_{400}+0.13W_{440}$$

$$+ 0.92W_{600} + 3.45W_{640})(\overset{\circ}{T}_{11} - \overset{\circ}{T}_{22})$$
$$- (0.015W_{420} + 2.52W_{620})(\overset{\circ}{T}_{11} + \overset{\circ}{T}_{22})$$
$$+ (0.60W_{420} + 5.03W_{620})\overset{\circ}{T}_{33} + o(\|w - w_{\text{iso}}\|); \tag{81}$$

here the stresses are in units of GPa. Substituting the texture coefficients measured by Johnson and Springer [29] into equations (80) and (81), we obtain

$$\frac{V_{31} - V_{32}}{\overline{V}_T} = -0.0171 - 0.032\overset{\circ}{T}_{11} + 0.060\overset{\circ}{T}_{22} - 0.026\overset{\circ}{T}_{33}, \tag{82}$$

and

$$\frac{V_{31} - V_{32}}{\overline{V}_T} = -0.0171 - 0.029\overset{\circ}{T}_{11} + 0.057\overset{\circ}{T}_{22} - 0.025\overset{\circ}{T}_{33}. \tag{83}$$

Equations (80)–(83) are close to those predicted by the first model (cf. [14, equations (73)–(76)]).

From the tables above we notice that the biggest discrepancy between the predictions of the two models is found in copper. As a consequence we also find a significant difference between the acoustoelastic birefringences predicted by the two models for copper. For completeness we report here the acoustoelasic birefringence for copper obtained from the first model,

$$\frac{V_{31} - V_{32}}{\overline{V}_T} = -19.40W_{420} + (-0.022 - 1.54W_{400} - 0.69W_{440}$$
$$+ 0.85W_{600} - 1.54W_{640})(\overset{\circ}{T}_{11} - \overset{\circ}{T}_{22})$$
$$- (0.23W_{420} + 2.32W_{620})(\overset{\circ}{T}_{11} + \overset{\circ}{T}_{22})$$
$$+ (1.29W_{420} + 4.64W_{620})\overset{\circ}{T}_{33}, \tag{84}$$

and the one obtained from the second model,

$$\frac{V_{31} - V_{32}}{\overline{V}_T} = -19.40W_{420} + (-0.023 - 0.37W_{400} - 0.11W_{440}$$
$$+ 0.60W_{600} + 2.24W_{640})(\overset{\circ}{T}_{11} - \overset{\circ}{T}_{22})$$
$$- (-0.12W_{420} + 1.63\, W_{620})(\overset{\circ}{T}_{11} + \overset{\circ}{T}_{22})$$
$$+ (0.57W_{420} + 3.27W_{620})\overset{\circ}{T}_{33} + o(\|w - w_{\text{iso}}\|). \tag{85}$$

At present we have no theoretical means to check on predictions pertaining to sixth-order tensors. For a material such as copper, where the two models lead to considerably different material constants in $\mathbb{D}(w)$ and $\widehat{\mathbb{D}}(w)$, the predicitions are practically useless.

For a material such as aluminum, where the predictions from the two models are close, there is still hope that they might indeed carry some value. Indeed we have pointed out previously [14, 39] that our first model seems to provide reasonably good predictions on the acoustoelastic coefficients of bulk waves in aluminum. As is explained elsewhere [34], predictions on acoustoelastic coefficients for Rayleigh waves in aluminum would provide a more stringent test. A recent experiment [40] conducted at Sandia National Laboratories (Livermore) showed that the predicted values of two Rayleigh-wave acoustoelastic coefficients from either model, which takes into account the effects of texture, are in good agreement with the measurements on two samples of 6061-T6 aluminum alloy.

Acknowledgement

The findings reported here were obtained in the course of work supported in part by grants from the National Science Foundation (Nos. CMS-9522829 and DMS-9803441) and by a DoD EPSCoR grant from AFOSR (No. F49620-98-1-0469).

Appendix

In the formulae below c_{11}, c_{12} and c_{44} are second-order elastic stiffnesses of the cubic single crystal in question, while $d_{111}, d_{112}, d_{144}, d_{155}, d_{123}$ and d_{456} are third-order stiffnesses. They refer to a Cartesian coordinate system whose coordinate axes agree with the three four-fold axes of cubic symmetry of the single crystal. For our first model, s_{11}, s_{12} and s_{44} are second-order elastic compliances.

$$\begin{aligned}
\beta_1 &= (-3c_{11} - 4c_{12} + 6c_{44} + d_{111} + 18d_{112} - 10d_{144} - 4d_{155} + 16d_{123})\frac{s_{11}}{35} \\
&\quad + (10c_{11} + 32c_{12} - 20c_{44} + 6d_{111} + 52d_{112} \\
&\quad - 4d_{144} - 24d_{155} + 12d_{123})\frac{s_{12}}{35} \\
&\quad + (-2c_{11} - 12c_{12} + 4c_{44} - 10d_{144} - 4d_{155} + 8d_{456})\frac{s_{44}}{35}, \\
\beta_2 &= (-6c_{11} + 6c_{12} - 2c_{44} + 2d_{111} + 8d_{112} + 22d_{144} + 20d_{155} - 10d_{123})\frac{s_{11}}{35} \\
&\quad + (20c_{11} - 20c_{12} + 44c_{44} + 12d_{111} - 8d_{112} + 20d_{144} \\
&\quad + 64d_{155} - 4d_{123})\frac{s_{12}}{35} \\
&\quad + (-4c_{11} + 4c_{12} - 20c_{44} + 8d_{144} - 8d_{155} - 12d_{456})\frac{s_{44}}{35}, \\
\beta_3 &= (8c_{11} + 20c_{12} - 16c_{44} + 2d_{111} + 8d_{112} + 8d_{144} - 8d_{155} - 10d_{123})\frac{s_{11}}{35} \\
&\quad + (-8c_{11} - 20c_{12} + 16c_{44} - 2d_{111} - 8d_{112} - 8d_{144} \\
&\quad + 8d_{155} + 10d_{123})\frac{s_{12}}{35} \\
&\quad + (3c_{11} + 18c_{12} - 6c_{44} + 15d_{144} + 6d_{155} - 12d_{456})\frac{s_{44}}{35},
\end{aligned}$$

$$
\begin{aligned}
\beta_4 &= (16c_{11} - 16c_{12} + 24c_{44} + 4d_{111} - 12d_{112} - 12d_{144} + 12d_{155} \\
&\quad + 8d_{123})\frac{s_{11}}{35} + (-16c_{11} + 16c_{12} - 24c_{44} - 4d_{111} + 12d_{112} + 12d_{144} \\
&\quad - 12d_{155} - 8d_{123})\frac{s_{12}}{35} \\
&\quad + (6c_{11} - 6c_{12} + 30c_{44} - 12d_{144} + 12d_{155} + 18d_{456})\frac{s_{44}}{35}, \\
\gamma_1 &= \frac{-32\sqrt{2}\pi^2}{385}(-49c_{11} + 5c_{12} + 10c_{44} - 15d_{111} + 12d_{112} \\
&\quad + 6d_{144} + 16d_{155} + 3d_{123})s_{11} \\
&\quad - \frac{64\sqrt{2}\pi^2}{385}(19c_{11} + 3c_{12} + 6c_{44} + 2d_{111} - 6d_{112} + 8d_{144} \\
&\quad + 14d_{155} + 4d_{123})s_{12} \\
&\quad - \frac{64\sqrt{2}\pi^2}{385}(4c_{11} + 7c_{12} + 14c_{44} + 3d_{144} + 8d_{155} + 6d_{456})s_{44}, \\
\gamma_2 &= \frac{16\sqrt{2}\pi^2}{385}(-5c_{11} + 5c_{12} + 32c_{44} - 4d_{111} + d_{112} - 5d_{144} + 27d_{155} \\
&\quad + 3d_{123})s_{11} + \frac{16\sqrt{2}\pi^2}{385}(-6c_{11} + 6c_{12} - 10c_{44} - 7d_{111} - d_{112} \\
&\quad + 27d_{144} + 17d_{155} + 8d_{123})s_{12} \\
&\quad - \frac{16\sqrt{2}\pi^2}{385}(3c_{11} - 3c_{12} + 5c_{44} - 6d_{144} + 6d_{155} - d_{456})s_{44}, \\
\gamma_3 &= \frac{8\sqrt{5}\pi^2}{385}(-13c_{11} + 13c_{12} + 4c_{44} - 6d_{111} + 7d_{112} + 9d_{144} \\
&\quad + 13d_{155} - d_{123})s_{11} \\
&\quad + \frac{8\sqrt{5}\pi^2}{385}(2c_{11} - 2c_{12} + 18c_{44} - 5d_{111} - 7d_{112} + 13d_{144} \\
&\quad + 31d_{155} + 12d_{123})s_{12} \\
&\quad + \frac{8\sqrt{5}\pi^2}{385}(c_{11} - c_{12} + 9c_{44} - 2d_{144} + 2d_{155} + 7d_{456})s_{44}, \\
\gamma_4 &= \frac{-4\sqrt{2}\pi^2}{385}(39c_{11} - 39c_{12} + 98c_{44} + 7d_{111} - 32d_{112} - 38d_{144} \\
&\quad + 60d_{155} + 25d_{123})s_{11} \\
&\quad + \frac{8\sqrt{2}\pi^2}{385}(25c_{11} - 25c_{12} + 38c_{44} + 9d_{111} - 16d_{112} - 30d_{144} \\
&\quad + 8d_{155} + 7d_{123})s_{12}
\end{aligned}
$$

$$
\begin{aligned}
&+\frac{8\sqrt{2}\pi^2}{385}(7c_{11}-7c_{12}+30c_{44}-14d_{144}+14d_{155}+16d_{456})s_{44},\\
\gamma_5 =\ &\frac{-64\sqrt{26}\pi^2}{1001}(-4c_{11}+4c_{12}+8c_{44}-d_{111}+3d_{112}-4d_{144}\\
&+4d_{155}-2d_{123})s_{11}\\
&+\frac{64\sqrt{26}\pi^2}{1001}(-4c_{11}+4c_{12}+8c_{44}-d_{111}+3d_{112}-4d_{144}\\
&+4d_{155}-2d_{123})s_{12}\\
&-\frac{128\sqrt{26}\pi^2}{1001}(c_{11}-c_{12}-2c_{44}-2d_{144}+2d_{155}-4d_{456})s_{44}.
\end{aligned}
$$

For our second model, the constants $\widetilde{\beta}_i$ and $\widetilde{\gamma}_p$ are given by the corresponding expression for β_i and for γ_p above, respectively, but the coefficients s_{11}, s_{12}, and s_{44} must be replaced by the constants $\tilde{s}_{11} = 1$, $\tilde{s}_{12} = 0$, and $\tilde{s}_{44} = 2$, respectively.

References

1. C. Truesdell, *The Mechanical Foundations of Elasticity and Fluid Dynamics*. Gordon and Breach, New York (1966).
2. C. Truesdell and W. Noll, *The Non-Linear Field Theories of Mechanics*. Springer, Berlin (1992).
3. C.-S. Man and W.Y. Lu, Towards an acoustoelastic theory for measurement of residual stress. *J. Elasticity* **17** (1987) 159–182.
4. D.S. Hughes and J.L. Kelly, Second-order elastic deformation of solids. *Phys. Rev.* **92** (1953) 1145–1149.
5. T.D. Murnaghan, *Finite Deformation of an Elastic Solid*. Wiley, New York (1951).
6. R.A. Toupin and B. Bernstein, Sound waves in deformed perfectly elastic materials. Acoustoelastic effect. *J. Acoust. Soc. Am.* **33** (1961) 216–225.
7. R.N. Thurston and K. Brugger, Third-order elastic constants and the velocity of small amplitude elastic waves in homogeneously stressed media. *Phys. Rev. A* **133** (1964) 1604–1610.
8. Y.-H. Pao, W. Sachse and H. Fukuoka, Acoustoelasticity and ultrasonic measurements of residual stress. In: W.P. Mason and R.N. Thurston (eds), *Physical Acoustics*, Vol. 17. Academic Press, Orlando (1984) pp. 61–143.
9. D.I. Crecraft, The measurement of applied and residual stresses in metals using ultrasonic waves. *J. Sound Vib.* **5** (1967) 173–192.
10. H.J. Bunge, *Texture Analysis in Materials Science*. Butterworth's, London (1982).
11. R.-J. Roe, Description of crystallite orientation in polycrystalline materials. III. General solution to pole figures. *J. Appl. Phys.* **36** (1965) 2024–2031.
12. C.M. Sayers, Ultrasonic velocities in anisotropic polycrystalline aggregates. *J. Phys. D* **15** (1982) 2157–2167.
13. G.C. Johnson, Acoustoelastic response of a polycrystalline aggregate with orthotropic texture. *ASME J. Appl. Mech.* **52** (1985) 659–663.
14. C.-S. Man and R. Paroni, On the separation of stress-induced and texture-induced birefringence in acoustoelasticity. *J. Elasticity* **45** (1996) 91–116.

15. R. Paroni and C.-S. Man, Constitutive equations of elastic polycrystalline materials. *Arch. Rational Mech. Anal.* **150** (1999) 153–177.
16. C. Truesdell and R. Toupin, The classical field theories. In: S. Flügge (ed.), *Encyclopedia of Physics*, Vol. III/1. Springer, Berlin (1960) pp. 226–793.
17. L.E. Malvern, *Introduction to the Mechanics of a Continuous Medium*. Prentice-Hall, Englewood Cliffs, NJ (1969).
18. P.R. Morris, Averaging fourth-rank tensors with weight functions. *J. Appl. Phys.* **40** (1969) 447–448.
19. H. Bross, Berechnung der elastischen konstanten dritter ordnung der alkalihalogenidkristalle. *Z. Physik* **175** (1963) 345–369.
20. R. Chang, Relationships between the nonlinear elastic constants of monocrystalline and polycrystalline solids of cubic symmetry. *Appl. Phys. Lett.* **11** (1967) 305–308.
21. G.R. Barsch, Relation between third-order elastic constants of single crystals and polycrystals. *J. Appl. Phys.* **39** (1968) 3780–3793.
22. G.C. Johnson, Acoustoelastic response of polycrystalline aggregates exhibiting transverse isotropy. *J. Nondestructive Evaluation* **3** (1982) 1–8.
23. T. Tokuoka and Y. Iwashimizu, Acoustical birefringence of ultrasonic waves in deformed isotropic elastic materials. *Internat. J. Solids Struct.* **4** (1968) 383–389.
24. T. Tokuoka and M. Saito, Elastic wave propagations and acoustical birefringence in stressed crystals. *J. Acoust. Soc. Am.* **45** (1968) 1241–1246.
25. P. Haupt, Y.-H. Pao and K. Hutter, Theory of incremental motion in a body with initial elasto-plastic deformation. *J. Elasticity* **28** (1992) 193–221.
26. D.C. Wallace, Thermoelastic theory of stressed crystals and higher-order elastic constants. In: H. Ehrenreich, F. Seitz and D. Turnbull (eds), *Solid State Physics*, Vol. 25. Academic Press, New York (1970) pp. 301–404.
27. C.M. Sayers and D.R. Allen, The influence of stress on the principal polarisation directions of ultrasonic shear waves in textured steel plates. *J. Phys. D* **17** (1984) 1399–1413.
28. G.C. Johnson and W.C. Springer, A comparison of predictions and measurements of the acoustoelastic response of a textured aggregate. In: D.O. Thompson and D.E. Chimenti (eds), *Review of Progress in Quantitative Nondestructive Evaluation*, Vol. 6B. Plenum, New York (1987) pp. 1495–1503.
29. G.C. Johnson and W.C. Springer, A comparison of measured and predicted second- and third-order elastic constants of a textured aggregate. *Internat. J. Solids Struct.* **25** (1989) 609–619.
30. W. Rudin, *Real and Complex Analysis*. McGraw-Hill, New York (1987).
31. W. Rudin, *Functional Analysis*. McGraw-Hill, New York (1973).
32. C.-S. Man, On the constitutive equations of some weakly-textured materials. *Arch. Rational Mech. Anal.* **143** (1998) 77–103.
33. M. Hirao, K. Aoki and H. Fukuoka, Texture of polycrystalline metals characterized by ultrasonic velocity measurements. *J. Acoust. Soc. Am.* **81** (1987) 1434–1440.
34. C.-S. Man, Effects of crystallographic texture on the acoustoelastic coefficients of polycrystals. *Nondestr. Test. Eval.* **15** (1999) 191–214.
35. C.-S. Man, Hartig's law and linear elasticity with initial stress. *Inverse Problems* **14** (1998) 313–319.
36. J.F. Thomas, Third order elastic constants of aluminum. *Phys. Rev.* **175** (1968) 955–962.
37. V.P.N. Sarma and P.J. Reddy, Third-order elastic constants of aluminum. *Phys. Status Solidi (a)* **10** (1972) 563–567.
38. R.F.S. Hearmon, Crystal and solid state physics. In: K.H. Hellwege and A.M. Hellwege (eds), *Numerical Data and Functional Relationships in Science and Technology*, Vol. 11. Springer, New York (1979).

39. C.-S. Man and R. Paroni, Explicit formulae showing the effects of texture on acoustoelastic coefficients. In: D.O. Thompson and D.E. Chimenti (eds), *Review of Progress in Quantitative Nondestructive Evaluation*, Vol. 16. Plenum, New York (1997) pp. 1683–1690.
40. C.-S. Man, W.Y. Lu and J. Li, Effects of crystallographic texture on the acoustoelastic coefficients for Rayleigh waves in aluminum. In: D.O. Thompson and D.E. Chimenti (eds), *Review of Progress in Quantitative Nondestructive Evaluation*, Vol. 18. Plenum, New York (1999) pp. 1879–1886.

Journal of Elasticity **59:** 175–193, 2000.

Constitutive Functions of Elastic Materials in Finite Growth and Deformation

YI-CHAO CHEN[1] and ANNE HOGER[2]
[1]*Department of Mechanical Engineering, University of Houston, Houston, TX 77204, U.S.A.*
[2]*Department of Applied Mechanics and Engineering Sciences, University of California, San Diego, LaJolla, CA 92093, U.S.A.*

Received 22 January 1999; in revised form 26 January 2000

Dedicated to Roger Fosdick.

Abstract. The constitutive functions of soft biological tissues during growth are studied. A growth, treated as addition (often non-uniform) of material points, results in deformation, residual stresses, and evolution of the constitutive functions. A theory based on the concept of equivalent material points is developed with the current configuration taken as the reference. The residual stresses developed in a spherical shell undergoing spherical growths are studied.

Mathematics Subject Classifications (1991): 73B05, 73G05, 73P20, 92C10.

Key words: biomechanics, growth, residual stresses.

1. Introduction

The description of the biological process of growth and remodeling is of interest both because of its importance as a fundamental mechanical process that occurs in normal development and in a number of pathological conditions, and because it offers an interesting and unusual application of continuum mechanics. It is known that mechanical quantities such as stress and strain can modulate growth, and growth of a tissue can, in turn, induce the development of residual stress. These are among the reasons that the phenomena of growth and remodeling have attracted the attention of researchers in both the mechanics and the biology communities. A number of earlier authors constructed kinematical descriptions of growth using the methods of continuum mechanics, but the work of Hsu [1], Cowin and Hegedus [2], Skalak [3], and Skalak et al. [4] are the initial efforts made to formulate mechanical theories to describe the growth of biological tissues. An excellent article by Taber [5] reviews the progress in the field through 1995.

The most general 3-dimensional theory of stress-dependent volumetric growth for soft biological tissues was first presented by Rodriguez et al. [6]. That theory introduced two fundamental ideas. The shape change of the unloaded tissue during growth can be described by a mapping whose gradient is analogous to a deformation gradient tensor. The first fundamental idea in [6] was that this gradient could be decomposed into the product of a "growth stretch tensor", which describes the addition of material at a point and the orientation of its deposition, and a tensor that represents the gradient of the elastic accommodation of the body to the new material. This elastic portion of the gradient ensures compatibility of the total growth deformation. The second fundamental idea was that the stress due to growth of a tissue was a function of only the elastic portion of the gradient of the mapping.

Because at the time when [6] was written it was not yet known how to construct constitutive equations for residually stressed finite elastic materials, the theory presented in [6] was illustrated only by problems in which the initial configuration was stress-free. In addition, their interpretation of the theory required knowledge of a locally stress-free reference state for the grown material. That such a zero-stress state exists is suggested by the destructive experiments that are commonly used to determine the residual stress in biological tissues and other materials. When a piece of residually stressed material is cut into progressively smaller pieces, the residual stress is relieved as the cuts are made.

Johnson and Hoger [7] stated and proved a mathematical description of this destructive testing process, which they used to motivate the concept of a virtual configuration of a residually stressed material. They applied the concept of the virtual configuration to derive constitutive equations for residually stressed elastic materials [7, 8]. The derivation of these constitutive equations requires that the response function of the corresponding natural material be locally invertible and satisfy certain smoothness requirements.

Recently, Hoger [9] expanded the theory of growth originally presented in [6] to take advantage of the availability of constitutive equations for finite elastic residually stressed materials. That work included an example of growth where the original configuration of the body supports a residual stress. It also contains a discussion of the growth stretch tensor, and identifies broad classes of growth based on the form taken by the growth stretch tensor.

The theory of growth presented by Rodriguez et al. and Hoger incorporates the use of a locally stress-free configuration of the material either explicitly [6] or through the use of the virtual configuration in the construction of a constitutive equation [9]. In both cases, the locally stress-free state is used as a reference configuration in which the mechanical properties of the natural material are known. In this paper we will not rely on the existence of a locally stress-free configuration, but, instead, will present a more general approach to developing the appropriate constitutive equation for a material comprising a residually stressed growing body.

To accomplish this, we will first focus on the issue of how to determine whether two material points are equivalent in the sense that their intrinsic mechanical properties are the same. The two points may be in different elastic bodies, or they may be the same material point in two configurations of the body; they may be in different stress states and may have deformation histories that are unknown. We engage this question by defining the constitutive functions with the current configuration taken as reference. The material properties are completely characterized by a density function that gives the density in the current state, and a response function that gives the stress in any configuration obtained by an elastic deformation out of the current state. No information on the histories of the deformation and stress of the material is needed. In particular, it is not necessary to identify a zero-stress configuration.

We then will use this framework to construct a theory of growth in which the growth tensor and the total deformation are defined on the current configuration. The forms of the density function and the response function evolve as the material grows and deforms.

There are two advantages of the theory of growth that we present in this paper over the theory of Rodriguez et al. [6] and Hoger [9]. The first advantage is due to the use of the current configuration as reference. The growth theory of [6] and [9] is Lagrangian: every function is referred to the original configuration. This typically presents no difficulty in the mechanics of solid materials, but if a body is growing, it is gaining material points, so there is no easily identified fixed reference configuration. One consequence of this difficulty is that the theory can only approximate continuous growth by modeling the growth as taking place in discrete increments [9]. The theory of growth we present in this paper is constructed with the current configuration as reference, thereby making it ideal for the description of continuous growth processes.

Second, neither Rodriguez et al. [6], nor Hoger [9] provided a theoretically sound foundation for the two central ideas in their theory of growth: the decomposition of the gradient of the total deformation; and the related assertion that the response of the material should depend only on the elastic portion of the decomposed gradient. In this paper we provide that foundation. We present a derivation of the evolution of the response function during growth and deformation, and determine explicitly how the evolution depends on the growth and on the total deformation. In particular, we provide a mathematically rigorous justification for the statement made in [6]: "Intuitively, residual stress arises from the part of the total growth deformation that is responsible for accommodating the newly grown tissue to prevent discontinuities in the grown state of the body."

In the next section we focus on the general issue of the characterization of the mechanical properties of a compressible material in the context of continuum mechanics. The density function and the response function are introduced with the current configuration taken as reference. Our goal in the section is to provide a rigorous definition for the statement that two material points have the same intrinsic material properties. We do this through the definitions of identical material points and equivalent material points, both of which are constructed using the density and response functions. The definition of equivalent material points requires the introduction of the equivalence transformation tensor, whose properties are examined. The relationship of this transformation tensor to the locally-stress-free configuration of [6] and the virtual configuration of [7–9] is also discussed.

In Section 3, we construct a theory of growth for a compressible elastic body. Only growth processes in which the intrinsic mechanical properties of the material do not change are considered. This allows the equivalence transformation tensor associated with the growth at a point to be defined through the requirement of material equivalence. The growth tensor, which describes the amount and the orientation of the material added at a point during a growth process, is defined. The dependence of the equivalence transformation tensor on the growth tensor and the total deformation gradient is then derived, and used to obtain the equations that explicitly describe the evolution of the density function and the response function during a growth process.

In Section 4, we outline the analogous theory of growth for incompressible elastic materials. The concepts of identical material points and equivalent material points are defined for incompressible materials; the appropriate equivalence transformation tensor is introduced; and the dependence of the Cauchy stress on the growth tensor and on the total deformation gradient is given.

We conclude the paper in Section 5 with an example in which we examine the spherically symmetric continuous growth of a spherical shell that is initially stress-free. The shell remains unloaded during the growth process. This growth boundary value problem is solved for the general class of homogeneous, isotropic, incompressible, elastic materials. We obtain expressions for the components of the residual stress generated by the growth process in terms of the growth tensor. In addition, explicit expressions for the time rates of change of the residual stress components are derived in terms of the growth rate. The section concludes with a discussion of the dependence of the growth-induced residual stress on the form of the growth tensor.

2. Characterization of Material. Equivalence of Material Points

Consider a body composed of a compressible elastic material that occupies the domain $\Omega \subset \Re^3$ at the current time. The body may undergo deformations due purely to external loads or deformations that include growth. It is usual in continuum mechanics to characterize the material of the body, especially the relation between the deformation and the stress, in terms of its mechanical properties in a fixed reference configuration. This approach is not convenient, however, if a fixed reference configuration cannot be easily identified, or if there is no fixed reference configuration in which the properties of the material are known.

An alternate approach is to use the current configuration Ω as the reference configuration, and to define the response function of the material in terms of the relation between deformations out of Ω and the stresses in the deformed configurations. If the body is subject to a time dependent deformation or growth process, then Ω will vary in time continuously, as will the characterization of the material.

For the present study, we will characterize the material by a density function* $\rho: \Omega \to \mathfrak{R}_+$ and a response function $\widehat{\mathbf{T}}$: $\mathrm{Lin}^+ \times \Omega \to \mathrm{Sym}$, where Lin^+ is the set of all linear transformations on $\mathfrak{R}^3$ that have positive determinant, and Sym the set of all symmetric linear transformations on $\mathfrak{R}^3$. These functions are assumed to possess the smoothness needed in the analysis. The value of the density function $\rho(\mathbf{x})$ gives the current density of the material at $\mathbf{x} \in \Omega$. We will term a deformation that is produced purely by external loads a *pure deformation* to distinguish it from a deformation in which growth occurs. Let the mapping $\mathbf{w}: \Omega \to \mathfrak{R}^3$ represent a pure deformation of the body out of the current configuration. Then the value of the response function $\widehat{\mathbf{T}}(\nabla\mathbf{w}(\mathbf{x}), \mathbf{x})$ gives the Cauchy stress tensor $\mathbf{T}(\mathbf{w})$ at $\mathbf{w} = \mathbf{w}(\mathbf{x})$. Note that the Cauchy stress $\mathbf{T}(\mathbf{x})$ at $\mathbf{x}$ in the current configuration Ω is given by

$$\mathbf{T}(\mathbf{x}) = \widehat{\mathbf{T}}(\mathbf{I}, \mathbf{x}), \tag{1}$$

where $\mathbf{I}$ is the identity tensor. If the body is not loaded in the current configuration, this stress is typically called the residual stress.

As a body deforms, the expressions for the density function and the response function at a point will evolve to reflect the changing reference configuration. We now examine how these functions evolve over a pure deformation. To be definite, let the configurations of the body before and after a pure deformation $\mathbf{w}$ be denoted by Ω and Ω_w, respectively, where $\Omega_w \equiv \mathbf{w}(\Omega)$. In addition, let $(\rho, \widehat{\mathbf{T}})$ and $(\rho_w, \widehat{\mathbf{T}}_w)$ be the density and response function pairs in Ω and Ω_w, respectively. By the conservation of mass, the density of the deformed body at $\mathbf{w} = \mathbf{w}(\mathbf{x})$ is given by $[\det\nabla\mathbf{w}(\mathbf{x})]^{-1}\rho(\mathbf{x})$.

Now consider a further pure deformation $\mathbf{z}: \Omega_w \to \mathfrak{R}^3$ of the body out of configuration Ω_w. The Cauchy stress at $\mathbf{z} = \mathbf{z}(\mathbf{w}(\mathbf{x}))$ is given by $\widehat{\mathbf{T}}(\nabla\mathbf{z}\nabla\mathbf{w}, \mathbf{x})$. Hence, the density function ρ_w and the response function $\widehat{\mathbf{T}}_w$ are given by

$$\rho_w(\mathbf{w}) = \det\bigl[\nabla\mathbf{w}(\mathbf{x})\bigr]^{-1}\rho(\mathbf{x}), \tag{2}$$

and

$$\widehat{\mathbf{T}}_w(\mathbf{F}, \mathbf{w}) = \widehat{\mathbf{T}}\bigl(\mathbf{F}\nabla\mathbf{w}(\mathbf{x}), \mathbf{x}\bigr), \tag{3}$$

where $\mathbf{w} = \mathbf{w}(\mathbf{x})$.

Up to this point we have obtained the relations between the density functions and the response functions for the points $\mathbf{x}$ and $\mathbf{w}$ that are related by a pure deformation. We now turn to the more general problem of describing the relation between the constitutive functions of any two points that have the same intrinsic material properties. By intrinsic properties of a material point, we mean the density and response functions at this point in a configuration with a prescribed stress state.

* Here we have chosen to include the density function in the characterization of a material, because mass change is an important aspect of growth, and because the evolution of density must be taken into account when studying a full dynamical problem of growth. The density function is, however, not needed in a static or quasistatic problem, such as the one presented in Section 5.

If a material has a stress-free natural state, it is convenient to use the properties of this underlying natural material as the intrinsic material properties. In any case, the intrinsic properties of a material do not change when the material undergoes pure deformations.

Two material points will be said to be *identical* if their density functions and their response functions are the same. Two such identical material points may belong to the same body or to different bodies, or may be the same material point at two different deformation or growth configurations.* In any case, it is not possible to distinguish between two identical material points with mechanical experiments.

During all pure deformations and for certain types of growth, the intrinsic properties of a material point remain unchanged, although the forms of the density function and the response function at the material point will typically change. In order to identify materials which share intrinsic material properties, we now introduce the concept of *equivalent material points*: two material points $\mathbf{x}_1$ and $\mathbf{x}_2$ will be said to be *equivalent* if there exists a pure deformation at $\mathbf{x}_1$ such that after this deformation, $\mathbf{x}_1$ is identical to $\mathbf{x}_2$. More precisely, let $(\rho_1, \widehat{\mathbf{T}}_1)$ and $(\rho_2, \widehat{\mathbf{T}}_2)$ be the density and response function pairs associated with the material points $\mathbf{x}_1$ and $\mathbf{x}_2$. These two material points are equivalent if there exists $\overline{\mathbf{F}} \in \text{Lin}^+$ such that

$$\rho_2(\mathbf{x}_2) = \left(\det\overline{\mathbf{F}}\right)^{-1} \rho_1(\mathbf{x}_1) \tag{4}$$

and

$$\widehat{\mathbf{T}}_2(\mathbf{F}, \mathbf{x}_2) = \widehat{\mathbf{T}}_1\left(\mathbf{F}\overline{\mathbf{F}}, \mathbf{x}_1\right) \quad \forall \mathbf{F} \in \text{Lin}^+. \tag{5}$$

The tensor $\overline{\mathbf{F}}$ will be termed the *equivalence transformation*.

By comparing equations (4) and (5) with equations (2) and (3), we observe that the equivalence transformation tensor can be thought of as the gradient of the homogeneous deformation under which the material point $\mathbf{x}_1$ becomes identical to the material point $\mathbf{x}_2$.

It is obvious that if two material points are identical, then they are equivalent with $\overline{\mathbf{F}} = \mathbf{I}$. It is also clear that a material point is equivalent to itself after any pure deformation. For example, let $\mathbf{w}\colon \Omega \to \mathfrak{R}^3$ be a pure deformation of Ω. The material point $\mathbf{x} \in \Omega$ is equivalent to $\mathbf{w} = \mathbf{w}(\mathbf{x})$ with

$$\overline{\mathbf{F}} = \nabla \mathbf{w}(\mathbf{x}). \tag{6}$$

It follows from equation (6) that if the material undergoes a further pure deformation $\mathbf{z}\colon \mathbf{w}(\Omega) \to \mathfrak{R}^3$, then the new equivalence transformation $\overline{\mathbf{F}}^*$ for the point $\mathbf{z} = \mathbf{z}(\mathbf{w}(\mathbf{x}))$ is given by the product of $\nabla\mathbf{z}$ and the equivalence transformation associated with $\mathbf{w}$:

$$\overline{\mathbf{F}}^* = \nabla(\mathbf{z} \circ \mathbf{w})(\mathbf{x}) = \nabla\mathbf{z}\nabla\mathbf{w} = \nabla\mathbf{z}\overline{\mathbf{F}}. \tag{7}$$

* In this latter case, one must redefine the "current times" of the two configurations accordingly.

If the stresses at two equivalent material points $\mathbf{x}_1$ and $\mathbf{x}_2$ are the same, then it follows from (1) and (5) that

$$\widehat{\mathbf{T}}_1(\mathbf{I}, \mathbf{x}_1) = \widehat{\mathbf{T}}_2\big(\mathbf{I}, \mathbf{x}_2\big) = \widehat{\mathbf{T}}_1(\overline{\mathbf{F}}, \mathbf{x}_1).$$

A solution of the above equation is $\overline{\mathbf{F}} = \mathbf{I}$. We shall assume that this solution is unique. Hence, if the stresses and densities at two equivalent material points are the same, these two material points are identical.

In the next section we shall use the concept of material equivalence developed above to study growth of soft biological tissues. Of course, growth does not happen in discrete increments, but is a continuous process. Using the current configuration as reference leads to a natural formulation for a continuous growth process, as displayed at the end of Section 3.

3. Growth

Changes in size and shape of a biological tissue can involve both growth, a change in mass, and remodeling, which includes changes in internal structure, density and material properties. Often growth and remodeling are linked, since the process of mass alteration can change the mechanical properties of the tissue. For example, the newly deposited tissue may have different material properties than the original tissue; or only one component of the tissue may grow, thus changing the mechanical properties of the tissue as a whole. Here we will focus exclusively on growth, which can involve both the addition and removal of material, sometimes occurring simultaneously. See Rodriguez et al. [6] and Hoger [9] for a full discussion of the kinematical description of growth.

In this work, we will restrict our attention to growth processes that meet the following two requirements. First, the material points must be dense during growth. This implies that in any arbitrary neighborhood in the grown body, there will always be material points that existed before the growth took place. The second requirement is that the intrinsic mechanical properties of the material should not change during growth. In other words, the new material has the same properties as does the original material. This implies that not only is a material point equivalent (in the sense of Section 2) to itself after growth, but also that a new material point added during growth is equivalent to an original material point in an arbitrarily small neighborhood of the new material point. Growth processes that meet these restrictions are special forms of volumetric growth, which takes place in the volume of a tissue rather than on the surface.

Kinematically, the addition of material alone can be described by a tensor valued function $\mathbf{G}\colon \Omega \to \mathrm{Lin}^+$, termed the growth tensor, that describes the amount and the orientation of material deposition [6, 9].

To set ideas, we first consider the simple case of a homogeneous elastic body with constant density and zero stress in the current configuration. Let the body undergo a homogeneous growth, with no applied external loads. For this case, the shape change of the grown body can be described by an affine mapping similar to a homogeneous deformation in continuum mechanics. The growth tensor in this simple case would correspond to the gradient of this mapping, which is uniform. The Cauchy stress in the grown body is maintained at zero, and the density of the body remains constant.

The situation just described is atypical. The growth tensor is not homogeneous in general, and the Cauchy stress in the body is usually inhomogeneous. To generalize the idea introduced in the above simple example, consider an inhomogeneous elastic body supporting an inhomogeneous stress in the current configuration Ω, and consider a small spherical neighborhood centered at point $\mathbf{x} \in \Omega$. Let the body undergo an inhomogeneous growth, and suppose that loads are applied on the grown body so that the stress at point $\mathbf{x}$ remains the same in the grown state as it was in Ω. Due to the growth, the sphere grows into a shape which can be approximated by an ellipsoid. The growth tensor $\mathbf{G}(\mathbf{x})$ describes the geometric change of the sphere to the ellipsoid.

The above argument is local. Globally, it is generally impossible to maintain, by applying the surface loads alone, the same inhomogeneous stress in the entire body during the growth. The shape change of a body subjected to an arbitrary growth process can be described by a mapping $\mathbf{y}\colon \Omega \to \mathfrak{R}^3$, which will be referred to as the *total deformation*. In general, the gradient of this total deformation will be different from the growth tensor. Hence, the total deformation cannot be obtained from the growth tensor alone. In fact, the growth tensor will in general not be the gradient of a vector field. However, at any material point where the gradient of the total deformation happens to be equal to the value of the growth tensor, the stress at that point will remain unchanged after the growth process.

A growth process can also be accompanied by an additional pure deformation due to a change of the applied loads on the body. Of course, in this case the additional pure deformation will contribute to the total deformation.

Let $\rho(\mathbf{x})$ and $\widehat{\mathbf{T}}(\mathbf{F}, \mathbf{x})$ denote the density and response functions for the material in the current configuration. And let $\rho_y(\mathbf{y})$ and $\widehat{\mathbf{T}}_y(\mathbf{F}, \mathbf{y})$ denote the density and response functions of the material at some later time after growth and deformation has occurred, as described by the total deformation $\mathbf{y}(x)$. We now examine how $\rho_y(\mathbf{y})$ and $\widehat{\mathbf{T}}_y(\mathbf{F}, \mathbf{y})$ are related to $\rho(\mathbf{x})$ and $\widehat{\mathbf{T}}(\mathbf{F}, \mathbf{x})$.

By the definition of material equivalence and by the requirement that a material point be equivalent to itself at any time during a growth or deformation, there exists an equivalence transformation $\overline{\mathbf{F}}\colon \Omega \times \mathrm{Lin}^+ \times \mathrm{Lin}^+ \to \mathrm{Lin}^+$, such that

$$\rho_y\big(\mathbf{y}(\mathbf{x})\big) = \big[\det\overline{\mathbf{F}}\big(\mathbf{x}, \mathbf{G}(\mathbf{x}), \nabla\mathbf{y}(\mathbf{x})\big)\big]^{-1}\rho(\mathbf{x}), \tag{8}$$

and

$$\widehat{\mathbf{T}}_y(\mathbf{F}, \mathbf{y}(\mathbf{x})) = \widehat{\mathbf{T}}\big(\mathbf{F}\overline{\mathbf{F}}\big(\mathbf{x}, \mathbf{G}(\mathbf{x}), \nabla\mathbf{y}(\mathbf{x})\big), \mathbf{x}\big) \quad \forall \mathbf{F} \in \mathrm{Lin}^+. \tag{9}$$

Here we have temporarily assumed that the equivalence transformation depends on the material point, the growth tensor, and the total deformation gradient.

We now determine the form of the equivalence transformation $\overline{\mathbf{F}}(\mathbf{x}, \mathbf{G}, \nabla\mathbf{y})$, and show that the form of $\overline{\mathbf{F}}$ is actually independent of the material. By the properties of the growth tensor, we observe that if the values of the total deformation gradient and the growth tensor are the same at a point, then the stress and density at this point do not change. Consequently, such a material point is identical to itself after the growth and deformation. That is,

$$\overline{\mathbf{F}}\big(\mathbf{x}, \mathbf{G}(\mathbf{x}), \mathbf{G}(\mathbf{x})\big) = \mathbf{I}. \tag{10}$$

If the material, after the total deformation $\mathbf{y}$, undergoes a further pure deformation $\mathbf{z}$: $\mathbf{y}(\Omega) \to \mathfrak{R}^3$, then the transformation tensor for the combined deformations can be obtained by considering $\mathbf{z}$ as a pure deformation imposed after the total deformation $\mathbf{y}$. The new equivalence transformation can be obtained by using equation (7):

$$\overline{\mathbf{F}}(\mathbf{x}, \mathbf{G}, \nabla\mathbf{z}\nabla\mathbf{y}) = \nabla\mathbf{z}\overline{\mathbf{F}}(\mathbf{x}, \mathbf{G}, \nabla\mathbf{y}) \quad \forall\mathbf{z}. \tag{11}$$

For a given growth tensor $\mathbf{G}$, the value of the total deformation gradient at a particular point $\mathbf{x} \in \Omega$ can be made equal to any given tensor in Lin^+ by applying appropriate loads to the body. Suppose that loads are chosen so that, at the point $\mathbf{x}$, $\nabla\mathbf{y}(\mathbf{x}) = \mathbf{G}(\mathbf{x})$. Then by (11),

$$\overline{\mathbf{F}}(\mathbf{x}, \mathbf{G}, \nabla\mathbf{z}\mathbf{G}) = \nabla\mathbf{z}\overline{\mathbf{F}}(\mathbf{x}, \mathbf{G}, \mathbf{G}) \quad \forall\mathbf{z},$$

which, with (10) and $\mathbf{F} = \nabla\mathbf{z}\mathbf{G}$, can be written as

$$\overline{\mathbf{F}}(\mathbf{x}, \mathbf{G}, \mathbf{F}) = \mathbf{F}\mathbf{G}^{-1} \quad \forall\mathbf{F} \in \mathrm{Lin}^+. \tag{12}$$

Equation (12) gives the explicit form of the equivalence transformation. In particular, it indicates that $\overline{\mathbf{F}}$ does not depend explicitly on either position or on the material. Rather, for a given growth tensor $\mathbf{G}$ and the total deformation $\mathbf{y}$ on Ω, the value of $\overline{\mathbf{F}}$ depends on $\mathbf{x}$ implicitly through $\mathbf{G}(\mathbf{x})$ and $\nabla\mathbf{y}(\mathbf{x})$.

Substitution of (12) into (8) and (9) yields

$$\rho_y\big(\mathbf{y}(\mathbf{x})\big) = \det\mathbf{G}(\mathbf{x})\big[\det\nabla\mathbf{y}(\mathbf{x})\big]^{-1}\rho(\mathbf{x}), \tag{13}$$

and

$$\widehat{\mathbf{T}}_y\big(\mathbf{F}, \mathbf{y}(\mathbf{x})\big) = \widehat{\mathbf{T}}\big(\mathbf{F}\nabla\mathbf{y}(\mathbf{x})\mathbf{G}^{-1}(\mathbf{x}), \mathbf{x}\big) \quad \forall\mathbf{F} \in \mathrm{Lin}^+.$$

These equations describe the changes in the density function and the response function after growth. In particular, the Cauchy stress in the grown state is given by

$$\mathbf{T}(\mathbf{y}) = \widehat{\mathbf{T}}_y(\mathbf{I}, \mathbf{y}) = \widehat{\mathbf{T}}\big(\nabla\mathbf{y}(\mathbf{x})\mathbf{G}^{-1}(\mathbf{x}), \mathbf{x}\big), \tag{14}$$

where $\mathbf{y} = \mathbf{y}(\mathbf{x})$.

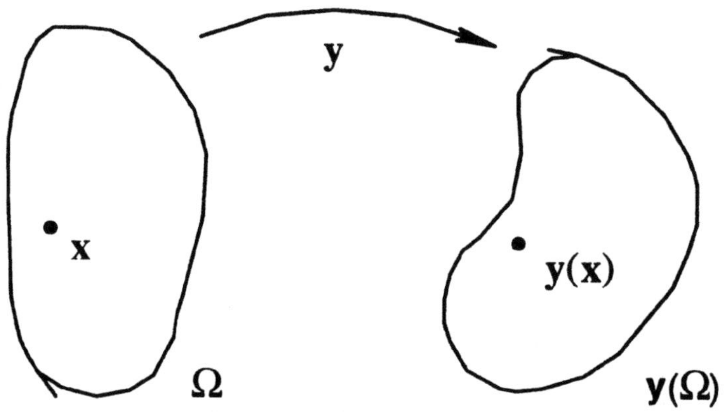

Figure 1. The body originally occupies domain Ω. In this configuration the material is characterized by the density function $\rho: \Omega \to \mathfrak{R}_+$ and the response function $\widehat{\mathbf{T}}: \mathrm{Lin}^+ \times \Omega \to \mathrm{Sym}$. The density of the material at $\mathbf{x} \in \Omega$ is given by $\rho(\mathbf{x})$, and the Cauchy stress $\mathbf{T}(\mathbf{x})$ in the configuration Ω is given by $\mathbf{T}(\mathbf{x}) = \widehat{\mathbf{T}}(\mathbf{I}, \mathbf{x})$. After a growth process, represented here by the total deformation $\mathbf{y}(\mathbf{x})$, the body occupies the region $\mathbf{y}(\Omega)$. The density $\rho_y(\mathbf{y})$ and the stress $\mathbf{T}(\mathbf{y})$ at point $\mathbf{y} \in \mathbf{y}(\Omega)$ are given by $\rho_y(\mathbf{y}) = \det\mathbf{G}(\mathbf{x})[\det\nabla\mathbf{y}(\mathbf{x})]^{-1}\rho(\mathbf{x})$ and $\mathbf{T}(\mathbf{y}) = \widehat{\mathbf{T}}(\nabla\mathbf{y}(\mathbf{x})\mathbf{G}^{-1}(\mathbf{x}), \mathbf{x})$.

The derivations leading to equations (12) and (14) provide a rigorous foundation for two suggestions originally proposed by Rodrigue et al. [6] and elaborated by Hoger [9]. Equation (12) can be rewritten as

$$\overline{\mathbf{F}}(\mathbf{x}, \mathbf{G}, \nabla\mathbf{y}) = \nabla\mathbf{y}\mathbf{G}^{-1},$$

or

$$\nabla\mathbf{y} = \overline{\mathbf{F}}\mathbf{G}.$$

This is the decomposition of the total deformation originally presented in [6, equation 9]. The decomposition was introduced in the context where $\mathbf{G}$ is defined as the growth from the original stress-free reference state to a new locally stress-free state, and $\overline{\mathbf{F}}$ is viewed as an elastic deformation that ensures the continuity of the body. In the present work, $\mathbf{G}$ corresponds to the growth of the body from the configuration Ω, which is not necessarily stress free. The introduction of the equivalence transformation $\overline{\mathbf{F}}$ is based on the requirement of material equivalence for growth rather than being assumed, as it was in [6, 7, 8, 9]. Of course, $\overline{\mathbf{F}}$ is generally not the gradient of a deformation of the body, as the notion of equivalence of material points is local. In [6, 7, 8, 9], it is conjectured that the stress which arises from the growth depends only on the elastic deformation that ensures the continuity of the body. The derivation of (14) provided here gives a rigorous proof of this conjecture.

A continuous growth process can be described by a time-dependent growth tensor $\mathbf{G}(\mathbf{x}, t)$ and a time-dependent total deformation $\mathbf{y}(\mathbf{x}, t)$. Equations (13) and (14) can be applied to this growth process, yielding

$$\rho_y(\mathbf{y}, t) = \det\mathbf{G}(\mathbf{x}, t)\left[\det\nabla\mathbf{y}(\mathbf{x}, t)\right]^{-1}\rho(\mathbf{x}),$$

and

$$\mathbf{T}(\mathbf{y}, t) = \widehat{\mathbf{T}}\left(\nabla\mathbf{y}(\mathbf{x}, t)\mathbf{G}^{-1}(\mathbf{x}, t), \mathbf{x}\right),$$

where $\mathbf{y} = \mathbf{y}(\mathbf{x}, t)$, and $\rho_y(\mathbf{y}, t)$ and $\mathbf{T}(\mathbf{y}, t)$ are the density and the stress at $\mathbf{y}$ in the grown configuration at time t.

4. Incompressible Materials

The concepts developed in Sections 2 and 3 can also be used for incompressible materials. In this section we present the parallel theory for the growth of an incompressible elastic body. For brevity, we shall present the basic equations with little or no derivation.

Let the body composed of an incompressible elastic material occupy the domain $\Omega \in \mathfrak{R}^3$ at the present time. As in the compressible case, the material is characterized by a density ρ function and a response function $\widehat{\mathbf{T}}$, but for incompressible materials $\widehat{\mathbf{T}}$: Unim $\times\, \Omega \to$ Sym, where Unim is the set of all linear transformations on $\mathfrak{R}^3$ whose determinant is unity. The value of the response function $\widehat{\mathbf{T}}$ gives the Cauchy stress only up to an arbitrary hydrostatic pressure. Let $\mathbf{w}$: $\Omega \to \mathfrak{R}^3$ with $\nabla\mathbf{w} \in$ Unim be a pure deformation of the body. The Cauchy stress in the deformed configuration is

$$\mathbf{T}(\mathbf{w}) = -p_w(\mathbf{w})\mathbf{I} + \widehat{\mathbf{T}}(\mathbf{F}, \mathbf{x}), \tag{15}$$

where $\mathbf{w} = \mathbf{w}(\mathbf{x})$, $\mathbf{F} = \nabla\mathbf{w}(\mathbf{x})$, and p_w: $\mathbf{w}(\Omega) \to \mathfrak{R}$ is the hydrostatic pressure required by the incompressibility constraint. The Cauchy stress $\mathbf{T}(\mathbf{x})$ in the current configuration Ω is given by

$$\mathbf{T}(\mathbf{x}) = -p(\mathbf{x})\mathbf{I} + \widehat{\mathbf{T}}(\mathbf{I}, \mathbf{x}),$$

where p: $\Omega \to \mathfrak{R}$ is in general different from p_w appearing in (15).

The notions of identical material points and of equivalent material points can be defined for incompressible bodies in a way similar to that used for compressible bodies. However, since for incompressible bodies the response function determines only those components of the Cauchy stress that are orthogonal to the identity tensor $\mathbf{I}$, only the orthogonal components of the response function need be considered when making the definitions. In the following, $\overline{\mathbf{T}}$ will denote the part of $\widehat{\mathbf{T}}$ that is orthogonal to $\mathbf{I}$:

$$\overline{\mathbf{T}} \equiv \widehat{\mathbf{T}} - \frac{1}{3}(\mathrm{tr}\widehat{\mathbf{T}})\mathbf{I}.$$

Consider two bodies Ω_1 and Ω_2 (which may represent two configurations of the same body), with associated density and response function pairs $(\rho_1, \widehat{\mathbf{T}}_1)$ and $(\rho_2, \widehat{\mathbf{T}}_2)$. Two material points $\mathbf{x}_1 \in \Omega_1$ and $\mathbf{x}_2 \in \Omega_2$ are said to be identical if

$$\rho_1(\mathbf{x}_1) = \rho_2(\mathbf{x}_2), \tag{16}$$

and

$$\overline{\mathbf{T}}_1(\mathbf{F}, \mathbf{x}_1) = \overline{\mathbf{T}}_2(\mathbf{F}, \mathbf{x}_2) \quad \forall \mathbf{F} \in \text{Unim}.$$

If instead, the two material points satisfy (16) and

$$\overline{\mathbf{T}}_1(\mathbf{F}\overline{\mathbf{F}}, \mathbf{x}_1) = \overline{\mathbf{T}}_2(\mathbf{F}, \mathbf{x}_2) \quad \forall \mathbf{F} \in \text{Unim} \tag{17}$$

for some equivalence transformation tensor $\overline{\mathbf{F}} \in$ Unim, then the points are said to be equivalent.

We consider the growth of an incompressible body that meets the requirements that the material points are dense during the growth, and that the intrinsic mechanical properties of the material do not change. Such a growth is described, in analogy to compressible materials, by a growth tensor $\mathbf{G}\colon \Omega \to \text{Lin}^+$, along with a total deformation $\mathbf{y}\colon \Omega \to \Re^3$ which provides the total shape change of the body.

Let $\rho_y(\mathbf{y})$ and $\widehat{\mathbf{T}}_y(\mathbf{F}, \mathbf{y})$ be the density and response functions of the material in the grown body. Since a material point is equivalent to itself after a total deformation, it follows from (16) and (17) that there exits an equivalence transformation tensor $\overline{\mathbf{F}}\colon \Omega \times \text{Lin}^+ \times \text{Lin}^+ \to \text{Unim}$ such that

$$\rho_y\big(\mathbf{y}(\mathbf{x})\big) = \rho(\mathbf{x}),$$

and

$$\overline{\mathbf{T}}_y\big(\mathbf{F}, \mathbf{y}(\mathbf{x})\big) = \overline{\mathbf{T}}\big(\mathbf{F}\overline{\mathbf{F}}\big(\mathbf{x}, \mathbf{G}(\mathbf{x}), \nabla\mathbf{y}(\mathbf{x})\big), \mathbf{x}\big) \quad \forall \mathbf{F} \in \text{Unim}. \tag{18}$$

The arguments that lead to (12) for compressible materials are equally valid for incompressible materials; so the representation (12) for the equivalence transformation holds for incompressible materials as well. Since $\overline{\mathbf{F}} \in$ Unim, equation (12) implies that

$$\det\mathbf{G}(\mathbf{x}) = \det\nabla\mathbf{y}(\mathbf{x}). \tag{19}$$

Substitution of (12) into (18) gives

$$\overline{\mathbf{T}}_y\big(\mathbf{F}, \mathbf{y}(\mathbf{x})\big) = \overline{\mathbf{T}}\big(\mathbf{F}\nabla\mathbf{y}(\mathbf{x})\mathbf{G}^{-1}(\mathbf{x}), \mathbf{x}\big) \quad \forall \mathbf{F} \in \text{Unim}.$$

The Cauchy stress in the grown state is then given by

$$\mathbf{T}(\mathbf{y}) = -p_y(\mathbf{y})\mathbf{I} + \widehat{\mathbf{T}}_y(\mathbf{I}, \mathbf{y}) = -p(\mathbf{x})\mathbf{I} + \widehat{\mathbf{T}}\big(\nabla\mathbf{y}(\mathbf{x})\mathbf{G}^{-1}(\mathbf{x}), \mathbf{x}\big),$$

where $\mathbf{y} = \mathbf{y}(\mathbf{x})$.

Continuous growth of an incompressible body can be described by a time-dependent growth tensor $\mathbf{G}(\mathbf{x}, t)$ and a time-dependent total deformation $\mathbf{y}(\mathbf{x}, t)$. For such a growth process the Cauchy stress becomes

$$\mathbf{T}(\mathbf{y}, t) = -p(\mathbf{x}, t)\mathbf{I} + \widehat{\mathbf{T}}\big(\nabla\mathbf{y}(\mathbf{x}, t)\mathbf{G}^{-1}(\mathbf{x}, t), \mathbf{x}\big), \tag{20}$$

where $\mathbf{y} = y(\mathbf{x}, t)$.

5. Growth of a spherical shell

In this section we examine the spherically symmetric growth of a spherical shell which, in the initial configuration, is composed of a stress-free, homogeneous, isotropic, incompressible, elastic material. The shell remains unloaded during the growth process. Here we investigate the generation of residual stresses by growth.

The initial spherical configuration Ω is defined, in spherical coordinates (R, Φ, Θ), by

$$A \geqslant R \geqslant B, \qquad 0 \geqslant \Phi \geqslant \pi, \qquad 0 \geqslant \Theta \geqslant 2\pi,$$

where A and B are the inner and outer radii of the shell, respectively. We consider the growth to be spherically symmetric and continuous, so in spherical coordinates the growth tensor $\mathbf{G}(R, t)$ has the component form

$$\mathbf{G}(R, t) = \begin{pmatrix} G_1(R, t) & 0 & 0 \\ 0 & G_2(R, t) & 0 \\ 0 & 0 & G_2(R, t) \end{pmatrix}. \tag{21}$$

The smooth growth functions G_1 and G_2, which represent the radial and circumferential growth of the shell, respectively, must satisfy the initial conditions

$$G_1(R, 0) = G_2(R, 0) = 1. \tag{22}$$

When the value of a growth function is greater than 1, it represents growth; a value less than 1 represents resorption [9], which is the removal of material. Following the nomenclature introduced by Hoger [9], we will refer to the case where $G_1 > 1$ and $G_2 = 1$ as fiber growth in the radial direction; the case where $G_1 = 1$ and $G_2 > 1$ will be termed area growth perpendicular to the radial direction; and the case where $G_1 = G_2 > 1$ is called isotropic growth.

Because we have restricted our attention to the situation where the spherical shell grows into a spherical shell, the total deformation $\mathbf{y}(\mathbf{x}, t)$ associated with the growth function is spherically symmetric, and its components in spherical coordinates are given by

$$r = r(R, t), \qquad \phi = \Phi, \qquad \theta = \Theta,$$

where the radial deformation $r(R, t)$ is a smooth function satisfying

$$r(R, 0) = R. \tag{23}$$

With a prime denoting the derivative with respect to R, the corresponding deformation gradient $\nabla\mathbf{y}(\mathbf{x},t)$ has the component form

$$\nabla\mathbf{y}(\mathbf{x},t)=\begin{pmatrix} r'(R,t) & 0 & 0 \\ 0 & \dfrac{r(R,t)}{R} & 0 \\ 0 & 0 & \dfrac{r(R,t)}{R} \end{pmatrix}. \tag{24}$$

Substitution of (21) and (24) into the incompressibility condition (19) yields

$$G_1G_2^2=\frac{r^2r'}{R^2}.$$

This equation can be integrated to give the deformation $r(R,t)$ in terms of the growth functions $G_1(R,t)$ and $G_2(R,t)$:

$$r(R,t)=\left[a^3(t)+\int_A^R 3\xi^2G_1(\xi,t)G_2^2(\xi,t)\,\mathrm{d}\xi\right]^{1/3}, \tag{25}$$

where

$$a(t)\equiv r(A,t)$$

is the radial position of the inner surface during the growth process.

Recall that we assumed the material in the initial configuration Ω to be homogeneous and isotropic. Thus, the response function $\widehat{\mathbf{T}}(F,x)$ has the following representation (see [10, Section 49]):

$$\widehat{\mathbf{T}}(\mathbf{F},\mathbf{x})=f_1(I_1,I_2)\mathbf{V}+f_2(I_1,I_2)\mathbf{V}^2, \tag{26}$$

where $\mathbf{V}\equiv(\mathbf{F}\mathbf{F}^{\mathrm{T}})^{1/2}$, I_1 and I_2 are the first and second principal invariants of $\mathbf{V}$, and f_1 and f_2 are two scalar functions of the principal invariants. By substituting (26) into (20), we obtain the Cauchy stress

$$\mathbf{T}=-p\mathbf{I}+f_1(I_1,I_2)\mathbf{V}+f_2(I_1,I_2)\mathbf{V}^2.$$

For the total deformation gradient (24) and the growth tensor (21), we have

$$\mathbf{V}=\begin{pmatrix} \dfrac{r'}{G_1} & 0 & 0 \\ 0 & \dfrac{r}{RG_2} & 0 \\ 0 & 0 & \dfrac{r}{RG_2} \end{pmatrix},$$

$$I_1=\frac{r'}{G_1}+\frac{2r}{RG_2},\qquad I_2=\frac{r^2}{R^2G_2^2}+\frac{2rr'}{RG_1G_2}. \tag{27}$$

The material is assumed to satisfy the Baker–Ericksen inequality, which asserts that the greater principal stress occurs in the direction of the greater principal stretch. For the present problem, the Baker–Ericksen inequality states

$$f_1 + (v_1 + v_2) f_2 > 0 \quad \text{if } v_1 \neq v_2, \tag{28}$$

where the principal stretches v_1 and v_2 are given by

$$v_1 = \frac{r'}{G_1}, \qquad v_2 = \frac{r}{RG_2}. \tag{29}$$

In the absence of body forces, the only nontrivial component of the equation of equilibrium is

$$\frac{\partial T_1}{\partial r} + \frac{2(T_1 - T_2)}{r} = 0, \tag{30}$$

where T_1 and T_2 are the radial and circumferential stresses, respectively, given by

$$T_1 = -p + v_1 f_1 + v_1^2 f_2, \qquad T_2 = -p + v_2 f_1 + v_2^2 f_2. \tag{31}$$

We shall examine the case where the spherical shell is unloaded during the growth, so the boundary conditions are

$$T_1\big(r(A,t),t\big) = 0, \qquad T_1\big(r(B,t),t\big) = 0. \tag{32}$$

Equation (30) can be written, with the help of (31), as

$$T_1' = \frac{2(v_2 - v_1)r'}{r}\big[f_1 + (v_1 + v_2) f_2\big],$$

which can be integrated to yield

$$\begin{aligned} T_1\big(r(R,t),t\big) &= \int_A^R \frac{2[v_2(\xi,t) - v_1(\xi,t)]r'(\xi,t)}{r(\xi,t)} \big\{f_1\big(I_1(\xi,t), I_2(\xi,t)\big) \\ &\quad + \big[v_1(\xi,t) + v_2(\xi,t)\big] f_2\big(I_1(\xi,t), I_2(\xi,t)\big)\big\} \mathrm{d}\xi. \end{aligned} \tag{33}$$

This solution satisfies the boundary condition $(32)_1$ on the inner surface. It also follows from (31) that

$$\begin{aligned} T_2\big(r(R,t),t\big) &= T_1\big(r(R,t),t\big) + \big[v_2(R,t) - v_1(R,t)\big] \\ &\quad \times \big\{f_1\big(I_1(R,t), I_2(R,t)\big) \\ &\quad + \big[v_1(R,t) + v_2(R,t)\big] f_2\big(I_1(R,t), I_2(R,t)\big)\big\}. \end{aligned} \tag{34}$$

Equations (33) and (34) give the stresses in the growing shell as a function of time. Since the shell is unloaded, these equations provide the residual stress due solely to the growth that evolves during the growth process.

A special case of isotropic growth is when the growth functions are uniform: $G_1(R,t) = G_2(R,t) = G(t)$. In this case, it is straightforward to verify that the solution is given by

$$r(R,t) = RG(t), \qquad I_1 = I_2 = 3, \qquad T_1 = T_2 = 0.$$

That is, this uniform isotropic growth gives rise to a pure dilatation of the shell and produces no residual stress.

For general growth functions, it is not possible to go any further than (33) and (34) in the determination of the residual stress without specifying the form of the constitutive functions f_1 and f_2. However, it is possible to derive explicit expressions for the time derivatives of the components of the residual stress in the initial state. This provides useful information on the residual stress distribution for configurations in which the shell has grown slightly.

First, observe from (22), (23) and (29) that

$$v_1(R,0) = v_2(R,0) = 1. \tag{35}$$

This implies, by (33) and (34), that

$$T_1(R,0) = T_2(R,0) = 0,$$

which simply states that initially the material is in its natural, or stress-free, state.

We shall use the following notation for the time derivative of a function $F(R,t)$ in the initial state:

$$\dot{F}(R) \equiv \left.\frac{\partial F(R,t)}{\partial t}\right|_{t=0}.$$

By differentiating (25) and applying (22) and (23), we find that

$$\begin{aligned}
\dot{r}(R) &= \frac{A^2\dot{a}}{R^2} + \frac{1}{R^2}\int_A^R \xi^2\big[\dot{G}_1(\xi) + 2\dot{G}_2(\xi)\big]\mathrm{d}\xi,\\
\dot{r}'(R) &= \dot{G}_1(R) + 2\dot{G}_2(R) - \frac{2A^2\dot{a}}{R^3} - \frac{2}{R^3}\int_A^R \xi^2\big[\dot{G}_1(\xi) + 2\dot{G}_2(\xi)\big]\mathrm{d}\xi.
\end{aligned}$$

Differentiation of (29), together with (22) and (23), gives

$$\begin{aligned}
\dot{v}_1(R) &= \dot{r}'(R) - \dot{G}_1(R)\\
&= 2\dot{G}_2(R) - \frac{2A^2\dot{a}}{R^3} - \frac{2}{R^3}\int_A^R \xi^2\big[\dot{G}_1(\xi) + 2\dot{G}_2(\xi)\big]\mathrm{d}\xi,
\end{aligned} \tag{36}$$

and

$$\begin{aligned}
\dot{v}_2(R) &= \frac{\dot{r}(R)}{R} - \dot{G}_2(R)\\
&= -\dot{G}_2(R) + \frac{A^2\dot{a}}{R^3} + \frac{1}{R^3}\int_A^R \xi^2\big[\dot{G}_1(\xi) + 2\dot{G}_2(\xi)\big]\mathrm{d}\xi.
\end{aligned} \tag{37}$$

Further, by (22), (23), (27), (33) and (35), we have

$$\dot{T}_1(R) = 2(f_1 + 2f_2) \int_A^R \frac{\dot{v}_2(\xi) - \dot{v}_1(\xi)}{\xi}\, d\xi. \tag{38}$$

Because we are evaluating the derivatives of the stress in the initial configuration, the functions f_1 and f_2 are evaluated at $(I_1, I_2) = (3, 3)$. By substituting (36) and (37) into (38) and evaluating one integral after changing the order of integration, we find that

$$\begin{aligned}\dot{T}_1(R) &= \frac{2(f_1 + 2f_2)}{R^3}\left\{\frac{R^3 - A^3}{A}\dot{a}\right.\\ &\quad \left. + \int_A^R \left[\frac{R^3 - \xi^3}{\xi}\dot{G}_1(\xi) - \frac{R^3 + 2\xi^3}{\xi}\dot{G}_2(\xi)\right] d\xi\right\}.\end{aligned} \tag{39}$$

The initial velocity $\dot{a}$ of the inner surface appearing in (39) can be determined by using the remaining traction free boundary condition $(32)_2$ on the outer surface. Indeed, evaluation of (39) at $R = B$ and use of $(32)_2$ give

$$\dot{a} = -\frac{A}{B^3 - A^3} \int_A^B \left[\frac{B^3 - \xi^3}{\xi}\dot{G}_1(\xi) - \frac{B^3 + 2\xi^3}{\xi}\dot{G}_2(\xi)\right] d\xi. \tag{40}$$

By substituting (40) back to (39) and simplifying the result, we arrive at

$$\begin{aligned}\dot{T}_1(R) &= -\frac{2(f_1 + 2f_2)}{(B^3 - A^3)R^3}\left\{\left(B^3 - R^3\right)\right.\\ &\quad \times \int_A^R \left[\frac{\xi^3 - A^3}{\xi}\dot{G}_1(\xi) + \frac{A^3 + 2\xi^3}{\xi}\dot{G}_2(\xi)\right] d\xi\\ &\quad \left. +\left(R^3 - A^3\right) \int_R^B \left[\frac{B^3 - \xi^3}{\xi}\dot{G}_1(\xi) - \frac{B^3 + 2\xi^3}{\xi}\dot{G}_2(\xi)\right] d\xi\right\}.\end{aligned} \tag{41}$$

Finally, differentiation of (34), together with (22), (23), (27), (35)–(37), (40) and (41), yields

$$\begin{aligned}\dot{T}_2(R) &= \dot{T}_1(R) + (f_1 + 2f_2)\left[\dot{v}_2(R) - \dot{v}_1(R)\right]\\ &= \frac{f_1 + 2f_2}{(B^3 - A^3)R^3}\left\{\left(B^3 + 2R^3\right)\right.\\ &\quad \times \int_A^R \left[\frac{\xi^3 - A^3}{\xi}\dot{G}_1(\xi) + \frac{A^3 + 2\xi^3}{\xi}\dot{G}_2(\xi)\right] d\xi\\ &\quad \left. -\left(A^3 + 2R^3\right) \int_R^B \left[\frac{B^3 - \xi^3}{\xi}\dot{G}_1(\xi) - \frac{B^3 + 2\xi^3}{\xi}\dot{G}_2(\xi)\right] d\xi\right\}\\ &\quad -3(f_1 + 2f_2)\dot{G}_2(R).\end{aligned} \tag{42}$$

Equations (41) and (42) give the residual stress rates in the initial state in terms of the growth rates. Since the shell is stress free in the initial state, these equations also give the first order approximations of the residual stress components that result from the growth. Some interesting observations can be made based on these equations.

Firstly, it is straight forward to verify that the condition that $\dot{G}_1$ and $\dot{G}_2$ are equal to a same constant implies that $\dot{T}_1 = \dot{T}_2 = 0$. This confirms the observation made earlier that a uniform growth results in no residual stress.

Secondly, the radial residual stress rate $\dot{T}_1$ depends on the growth rate "globally" in the sense that the values of $\dot{T}_1$ depends on the values of $\dot{G}_1$ and $\dot{G}_2$ in the entire thickness of the shell. On the other hand, for the circumferential stress rate $\dot{T}_2$, besides a global dependence, there is also a "local" dependence on $\dot{G}_2$, as indicated by the last term in (42). This local dependence is due to the circumferential constraint provided by the continuity condition.

Thirdly, we observe from (28) and (35) that the factor f_1+2f_2 appearing in (41) and (42) is positive. It then follows from (41) that any radial fiber growth (in which $\dot{G}_1 > 0$, $\dot{G}_2 = 0$) will produce a compressive radial residual stress component in the entire shell, while a radial fiber resorption will result in a tensile radial residual stress component. This observation seems to be consistent with our intuition that growth leads to compressive residual stresses when the material is constrained in the growth direction in some way. On the other hand, it is observed from (42) that a fiber growth can lead to either tensile or compressive circumferential residual stresses, depending on the distribution of $\dot{G}_1$.

Finally, the effects of area growths on the residual stress are somewhat different, and perhaps more interesting. As can be seen from (41), an area growth perpendicular to the radius (in which $\dot{G}_2 > 0$, $\dot{G}_1 = 0$) can lead to either tensile or compressive radial residual stresses, depending again on the distribution of $\dot{G}_2$. On the other hand, equation (42) shows that the local contribution of an area growth to the circumferential residual stress is compressive, while the global contribution is tensile. This also is consistent with our intuition: consider a given spherical surface in the spherical shell. If an area growth takes place in this surface but nowhere else, a compressive circumferential residual stress will develop in the surface, while the rest of the shell will be in tension circumferentially. If, on the other hand, an area growth takes place everywhere but this spherical surface, the surface will be in tension while the rest of the shell in compression.

Acknowledgements

The authors wish to thank the Institute for Mechanics and Materials at UC, San Diego, for supporting a scientific visit of Y.-C. Chen, during which this work was initiated. A. Hoger wishes to acknowledge the support of NSF through grant CMS-96-34903.

References

1. F. Hsu, The influences of mechanical loads on the form of a growing elastic body. *J. Biomechanics* **1** (1968) 303–311.
2. S.C. Cowin and D.H. Hegedus, Bone remodeling I: Theory of adaptive elasticity. *J. Elasticity* **6** (1976) 313–326.
3. R. Skalak, Growth as a finite displacement field. In: D.E. Carlson and R.T. Shield (eds), *Proc. of IUTAM Symp. on Finite Elasticity*, Martinus Nijhoff, The Hague (1981) pp. 348–355.
4. R. Skalak, G. Dasgupta, M. Moss, E. Otten, P. Dullemeijer and H. Vilmann, Analytical description of growth. *J. Theor. Biol.* **94** (1982) 555–577.
5. L.A. Taber, Biomechanics of growth, remodeling, and morphogenesis. *ASME Appl. Mech. Rev.* **48** (1995) 487–545.
6. E.K. Rodriguez, A. Hoger and A.D. McCulloch, Stress-dependent finite growth in soft elastic tissues. *J. Biomechanics* **27** (1994) 455–467.
7. B.E. Johnson and A. Hoger, The use of a virtual configuration in formulating constitutive equations for residually stressed elastic materials. *J. Elasticity* **41** (1995) 177–215.
8. A. Hoger, Virtual configurations and constitutive equations for residually stressed bodies with material symmetry. *J. Elasticity* **48** (1997) 125–144.
9. A. Hoger, An incremental mechanical theory of growth for soft biological tissues. Unpublished notes, manuscript submitted. See also: A. Hoger, "An incremental theory of volumetric growth for soft biological tissues," *1999 ASME Mechanics and Materials Conference*, June 27–30, 1999, University of Virginia, Blacksburg, Virginia.
10. C. Truesdell and W. Noll, *Non-Linear Field Theories of Mechanics*, 2nd edn. Springer, Berlin (1992).

Journal of Elasticity **59:** 195–211, 2000.

Toughening by Phase Boundary Propagation

N.K. SIMHA
Department of Mechanical Engineering, University of Miami, P.O. Box 248294, Coral Gables, FL 33124-0624, U.S.A. E-mail: nsimha@miami.edu

Received 10 February 1999; in revised form 14 February 2000

Dedicated to Roger Fosdick on the occasion of his 60th birthday.

Abstract. Transformation toughening has enhanced the fracture toughness of certain Zirconia-Toughened Ceramics (ZTC) by factors of 2–4. The primary explanation of toughening, by McMeeking and Evans [1] and Budiansky et al. [2], suggests that the main source of toughening is the energy stored by the transformed inclusions in the wake of a propagating crack. In the case of supercritical ZTC where the boundary of the transformed zone is a phase boundary (surface of strain discontinuity), this paper suggests an additional source of toughening – that due to propagation of the phase boundary. By extending the J-integral to cracked bodies containing surfaces of strain discontinuity, the transformation toughening for a steady Mode I crack is evaluated.

Key words: transformation toughening, phase boundary propagation, zirconia-toughened ceramics, shape-memory alloys, J-integral.

1. Introduction

Transformation toughening has enhanced the fracture toughness of certain Zirconia-Toughened Ceramics (ZTC) by factors of 2–4 and is attributed to the tetragonal to monoclinic transformation of zirconia inclusions induced by the crack-tip stresses. Since the transformation is dissipative, not all the energy supplied by a loading device is dissipated at the crack tip, and consequently there is toughening. The region $\mathcal{A}$ surrounding the crack where inclusions undergo transformation is called the transformation zone (see Figure 1). In certain types of ZTC it is observed that all the zirconia inclusions inside the zone undergo transformation [3, 4]. These ceramics are called super-critical ZTC [2], and examples include MgO- and CaO-partially stabilized zirconia [5]. By attributing the toughening to the energy stored in the wake of the transformation zone, McMeeking and Evans [1] as well as Budiansky et al. [2] estimated the transformation toughening. However, measurements are usually between two to three times the theoretically predicted values [5], which suggests that there may be other sources of toughening.

In this paper we explore an alternate cause of the toughening in supercritical ZTC. Notice that the boundary of the transformation zone Σ is a coherent phase

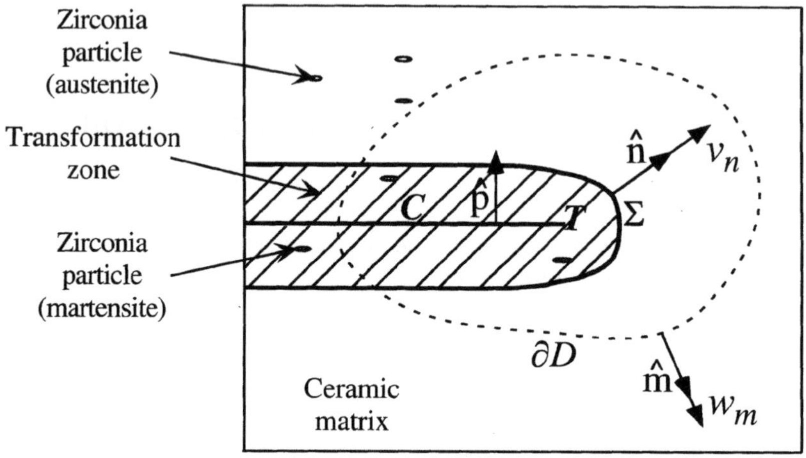

Figure 1. Transformation zone $\mathcal{A}$ (hatched area) around a crack C.

boundary (see Figure 1). It has been shown that phase boundary propagation is a dissipative process [6 – 8], and we argue that this is exactly the contribution that was neglected by the micromechanics calculations for supercritical ceramics [1, 2]. This paper will include the dissipation due to the propagating transformation front and evaluate the total toughening.

In simple terms, toughening can be evaluated using the J-integral on two contours: one contour close to the crack-tip and a second far away from the crack-tip. Since a propagating phase boundary will be present between the contours the J-integral is no longer path-independent. The main calculation in this paper is to find the relation between the two J-integrals, and the concept of configurational forces due to Gurtin [8] provides a convenient framework. It has been used to study crack propagation [9, 10] and phase boundary propagation [8, 11, 12]. A synthesis of these two calculations will be used to calculate the toughening in Section 2. Following [2] a constitutive model for supercritical ZTC that is nonlinear in dilatation but linear in the deviatoric strain will be constructed in Section 3.1. Although such a model may not account for shear effects [13], it is sufficient to demonstrate toughening due to phase boundary propagation. The toughening is evaluated in Section 3.2 by considering steady-state propagation of a Mode I crack. We end with a discussion of our results and with a comment on the extension of these results to polycrystalline shape-memory alloys (Section 4).

2. Dissipation Due to Cracks and Phase Boundaries

2.1. KINEMATICS

Consider a body in its reference configuration $\Omega \in \mathcal{R}^2$; the body contains a crack C and a coherent phase boundary Σ (see Figure 1). The crack and phase boundary are considered to be curved interfaces in the body. Zirconia inclusions inside the transformation zone $\mathcal{A}$ are in the martensite phase, whereas outside they are in the austenite phase. The unit normal to the phase boundary is $\hat{\mathbf{n}}$ and the unit normal to the crack is $\hat{\mathbf{p}}$.

Phase transformation is modeled by allowing the interface Σ to evolve in the reference configuration, and thus the position of the interface Σ in the reference depends on the time t. Let the interface $\Sigma(t)$ be parametrized as

$$\Sigma(t) = \left\{\mathbf{x} \in \mathcal{R}^2 \colon \mathbf{x} = \bar{\mathbf{x}}(\xi, t)\right\}, \tag{2.1}$$

where ξ is a scalar. The interface velocity in the reference $\mathbf{v}$ and its normal component v_n are given by

$$\mathbf{v} = \frac{\partial \bar{\mathbf{x}}(\cdot, t)}{\partial t} \quad \text{and} \quad v_n = \mathbf{v} \cdot \hat{\mathbf{n}}, \tag{2.2}$$

respectively; only the normal component of the interface velocity is independent of the parametrization (2.1).

Although the crack remains fixed, its tip T moves in the reference configuration; let the crack-tip velocity be $\mathbf{v}_T$. Notice that the bulk fields such as the stress may be singular at the crack-tip. To tackle this, we exclude a small region around the singular point (following classical potential theory) and perform the required analysis in the punctured domain; the relevant equations are obtained in the limit as the region vanishes. A similar approach has been used in [9, 10, 11, 12]. The region we choose is a disk $\mathcal{B}_r$ centered at the crack-tip T and with radius $r > 0$. The normal to its boundary $\partial\mathcal{B}_r$ is denoted by $\hat{\mathbf{m}}$ (see Figure 2). Since the disk has to contain the moving crack-tip for all values of $r > 0$, it also moves in the

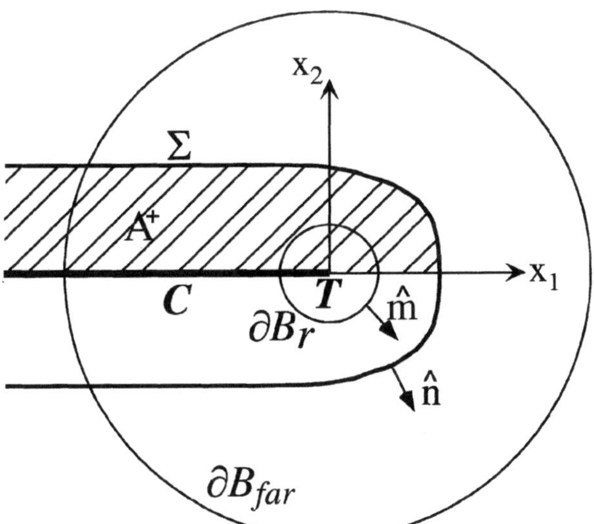

Figure 2. The region $\mathcal{D}$ for calculating the toughness lies between the two circles ∂B_r and ∂B_{far} and excludes the crack.

reference. Let the reference velocity field of the circle $\partial\mathcal{B}_r$ be denoted by $\mathbf{w}$. We note that $\mathbf{w} \to \mathbf{v}_T$ in the limit $r \to 0$.

Let the deformation be $\mathbf{y}$: $\Omega \times I \to \mathcal{R}^2$ where I is a time interval. The deformation gradient is $\mathbf{F} = \nabla_x \mathbf{y}(\mathbf{x}, t)$ and the material velocity is $\dot{\mathbf{y}}(\mathbf{x}, t) = \partial \mathbf{y}(\mathbf{x}, t)/\partial t$. The deformation is sufficiently smooth so that the deformation gradient and material velocity are smooth everywhere in Ω except at the phase boundary, the crack and the crack-tip. Since we take the phase boundary to be coherent, the following conditions are satisfied at Σ [11]:

$$[[\mathbf{F}]] = [[\mathbf{F}]]\hat{\mathbf{n}} \otimes \hat{\mathbf{n}} \quad \text{and} \quad [[\dot{\mathbf{y}}]] = -v_n[[\mathbf{F}]]\hat{\mathbf{n}}, \tag{2.3}$$

where the symbol $[[\mathbf{a}]] = (\mathbf{a}^+ - \mathbf{a}^-)$ denotes the jump across the interface Σ and $\mathbf{a}^+$ is the limit of the parameter $\mathbf{a}$ at the interface Σ from outside the transformed zone $\mathcal{A}$, while $\mathbf{a}^-$ is the limit from the inside.

The deformation is continuous from either side of the crack, but the two limits may not be equal. However, the deformation is taken to be continuous at the crack-tip, which allows us to define the velocity of the crack-tip in the current configuration. If the deformation $\mathbf{y}(\mathbf{x}, t) \to \mathbf{y}(T(t), t)$ as $\mathbf{x} \to T$, then $\mathbf{y}(T(t), t)$ is differentiable in t and the crack-tip velocity in the current configuration is $\dot{\mathbf{y}} + \mathbf{F}\mathbf{v}_T$ [9].

2.2. BALANCE LAWS

In the configurational forces approach two systems of forces are introduced: the classical deformational forces that act in the current configuration such as gravity and a new system of forces called configurational forces that act in the reference configuration. The configurational forces are responsible for kinematic changes in the reference such as the propagation of the crack-tip T and of the phase boundary Σ. Consequently there are separate balance laws for the deformational and configurational forces.

Let $\mathcal{D} \subset \Omega$ be a part of the body that contains the crack-tip and a part of the transformation zone (see Figure 1). We denote the part of the phase-boundary that lies inside $\mathcal{D}$ by Σ_D and the part of the crack by C_D. For evaluating the dissipation due to crack propagation in supercritical ZTC we neglect inertia and heat transfer and consider the following mechanical setting. In the bulk we choose a deformational stress $\mathbf{S}$, a configurational stress $\mathbf{C}$ and a configurational force $\mathbf{f}$. Additional configurational forces will be considered on the phase boundary $\mathbf{f}_\Sigma$ and at the crack-tip $\mathbf{f}_T$. Deformational forces are ignored.

The deformational contact force arises from the bulk stress $\mathbf{S}$ acting on the boundary $\partial\mathcal{D}(t)$. The balance of deformational forces requires that

$$\int_{\partial\mathcal{D}(t)} \mathbf{S}\hat{\mathbf{n}}\, da = \mathbf{0}$$

for every $\mathcal{D} \subset \Omega$ and $t \in I$. The configurational contact force arises from the bulk stress $\mathbf{C}$ acting on $\partial\mathcal{D}(t)$ and the balance of configurational forces requires that

$$\int_{\mathcal{D}(t)} \mathbf{f}\,\mathrm{d}a + \int_{\Sigma_D} \mathbf{f}_\Sigma\,\mathrm{d}l + \int_{\partial\mathcal{D}(t)} \mathbf{C}\widehat{\mathbf{m}}\,\mathrm{d}l + \mathbf{f}_T = \mathbf{0}$$

for every $\mathcal{D}(t) \subset \Omega$ and each time $t \in I$.

Since the bulk stresses can be singular at the crack-tip, we need to be careful in localizing these balance laws. We remove a circle $\mathcal{B}_r$ of radius r centered at the crack-tip, apply the usual divergence theorem in the domain $\mathcal{D}(t)\backslash\mathcal{B}_r$ and then take the limit $r \to 0$. See equations (A.1) and (A.2) in Appendix A for details. Localizing the deformational and configurational force balances, we obtain

$$\nabla\cdot\mathbf{S} = \mathbf{0}, \qquad \nabla\cdot\mathbf{C}+\mathbf{f}=\mathbf{0} \quad \text{in } \Omega, \tag{2.4}$$

$$[[\mathbf{S}]]\hat{\mathbf{n}} = \mathbf{0}, \qquad [[\mathbf{C}]]\hat{\mathbf{n}}+\mathbf{f}_\Sigma=\mathbf{0} \quad \text{in } \Sigma, \tag{2.5}$$

$$[[\mathbf{S}]]\hat{\mathbf{p}} = \mathbf{0}, \qquad [[\mathbf{C}]]\hat{\mathbf{p}}=\mathbf{0} \quad \text{in } C, \tag{2.6}$$

$$\lim_{r\to 0}\int_{\partial\mathcal{B}_r} \mathbf{S}\hat{\mathbf{n}}\,\mathrm{d}l = \mathbf{0}, \qquad \lim_{r\to 0}\int_{\partial\mathcal{B}_r} \mathbf{C}\hat{\mathbf{n}}\,\mathrm{d}l + \mathbf{f}_T = \mathbf{0} \quad \text{at } T. \tag{2.7}$$

Equation $(2.4)_1$ in the bulk is the familiar equilibrium equation of continuum mechanics. On the phase boundary, equation $(2.5)_1$ ensures traction continuity, whereas equation $(2.6)_1$ is trivially satisfied since the crack faces are traction free. The limiting value of the singular bulk stress vanishes at the tip, $(2.7)_1$.

2.3. DISSIPATION

In the configurational forces approach the dissipation of a control volume $\mathcal{D}(t) \subset \Omega$ that evolves in the reference is used to find the driving forces on the phase boundary and crack-tip. The boundary $\partial\mathcal{D}(t)$ has an external normal $\widehat{\mathbf{m}}$ and suppose that it is described by the parameterization

$$\partial\mathcal{D}(t) = \left\{\mathbf{x} \in \Omega\colon \mathbf{x} = \tilde{\mathbf{x}}(\delta, t)\right\}, \tag{2.8}$$

where δ is a scalar parameter. The velocity of $\partial\mathcal{D}$ in the reference, $\mathbf{w}$, and its normal velocity, w_m, are given by

$$\mathbf{w} = \frac{\partial\tilde{\mathbf{x}}(\cdot, t)}{\partial t} \quad \text{and} \quad w_m = \mathbf{w}\cdot\widehat{\mathbf{m}},$$

respectively; only the normal component w_m is independent of the parametrization (2.8). The position of the boundary $\partial\mathcal{D}(t)$ in the deformed configuration can be obtained through the composition $\mathbf{y}(\tilde{\mathbf{x}}(\cdot, t), t)$. Differentiating this we get the velocity in the deformed configuration as

$$\frac{\partial\mathbf{y}(\tilde{\mathbf{x}}(\cdot, t), t)}{\partial t} = \dot{\mathbf{y}} + \mathbf{F}\mathbf{w}.$$

Table I. Velocities in the reference and current configurations

Region	Reference vel.	Current vel.
Σ	$\mathbf{v}$	$\dot{\mathbf{y}} + \mathbf{F}\mathbf{v}$
T	$\mathbf{v}_T$	$\dot{\mathbf{y}} + \mathbf{F}\mathbf{v}_T$
$\partial \mathcal{B}_r$	$\mathbf{w}$	$\dot{\mathbf{y}} + \mathbf{F}\mathbf{w}$
$\partial \mathcal{D}$	$\mathbf{w}$	$\dot{\mathbf{y}} + \mathbf{F}\mathbf{w}$

In a similar manner we can find the velocities of the phase boundary Σ, crack-tip T and the disk $\mathcal{B}_r$ in the deformed configuration (see Table I).

Working. The important difference between the deformational and configurational forces is their conjugate velocities: deformational forces do work on velocities in the deformed configuration, whereas the configurational forces perform work on reference velocities. Thus the external working on the region $\mathcal{D}(t)$ is

$$\mathcal{W}\big(\mathcal{D}(t)\big) = \int_{\partial \mathcal{D}(t)} \big[\mathbf{S}\widehat{\mathbf{m}} \cdot (\dot{\mathbf{y}} + \mathbf{F}\mathbf{w}) + \mathbf{C}\widehat{\mathbf{m}} \cdot \mathbf{w}\big] \mathrm{d}l. \tag{2.9}$$

The configurational forces $\mathbf{f}$, $\mathbf{f}_\Sigma$ and $\mathbf{f}_T$ are internal, and hence do not contribute to the working. The working due to the configurational stress accounts for the material entering or leaving the evolving control volume $\mathcal{D}(t)$.

The working (2.9) is required to be independent of parameterizations of the curve $\partial \mathcal{D}(t)$. Since only the normal component of the curve velocity ($\mathbf{w} \cdot \widehat{\mathbf{m}}$) is independent of the parameterization (2.8), it is necessary that [8]

$$\mathbf{C} = \pi \mathbf{I} - \mathbf{F}^{\mathrm{T}}\mathbf{S}, \tag{2.10}$$

where π is a scalar field and $\mathbf{I}$ denotes the identity.

In the next few steps we use the representation (2.10) for the configurational stress to simplify the working (2.9) as follows:

$$\mathcal{W} = \int_{\partial \mathcal{D}} \pi w_m \, \mathrm{d}l + \int_{\partial \mathcal{D}} \mathbf{S}\widehat{\mathbf{m}} \cdot \dot{\mathbf{y}} \, \mathrm{d}l.$$

Then applying the divergence theorem (A.1) to the second integral we obtain

$$\begin{aligned} \mathcal{W} = & \int_{\partial \mathcal{D}} \pi w_m \, \mathrm{d}l + \int_{\mathcal{D}} (\nabla \cdot \mathbf{S} + \mathbf{S} \cdot \dot{\mathbf{F}}) \, \mathrm{d}a + \lim_{r \to 0} \int_{\partial \mathcal{B}_r} \mathbf{S}\widehat{\mathbf{m}} \cdot \dot{\mathbf{y}} \, \mathrm{d}l \\ & + \int_{\Sigma_D} \big([[\mathbf{S}]]\hat{\mathbf{n}} \cdot \langle \dot{\mathbf{y}} \rangle + \langle \mathbf{S} \rangle \hat{\mathbf{n}} \cdot [[\dot{\mathbf{y}}]] \big) \mathrm{d}l \\ & + \int_{C_D} \big([[\mathbf{S}]]\hat{\mathbf{p}} \cdot \langle \dot{\mathbf{y}} \rangle + \langle \mathbf{S} \rangle \hat{\mathbf{p}} \cdot [[\dot{\mathbf{y}}]] \big) \mathrm{d}l, \end{aligned}$$

where $\langle \mathbf{a} \rangle = (\mathbf{a}^+ + \mathbf{a}^-)/2$. Upon using the balance laws $(2.4)_1$, $(2.5)_1$, $(2.6)_1$ along with the compatibility condition $(2.3)_2$, the working reduces to

$$\begin{aligned} \mathcal{W} &= \int_{\partial\mathcal{D}} \pi w_m \,\mathrm{d}l + \int_{\mathcal{D}} \mathbf{S}\cdot\dot{\mathbf{F}}\,\mathrm{d}a + \lim_{r\to 0}\int_{\partial\mathcal{B}_r} \mathbf{S}\widehat{\mathbf{m}}\cdot\dot{\mathbf{y}}\,\mathrm{d}l \\ &\quad - \int_{\Sigma_D} v_n\langle \mathbf{S}\rangle\hat{\mathbf{n}}\cdot[[\mathbf{F}]]\hat{\mathbf{n}}\,\mathrm{d}l. \end{aligned} \tag{2.11}$$

Energy. Let the bulk energy per unit reference area be ϕ. Then the energy of the region $\mathcal{D}(t)$ is $\mathcal{E}(\mathcal{D}(t)) = \int_{\mathcal{D}(t)} \phi\,\mathrm{d}a$. The transport identity (A.3) accounts for the phase boundary and the possibility of singular bulk fields at the crack-tip. Applying it to the evolving region $\mathcal{D}(t)$, we get

$$\begin{aligned} \frac{\mathrm{d}\mathcal{E}(\mathcal{D}(t))}{\mathrm{d}t} &= \int_{\partial\mathcal{D}(t)} \phi w_m\,\mathrm{d}l + \int_{\mathcal{D}(t)} \dot{\phi}\,\mathrm{d}a - \int_{\Sigma_D} [[\phi]]v_n\,\mathrm{d}l \\ &\quad - \lim_{r\to 0}\int_{\partial\mathcal{B}_r} \phi w_m\,\mathrm{d}a. \end{aligned} \tag{2.12}$$

Dissipation. The dissipation Γ of the region $\mathcal{D}(t)$ is defined as

$$\Gamma\big(\mathcal{D}(t)\big) = \mathcal{W}\big(\mathcal{D}(t)\big) - \frac{\mathrm{d}\mathcal{E}\big(\mathcal{D}(t)\big)}{\mathrm{d}t}, \tag{2.13}$$

and a version of the second law of thermodynamics appropriate for this mechanical setting is that the dissipation Γ be non-negative, i.e.,

$$\Gamma\big(\mathcal{D}(t)\big) \geqslant 0 \quad \text{for every } \mathcal{D}(t) \subset \Omega,\ t \in I. \tag{2.14}$$

Substituting equations (2.11) and (2.12) in (2.13) we write the dissipation as

$$\begin{aligned} \Gamma\big(\mathcal{D}(t)\big) &= \int_{\partial\mathcal{D}(t)} (\pi - \phi) w_m\,\mathrm{d}l + \int_{\mathcal{D}(t)} \big[\mathbf{S}\cdot\dot{\mathbf{F}} - \dot{\phi}\big]\mathrm{d}a \\ &\quad + \int_{\Sigma_D} \big([[\phi]] - \langle\mathbf{S}\rangle\hat{\mathbf{n}}\cdot[[\mathbf{F}]]\hat{\mathbf{n}}\big)v_n\,\mathrm{d}l + \Gamma_1 \geqslant 0 \end{aligned} \tag{2.15}$$

$$\Gamma_1 = \lim_{r\to 0}\int_{\partial\mathcal{B}_r} \big[\mathbf{S}\widehat{\mathbf{m}}\cdot\dot{\mathbf{y}} + \phi w_m\big]\mathrm{d}l, \tag{2.16}$$

where we have collected certain terms as Γ_1 for further simplification.

We now show that $\pi = \phi$. Consider a second region $\mathcal{D}'(t)$ such that it coincides with the region $\mathcal{D}(t)$ at time t and the boundary $\partial\mathcal{D}'(t)$ has a normal velocity $w'_m \neq w_m$. We obtain an expression similar to (2.15) for $\Gamma(\mathcal{D}'(t))$. The dissipation inequality (2.14) must be satisfied for both regions, which implies that $\pi = \phi$. Consequently, equation (2.10) implies that the bulk configurational stress is [8]

$$\mathbf{C} = \phi\mathbf{I} - \mathbf{F}^{\mathrm{T}}\mathbf{S}, \tag{2.17}$$

which is nothing but the Eshelby or energy-momentum tensor [14].

By adding and subtracting the term $\mathbf{S}\widehat{\mathbf{m}} \cdot \mathbf{F}\mathbf{w}$ to the integrand, we rewrite Γ_1 as

$$\Gamma_1 = \lim_{r\to 0} \int_{\partial \mathcal{B}_r} \left[\mathbf{S}\widehat{\mathbf{m}} \cdot (\dot{\mathbf{y}} + \mathbf{F}\mathbf{w}) + \mathbf{w} \cdot \left(\phi\mathbf{I} - \mathbf{F}^{\mathrm{T}}\mathbf{S}\right)\widehat{\mathbf{m}}\right] \mathrm{d}l.$$

Since $\mathbf{w} \to \mathbf{v}_T$ and $\dot{\mathbf{y}} + \mathbf{F}\mathbf{w} \to \dot{\mathbf{y}} + \mathbf{F}\mathbf{v}_T$ as $r \to 0$, it can be shown that [9]

$$\Gamma_1 = (\dot{\mathbf{y}} + \mathbf{F}\mathbf{v}_T) \cdot \lim_{r\to 0} \int_{\partial \mathcal{B}_r} \mathbf{S}\widehat{\mathbf{m}}\, \mathrm{d}l + \mathbf{v}_T \cdot \lim_{r\to 0} \int_{\partial \mathcal{B}_r} \left(\phi\mathbf{I} - \mathbf{F}^{\mathrm{T}}\mathbf{S}\right)\widehat{\mathbf{m}}\, \mathrm{d}l. \tag{2.18}$$

Finally, upon using the balance law $(2.7)_1$ to simplify the relation (2.18), the dissipation inequality (2.15) reduces to

$$\begin{aligned} \Gamma(\mathcal{D}) = & \int_{\mathcal{D}} \left[\mathbf{S} \cdot \dot{\mathbf{F}} - \dot{\phi}\right] \mathrm{d}a + \int_{\Sigma_D} \left([[\phi]] - \langle\mathbf{S}\rangle\hat{\mathbf{n}} \cdot [[\mathbf{F}]]\hat{\mathbf{n}}\right) v_n\, \mathrm{d}l \\ & + \mathbf{v}_T \cdot \lim_{r\to 0} \int_{\partial \mathcal{B}_r} \left(\phi\mathbf{I} - \mathbf{F}^{\mathrm{T}}\mathbf{S}\right)\widehat{\mathbf{m}}\, \mathrm{d}l \geqslant 0. \end{aligned} \tag{2.19}$$

If we view the normal velocity of the phase boundary v_n and the velocity of the crack-tip $\mathbf{v}_T$ as internal variables, then the terms that are conjugate to these variables in the dissipation (2.19) can be identified as the driving forces [7]. Consequently the driving force d_Σ on the phase boundary and the crack-tip driving force $\mathbf{d}_T$ are given by

$$d_\Sigma := [[\phi]] - \langle\mathbf{S}\rangle\hat{\mathbf{n}} \cdot [[\mathbf{F}]]\hat{\mathbf{n}}, \tag{2.20}$$

$$\mathbf{d}_T := \lim_{r\to 0} \int_{\partial \mathcal{B}_r} (\phi\mathbf{I} - \mathbf{F}^{\mathrm{T}}\mathbf{S})\widehat{\mathbf{m}}\, \mathrm{d}l. \tag{2.21}$$

The interface driving force (2.20) has been derived by [6, 7, 8] and the crack-tip driving force (2.21) was obtained by [9, 15, 16].

2.4. CRACK-TIP DISSIPATION AND TOUGHENING

Consider a homogeneous hyperelastic specimen with a crack. By measuring the load-displacement curve and tracking the crack-length the energy dissipated can be measured. This is exactly equal to the energy dissipated at the crack-tip, since there are no other dissipative sources in the specimen. On the other hand, for a specimen of supercritical ZTC the measured dissipation need not be equal to the dissipation at the crack-tip, and we refer to the difference between the measured and the crack-tip dissipation as the toughening.

Following Gurtin and Podio-Guidugli [9], we define the crack-tip dissipation Γ_T as the limiting value of the dissipation of the region $\mathcal{B}_r$ in the limit as the radius $r \to 0$. By definition, the dissipation of the region $\mathcal{B}_r$ is

$$\Gamma(\mathcal{B}_r) = \int_{\partial \mathcal{B}_r} \left[\mathbf{S}\widehat{\mathbf{m}} \cdot (\dot{\mathbf{y}} + \mathbf{F}\mathbf{w}) + \mathbf{C}\widehat{\mathbf{m}} \cdot \mathbf{w}\right] \mathrm{d}l - \frac{\mathrm{d}}{\mathrm{d}t} \int_{\mathcal{B}_r} \phi\, \mathrm{d}a.$$

Suppose that the energy density ϕ satisfies:
(a) the derivative of ϕ following the crack tip is integrable on the disk $\mathcal{B}_r$, and
(b) $[[\phi]]\hat{\mathbf{p}} \cdot \mathbf{v}_T$ is integrable on the crack C, where $[[\phi]]$ denotes the jump across the crack.

The linear elastic energy, for instance, satisfies these conditions. For such energies, Gurtin and Podio-Guidugli [9, 2.20] show that

$$\lim_{r\to 0} \frac{\mathrm{d}}{\mathrm{d}t} \int_{\mathcal{B}_r} \phi \, \mathrm{d}a = 0 \, .$$

Next, on using the arguments that lead to equation (2.18), the crack-tip dissipation reduces to

$$\begin{aligned}\Gamma_T : &= \lim_{r\to 0} \Gamma(\mathcal{B}_r) = (\dot{\mathbf{y}} + \mathbf{F}\mathbf{v}_T) \cdot \lim_{r\to 0} \int_{\partial\mathcal{B}_r} \mathbf{S}\widehat{\mathbf{m}} \, \mathrm{d}l \\ &\quad + \mathbf{v}_T \cdot \lim_{r\to 0} \int_{\partial\mathcal{B}_r} (\phi \mathbf{I} - \mathbf{F}^{\mathrm{T}}\mathbf{S})\widehat{\mathbf{m}} \, \mathrm{d}l.\end{aligned}$$

Finally, using the balance law $(2.7)_1$, we see that the crack-tip dissipation is given by

$$\Gamma_T = \mathbf{v}_T \cdot \lim_{r\to 0} \int_{\partial\mathcal{B}_r} (\phi \mathbf{I} - \mathbf{F}^{\mathrm{T}}\mathbf{S})\widehat{\mathbf{m}} \, \mathrm{d}l = \mathbf{v}_T \cdot \mathbf{d}_T \geqslant 0. \tag{2.22}$$

In fracture mechanics the energy dissipated per unit crack extension is commonly used, and it can be written as

$$J_T = \hat{\mathbf{e}}_1 \cdot \lim_{r\to 0} \int_{\partial\mathcal{B}_r} (\phi \mathbf{I} - \mathbf{F}^{\mathrm{T}}\mathbf{S})\widehat{\mathbf{m}} \, \mathrm{d}l = \hat{\mathbf{e}}_1 \cdot \mathbf{d}_T \geqslant 0, \tag{2.23}$$

where the unit vector $\hat{\mathbf{e}}_1 = \mathbf{v}_T/|\mathbf{v}_T|$ lies along the direction of crack growth. The above integral is the J-integral of fracture mechanics [16]. We note that it was derived without using any specific form for the energy ϕ, hence it is valid even for non-linear hyperelastic materials.

We now define toughening by relating the crack-tip dissipation to the total dissipation. Let $\mathcal{D}$ be the region between the two circles $\partial\mathcal{B}_r$ and $\partial\mathcal{B}_{\mathrm{far}}$ (see Figure 2); note that $\mathcal{D}$ does not include the crack or the disk $\mathcal{B}_r$. Thus, the statement of configurational force balance for region $\mathcal{D}$ is

$$\int_{\mathcal{D}} \mathbf{f} \, \mathrm{d}a + \int_{\Sigma_D} \mathbf{f}_\Sigma \, \mathrm{d}l + \int_{\partial\mathcal{D}} \mathbf{C}\widehat{\mathbf{m}} \, \mathrm{d}l = 0.$$

The integral of the configurational contact stress can be decomposed as follows:

$$\int_{\partial\mathcal{D}} \mathbf{C}\widehat{\mathbf{m}} \, \mathrm{d}l = \int_{\partial\mathcal{B}_{\mathrm{far}}} \mathbf{C}\widehat{\mathbf{m}} \, \mathrm{d}l - \int_{\partial\mathcal{B}_r} \mathbf{C}\widehat{\mathbf{m}} \, \mathrm{d}l - \int_{C_D} [[\mathbf{C}]]\hat{\mathbf{p}} \, \mathrm{d}l.$$

The balance law $(2.6)_2$ implies that $[[\mathbf{C}]]\hat{\mathbf{p}} = 0$. Thus we obtain

$$\int_{\partial \mathcal{B}_{\text{far}}} \mathbf{C}\widehat{\mathbf{m}}\, dl - \int_{\partial \mathcal{B}_r} \mathbf{C}\widehat{\mathbf{m}}\, dl = -\int_{\mathcal{D}} \mathbf{f}\, da - \int_{\Sigma_D} \mathbf{f}_\Sigma\, dl. \tag{2.24}$$

Remembering the balance law $(2.5)_2$, we note that $\mathbf{f}_\Sigma \cdot \hat{\mathbf{n}} = -d_\Sigma$. However, since tangential motions of the phase boundary are indistinguishable from reparametrization, the tangential component of the interface configurational force is indeterminate [8]. Hence, using $\mathbf{f}_\Sigma = -d_\Sigma \hat{\mathbf{n}}$, representation (2.17) and taking the limit as $r \to 0$, we rewrite (2.24) as

$$\hat{\mathbf{e}}_1 \cdot \int_{\partial \mathcal{B}_{\text{far}}} (\phi \mathbf{I} - \mathbf{F}^{\text{T}}\mathbf{S})\widehat{\mathbf{m}}\, dl - \hat{\mathbf{e}}_1 \cdot \lim_{r \to 0} \int_{\partial \mathcal{B}_r} (\phi \mathbf{I} - \mathbf{F}^{\text{T}}\mathbf{S})\widehat{\mathbf{m}}\, dl$$
$$= -\hat{\mathbf{e}}_1 \cdot \int_{\mathcal{D}} \mathbf{f}\, da + \int_{\Sigma_D} d_\Sigma (\hat{\mathbf{n}} \cdot \hat{\mathbf{e}}_1)\, dl. \tag{2.25}$$

For homogeneous hyperelastic materials there is no phase boundary and $\mathbf{f} = 0$ [8, 11], consequently equation (2.25) would imply the path-independence of the J-integral (in which case the limit in (2.21) and (2.23) is unnecessary).

3. Supercritical ZTC

3.1. CONSTITUTIVE MODEL

We now construct a geometrically linear double-well energy for supercritical ZTC. Let $\mathbf{u}$ denote the displacement and $\mathbf{E} = (\nabla \mathbf{u} + \nabla \mathbf{u}^{\text{T}})/2$ the usual linear strain. The strain can be decomposed as

$$\mathbf{E} = \theta \mathbf{I} + \mathbf{e}; \quad \theta = Tr\mathbf{E} \quad \text{and} \quad \mathbf{e} = \mathbf{E} - \frac{\theta}{3}\mathbf{I}, \tag{3.1}$$

where θ is the dilatation, $\mathbf{I}$ is the identity tensor and $\mathbf{e}$ denotes the deviatoric part of the strain. Similarly, we decompose the (Cauchy) stress into hydrostatic and deviatoric parts as

$$\mathbf{S} = \sigma \mathbf{I} + \mathbf{s}; \quad \sigma = \frac{Tr\mathbf{S}}{3} \quad \text{and} \quad \mathbf{s} = \mathbf{S} - \sigma \mathbf{I}, \tag{3.2}$$

where σ is the mean-stress and $\mathbf{s}$ is the deviatoric stress.

Following [2], we choose the following form for the (Helmholtz) free energy:

$$\phi = \hat{\phi} + \mu \mathbf{e} \cdot \mathbf{e}, \quad \text{where} \quad \hat{\phi} = \begin{cases} \dfrac{\kappa \theta^2}{2} + \hat{\phi}_a, & -\theta_1 \leqslant \theta \leqslant \theta_1, \\ \dfrac{\kappa (\theta - \theta_t)^2}{2} + \hat{\phi}_m, & \theta_t - \theta_2 \leqslant \theta, \end{cases} \tag{3.3}$$

where the scalars $\hat{\phi}_m > \hat{\phi}_a \geqslant 0$, $\theta_1 > 0$, $\theta_2 > 0$ and the transformation dilatation $\theta_t > 0$. The energy has two wells in the direction of the dilatation but is linear in the space of the deviatoric strains. The austenite well corresponds to the dilatations for $-\theta_1 \leqslant \theta \leqslant \theta_1$, whereas the martensite well corresponds to the dilatation range $\theta_t - \theta_2 \leqslant \theta$. The elastic moduli of the two phases are taken to be the same. The bulk modulus is denoted by κ and the shear modulus by μ.

If we consider a body subjected to dead loads and restrict ourselves to homogenous deformations, then equilibrium corresponds to minimizing the Gibbs potential $\psi = \phi - \mathbf{S} \cdot \mathbf{E}$ where $\mathbf{S}$ is the stress corresponding to the dead loads. The decompositions (3.1), (3.2) and energy (3.3) imply that the Gibbs potential can also be decomposed as follows:

$$\psi = \hat{\psi} + (\mu \mathbf{e} \cdot \mathbf{e} - \mathbf{s} \cdot \mathbf{e}), \quad \text{where} \quad \hat{\psi} := \hat{\phi} - \sigma\theta. \tag{3.4}$$

Differentiating the energy (3.3) with respect to strain, we obtain the stress–strain relations

$$\mathbf{s} = 2\mu\mathbf{e} \quad \text{and} \quad \sigma = \begin{cases} \kappa\theta, & -\theta_1 \leqslant \theta \leqslant \theta_1, \\ \kappa(\theta - \theta_t), & \theta_t - \theta_2 \leqslant \theta. \end{cases} \tag{3.5}$$

For certain values of the mean stress, the corresponding dilatation can either lie in the austenite or in the martensite phase. The Maxwell stress is the applied stress corresponding to the situation when the two phases have the same (Gibbs) energy. From equations (3.3)–(3.5), it can be shown that the Maxwell stress σ^m is given by

$$\sigma^m = \frac{\hat{\phi}_m - \hat{\phi}_a}{\theta_t}. \tag{3.6}$$

Since the energy (3.3) is non-linear only in dilatation, it follows that the Maxwell stress is a mean-stress.

In addition, we need a nucleation criterion for stress-induced phase transformation. Following [17], we choose

$$\text{austenite} \rightarrow \text{martensite} \quad \sigma \nearrow \sigma_1 = \sigma^m + \sigma^*, \tag{3.7}$$

$$\text{martensite} \rightarrow \text{austenite} \quad \sigma \searrow \sigma_2 = \sigma^m - \sigma^*. \tag{3.8}$$

Since few, if any, particles in the wake of a crack transform back to the austenite phase in supercritical zirconia, we take $\sigma_2 < 0$, whereas $\sigma_1 > 0$. This implies that $\sigma^m < \sigma^1/2$.

We now find the conditions for the existence of a phase boundary separating austenite and martensite in a material described by the energy (3.3). The following conditions need to be satisfied at the phase boundary:

(a) $[[\mathbf{S}]]\hat{\mathbf{n}} = \mathbf{0}$, and

(b) $[[\nabla \mathbf{u}]] = \mathbf{a} \otimes \hat{\mathbf{n}}$ for some nonzero vector $\mathbf{a}$.

In this geometrically linear setting, (a) corresponds to the traction continuity (2.5) and (b) to the Hadamard compatibility (2.3). It can be shown that these two conditions and the stress–strain relation (3.5) imply the following [18]:

$$\mathbf{a} = [[\theta]]\hat{\mathbf{n}}, \qquad [[\theta]] = -\frac{(1+\nu)\theta_t}{3(1-\nu)} \quad \text{and} \quad \frac{[[\sigma]]}{[[\theta]]} = -\frac{4\mu}{3}, \tag{3.9}$$

where ν is the Poisson's ratio.

For later use, we list the stress–strain relations corresponding to the martensite branch. The stress–strain relations are given by

$$\mathbf{S} = \kappa(\theta - \theta_t)\mathbf{I} + 2\mu\left(\mathbf{E} - \frac{\theta}{3}\mathbf{I}\right), \tag{3.10}$$

$$\mathbf{E} = \frac{1}{2\mu}(\mathbf{S} - \sigma\mathbf{I}) + \frac{1}{3\kappa}\sigma\mathbf{I} + \frac{\theta_t}{3}\mathbf{I}. \tag{3.11}$$

The bulk and shear moduli are related to the Young's modulus E and Poisson's ratio ν through the relations

$$\kappa = \frac{E}{3(1-2\nu)} \quad \text{and} \quad \mu = \frac{E}{2(1+\nu)}. \tag{3.12}$$

3.2. TOUGHENING

We now model the situation depicted in Figure 1 and obtain the toughening due to phase boundary propagation in supercritical ZTC. Consider a Mode-I crack specimen under plane-strain conditions. Then the equilibrium stresses in the vicinity of the crack-tip are given by the standard relations [19], and the corresponding mean-stress is

$$\sigma = \frac{2(1+\nu)}{3\sqrt{2\pi r}} K \cos\left(\frac{\beta}{2}\right), \tag{3.13}$$

where K is the applied stress-intensity factor and (r, β) are polar co-ordinates with origin at the crack-tip.

Suppose that the crack propagates in the x_1 direction in a steady manner (see Figure 2). To find the stress loading path at a point P due to the propagating crack-tip, we can fix the crack-tip and instead decrease the x_1 coordinate of the point. Thus when the point is in front of the crack and far from the crack-tip, the stresses are zero and the material is in the austenite phase. As the crack approaches closer, the mean-stress increases on the austenite branch, and when it reaches the critical value σ_1 the material at point P transforms to the martensite phase (3.7). The jumps in the dilatation and mean-stress are governed by the relations (3.9). As x_1 is further decreased, the mean-stress first increases, and then decreases. Finally, when x_1 is in the wake of the crack, the stresses become vanishingly small, however, the material at point P remains in the martensite phase.

The front boundary of the transformation zone is determined by the contour line of the mean-stress with value $\sigma = \sigma_1$ and since there is no reverse transformation the transformation zone is obtained by translating this contour backwards. The transformation zone is symmetric about the crack and the upper half can be specified as [2]

$$r \leqslant \begin{cases} \dfrac{8h}{3\sqrt{3}} \cos^2\left(\dfrac{\beta}{2}\right), & 0 \leqslant \beta \leqslant \dfrac{\pi}{3}, \\ \dfrac{h}{\sin(\beta)}, & \dfrac{\pi}{3} \leqslant \beta < \pi, \end{cases} \tag{3.14}$$

where (r, β) are polar coordinates with the origin at the crack-tip and the half-height of the transformation zone is given by

$$h = \frac{(1+\nu)^2}{4\pi\sqrt{3}}\left(\frac{K}{\sigma_1}\right)^2. \tag{3.15}$$

The toughening due to the phase transformation in supercritical ZTC can be calculated using (2.25). Note that the situation here corresponds to homogeneous materials inside and outside the transformation zone, consequently the configurational body force $\mathbf{f} = 0$ [8, 11]. In this geometrically linear setting the driving force on the phase boundary (2.20) and the energy-release rate per unit crack extension (2.23) (see [15]) are given by

$$d_\Sigma = [[\phi]] - \langle \mathbf{S}\hat{\mathbf{n}} \rangle \cdot [[\mathbf{E}]]\hat{\mathbf{n}}, \tag{3.16}$$

$$J_T = \hat{\mathbf{e}}_1 \cdot \lim_{r\to 0} \int_{\partial \mathcal{B}_r} (\phi \mathbf{I} - (\nabla \mathbf{u})^{\mathrm{T}}\mathbf{S})\widehat{\mathbf{m}}\, \mathrm{d}l. \tag{3.17}$$

Moreover, the integral over the boundary $\partial\mathcal{B}_{\text{far}}$ can be split into two integrals:
(i) one over the wake of the transformation zone, i.e., $\partial\mathcal{B}_{\text{far}} \cap \mathcal{A}$ and
(ii) over the rest of the boundary, i.e., $\partial\mathcal{B}_{\text{far}} \backslash \mathcal{A}$.
Typically, the height of the transformation zone in supercritical ZTC is on the order of micrometers. Hence, it is reasonable that part (ii) corresponds to the energy release rate, J_{far}, that can be measured in experiments [2]. Consequently, the expression (2.25) can be written as

$$J_{\text{far}} - J_T = -\hat{\mathbf{e}}_1 \cdot \int_{\partial\mathcal{B}_{\text{far}}\cap\mathcal{A}} (\phi\mathbf{I} - (\nabla\mathbf{u})^{\mathrm{T}}\mathbf{S})\widehat{\mathbf{m}}\, \mathrm{d}l + \int_{\Sigma_D} d_\Sigma (\hat{\mathbf{n}}\cdot\hat{\mathbf{e}}_1)\, \mathrm{d}l. \tag{3.18}$$

Next we evaluate the integrals in (3.18). Although the crack-tip stresses vanish in the wake, the transformation strain results in residual stresses and strains characterized by
(1) $S_{22} = S_{21} = 0$, due to the requirement that the crack is traction free,
(2) $E_{11} = 0$, due to the constraints of the elastic half-space above the wake, and
(3) $E_{13} = E_{23} = E_{33} = 0$, due to plane-strain [2].
Using these in the stress–strain relations (3.10) and (3.11), we get the only nonzero stresses and strains to be

$$S_{11} = S_{33} = -\frac{E\theta_t}{3(1-\nu)}, \qquad E_{22} = \frac{\theta_t(1+\nu)}{3(1-\nu)}. \tag{3.19}$$

For the residual fields in the wake, the term $(\nabla\mathbf{u})\hat{\mathbf{e}}_1 \cdot \mathbf{S}\widehat{\mathbf{m}} = 0$. Then, noting that $\widehat{\mathbf{m}} = -\hat{\mathbf{e}}_1$, we write the first integral in (3.18) as

$$-2\hat{\mathbf{e}}_1 \cdot \int_{\partial\mathcal{B}_{\text{far}}\cap\mathcal{A}^+} \phi\widehat{\mathbf{m}}\, \mathrm{d}l = 2\phi\int_0^h \mathrm{d}l = 2h\left[\hat{\phi}_m + \frac{E(5-7\nu)\theta_t^2}{27(1-\nu)^2}\right], \tag{3.20}$$

where the constant energy in the wake has been evaluated using (3.3), (3.12) and (3.19).

The driving force on the phase boundary can be calculated using (3.3), (3.5), (3.9) and (3.16) as

$$d_\Sigma = (\langle\sigma\rangle - \sigma^m)\theta_t,$$

where the average stress at the phase boundary is given by $\langle\sigma\rangle = \kappa(\langle\theta\rangle - 2\theta_t)$. Since the transformation condition (3.7) requires $\sigma^+ = \sigma_1$, we can use (3.5), (3.9) and (3.12) to obtain

$$d_\Sigma = \left(\sigma_1 - \sigma^m - \frac{E\theta_t}{9(1-\nu)}\right)\theta_t.$$

Thus, the driving force on the phase boundary is constant and, hence, the second integral in equation (3.18) becomes

$$\begin{aligned}\int_{\Sigma_D} d_\Sigma(\hat{\mathbf{n}}\cdot\hat{\mathbf{e}}_1)\mathrm{d}l &= d_\Sigma \int_0^{\pi/3} (\hat{\mathbf{n}}(\beta)\cdot\hat{\mathbf{e}}_1) r\,\mathrm{d}\beta \\ &= \left(\sigma_1 - \sigma^m - \frac{E\theta_t}{9(1-\nu)}\right)\theta_t\left(\frac{76h}{45\sqrt{3}}\right).\end{aligned} \tag{3.21}$$

To evaluate the integral above we have used the following expression for the normal to the boundary of the transformation zone:

$$\hat{\mathbf{n}} = \begin{cases} \cos\left(\frac{3\beta}{2}\right)\hat{\mathbf{e}}_2 + \sin\left(\frac{3\beta}{2}\right)\hat{\mathbf{e}}_2, & 0 \leqslant \beta \leqslant \frac{\pi}{3}, \\ \hat{\mathbf{e}}_2, & \frac{\pi}{3} \leqslant \beta < \pi, \end{cases}$$

which can be calculated from (3.14).

Consequently, the difference between the far-field and crack-tip energy release rates (3.18) can be written using (3.20, 3.21) as

$$\begin{aligned} J_{\mathrm{far}} - J_T &= \frac{76}{45\sqrt{3}} h\sigma_1\theta_t + \frac{2hE\theta_t^2}{27(1-\nu)}\left[\frac{5-7\nu}{1-\nu} - \frac{38}{15\sqrt{3}}\right] \\ &\quad + 2h\sigma_1\theta_t\left[\frac{\hat{\phi}_m}{\sigma_1\theta_t} - \frac{38}{45\sqrt{3}}\frac{\sigma_m}{\sigma_1}\right]. \end{aligned} \tag{3.22}$$

In contrast Budiansky et al. [2] consider the energy ϕ to be the area under the stress–strain curve. They evaluate $J_{\mathrm{far}} - J_T$ for subcritical ZTC where there is no phase boundary, note that the phase boundary appears in the limit $\kappa = -4\mu/3$, and by taking this limit they evaluate the toughening in super-critical ZTC as

$$J_{\mathrm{far}} - J_T = 2h\sigma_1\theta_t. \tag{3.23}$$

In fracture experiments typically h and K are measured. The critical stress σ_1 can then be calculated (3.15); by using known values for K_{tip} and the relation $J = K^2(1-\nu^2)/E$ the measured toughening can be found. These are sufficient to evaluate the toughening predicted by (3.23). But for the present model we need the values of $\xi = \hat{\phi}_m/(\sigma_1\theta_t)$ and σ_m/σ_1. (Note that $\sigma_m/\sigma_1 < 0.5$; see sentences following (3.8).) The energy ϕ_m is related to the heat of formation of monoclinic zirconia. Since measurements of σ_m and ϕ_m are typically not reported for specimens used in the fracture tests, we set $\sigma_m/\sigma_1 = 0.1$ and treat ξ as a parameter. This allows us to perform a preliminary comparison with the experimental data reported in [2]. One set of data is $E = 200$ GPa, $\theta_t = 0.0174$, $h := 0.6$ microns, $K - K_{\text{tip}} = 2.3$ MPa $\sqrt{m}$ and $\nu = 0.29$. Taking $K = 4$ MPa $\sqrt{m}$, we find the measured toughening corresponds to $J_{\text{far}} - J_T = 60.04$ N/m. The toughness predicted by (3.23) is 29.81 N/m. The first two terms of (3.22) give 24.84 N/m. The last term in (3.22) depends on ξ. For $\xi = 0.5$ the total toughening predicted by (3.22) is 38.3 N/m, and for $\xi = 1$ the toughening is 53.21 N/m.

4. Discussion

The J-integral has been extended to the situation where surfaces of strain discontinuity exist in the vicinity of a crack. Using this we have evaluated the steady-state toughening for supercritical ZTC by attributing the enhancement in toughness to both the energy stored in the wake and the energy dissipated by the propagating transformation front. Preliminary comparisons indicate that the toughness predicted by the present model is larger than the toughness predicted by models that attribute toughening only to the energy stored in the wake [1, 2]. However, the present model requires two additional parameters: the latent heat ϕ_m and the Maxwell stress σ_m. Since fracture experiments do not typically report these thermomechanical properties, definitive comparisons of the toughness predictions with experimental data remain elusive. Nevertheless, the present work conclusively demonstrates that phase boundary dissipation contributes to toughening in supercritical ZTC.

We propose that the model developed here also applies to pseudoelastic crack propagation in polycrystalline shape-memory alloys. If we interpret Σ_D as the collection of the austenite/martensite interface and all the martensite/martensite twin interfaces, then equation (3.18) provides the toughening. Of course, calculating this toughening is complicated by the need to develop constitutive models for polycrystalline shape-memory alloys and to determine the exact geometry of the austenite/martensite and martensite/martensite interfaces in the neighbourhood of the crack.

Acknowledgement

I thank Lev Truskinovsky for useful discussions and for introducing me to transformation toughening and phase boundary propagation. This work was supported by the University of Miami Summer Awards.

Appendix A. Integral Identities

Since the bulk fields can be singular at the crack-tip, the area integral over a region $\mathcal{D}$ that contains the crack-tip is calculated as

$$\int_{\mathcal{D}} \mathbf{b}\,\mathrm{d}a := \lim_{r\to 0}\int_{\mathcal{D}\backslash\mathcal{B}_r} \mathbf{b}\,\mathrm{d}a,$$

where $\mathcal{B}_r$ is a disk or radius r centered at the crack-tip.

Upon using this definition for area integrals and using the usual divergence theorem, one can obtain (see [11])

$$\begin{aligned}\int_{\mathcal{D}} \nabla\cdot\mathbf{b}\,\mathrm{d}a &= \int_{\partial\mathcal{D}} \mathbf{b}\cdot\widehat{\mathbf{m}}\,\mathrm{d}a - \lim_{r\to 0}\int_{\partial\mathcal{B}_r} \mathbf{b}\cdot\widehat{\mathbf{m}}\,\mathrm{d}l \\ &\quad - \int_{\Sigma_D} [[\mathbf{b}]]\cdot\hat{\mathbf{n}}\,\mathrm{d}l - \int_{C_D} [[\mathbf{b}]]\cdot\hat{\mathbf{p}}\,\mathrm{d}l. \end{aligned} \tag{A.1}$$

Then by taking $\mathbf{b} = \mathbf{B}^{\mathrm{T}}\mathbf{a}$ where $\mathbf{a}$ is some constant vector, we get

$$\begin{aligned}\int_{\mathcal{D}} \nabla\cdot\mathbf{B}\,\mathrm{d}a &= \int_{\partial\mathcal{D}} \mathbf{B}\widehat{\mathbf{m}}\,\mathrm{d}a - \lim_{r\to 0}\int_{\partial\mathcal{B}_r} \mathbf{B}\widehat{\mathbf{m}}\,\mathrm{d}l \\ &\quad - \int_{\Sigma_D} [[\mathbf{B}]]\hat{\mathbf{n}}\,\mathrm{d}l - \int_{C_D} [[\mathbf{B}]]\hat{\mathbf{p}}\,\mathrm{d}l. \end{aligned} \tag{A.2}$$

Again, to account for the possiblity of singular bulk fields at the crack-tip, we define

$$\frac{\mathrm{d}}{\mathrm{d}t}\int_{\mathcal{D}} \phi\,\mathrm{d}a := \lim_{r\to 0}\left[\frac{\mathrm{d}}{\mathrm{d}t}\int_{\mathcal{D}\backslash\mathcal{B}_r} \phi\,\mathrm{d}a\right].$$

Then the transport identity for a moving control volume $\mathcal{D}$ that contains a moving phase boundary Σ_D and a propagating crack-tip is [11]

$$\begin{aligned}\frac{\mathrm{d}}{\mathrm{d}t}\int_{\mathcal{D}(t)} \phi(\mathbf{x},t)\,\mathrm{d}a &= \int_{\mathcal{D}(t)} \frac{\partial\phi(\cdot,t)}{\partial t}\,\mathrm{d}a + \int_{\partial\mathcal{D}(t)} \phi(\mathbf{w}\cdot\widehat{\mathbf{m}})\mathrm{d}l \\ &\quad - \int_{\Sigma_D(t)} [[\phi(\mathbf{x},t)]]v_n\,\mathrm{d}l - \lim_{r\to 0}\int_{\partial\mathcal{B}_r(t)} \phi(\mathbf{w}\cdot\widehat{\mathbf{m}})\mathrm{d}l. \end{aligned} \tag{A.3}$$

References

1. R.M. McMeeking and A.G. Evans, Mechanics of transformation-toughening in brittle materials. *J. Am. Ceram. Soc.* **65**(5) (1982) 242–246.
2. B. Budiansky, J.W. Hutchinson and J.C. Lambropoulos, Continuum theory of dilatant transformation toughening in ceramics. *Internat. J. Solids Struct.* **19**(4) (1983) 337–355.

3. R.C. Garvie, R.H.J. Hannink and M.V. Swain, X-ray analysis of the transformed zone in partially stabilized zirconia (PSZ). *J. Matt. Sci. Lett.* **1** (1982) 437–440.
4. M.V. Swain, R.H.J. Hannink and R.C. Garvie, The influence of precipitate size and temperature on the fracture toughness of calcia- and magnesia-partially stabilized zirconia. A.G. Evans, D.P.H. Hasselman and F.F. Lange (eds), *Fracture Mechanics of Ceramics*, Vol. 6. Plenum Press, New York (1983).
5. A.G. Evans and R.M. Cannon, Overview no. 48:toughening in brittle solids by martensitic transformations. *Acta Metall.* **34**(5) (1986) 761–800.
6. L. Truskinovsky, Equilibrium interphase boundaries. *Dokl. Akad. Nauk SSSR* **275**(2) (1982) 306–10.
7. R. Abeyaratne and J.K. Knowles, On the driving traction acting on a surface of strain discontiunity in a continuum. *J. Mech. Phys. Solids* **38** (1990) 345–60.
8. M.E. Gurtin, The nature of configurational forces. *Arch. Rational Mech. Anal.* 131 (1995) 67–100.
9. M.E. Gurtin and P. Podio-Guidugli, Configurational forces and the basic laws for crack propagation. *J. Mech. Phys. Solids* **44**(6) (1996) 905–927.
10. M.E. Gurtin and M.M. Shvartsman, Configurational forces and the dynamics of planar cracks in 3-dimensional bodies. *J. Elasticity* **48**(2) (1997) 167–191.
11. N.K. Simha and K. Bhattacharya, Kinetics of phase boundaries with edges and junctions. *J. Mech. Phys. Solids* **46** (1998) 2323–2359.
12. N.K. Simha and K. Bhattacharya, Kinetics of phase boundaries with edges and junctions in a three-dimensional multi-phase body. *J. Mech. Phys. Solids* (2000) in press.
13. N.K. Simha and L. Truskinovsky, Shear induced transformation toughening in ceramics. *Acta Metall. Mater.* **42**(11) (1994) 3827–36.
14. J.D. Eshelby, Energy relations and the energy-momentum tensor in continuum mechanics. In: M. Kanninen, W. Adler, A. Rosenfield and R. Jaffee (eds). *Inelastic Behavior of Solids*. McGraw-Hill, New York (1970).
15. G.A. Maugin, *Material Inhomogenieties in Elasticity*. Chapman and Hall, London (1993).
16. J.R. Rice, A path independent integral and the approximate analysis of strain concentration by notches and cracks. *J. Appl. Mech.* **35** (1968) 379–86.
17. R. Abeyaratne, S.-J. Kim and J.K. Knowles, A one-dimensional continuum model for shape-memory alloys. *Internat. J. Solids Struct.* **31**(16) (1994) 2229–2249.
18. L. Truskinovsky, private communication (1992).
19. J.R. Rice, The mathematical theory of fracture. *Fracture* **2** (1968) 357–372.

Journal of Elasticity **59:** 213–225, 2000.

Finite Amplitude Transverse Waves in Special Incompressible Viscoelastic Solids

MICHAEL A. HAYES[1] and GIUSEPPE SACCOMANDI[2]
[1]*Department of Mathematical Physics, University College Dublin, Dublin 4, Ireland*
[2]*Dipartimento di Ingegneria dell'Innovazione, Università di Lecce, 73100 Lecce, Italy*

Received 16 February 1999; in revised form 5 January 2000

Dedicated with esteem to Professor Roger Fosdick.

Abstract. We consider the propagation of finite amplitude plane transverse waves in a class of homogeneous isotropic incompressible viscoelastic solids. It is assumed that the Cauchy stress may be written as the sum of an elastic part and a dissipative viscoelastic part. The elastic part is of the form of the stress corresponding to a Mooney–Rivlin material, whereas the dissipative part is a linear combination of $\mathbf{A}_1$, $\mathbf{A}_1^2$ and $\mathbf{A}_2$, where $\mathbf{A}_1$, $\mathbf{A}_2$ are the first and second Rivlin–Ericksen tensors. The body is first subject to a homogeneous static deformation. It is seen that two finite amplitude transverse plane waves may propagate in every direction in the deformed body. It is also seen that a finite amplitude circularly polarized wave may propagate along either $\mathbf{n}^+$ or $\mathbf{n}^-$, where $\mathbf{n}^+$, $\mathbf{n}^-$ are the normals to the planes of the central circular section of the ellipsoid $\mathbf{x} \cdot \mathbf{B}^{-1}\mathbf{x} = 1$. Here $\mathbf{B}$ is the left Cauchy–Green strain tensor corresponding to the finite static homogeneous deformation.

1. Introduction

The aim of this paper is to consider the propagation of finite amplitude shear waves in homogeneous isotropic incompressible viscoelastic materials which are maintained in a state of finite static homogeneous deformation. This is a topic of actual interest in many applications as for example in geophysics where the nonlinear elastic and dissipative behavior of rocks has been recently observed in many experiments [23], in the design of automotive elastomeric components [21] and in the study of seismic isolation devices [18].

Here we consider a special constitutive assumption for which the Cauchy stress $\mathbf{T}$ can be divided into an elastic part $\mathbf{T}^E$ and in a dissipative part $\mathbf{T}^D$. This strong, but interesting, hypothesis has been recently investigated in several papers by Fosdick and coworkers [15, 16]. In these papers the authors address fundamental questions about the thermodynamics and the stability of these materials and they study the nonlinear oscillations of a sphere. The practical use of these models in modeling vibration damping can be found in further studies by Fosdick et al. [17].

The additivity hypothesis for the Cauchy stress may be traced back to Coleman and Noll who in [12] introduce the theory of *Finite Linear Viscoelasticity*. This assumption has been reproposed in several papers and also tested experimentally (see, for example, [20]); today we know that for unfilled polymers this framework leads to quite *good* mathematical models [19].

Using the method considered by Boulanger and Hayes in [4, 6, 7, 8] here we give a complete investigation of transverse waves when the viscoelastic material is maintained in a state of arbitrary static finite deformation and we will exhibit some simple exact solutions. To the best of our knowledge the only exact solutions in nonlinear viscoelasticity that we have to record are the universal solutions found by Carroll [9] and Fosdick [14] and the wave solutions found by Carroll [10, 11].

Some general results about nonlinear wave propagation in viscoelasticity may be found in the reviews and books [1, 13, 22, 24], but these results are mainly concerned with propagation of singular surfaces, qualitative analysis and the derivation of a nonlinear evolution equation by means of asymptotic or spectral methods. They are not strictly related to our work.

2. Basic Equations

The elastic part of our model is characterized by the Mooney–Rivlin strain-energy density, W, which measured per unit volume in the undeformed state is given by

$$2W = C(I-3) + E(II-3), \tag{1}$$

where C, E are material constants and I, II denote the principal invariants of the left Cauchy–Green strain tensor $\mathbf{B} = \mathbf{F}\mathbf{F}^{\mathrm{T}}$:

$$I = tr\,\mathbf{B}, \qquad 2II = (tr\,\mathbf{B})^2 - tr\left(\mathbf{B}^2\right). \tag{2}$$

The components of the gradient of deformation, $\mathbf{F}$, in a rectangular Cartesian coordinate system are

$$F_{ij} = \frac{\partial x_i}{\partial X_j}, \tag{3}$$

where x_i $(i = 1, 2, 3)$ are the coordinates at time t of the point whose coordinates are X_j $(j = 1, 2, 3)$ in the undeformed reference configuration. Here we are considering only isochoric deformations. For these

$$\det \mathbf{F} = 1, \tag{4}$$

in which case $II = tr\,\mathbf{B}^{-1}$.

We assume that

$$C \geqslant 0, \quad E > 0, \quad \text{or} \quad C > 0, \quad E \geqslant 0, \tag{5}$$

which are equivalent to the strong ellipticity condition for Mooney–Rivlin materials [4]. If $E = 0$, the strain-energy has the *neo-Hookean* form.

We assume that the dissipative part of the stress is given by

$$\mathbf{T}^{\mathrm{D}} = \mu\mathbf{A}_1 + \alpha_1\mathbf{A}_2 + \alpha_2\mathbf{A}_1^2, \tag{6}$$

where $\mathbf{A}_1$ and $\mathbf{A}_2$ are the first two Rivlin–Ericksen tensors given by

$$\mathbf{A}_1 = \mathbf{L} + \mathbf{L}^{\mathrm{T}},$$

and

$$\mathbf{A}_2 = \dot{\mathbf{A}}_1 + \mathbf{A}_1\mathbf{L} + \mathbf{L}^{\mathrm{T}}\mathbf{A}_1,$$

with $\mathbf{L} = \dot{\mathbf{F}}\mathbf{F}^{-1}$ (spatial velocity gradient). Moreover, μ is the classical, constant Newtonian viscosity, and α_1 and α_2 are two constant coefficients supposed to represent the non-Newtonian effect. In this case the constitutive equation for the Cauchy stress tensor, $\mathbf{T}$, is given by

$$\mathbf{T} = -\,p\mathbf{1} + C\mathbf{B} - E\mathbf{B}^{-1} + \mu\mathbf{A}_1 + \alpha_1\mathbf{A}_2 + \alpha_2\mathbf{A}_1^2, \tag{7}$$

where p is the indeterminate pressure introduced by the incompressibility constraint (4). The case $\alpha_1 = 0$, $\alpha_2 = 0$, $E = 0$, a blend between a neo-Hookean material and a Navier–Stokes fluid

$$\mathbf{T} = -p\mathbf{1} + C\mathbf{B} + \mu\mathbf{A}_1, \tag{8}$$

has been considered by Beatty and Zhou [3].

The thermodynamic analysis of (7) can be derived from the work of Fosdick and Yu [15]. Using the Clausius-Duhem inequality and the requirement that the Helmholtz free energy function is a minimum at equilibrium, they derived the following restrictions on the constitutive parameters:

$$\mu \geqslant 0, \quad \alpha_1 + \alpha_2 = 0, \quad \alpha_1 \geqslant 0.$$

We recall that the equations of motion, in the absence of body forces, are

$$\rho\ddot{\mathbf{x}} = \operatorname{div}\mathbf{T}, \tag{9}$$

where ρ is the constant mass density of the material.

3. Transverse Waves

Let us consider a finite static isochoric homogeneous deformation defined by

$$\mathbf{x} = \mathbf{F}\mathbf{X}, \tag{10}$$

where F_{iA} are constant and $\det\mathbf{F} = 1$. For this deformation both $\mathbf{B}$ and $\mathbf{B}^{-1}$ are constants, $\mathbf{A}_1 = \mathbf{A}_2 = 0$, and

$$\mathbf{T} = -p\mathbf{1} + C\mathbf{B} - E\mathbf{B}^{-1}. \tag{11}$$

Moreover, the equilibrium equations give $p = p^*$, a constant.

We superpose on this static deformation a plane transverse wave, i.e., a wave propagating in the direction of the unit vector $\mathbf{n}$, and polarized in the direction of the unit vector $\mathbf{a}$, such that $\mathbf{n} \perp \mathbf{a}$.

Thus, we seek solutions of the equations of motion in the form

$$\bar{\mathbf{x}} = \mathbf{x} + \mathbf{a} f(\eta, t), \tag{12}$$

where

$$\eta = \mathbf{n} \cdot \mathbf{x} = n_i F_{iA} X_A. \tag{13}$$

To simplify the computations we introduce the orthonormal triad

$$\mathbf{a}, \quad \mathbf{n}, \quad \mathbf{b} = \mathbf{n} \wedge \mathbf{a};$$

and we call $(\bar{\xi}, \bar{\eta}, \bar{\zeta})$ and (ξ, η, ζ), respectively, the components of $\bar{\mathbf{x}}$ and $\mathbf{x}$ in this triad. In such a way we rewrite (12) as

$$\bar{\eta} = \eta, \qquad \bar{\xi} = \xi + f(\eta, t), \qquad \bar{\zeta} = \zeta. \tag{14}$$

The stress field corresponding to (12) is given by

$$\overline{\mathbf{T}} = -\bar{p}\mathbf{1} + C\overline{\mathbf{B}} - E\overline{\mathbf{B}}^{-1} + \mu\overline{\mathbf{A}}_1 + \alpha_1\overline{\mathbf{A}}_2 + \alpha_2\overline{\mathbf{A}}_1^2. \tag{15}$$

Here $-\bar{p}\mathbf{I}$ is the pressure term associated with the final state of deformation. Because we are dealing with plane waves propagating in the direction $\mathbf{n}$, we assume that $\bar{p} = \bar{p}(\eta, t)$. Then all the stress components $\overline{T}_{ij}$ will be functions of η and t, only, and the components of the traction across the planes η =constant will be functions of t alone. Also, we may write $\bar{p} = p^* + p(\eta, t)$, thus incorporating p^* corresponding to the finite static deformation, while $p(\eta, t)$ is essentially due to the motion.

Also, $\overline{\mathbf{F}} = (\partial\bar{\mathbf{x}}/\partial\mathbf{X})$, the gradient of the deformation can be computed as

$$\overline{\mathbf{F}} = (\mathbf{1} + f_\eta \mathbf{a} \otimes \mathbf{n})\mathbf{F}, \tag{16}$$

and its inverse $\overline{\mathbf{F}}^{-1}$ is given by

$$\overline{\mathbf{F}}^{-1} = \mathbf{F}^{-1}(\mathbf{1} - f_\eta \mathbf{a} \otimes \mathbf{n}). \tag{17}$$

The corresponding left Cauchy–Green strain tensor is

$$\overline{\mathbf{B}} = (\mathbf{1} + f_\eta \mathbf{a} \otimes \mathbf{n})\mathbf{B}(\mathbf{1} + f_\eta \mathbf{n} \otimes \mathbf{a}), \tag{18}$$

its inverse

$$\overline{\mathbf{B}}^{-1} = (\mathbf{1} - f_\eta \mathbf{n} \otimes \mathbf{a})\mathbf{B}^{-1}(\mathbf{1} - f_\eta \mathbf{a} \otimes \mathbf{n}), \tag{19}$$

so that the principal invariants of $\overline{\mathbf{B}}$ are given by

$$\begin{aligned}\overline{I} &= I + 2f_\eta \mathbf{n}\cdot\mathbf{Ba} + f_\eta^2 \mathbf{n}\cdot\mathbf{Bn},\\ \overline{II} &= II - 2f_\eta \mathbf{n}\cdot\mathbf{B}^{-1}\mathbf{a} + f_\eta^2 \mathbf{a}\cdot\mathbf{B}^{-1}\mathbf{a}.\end{aligned} \tag{20}$$

Also, the spatial velocity gradient $\overline{\mathbf{L}} = \dot{\overline{\mathbf{F}}}\,\overline{\mathbf{F}}^{-1}$, in the final state of deformation, is given by

$$\overline{\mathbf{L}} = f_{\eta t}\mathbf{a}\otimes\mathbf{n}. \tag{21}$$

The corresponding first and second Rivlin-Ericksen tensors $\overline{\mathbf{A}}_1, \overline{\mathbf{A}}_2$ are given by

$$\begin{aligned}\overline{\mathbf{A}}_1 &= f_{\eta t}(\mathbf{a}\otimes\mathbf{n} + \mathbf{n}\otimes\mathbf{a}),\\ \overline{\mathbf{A}}_2 &= f_{\eta tt}(\mathbf{a}\otimes\mathbf{n} + \mathbf{n}\otimes\mathbf{a}) + 2f_{\eta t}^2\mathbf{n}\otimes\mathbf{n}.\end{aligned} \tag{22}$$

We note that $\mathbf{L} = \mathbf{0}$ for the basic static deformation (10), so that $\mathbf{A}_1 = \mathbf{A}_2 = \mathbf{0}$ in equation (7).

Then, on using the orthogonal triad already introduced, the equilibrium equations are

$$\begin{aligned}0 &= \frac{\partial \overline{T}_{\eta\eta}}{\partial\bar{\eta}},\\ \rho f_{tt} &= \frac{\partial \overline{T}_{\xi\eta}}{\partial\bar{\eta}},\\ 0 &= \frac{\partial \overline{T}_{\zeta\eta}}{\partial\bar{\eta}}.\end{aligned} \tag{23}$$

4. Propagation Condition

To write down the equation of motion, we first note that

$$\begin{aligned}\overline{T}_{\eta\eta} &= \mathbf{n}\cdot\overline{\mathbf{T}}\mathbf{n}\\ &= -p + \mathbf{n}\cdot\mathbf{Tn} + (2\alpha_1+\alpha_2)f_{\eta t}^2 - E\big(f_\eta^2\mathbf{a}\cdot\mathbf{B}^{-1}\mathbf{a} - 2f_\eta\mathbf{n}\cdot\mathbf{B}^{-1}\mathbf{a}\big),\\ \overline{T}_{\xi\eta} &= \mathbf{a}\cdot\overline{\mathbf{T}}\mathbf{n}\\ &= \mathbf{a}\cdot\mathbf{Tn} + \big(C\mathbf{n}\cdot\mathbf{Bn} + E\mathbf{a}\cdot\mathbf{B}^{-1}\mathbf{a}\big)f_\eta + \mu f_{\eta t} + \alpha_1 f_{\eta tt},\\ \overline{T}_{\zeta\eta} &= \mathbf{b}\cdot\overline{\mathbf{T}}\mathbf{n} = \mathbf{b}\cdot\mathbf{Tn} + \big(E\mathbf{b}\cdot\mathbf{B}^{-1}\mathbf{a}\big)f_\eta.\end{aligned} \tag{24}$$

From $(23)_1$, it now follows that $p = p(\eta, t)$ is given by

$$p = (2\alpha_1+\alpha_2)f_{\eta t}^2 - E\big(f_\eta^2\mathbf{a}\cdot\mathbf{B}^{-1}\mathbf{a} - 2f_\eta\mathbf{n}\cdot\mathbf{B}^{-1}\mathbf{a}\big) + p_0(t), \tag{25}$$

where $p_0(t)$ is an arbitrary function of time. Then equations $(23)_2$ and $(23)_3$ give

$$\rho f_{tt} = \big(C\mathbf{n}\cdot\mathbf{Bn} + E\mathbf{a}\cdot\mathbf{B}^{-1}\mathbf{a}\big)f_{\eta\eta} + \mu f_{\eta\eta t} + \alpha_1 f_{\eta\eta tt}, \tag{26}$$

and

$$E\big(\mathbf{b}\cdot\mathbf{B}^{-1}\mathbf{a}\big)f_{\eta\eta} = 0. \tag{27}$$

From equation (27), on assuming $E \neq 0$ and $f_{\eta\eta} \neq 0$, we obtain the propagation condition

$$\mathbf{b} \cdot \mathbf{B}^{-1}\mathbf{a} = 0. \tag{28}$$

This condition is the same as in the purely elastic case for Mooney–Rivlin materials [4]. It means that the vectors **a** and **b** are conjugate with respect to the ellipsoid $\mathbf{x} \cdot \mathbf{B}^{-1}\mathbf{x} = 1$. Because **a** and **b** are orthogonal, it follows that they must be along the principal axes of the elliptical section of the ellipsoid by the plane $\mathbf{n} \cdot \mathbf{x} = 0$. Hence, for every propagation direction **n** two linearly polarized waves may propagate. The two possible directions **a**, **b** of polarization are along the principal axes of the elliptical section of the $\mathbf{B}^{-1}$ ellipsoid by the plane $\mathbf{n} \cdot \mathbf{x} = 0$.

Equation (26) is linear and obviously, under some restrictions, damped and attenuated planes waves are possible solutions. An equation similar to (26) appears in the study of continuum models with inherent characteristic length recently proposed by Rubin et al. [25].

4.1. DAMPED PLANE WAVES

We first consider the classical solution for damped waves propagating along **n** with amplitude **a**:

$$\bar{\mathbf{x}} = \mathbf{x} + \mathbf{a}\exp(-kt)\sin(\eta - ct). \tag{29}$$

This solution is possible when

$$k = \frac{\mu}{2(\rho + \alpha_1)} \tag{30}$$

and

$$c^2(\mathbf{n};\ \mathbf{a}) = \frac{(C\mathbf{n} \cdot \mathbf{B}\mathbf{n} + E\mathbf{a} \cdot \mathbf{B}^{-1}\mathbf{a})}{(\rho + \alpha_1)} - \frac{\mu^2}{4(\rho + \alpha_1)^2}, \tag{31}$$

where the notation $c(\mathbf{n}, \mathbf{a})$ is used for the phase speed of the wave propagating along **n** with amplitude **a**. Similarly the speed $c(\mathbf{n}; \mathbf{b})$ for the wave propagating along **n** with amplitude **b** is given by

$$c^2(\mathbf{n}; \mathbf{b}) = \frac{(C\mathbf{n} \cdot \mathbf{B}\mathbf{n} + E\mathbf{b} \cdot \mathbf{B}^{-1}\mathbf{b})}{(\rho + \alpha_1)} - \frac{\mu^2}{4(\rho + \alpha_1)^2}. \tag{32}$$

Clearly

$$c^2(\mathbf{n}; \mathbf{a}) - c^2(\mathbf{n}; \mathbf{b}) = E\frac{(\mathbf{a} \cdot \mathbf{B}^{-1}\mathbf{a} - \mathbf{b} \cdot \mathbf{B}^{-1}\mathbf{b})}{(\rho + \alpha_1)}, \tag{33}$$

and the two speeds are equal only if $E = 0$ or if

$$\mathbf{a} \cdot \mathbf{B}^{-1}\mathbf{a} = \mathbf{b} \cdot \mathbf{B}^{-1}\mathbf{b}. \tag{34}$$

Now let **i**, **j**, **k** be unit vectors along the principal axes of **B** and $\lambda_1, \lambda_2, \lambda_3$ the principal stretches, ordered as $\lambda_1^2 > \lambda_2^2 > \lambda_3^2$. In this case $\mathbf{n} \cdot \mathbf{B}\mathbf{n}$ assumes its largest value when $\mathbf{n} = \mathbf{i}$ and $\mathbf{a} \cdot \mathbf{B}^{-1}\mathbf{a}$ its largest value when $\mathbf{a} = \mathbf{k}$. Thus, on noting the inequalities (5), the fastest transverse wave propagates in the direction of the greatest stretch λ_1 ($\mathbf{n} = \mathbf{i}$), with amplitude in the direction of the least stretch λ_3 ($\mathbf{a} = \mathbf{k}$) and with speed $c_{\max}$ given by

$$c_{\max}^2 = \frac{\lambda_3^{-2}(C\lambda_1^2\lambda_3^2 + E)}{(\rho + \alpha_1)} - \frac{\mu^2}{4(\rho + \alpha_1)^2}. \tag{35}$$

Also, the slowest transverse wave propagates in the direction of the least stretch ($\mathbf{n} = \mathbf{k}$) with amplitude in the direction of the greatest stretch ($\mathbf{a} = \mathbf{i}$) and with speed $c_{\min}$ given by

$$c_{\min}^2 = \frac{\lambda_1^{-2}(C\lambda_1^2\lambda_3^2 + E)}{(\rho + \alpha_1)} - \frac{\mu^2}{4(\rho + \alpha_1)^2}. \tag{36}$$

This is the same situation as in the elastic case [4], but here the inequalities (5) do not ensure the reality of the propagation speed. Obviously, for given values of the material constants, there is always a deformation sufficiently *large* to ensure the possibility of propagation of transverse waves with speed $c_{\max}$.

We note that

$$\lambda_3^2\left(c_{\max}^2 + \frac{\mu^2}{4(\rho + \alpha_1)^2}\right) = \lambda_1^2\left(c_{\min}^2 + \frac{\mu^2}{4(\rho + \alpha_1)^2}\right), \tag{37}$$

a result which is independent of the *elastic* coefficients C and E.

Now the tensor $\mathbf{B}^{-1}$ may be written as [5]

$$\begin{aligned} \mathbf{B}^{-1} &= \lambda_1^{-2}\mathbf{i}\otimes\mathbf{i} + \lambda_2^{-2}\mathbf{j}\otimes\mathbf{j} + \lambda_3^{-2}\mathbf{k}\otimes\mathbf{k} \\ &= \lambda_2^{-2}\mathbf{I} - \frac{1}{2}(\lambda_3^{-2} - \lambda_1^{-2})(\mathbf{n}^+\otimes\mathbf{n}^- + \mathbf{n}^-\otimes\mathbf{n}^+), \end{aligned} \tag{38}$$

where

$$\mathbf{n}^{\pm} = \sqrt{\frac{\lambda_2^{-2} - \lambda_1^{-2}}{\lambda_3^{-2} - \lambda_1^{-2}}}\,\mathbf{i} \pm \sqrt{\frac{\lambda_3^{-2} - \lambda_2^{-2}}{\lambda_3^{-2} - \lambda_1^{-2}}}\,\mathbf{k}. \tag{39}$$

The unit vectors $\mathbf{n}^{\pm}$ are normal to the planes of the central circular sections of the $\mathbf{B}^{-1}$-ellipsoid: $\mathbf{x} \cdot \mathbf{B}^{-1}\mathbf{x} = 1$.

Now for a given direction of propagation **n**, the polarization directions **a**, **b** characterized by $\mathbf{n} \cdot \mathbf{a} = \mathbf{n} \cdot \mathbf{b} = 0$ and $\mathbf{a} \cdot \mathbf{b} = \mathbf{a} \cdot \mathbf{B}^{-1}\mathbf{b} = 0$ may be obtained in terms of **n** [4]. The unit vectors are along $\mathbf{a}^{\times}$ and $\mathbf{b}^{\times}$, respectively, given by [6]

$$\left.\begin{matrix}\mathbf{a}^{\times} \\ \mathbf{b}^{\times}\end{matrix}\right\} = \frac{1}{\delta_+}\mathbf{n}\wedge\mathbf{n}^+ \pm \frac{1}{\delta_-}\mathbf{n}\wedge\mathbf{n}^-, \tag{40}$$

with

$$\delta_{\pm}^2 = (\mathbf{n} \wedge \mathbf{n}^{\pm}) \cdot (\mathbf{n} \wedge \mathbf{n}^{\pm}) = \lambda_2^2 (\mathbf{n} \wedge \mathbf{n}^{\pm}) \cdot \mathbf{B}^{-1} (\mathbf{n} \wedge \mathbf{n}^{\pm}). \tag{41}$$

Note that $\delta_{\pm} = \sin \varphi_{\pm}$, where $\varphi_{\pm}$ denotes the angle between the propagation direction and $\mathbf{n}^{\pm}$.

It has been shown [4] that

$$\begin{aligned} 2\mathbf{a} \cdot \mathbf{B}^{-1}\mathbf{a} &= \lambda_3^{-2} + \lambda_1^{-2} + (\lambda_3^{-2} - \lambda_1^{-2}) \cos(\varphi^+ - \varphi^-), \\ 2\mathbf{b} \cdot \mathbf{B}^{-1}\mathbf{b} &= \lambda_3^{-2} + \lambda_1^{-2} + (\lambda_3^{-2} - \lambda_1^{-2}) \cos(\varphi^+ + \varphi^-). \end{aligned} \tag{42}$$

Hence, from (33)

$$c^2(\mathbf{n}; \mathbf{a}) - c^2(\mathbf{n}; \mathbf{b}) = E(\lambda_3^{-2} - \lambda_1^{-2}) \frac{\sin \varphi^+ \sin \varphi^-}{(\rho + \alpha_1)}. \tag{43}$$

Thus, the difference between the squared wave speeds for any direction $\mathbf{n}$ is proportional to the product $\sin \varphi^+ \sin \varphi^-$. This was shown by Boulanger and Hayes [4] to be the case for the two waves propagating in the direction $\mathbf{n}$ in the Mooney–Rivlin material. They noted that there is a similar situation for the two electromagnetic waves propagating in any direction $\mathbf{n}$ in a biaxial crystal.

For a given $\mathbf{n}$ the speeds are equal, provided $E = 0$ or if $\varphi^+ = 0, \pi$ or $\varphi^- = 0, \pi$. Thus in general ($E \neq 0$), there are two and only two directions, namely $\mathbf{n}^+$ and $\mathbf{n}^-$, for which the speeds of propagation are equal.

For propagation along $\mathbf{n}^+$ (or $\mathbf{n}^-$), we note that $\mathbf{n}^{\pm} \cdot \mathbf{B}^{-1}\mathbf{n}^{\pm} = \lambda_2^{-4}$, $\mathbf{a} \cdot \mathbf{B}^{-1}\mathbf{a} = \lambda_2^{-2}$ for all $\mathbf{a}$ orthogonal to $\mathbf{n}^+$ (or $\mathbf{n}^-$). Thus, in this case equation (30) is unchanged and equation (31) gives

$$c^{*2} = \frac{C + \lambda_2^2 E}{\lambda_2^4 (\rho + \alpha_1)} - \frac{\mu^2}{4(\rho + \alpha_1)^2}, \tag{44}$$

where c^* is the speed of propagation along $\mathbf{n}^+$ or $\mathbf{n}^-$.

4.2. ATTENUATED WAVES

Also attenuated waves

$$\bar{\mathbf{x}} = \mathbf{x} + \mathbf{a} \exp(-m\eta) \cos(\omega t - n\eta), \tag{45}$$

are possible solutions of (26) provided m, n satisfy

$$n^2 - m^2 = \frac{\rho\omega^2[(C\mathbf{n} \cdot \mathbf{B}\mathbf{n} + E\mathbf{a} \cdot \mathbf{B}^{-1}\mathbf{a}) - \alpha_1\omega^2]}{[\alpha_1\omega^2 - (C\mathbf{n} \cdot \mathbf{B}\mathbf{n} + E\mathbf{a} \cdot \mathbf{B}^{-1}\mathbf{a})]^2 + \mu^2\omega^2} \tag{46}$$

and

$$2nm = \frac{\rho\mu\omega^3}{[\alpha_1\omega^2 - (C\mathbf{n} \cdot \mathbf{B}\mathbf{n} + E\mathbf{a} \cdot \mathbf{B}^{-1}\mathbf{a})]^2 + \mu^2\omega^2}. \tag{47}$$

It may be seen that possible solutions to (46) and (47) for given $\mathbf{n}$ are

$$n = \frac{\sqrt{\rho}\omega\cos(\Delta/2)}{(H(\mathbf{n};\mathbf{a})^2 + \mu^2\omega^2)^{1/4}}, \qquad m = \frac{\sqrt{\rho}\omega\sin(\Delta/2)}{(H(\mathbf{n};\mathbf{a})^2 + \mu^2\omega^2)^{1/4}}. \tag{48}$$

where

$$H(\mathbf{n};\mathbf{a}) = \left(C\mathbf{n}\cdot\mathbf{B}\mathbf{n} + E\mathbf{a}\cdot\mathbf{B}^{-1}\mathbf{a}\right) - \alpha_1\omega^2 \tag{49}$$

and

$$\tan\Delta = \frac{\mu\omega}{H(\mathbf{n};\mathbf{a})}.$$

The corresponding wave speed of the attenuated wave is given by

$$c^2(\mathbf{n};\mathbf{a}) = \frac{\omega^2}{n^2} = \frac{2(H^2(\mathbf{n};\mathbf{a}) + \mu^2\omega^2)^{3/2}}{\rho[(H^2(\mathbf{n};\mathbf{a}) + \mu^2\omega^2)^{1/2} + H(\mathbf{n};\mathbf{a})]}. \tag{50}$$

If n_1, m_1 are solutions corresponding to $\mathbf{n}$ and $\mathbf{b}$ where $\mathbf{a}\cdot\mathbf{b} = \mathbf{a}\cdot\mathbf{B}^{-1}\mathbf{b} = 0$, then we have to consider

$$H(\mathbf{n};\mathbf{b}) = \left(C\mathbf{n}\cdot\mathbf{B}\mathbf{n} + E\mathbf{b}\cdot\mathbf{B}^{-1}\mathbf{b}\right) - \alpha_1\omega^2,$$

and

$$\begin{aligned} n_1 &= \frac{\sqrt{\rho}\omega\cos(\Delta_1/2)}{(H^2(\mathbf{n};\mathbf{b}) + \mu^2\omega^2)^{1/4}}, \\ m_1 &= \frac{\sqrt{\rho}\omega\sin(\Delta_1/2)}{(H^2(\mathbf{n};\mathbf{b}) + \mu^2\omega^2)^{1/4}}, \\ \tan\Delta_1 &= \frac{\mu\omega}{H(\mathbf{n};\mathbf{b})}, \end{aligned} \tag{51}$$

$$c^2(\mathbf{n};\mathbf{b}) = \frac{\omega^2}{n_1^2} = \frac{2(H^2(\mathbf{n};\mathbf{b}) + \mu^2\omega^2)^{3/2}}{\rho[(H^2(\mathbf{n};\mathbf{b}) + \mu^2\omega^2)^{1/2} + H(\mathbf{n};\mathbf{b})]}. \tag{52}$$

It may be shown that if we demand $n_1 = n$, $m_1 = m$, then we must have $\mathbf{a}\cdot\mathbf{B}^{-1}\mathbf{a} = \mathbf{b}\cdot\mathbf{B}^{-1}\mathbf{b}$, which is only possible if $\mathbf{n} = \mathbf{n}^{\pm}$, so propagation is along an *acoustic axis* of $\mathbf{B}^{-1}$. In this case $H = H^*$, is given by

$$H^* = \lambda_2^{-4}C + \lambda_2^{-2}E - \alpha_1\omega^2. \tag{53}$$

5. Superposition of Waves. Circularly Polarized Waves

Here we consider the possibility of superposing two finite amplitude transverse waves, propagating in the direction $\mathbf{n}$, with amplitudes along $\mathbf{a}$ and $\mathbf{b}$, where $\mathbf{a}\cdot\mathbf{b} = \mathbf{a}\cdot\mathbf{B}^{-1}\mathbf{b} = 0$. It is seen that such solutions are indeed possible both for damped plane waves and for attenuated waves. In particular for propagation along $\mathbf{n}^{\pm}$, the *acoustic axes* of $\mathbf{B}$, it is seen that circularly polarized waves of either handedness may propagate.

5.1. SUPERPOSITION

Let

$$\bar{\mathbf{x}} = \mathbf{x} + \mathbf{a}f(\eta, t) + \mathbf{b}g(\eta, t), \qquad \mathbf{a}\cdot\mathbf{b} = \mathbf{a}\cdot\mathbf{B}^{-1}\mathbf{b} = 0. \tag{54}$$

As before, we assume that $-\bar{p}\mathbf{I}$, the pressure term associated with the final state of deformation, is such that $\bar{p} = p^* + p(\eta, t)$, where p^* corresponds to the finite static deformation. Then, all the stress components $\overline{T}_{ij}$ will be functions of η and t, only and the components of traction across the planes $\eta =$ constant will be functions of t alone.

In the system of coordinates along **n, a, b,** the motion (54) may be written

$$\bar{\eta} = \eta, \qquad \bar{\xi} = \xi + f(\eta, t), \qquad \bar{\zeta} = \zeta + g(\eta, t).$$

Then,

$$\begin{aligned}
\overline{\mathbf{F}} &= (\mathbf{1} + f_\eta\mathbf{a}\otimes\mathbf{n} + g_\eta\mathbf{b}\otimes\mathbf{n})\mathbf{F},\\
\overline{\mathbf{F}}^{-1} &= \mathbf{F}^{-1}(\mathbf{1} - f_\eta\mathbf{a}\otimes\mathbf{n} - g_\eta\mathbf{b}\otimes\mathbf{n}),\\
\overline{\mathbf{B}} &= (\mathbf{1} + f_\eta\mathbf{a}\otimes\mathbf{n} + g_\eta\mathbf{b}\otimes\mathbf{n})\mathbf{B}(\mathbf{1} + f_\eta\mathbf{n}\otimes\mathbf{a} + g_\eta\mathbf{n}\otimes\mathbf{b}),\\
\overline{\mathbf{B}}^{-1} &= (\mathbf{1} - f_\eta\mathbf{n}\otimes\mathbf{a} - g_\eta\mathbf{n}\otimes\mathbf{b})\mathbf{B}^{-1}(\mathbf{1} - f_\eta\mathbf{a}\otimes\mathbf{n} - g_\eta\mathbf{b}\otimes\mathbf{n}).
\end{aligned}$$

Here $\mathbf{F}$ and $\mathbf{B} = \mathbf{F}\mathbf{F}^{\mathrm{T}}$ are the deformation and the strain tensor corresponding to the static finite strain homogeneous deformation (10). Because

$$\det(\mathbf{1} + f_\eta\mathbf{a}\otimes\mathbf{n} + g_\eta\mathbf{b}\otimes\mathbf{n}) = 1,$$

it follows that $\overline{III} = \det\overline{\mathbf{B}} = 1$, so that the incompressibility constraint is satisfied. Also, the principal invariants are

$$\begin{aligned}
\overline{I} &= tr\,\overline{\mathbf{B}} = I + 2(f_\eta\mathbf{n}\cdot\mathbf{B}\mathbf{a} + g_\eta\mathbf{n}\cdot\mathbf{B}\mathbf{b}) + \left(f_\eta^2 + g_\eta^2\right)\mathbf{n}\cdot\mathbf{B}\mathbf{n},\\
\overline{II} &= tr\,\overline{\mathbf{B}}^{-1}\\
&= II - 2\left(f_\eta\mathbf{n}\cdot\mathbf{B}^{-1}\mathbf{a} + g_\eta\mathbf{n}\cdot\mathbf{B}^{-1}\mathbf{b}\right) + f_\eta^2\mathbf{a}\cdot\mathbf{B}^{-1}\mathbf{a} + g_\eta^2\mathbf{b}\cdot\mathbf{B}^{-1}\mathbf{b}.
\end{aligned}$$

Also,

$$\begin{aligned}
\overline{\mathbf{A}}_1 &= f_{\eta t}(\mathbf{a}\otimes\mathbf{n} + \mathbf{n}\otimes\mathbf{a}) + g_{\eta t}(\mathbf{b}\otimes\mathbf{n} + \mathbf{n}\otimes\mathbf{b}),\\
\overline{\mathbf{A}}_2 &= f_{\eta tt}(\mathbf{a}\otimes\mathbf{n} + \mathbf{n}\otimes\mathbf{a}) + g_{\eta tt}(\mathbf{b}\otimes\mathbf{n} + \mathbf{n}\otimes\mathbf{b})\\
&\quad + 2\left(f_{\eta t}^2 + g_{\eta t}^2\right)\mathbf{n}\otimes\mathbf{n}.
\end{aligned} \tag{55}$$

5.2. EQUATIONS OF MOTION

The components of $\overline{\mathbf{T}}$ may be obtained by using $\mathbf{b}\cdot\mathbf{B}^{-1}\mathbf{a} = 0$. The equation of motion are again equations (23). From equation $(23)_1$ we have

$$\begin{aligned}
p(\eta, t) &= p_0(t) + (2\alpha_1 + \alpha_2)\left(f_{\eta t}^2 + g_{\eta t}^2\right)\\
&\quad - E\left(f_\eta^2\mathbf{a}\cdot\mathbf{B}^{-1}\mathbf{a} - 2f_\eta\mathbf{n}\cdot\mathbf{B}^{-1}\mathbf{a}\right)\\
&\quad - E\left(g_\eta^2\mathbf{b}\cdot\mathbf{B}^{-1}\mathbf{b} - 2g_\eta\mathbf{n}\cdot\mathbf{B}^{-1}\mathbf{b}\right).
\end{aligned} \tag{56}$$

Also,

$$\begin{aligned}\overline{T}_{\eta\eta} &= \mathbf{n}\cdot\overline{\mathbf{T}}\mathbf{n} = -p_0 + \mathbf{n}\cdot\mathbf{T}\mathbf{n},\\ \overline{T}_{\xi\eta} &= \mathbf{a}\cdot\overline{\mathbf{T}}\mathbf{n}\\ &= \mathbf{a}\cdot\mathbf{T}\mathbf{n} + \mu f_{\eta t} + \alpha_1 f_{\eta tt} + \big(C\mathbf{n}\cdot\mathbf{B}\mathbf{n} + E\mathbf{a}\cdot\mathbf{B}^{-1}\mathbf{a}\big)f_\eta,\\ \overline{T}_{\zeta\eta} &= \mathbf{b}\cdot\overline{\mathbf{T}}\mathbf{n}\\ &= \mathbf{b}\cdot\mathbf{T}\mathbf{n} + \mu g_{\eta t} + \alpha_1 g_{\eta tt} + \big(C\mathbf{n}\cdot\mathbf{B}\mathbf{n} + E\mathbf{b}\cdot\mathbf{B}^{-1}\mathbf{b}\big)g_\eta.\end{aligned} \tag{57}$$

The equations of motion are reduced to

$$\rho f_{tt} = \big(C\mathbf{n}\cdot\mathbf{B}\mathbf{n} + E\mathbf{a}\cdot\mathbf{B}^{-1}\mathbf{a}\big)f_{\eta\eta} + \mu f_{\eta\eta t} + \alpha_1 f_{\eta\eta tt} \tag{58}$$

and

$$\rho g_{tt} = \big(C\mathbf{n}\cdot\mathbf{B}\mathbf{n} + E\mathbf{b}\cdot\mathbf{B}^{-1}\mathbf{b}\big)g_{\eta\eta} + \mu g_{\eta\eta t} + \alpha_1 g_{\eta\eta tt}. \tag{59}$$

Then (54) is a solution of the equations of motion if p is given by (56) and $f(\eta, t)$ and $g(\eta, t)$ satisfy the two uncoupled equations (58) and (59). Thus, (54) represents the superposition of two non-interacting waves propagating along **n** and polarized along **a** and **b**.

In the particular case of a pair of damped waves, the motion is described by

$$\begin{aligned}\bar{\mathbf{x}} &= \mathbf{x} + A\mathbf{a}\exp(-kt)\sin\big(\eta - c(\mathbf{n};\mathbf{a})t\big)\\ &\quad + B\mathbf{b}\exp(-kt)\cos\big(\eta - c(\mathbf{n};\mathbf{b})t\big),\end{aligned} \tag{60}$$

where k is given by (30) and $c(\mathbf{n};\mathbf{a})$, $c(\mathbf{n};\mathbf{b})$ by (31), (32) and A and B are arbitrary constants.

It has been seen that in general ($E \neq 0$), $c(\mathbf{n};\mathbf{a}) = c(\mathbf{n};\mathbf{b})$ provided only that $\mathbf{n} = \mathbf{n}^{\pm}$. In this case it is possible to have a circularly polarized wave propagating along $\mathbf{n}^+$ (or $\mathbf{n}^-$). The motion is described by

$$\bar{\mathbf{x}} = \mathbf{x} + A\exp(-kt)\big(\mathbf{a}\sin(\eta - c^*t) + \mathbf{b}\cos(\eta - c^*t)\big), \tag{61}$$

where A is an arbitrary constant, c^* is given by (44) and **a**, **b**, $\mathbf{n}^+$ (or $\mathbf{n}^-$) form an orthonormal triad.

Clearly, by choice of suitable A and B, it is possible to have elliptically polarized damped finite waves of either handedness propagate along $\mathbf{n}^+$ or $\mathbf{n}^-$.

Also, in the particular case of a pair of attenuated waves propagating along **n**, the motion is given by

$$\begin{aligned}\bar{\mathbf{x}} &= \mathbf{x} + A\mathbf{a}\exp(-m\eta)\sin(\omega t - n\eta)\\ &\quad + B\mathbf{b}\exp(-m_1\eta)\cos(\omega t - n_1\eta),\end{aligned} \tag{62}$$

where n, m and n_1, m_1 are given by (48) and (51). Here A and B are arbitrary constants. It has been noted that in order to have $n_1 = n$, $m_1 = m$, we must have $\mathbf{a}\cdot\mathbf{B}^{-1}\mathbf{a} = \mathbf{b}\cdot\mathbf{B}^{-1}\mathbf{b}$, so that $\mathbf{n} = \mathbf{n}^{\pm}$. In this case a possible solution is the circularly polarized attenuated wave

$$\bar{\mathbf{x}} = \mathbf{x} + A\exp(-m\eta)\big(\mathbf{a}\sin(\omega t - n\eta) + \mathbf{b}\cos(\omega t - n\eta)\big), \tag{63}$$

where A is an arbitrary constant and m, n are given by (48) with H replaced by the H^* given by (53), and $\mathbf{a}$, $\mathbf{b}$, $\mathbf{n}^+$ (or $\mathbf{n}^-$) form an orthonormal triad.

Clearly, by suitable choice of A and B, it is possible to have elliptically polarized attenuated finite amplitude waves of either handeness propagating along $\mathbf{n}^+$ or $\mathbf{n}^-$.

Acknowledgments

M.A.H. is grateful to Enterprise Ireland for support. G.S. is partially supported by Gruppo Nazionale di Fisica Matematica of Italian INDAM and Progetto Giovani Ricercatori of Italian MURST through a grant of the University of Perugia.

References

1. S.S. Antman, *Nonlinear Problems in Elasticity*. Springer, New York (1995).
2. M.F. Beatty, Topics in finite elasticity: Hyperelasticity of rubber, elastomers, and biological tissues – with examples. *Appl. Mech. Rev.* **40** (1987) 1699–1733. See also Introduction to Nonlinear Elasticity in: M.M. Carroll and M.A. Hayes (eds), *Nonlinear Effects in Fluids and Solids*. Plenum, New York (1996) pp. 16–112.
3. M.F. Beatty and Z. Zhou, Universal motions for a class of viscoelastic materials of differential type. *Continuum Mech. Thermodyn.* **3** (1991) 169–171.
4. Ph. Boulanger and M. Hayes, Finite amplitude waves in deformed Mooney–Rivlin materials. *Quart. J. Mech. Appl. Math.* **45** (1992) 575–593.
5. Ph. Boulanger and M. Hayes , *Bivectors and Waves in Mechanics and Optics*. Chapman & Hall, London (1993).
6. Ph. Boulanger, M. Hayes and C. Trimarco, Finite amplitude waves in deformed Hadamard materials. *Geophys. J. Internat.* **118** (1994) 447–458.
7. Ph. Boulanger and M. Hayes, Further properties of finite amplitude waves in deformed Mooney–Rivlin materials. *Quart. J. Mech. Appl. Math.* **48** (1995) 427–464.
8. Ph. Boulanger and M. Hayes, Wave propagation in sheared rubber. *Acta Mech.* **22** (1997) 75–87.
9. M.M. Carroll, Controllable deformations in incompressible simple elastic materials. *Internat. J. Engrg. Sci.* **5** (1967) 515–525.
10. M.M. Carroll, Unsteady homothermal motions of fluids and isotropic solids. *Arch. Rational Mech. Anal.* **53** (1974) 218.
11. M.M. Carroll, Plane circular shearing of incompressible fluids and solids. *Quart. J. Mech. Appl. Math.* **30** (1976) 223–234.
12. B.D. Coleman and W. Noll, Foundations of linear viscoelasticity. *Rev. Modern Phys.* **33** (1961) 239–249.
13. J. Engelbrecht, Qualitative aspects of nonlinear wave motion. In: J.L.Wegner and F.R. Norwood (eds), *Nonlinear Waves in Solids*. Appl. Mech. Rev. **46** (1993) 509–518.
14. R.L. Fosdick, Dynamically possible motions of isotropic incompressible simple materials. *Arch. Rational Mech. Anal.* **29** (1968) 272–288.
15. R.L. Fosdick and J.-H. Yu, Thermodynamics, stability and nonlinear oscillations of viscoelastic solids-I. *Internat. J. Nonlinear Mech.* **31** (1996) 495–516.
16. R.L. Fosdick and J.-H. Yu, Thermodynamics, stability and non-linear oscillations of viscoelastic solids-II. *Int. J. Nonlinear Mech.* **33** (1998) 165–188.
17. R.L. Fosdick, Y. Ketema and J.-H. Yu, Vibration damping through the use of materials with memory. *Internat. J. Solids Struct.* **35** (1998) 403–420.

18. J.M. Kelly, *Earthquake-resistant Design with Rubber*. Springer, London (1993).
19. J.L. Leblanc, Dynamic strain methods to probe the morphology of uncured rubber compounds. *Kautschuk Gummi Kunststoffe* **49** (1996) 258–266.
20. K.N. Morman, An adaptation of finite linear viscoelasticity theory for rubber-like viscoelasticity by use of a generalized strain measure. *Rheol. Acta* **27** (1988) 3–14.
21. K.N. Morman and Y.T. Pan, Application of finite element analysis in the design of automotive elastomeric components. *Rubber Chem. Technol.* **61** (1988) 503–533.
22. J.W. Nunziato, E.K. Walsh, F.W. Schuler and L.M. Barker, Wave propagation in nonlinear viscoelastic media. In: *Flügge's Handbuch der Physik.*, VIa/4. Springer, Berlin (1974).
23. K. Rasolofosan, B. Zinszner and P.A. Johnson, Propagation des ondes elasticques dans le materiaux non linéaires. *Revue de l'Institut Français du Pétrole* **52** (1997) 585–608.
24. M. Renardy, W.J. Hrusa and A.J. Nohel, *Mathematical problems of viscoelasticity*. Longman, New York (1987).
25. M.B. Rubin, P. Rosenau and O. Gottlieb, Continuum model of dispersion caused by an inherent material characteristic length. *J. Appl. Phys.* **77** (1995) 4054–4063.

Journal of Elasticity **59:** 227–236, 2000.

On Infinitesimal Shear

PH. BOULANGER[1] and M. HAYES[2]
[1]*Département de Mathématique, Université Libre de Bruxelles, Campus Plaine C.P.218/1, 1050 Bruxelles, Belgium. E-mail: phboul@ulb.ac.be*
[2]*Department of Mathematical Physics, University College Dublin, Belfield, Dublin 4, Ireland. E-mail: Michael.Hayes@ucd.ie*

Received 23 February 1999; in revised form 30 November 1999

Dedicated to Roger Fosdick on the occasion of his 60th birthday.

Abstract. The setting for this note is the theory of infinitesimal strain in the context of classical linearized elasticity. As a body is subjected to a deformation the angle between a pair of material line elements through a typical point P is changed. The decrease in angle is called the **shear** of this pair of elements.

Here, we determine all pairs of material line elements at P which are unsheared in a deformation. It is seen, in general, that corresponding to any given material line element in a given plane through P, there is one corresponding "companion" material line element such that the given element and its conjugate are unsheared in the deformation. There are two exceptions. If the plane through P is a plane of central circular section of the strain ellipsoid, then every material line element through P in this plane has an infinity of companion elements in this plane – all pairs of material line elements in the plane(s) of central circular section of the strain ellipsoid are unsheared. If the plane through P is not a plane of central circular section of the strain ellipsoid, then there are two exceptional material line elements through P such that neither of them has a companion material line element forming an unsheared pair with it. The directions of these exceptional elements in the plane are called "limiting directions". It is seen that it is the pair of elements along the limiting directions in a plane which suffer the maximum shear in that plane.

A geometrical construction is presented for the determination of the extensional strains along the pairs of elements which are unsheared. Also, it is shown that knowing one unsheared pair in a plane and their extensions is sufficient to determine the principal extensions and the principal axes in this plane. Expressions for all unsheared pairs in a given plane are given in terms of the normals to the planes of central circular sections of the strain ellipsoid.

Finally, for a given material line element, a formula is derived for the determination of all other material line elements which form an unsheared pair with the given element.

1. Introduction

The purpose of this paper is to consider the shear of pairs of material line elements emanating from a point P in an elastic body which is deformed. The context is the classical theory of infinitesimal strain. Central to the determination of shear is the strain tensor $\mathbf{e}$ (say). In general this is not positive definite, but a positive definite tensor $\mathbf{E} = \mathbf{e} + \lambda\mathbf{1}$ (where λ is suitably large) may be associated with it [1], in which case the corresponding quadric is an ellipsoid $\mathcal{E}$ (say). The principal axes of $\mathcal{E}$ are the principal axes of strain. The ellipsoid $\mathcal{E}$ is the vehicle which makes visualisation of the problem immediate and direct.

We solve the problem of the determination of all pairs of material line elements passing through P which suffer no shear. For a typical plane π through P, it turns out that there is an infinity of pairs of unsheared material line elements. Generally, every material line element $\mathcal{L}$ has just one "companion" element $\mathcal{L}'$ such that the pair $(\mathcal{L}, \mathcal{L}')$ is unsheared. However, there are two exceptions. If the plane π is a plane of central circular section of $\mathcal{E}$, than **all** pairs of material line elements in π are unsheared [4], so that each line element $\mathcal{L}$ has an infinity of "companion elements". If π is not a plane of central circular section of $\mathcal{E}$, then there are two special material line elements in π for neither of which is there a companion element in π forming an unsheared pair. We call the directions of these two special material line elements "limiting directions". They have the property that elements along them suffer the maximum shear in π.

Knowing the principal axes and principal extensions in the plane π, it is seen that the extensions of material elements forming an unsheared pair may be determined geometrically. It is also seen that if the extensions along the elements of one unsheared pair in π are known, then the principal axes and principal extensions in π may be determined, and hence, all unsheared pairs in π and their extensions may be found.

We also present (Section 4) an analytical expression for the determination of all unsheared pairs in a given plane π, using the normals to the planes of central circular section of the ellipsoid $\mathcal{E}$. Finally (Section 5), we consider one given material line element at P and determine all those "companion" material elements at P which form an unsheared pair with the given element.

2. Basic Equations

If $\Theta(\neq 0, \pi)$ is the angle between a pair of material line elements $\mathcal{L}$ and $\mathcal{L}'$ along the unit vectors $\mathbf{m}$ and $\mathbf{s}$ at a point P in an undeformed body, then $\Delta(\mathbf{m}, \mathbf{s})$, the decrease in the angle between this pair of elements as a result of the deformation is given (e.g., [6, 7]) by

$$\Delta(\mathbf{m}, \mathbf{s}) \sin\Theta = 2e_{ij}m_i s_j - \{e_{(\mathbf{m})} + e_{(\mathbf{s})}\}\cos\Theta. \tag{2.1}$$

Here $e_{ij} = (1/2)(\partial u_i/\partial X_j + \partial u_j/\partial X_i)$, where $u_i = x_i - X_i$ are the displacement components of the particle initially at $\mathbf{X}$ but currently at $\mathbf{x}$, are the components of the infinitesimal strain tensor at P, and for every unit vector $\mathbf{h}$, $e_{(\mathbf{h})}$ is defined by

$$e_{(\mathbf{h})} = e_{ij}h_i h_j. \tag{2.2}$$

In the special case when $\mathbf{m}$ and $\mathbf{s}$ are orthogonal, equation (2.1) reduces to the familiar expression for orthogonal shear:

$$\Delta(\mathbf{m}, \mathbf{s}) = 2e_{ij}m_i s_j, \quad \mathbf{m}\cdot\mathbf{s} = 0. \tag{2.3}$$

Let the eigenvalues of e_{ij} be denoted by e_α with $e_1 \geqslant e_2 \geqslant e_3$. Let the corresponding orthogonal unit eigenvectors be $\mathbf{\Omega}_\alpha$. In general the eigenvalues are not all positive. However, a positive definite tensor $\mathbf{E}$ may be defined through [1]

$$E_{ij} = e_{ij} + \lambda\delta_{ij}. \tag{2.4}$$

Here the scalar λ is chosen sufficiently large and positive so that the eigenvalues of $\mathbf{E}$, which are $E_\alpha = e_\alpha + \lambda$, are all positive. Similarly to $e_{(\mathbf{h})}$, $E_{(\mathbf{h})}$ is defined by

$$E_{(\mathbf{h})} = E_{ij}h_ih_j = e_{(\mathbf{h})} + \lambda. \tag{2.5}$$

The quadric surface $x_iE_{ij}x_j = 1$ associated with the tensor $\mathbf{E}$ is an ellipsoid which we denote by $\mathcal{E}$. Its principal axes are the principal axes of strain at P. Any plane π through P will cut $\mathcal{E}$ in an ellipse which we denote by $\mathcal{S}$. In the special cases when the ellipse $\mathcal{S}$ is a circle we denote it by $\mathcal{C}$. If all three principal strains e_α are different from each other there are two such central circular sections $\mathcal{C}^\pm$ (say). There is just one such if $\mathcal{E}$ is a spheroid, so that two and only two of the principal strains are equal. Of course if all three principal strains are equal then $\mathcal{E}$ is a sphere and every section by π is a circle.

When $e_1 > e_2 > e_3$, the unit normals $\mathbf{h}^\pm$ (say) to the planes of $\mathcal{C}^\pm$ are given by (e.g., [2, p. 90])

$$(e_1 - e_3)^{1/2}\mathbf{h}^\pm = (e_1 - e_2)^{1/2}\mathbf{\Omega}_1 \pm (e_2 - e_3)^{1/2}\mathbf{\Omega}_3, \tag{2.6}$$

and the tensor $\mathbf{E}$ may be written (see, for instance, [2, p. 54])

$$\mathbf{E} = (\lambda + e_2)\mathbf{1} + \frac{1}{2}(e_1 - e_3)\left(\mathbf{h}^+ \otimes \mathbf{h}^- + \mathbf{h}^- \otimes \mathbf{h}^+\right). \tag{2.7}$$

When $e_1 = e_2 > e_3$ or $e_1 > e_2 = e_3$, the unit normal to the plane of central circular section is $\mathbf{\Omega}_3$ or $\mathbf{\Omega}_1$, respectively.

To be definite we suppose that we consider material elements lying in the plane π at P. We suppose also that the ellipse $\mathcal{S}$ in which π cuts the ellipsoid $\mathcal{E}$ has major axis along the unit vector $\mathbf{I}$ and minor axis along the unit vector $\mathbf{J}$:

$$\mathbf{I} \cdot \mathbf{E}\mathbf{J} = \mathbf{I} \cdot \mathbf{e}\mathbf{J} = 0, \qquad e_{(\mathbf{J})} > e_{(\mathbf{I})}. \tag{2.8}$$

We seek all pairs of unit vectors $(\mathbf{m}, \mathbf{s})$ in the plane π such that material elements along $\mathbf{m}$ and $\mathbf{s}$ at P are unsheared in the deformation.

3. Unsheared Pairs for a Given Θ

We recall [1] that the expression (2.1) of Love may be greatly simplified by writing $\mathbf{m}$ and $\mathbf{s}$ in terms of the orthogonal unit vectors $\mathbf{c}$ and $\mathbf{d}$ which are along the internal and external bisectors of the angle Θ between $\mathbf{m}$ and $\mathbf{s}$:

$$\mathbf{m} = \mathbf{c}\cos(\Theta/2) - \mathbf{d}\sin(\Theta/2), \qquad \mathbf{s} = \mathbf{c}\cos(\Theta/2) + \mathbf{d}\sin(\Theta/2). \tag{3.1}$$

Then, from (2.1),

$$\Delta(\mathbf{m}, \mathbf{s}) = \{e_{(\mathbf{c})} - e_{(\mathbf{d})}\} \sin\Theta = e_{ij}(c_i + d_i)(c_j - d_j)\sin\Theta. \tag{3.2}$$

Hence, $\Delta(\mathbf{m}, \mathbf{s}) = 0$ if the orthogonal unit vectors **c** and **d** are such that

$$e_{(\mathbf{c})} = e_{(\mathbf{d})}. \tag{3.3}$$

This means that the radii to $\mathcal{E}$ along the orthogonal unit vectors **c** and **d** are equal in length.

Clearly, if $\mathcal{S}$, the ellipse in which the plane π cuts the ellipsoid $\mathcal{E}$, is the circle $\mathcal{C}$, then (3.3) is satisfied for **all** orthogonal pairs of elements lying along the radii of $\mathcal{C}$ and so $\Delta(\mathbf{m}, \mathbf{s}) = 0$ for all pairs of elements lying along radii of $\mathcal{C}$. Hence, we recover the result [4] that **all** pairs of material elements lying along the radii of the central circular section are unsheared.

Turning now to the general case when $\mathcal{S}$ is not a circle, it follows that, in order to satisfy equation (3.3), **c** and **d** must be equally inclined at $(\pi/4)$ to **J**, so that $\mathbf{c} = (\mathbf{I}+\mathbf{J})/\sqrt{2}$, $\mathbf{d} = (\mathbf{J}-\mathbf{I})/\sqrt{2}$. Hence, **m** and **s** are equally inclined at $(\Theta/2)$ to **c**, the internal bisector of the angle between **I** and **J** [1], so that

$$\begin{aligned}
\mathbf{m} &= \mathbf{I}\sin\left(\frac{\pi}{4}+\frac{\Theta}{2}\right) + \mathbf{J}\cos\left(\frac{\pi}{4}+\frac{\Theta}{2}\right) \\
&= \frac{(\mathbf{I}-\mathbf{J})\sin(\Theta/2) + (\mathbf{I}+\mathbf{J})\cos(\Theta/2)}{\sqrt{2}}, \\
\mathbf{s} &= \mathbf{I}\cos\left(\frac{\pi}{4}+\frac{\Theta}{2}\right) + \mathbf{J}\sin\left(\frac{\pi}{4}+\frac{\theta}{2}\right) \\
&= \frac{-(\mathbf{I}-\mathbf{J})\sin(\Theta/2) + (\mathbf{I}+\mathbf{J})\cos(\Theta/2)}{\sqrt{2}}.
\end{aligned} \tag{3.4}$$

As Θ is varied, an infinite set of unsheared pairs in the plane π is obtained. However, the directions of **m** and **s** coalesce along $(\mathbf{I} \pm \mathbf{J})$ as Θ approaches 0 or π. Thus, when $\Theta = 0, \pi$ we no longer obtain an unsheared pair. Accordingly, we now consider pairs of material elements with one element along $\mathbf{I}+\mathbf{J}$ or $\mathbf{I}-\mathbf{J}$, a bisector of the right angle between the principal axes of $\mathcal{S}$.

Let

$$\mathbf{s}' = \frac{\mathbf{I}+\mathbf{J}}{\sqrt{2}}, \qquad \mathbf{m}' = \mathbf{I}\sin\left(\frac{\pi}{4}+\Theta\right) + \mathbf{J}\cos\left(\frac{\pi}{4}+\Theta\right), \tag{3.5}$$

where the angle between $\mathbf{s}'$ and $\mathbf{m}'$ is Θ. Then, from equation (2.1), the decrease in the angle $\Delta(\mathbf{m}', \mathbf{s}')$ between the pair of material elements along $\mathbf{s}'$ and $\mathbf{m}'$ as a result of the deformation is given by

$$\Delta(\mathbf{m}', \mathbf{s}') = \{e_{(\mathbf{I})} - e_{(\mathbf{J})}\}\sin^2\Theta. \tag{3.6}$$

Because π is not a plane of central circular section of $\mathcal{E}$, it follows by (2.8) that $e_{(\mathbf{J})} > e_{(\mathbf{I})}$, so that $\Delta(\mathbf{m}', \mathbf{s}') \neq 0$. Thus, $\mathbf{s}'$ given by equation $(3.5)_1$ has no companion element in π forming an unsheared pair with it. Similarly, for $\mathbf{s}' = (\mathbf{I}-\mathbf{J})/\sqrt{2}$ there is no companion element in π forming a pair of unsheared elements.

We conclude that if π is a plane of central circular section of $\mathcal{E}$, then all pairs of material line elements in π are pairs of unsheared elements. If π intersects $\mathcal{E}$ in an ellipse, $\mathcal{S}$, then, apart from two exceptions, every line element $\mathcal{L}$ in π has just one companion $\mathcal{L}'$ in π such that $\mathcal{L}$ and $\mathcal{L}'$ form an unsheared pair in π. The two exceptions are elements along the bisectors of the angle between the principal axes of the ellipse $\mathcal{S}$. We call the directions of these exceptional elements along the bisectors "limiting directions".

REMARK 1 (Geometrical construction for determination of extensions of elements along unsheared pairs). From (3.4), it is easily checked that

$$e_{(\mathbf{s})} - e_{(\mathbf{m})} = \{e_{(\mathbf{J})} - e_{(\mathbf{I})}\} \sin\Theta, \tag{3.7}$$

$$e_{(\mathbf{s})} + e_{(\mathbf{m})} = e_{(\mathbf{J})} + e_{(\mathbf{I})} = 2e_{(\mathbf{c})} = 2e_{(\mathbf{d})}. \tag{3.8}$$

We note that for any two orthogonal unit vectors **a** and **b** (say), we have $e_{(\mathbf{a})}+e_{(\mathbf{b})} = e_{(\mathbf{I})} + e_{(\mathbf{J})} = \operatorname{tr}\mathbf{e} - e_{(\mathbf{n})}$, where $\mathbf{n} = \mathbf{I}\times\mathbf{J}$. Hence, the sum of the extensions $e_{(\mathbf{m})}$, $e_{(\mathbf{s})}$ of material elements along an unsheared pair $(\mathbf{m},\mathbf{s})$ is equal to the sum of the extensions along any orthogonal pair of material elements in the plane of **m** and **s**:

$$e_{(\mathbf{s})} + e_{(\mathbf{m})} = e_{(\mathbf{a})} + e_{(\mathbf{b})}, \qquad \mathbf{a}\cdot\mathbf{b} = 0. \tag{3.9}$$

Now, it follows from (3.7) and (3.8) that the extensions $e_{(\mathbf{m})}$ and $e_{(\mathbf{s})}$ are given by

$$\begin{aligned} 2e_{(\mathbf{m})} &= \{e_{(\mathbf{J})} + e_{(\mathbf{I})}\} - \{e_{(\mathbf{J})} - e_{(\mathbf{I})}\}\sin\Theta, \\ 2e_{(\mathbf{s})} &= \{e_{(\mathbf{J})} + e_{(\mathbf{I})}\} + \{e_{(\mathbf{J})} - e_{(\mathbf{I})}\}\sin\Theta. \end{aligned} \tag{3.10}$$

Thus, for each given angle Θ, $e_{(\mathbf{m})}$ and $e_{(\mathbf{s})}$, the extensions of material elements along an unsheared pair $(\mathbf{m},\mathbf{s})$, may be obtained geometrically by the construction of Figure 1.

REMARK 2 (Determination of the principal axes of $\mathcal{S}$). Suppose that in the plane π the pair of material elements $\mathcal{L}$ and $\mathcal{L}'$ along **m** and **s** are known to form an unsheared pair subtending an angle Θ. Suppose also that the extensions $e_{(\mathbf{m})}$ and $e_{(\mathbf{s})}$ are known and **m** and **s** are labelled so that $e_{(\mathbf{m})} < e_{(\mathbf{s})}$. From equations (3.7) and (3.8), the principal extensions in π, $e_{(\mathbf{I})}$ and $e_{(\mathbf{J})}$ are given by

$$\begin{aligned} 2e_{(\mathbf{I})} &= (1 - \csc\Theta)e_{(\mathbf{s})} + (1 + \csc\Theta)e_{(\mathbf{m})}, \\ 2e_{(\mathbf{J})} &= (1 + \csc\Theta)e_{(\mathbf{s})} + (1 - \csc\Theta)e_{(\mathbf{m})}. \end{aligned} \tag{3.11}$$

The directions of the principal axes of $\mathcal{S}$, the ellipse in which the plane π cuts $\mathcal{E}$, are at $\pi/4$ to the bisectors of the angle Θ between **m** and **s**. The major and minor axes are along **I** and **J**. Consideration of the ellipse $\mathcal{S}$ shows that because $e_{(\mathbf{m})} < e_{(\mathbf{s})}$ and $e_{(\mathbf{I})} < e_{(\mathbf{J})}$, then in the first quadrant the directions are in the following order, **I, m, s, J**.

Finally, note that if the material elements of the unsheared pair in π along **m** and **s** are such that $e_{(\mathbf{m})} = e_{(\mathbf{s})}$, then it follows that π is a plane of central circular

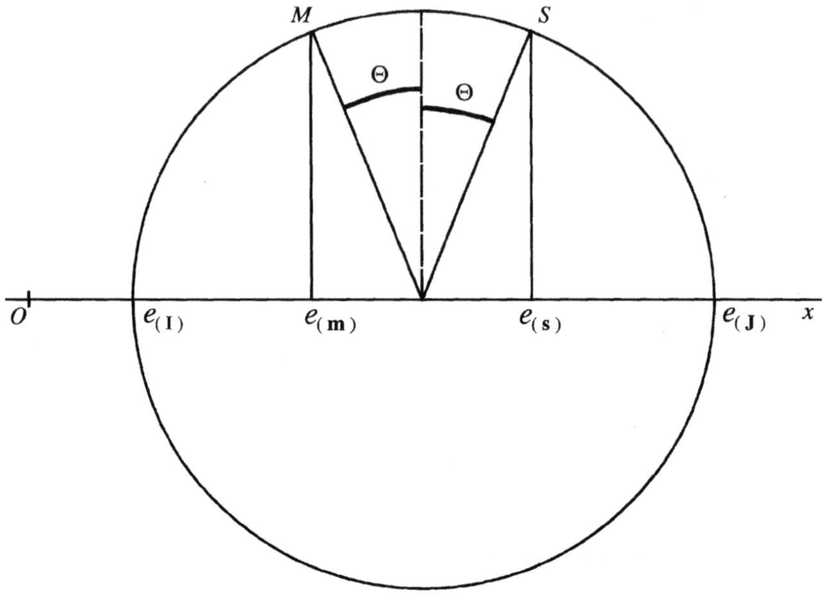

Figure 1. Construction of $e_{(\mathbf{m})}$ and $e_{(\mathbf{s})}$: on an axis Ox, represent the values of $e_{(\mathbf{I})}$ and $e_{(\mathbf{J})}$, and draw the circle with diameter $e_{(\mathbf{I})}$–$e_{(\mathbf{J})}$: the x-coordinates of the points M and S give the corresponding values of the extensions $e_{(\mathbf{m})}$ and $e_{(\mathbf{s})}$, respectively.

section $\mathcal{C}^{\pm}$ of $\mathcal{E}$. This is because, on using equation (3.4) for the unsheared pair $(\mathbf{m}, \mathbf{s})$, it follows that

$$\begin{aligned} e_{(\mathbf{m})} &= e_{(\mathbf{I})} \sin^2\left(\frac{\pi}{4} + \frac{\Theta}{2}\right) + e_{(\mathbf{J})} \cos^2\left(\frac{\pi}{4} + \frac{\Theta}{2}\right), \\ e_{(\mathbf{s})} &= e_{(\mathbf{I})} \cos^2\left(\frac{\pi}{4} + \frac{\Theta}{2}\right) + e_{(\mathbf{J})} \sin^2\left(\frac{\pi}{4} + \frac{\Theta}{2}\right). \end{aligned} \tag{3.12}$$

Then, $e_{(\mathbf{m})} = e_{(\mathbf{s})}$ gives $(e_{(\mathbf{I})} - e_{(\mathbf{J})}) \sin\Theta = 0$. Thus, because $\Theta \neq 0, \pi$, it follows that $e_{(\mathbf{I})} = e_{(\mathbf{J})}$ and so π is the plane of $\mathcal{C}^+$ or $\mathcal{C}^-$.

Thus, in a plane π, knowing just one unsheared pair and their extensions is sufficient to determine the principal extensions and the principal axes of $\mathcal{S}$, the ellipse in which the plane π cuts $\mathcal{E}$, and hence to determine all shears and extensions in π.

REMARK 3 (Maximum shear). From equation (3.2), for given Θ, the shear $\Delta(\mathbf{m}, \mathbf{s})$ of the pair of material elements along $(\mathbf{m}, \mathbf{s})$ subtending the (acute) angle Θ in π, takes its maximum value $|e_{(\mathbf{I})} - e_{(\mathbf{J})}| \sin\Theta$ when $\mathbf{c}$ is along $\mathbf{I}$ and $\mathbf{d}$ is along $\mathbf{J}$, and so $\mathbf{m}$ and $\mathbf{s}$ are equally disposed at $\Theta/2$ to $\mathbf{I}$. For given Θ, the shear takes its maximum value $|e_1 - e_3| \sin\Theta$ when $\mathbf{m}$ and $\mathbf{s}$ lie in the principal $\mathbf{\Omega}_1$–$\mathbf{\Omega}_3$ plane and are equally disposed at $\Theta/2$ to $\mathbf{\Omega}_1$. Clearly, for arbitrary Θ, the maximum shear for pairs of material elements in the plane π is $|e_{(\mathbf{I})} - e_{(\mathbf{J})}|$ and occurs for $\Theta = \pi/2$, in which case $\mathbf{m}$ and $\mathbf{s}$ are equally disposed at $\pi/4$ to $\mathbf{I}$ and $\mathbf{J}$.

The overall maximum shear is $|e_1 - e_3|$, occurs for $\Theta = \pi/2$ with $\mathbf{m}$ and $\mathbf{s}$ equally disposed at $\pi/4$ to $\mathbf{\Omega}_1$ and $\mathbf{\Omega}_3$ in the $\mathbf{\Omega}_1$–$\mathbf{\Omega}_3$ plane.

REMARK 4 (Universal relations). In passing, we note some universal relations amongst the shears of pairs of elements in the plane π. These relations are valid irrespective of the material or the deformation.

From equations (3.2), and (3.4), we have the universal relation

$$\Delta\left(\frac{\mathbf{I}+\mathbf{J}}{\sqrt{2}}, \mathbf{I}\sin\left(\frac{\pi}{4}+\Theta\right)+\mathbf{J}\cos\left(\frac{\pi}{4}+\Theta\right)\right)$$
$$= \sin\Theta\Delta\left(\mathbf{I}\cos\frac{\Theta}{2}-\mathbf{J}\sin\frac{\Theta}{2},\ \mathbf{I}\cos\frac{\Theta}{2}+\mathbf{J}\sin\frac{\Theta}{2}\right).$$

Also, from equation (3.6), we have the further universal relation

$$\sin^2\Phi\Delta\left(\frac{\mathbf{I}+\mathbf{J}}{\sqrt{2}},\ \mathbf{I}\sin\left(\frac{\pi}{4}+\Theta\right)+\mathbf{J}\cos\left(\frac{\pi}{4}+\Theta\right)\right)$$
$$= \sin^2\Theta\Delta\left(\frac{\mathbf{I}+\mathbf{J}}{\sqrt{2}},\ \mathbf{I}\sin\left(\frac{\pi}{4}+\Phi\right)+\mathbf{J}\cos\left(\frac{\pi}{4}+\Phi\right)\right).$$

4. Results in Terms of the Normals to the Central Circular Sections of $\mathcal{E}$

For $\Theta \neq 0, \pi$, equation (3.4) gives the unsheared pairs $(\mathbf{m}, \mathbf{s})$ in the plane π, where $\mathbf{I}$ and $\mathbf{J}$ are the unit vectors along the major and minor axes of $\mathcal{S}$, the ellipse in which the plane π cuts the ellipsoid $\mathcal{E}$. Thus, knowing $\mathbf{I}$ and $\mathbf{J}$ means that all unsheared pairs in π may be determined explicitly. But, as shown in [3], $\mathbf{I}$ and $\mathbf{J}$ may be determined in terms of the unit normal $\mathbf{n} = \mathbf{I} \times \mathbf{J}$ to the plane π. It follows that the unsheared pairs $(\mathbf{m}, \mathbf{s})$ in the plane π may be determined in terms of $\mathbf{n}$. The unit normals to the planes of central circular section of the ellipsoid $\mathcal{E}$ play a crucial role in this determination. Here, we consider the general case when all three principal strains are different from each other.

Two cases have to be considered separately [3]:

(a) when $\mathbf{n}$ is not coplanar with $\mathbf{h}^+$ and $\mathbf{h}^-$: $\mathbf{n}\cdot\mathbf{h}^+\times\mathbf{h}^- \neq 0$,

(b) when $\mathbf{n}$ is coplanar with $\mathbf{h}^+$ and $\mathbf{h}^-$: $\mathbf{n}\cdot\mathbf{h}^+\times\mathbf{h}^- = 0$.

Case (a) : $\mathbf{n}\cdot\mathbf{h}^+\times\mathbf{h}^- \neq 0$. Here, $\mathbf{n}$ is not coplanar with $\mathbf{h}^+$ and $\mathbf{h}^-$, or equivalently, with $\mathbf{\Omega}_1$ and $\mathbf{\Omega}_3$. Let ν be the angle between $\mathbf{n}\times\mathbf{h}^+$ and $\mathbf{n}\times\mathbf{h}^-$, and let $\widehat{\mathbf{n}\times\mathbf{h}^+}$ and $\widehat{\mathbf{n}\times\mathbf{h}^-}$ denote unit vectors along $\mathbf{n}\times\mathbf{h}^+$ and $\mathbf{n}\times\mathbf{h}^-$, respectively:

$$\widehat{\mathbf{n}\times\mathbf{h}^\pm} = \frac{\mathbf{n}\times\mathbf{h}^\pm}{|\mathbf{n}\times\mathbf{h}^\pm|}, \tag{4.1}$$

$$\cos\nu = \widehat{\mathbf{n}\times\mathbf{h}^+}\cdot\widehat{\mathbf{n}\times\mathbf{h}^-}. \tag{4.2}$$

Then [3], the unit vectors **I** and **J** along the principal axes of the ellipse $\mathcal{S}$ are given by

$$
\begin{aligned}
2\cos(\nu/2)\mathbf{I} &= \widehat{\mathbf{n}\times\mathbf{h}^+} + \widehat{\mathbf{n}\times\mathbf{h}^-},\\
2\sin(\nu/2)\mathbf{J} &= \widehat{\mathbf{n}\times\mathbf{h}^+} - \widehat{\mathbf{n}\times\mathbf{h}^-}.
\end{aligned}
\tag{4.3}
$$

Substituting these expressions for **I** and **J** into (3.4) yields

$$
\begin{aligned}
\sin\nu\,\mathbf{m} &= \widehat{\mathbf{n}\times\mathbf{h}^+}\cos\left(\frac{\pi}{4}+\frac{\Theta-\nu}{2}\right) - \widehat{\mathbf{n}\times\mathbf{h}^-}\cos\left(\frac{\pi}{4}+\frac{\Theta+\nu}{2}\right),\\
\sin\nu\,\mathbf{s} &= \widehat{\mathbf{n}\times\mathbf{h}^+}\sin\left(\frac{\pi}{4}+\frac{\Theta+\nu}{2}\right) - \widehat{\mathbf{n}\times\mathbf{h}^-}\sin\left(\frac{\pi}{4}+\frac{\Theta-\nu}{2}\right).
\end{aligned}
\tag{4.4}
$$

As Θ is varied, this gives all the unsheared pairs in the plane π with normal **n**, not in the plane of $\mathbf{\Omega}_1$ and $\mathbf{\Omega}_3$.

Case (b): $\mathbf{n}\cdot\mathbf{h}^+\times\mathbf{h}^- = 0$. Here, **n** is in the plane of $\mathbf{h}^+$ and $\mathbf{h}^-$, that is, in the principal $\mathbf{\Omega}_1$–$\mathbf{\Omega}_3$ plane. It is also assumed that **n** is not along $\mathbf{h}^+$ or $\mathbf{h}^-$, because when this is the case all pairs of material elements in π are unsheared [4]. Then [3], $\nu = 0$ or π, $\widehat{\mathbf{n}\times\mathbf{h}^+} = \pm\widehat{\mathbf{n}\times\mathbf{h}^-}$, and the unit vectors **I** and **J** along the principal axes of the ellipse $\mathcal{S}$ are given by

$$
\mathbf{I} = \widehat{\mathbf{n}\times\mathbf{h}^+},\qquad \mathbf{J} = \mathbf{n}\times\left(\widehat{\mathbf{n}\times\mathbf{h}^+}\right).
\tag{4.5}
$$

Substituting this into (3.4) yields

$$
\begin{aligned}
\mathbf{m} &= \widehat{\mathbf{n}\times\mathbf{h}^+}\sin\left(\frac{\pi}{4}+\frac{\Theta}{2}\right) + \mathbf{n}\times\left(\widehat{\mathbf{n}\times\mathbf{h}^+}\right)\cos\left(\frac{\pi}{4}+\frac{\Theta}{2}\right),\\
\mathbf{s} &= \widehat{\mathbf{n}\times\mathbf{h}^+}\cos\left(\frac{\pi}{4}+\frac{\Theta}{2}\right) + \mathbf{n}\times\left(\widehat{\mathbf{n}\times\mathbf{h}^+}\right)\sin\left(\frac{\pi}{4}+\frac{\Theta}{2}\right).
\end{aligned}
\tag{4.6}
$$

As Θ is varied, this gives all the unsheared pairs in the plane π with normal **n** in the plane of $\mathbf{\Omega}_1$ and $\mathbf{\Omega}_3$.

5. Unsheared Pairs One of Whose Elements Is Given

Let **s** be given. Consider now a plane π containing **s** and let **n** be the unit normal to this plane: $\mathbf{n}\cdot\mathbf{s} = 0$. We seek the direction **m** "companion to **s**" such that the pair of line elements $\mathcal{L}$ and $\mathcal{L}'$ in π along **m** and **s** is unsheared. By varying the plane π through the given **s**, all unsheared pairs (**m**, **s**) will thus be obtained.

Now

$$
e_{ij}m_i s_j = e_{(\mathbf{c})}\cos^2(\Theta/2) - e_{(\mathbf{d})}\sin^2(\Theta/2),
\tag{5.1}
$$

where Θ is the angle between **m** and **s**, and **c** and **d** are orthogonal unit vectors along the bisectors, internal and external, of the angle between **m** and **s**. From equation (3.2), if the pair of elements along $\mathbf{m}$ and $\mathbf{s}$ is unsheared, then $e_{(\mathbf{c})} = e_{(\mathbf{d})}$, so that

$$2e_{ij}m_i s_j = 2e_{(\mathbf{c})}\cos\Theta = \{\operatorname{tr}\mathbf{e} - e_{(\mathbf{n})}\}\cos\Theta = \{\operatorname{tr}\mathbf{e} - e_{(\mathbf{n})}\}m_i s_i. \tag{5.2}$$

Hence,

$$m_i[2e_{ij}s_j - \{\operatorname{tr}\mathbf{e} - e_{(\mathbf{n})}\}s_i] = 0. \tag{5.3}$$

Also, because $\mathbf{m}$ is in the plane π,

$$m_i n_i = 0. \tag{5.4}$$

Hence,

$$\mathbf{m} = \alpha\mathbf{n} \times \{2\mathbf{es} - [\operatorname{tr}\mathbf{e} - e_{(\mathbf{n})}]\mathbf{s}\}, \tag{5.5}$$

where α is an appropriate scalar such that $\mathbf{m}\cdot\mathbf{m} = 1$. Also, because $\mathbf{n}$, $\mathbf{s}$ and $\mathbf{n}\times\mathbf{s}$ form an orthonormal triad, we have

$$\operatorname{tr}\mathbf{e} = e_{(\mathbf{n})} + e_{(\mathbf{s})} + e_{(\mathbf{n}\times\mathbf{s})}, \tag{5.6}$$

$$\mathbf{es} = (\mathbf{n}\cdot\mathbf{es})\mathbf{n} + e_{(\mathbf{s})}\mathbf{s} + (\mathbf{n}\times\mathbf{s}\cdot\mathbf{es})\mathbf{n}\times\mathbf{s}, \tag{5.7}$$

and hence (5.5) may also be written,

$$\mathbf{m} = \alpha[2(\mathbf{n}\times\mathbf{es})\cdot\mathbf{s}]\mathbf{s} + \alpha[e_{(\mathbf{s})} - e_{(\mathbf{n}\times\mathbf{s})}]\mathbf{n}\times\mathbf{s} = \alpha[2(\mathbf{n}\times\mathbf{Es})\cdot\mathbf{s}]\mathbf{s} + \alpha[E_{(\mathbf{s})} - E_{(\mathbf{n}\times\mathbf{s})}]\mathbf{n}\times\mathbf{s}. \tag{5.8}$$

Thus, if $\mathbf{s}$ is given, then equation (5.5), or equivalently (5.8), determines in general a unique companion $\mathbf{m}$ to $\mathbf{s}$ in the plane π containing $\mathbf{s}$ such that the pair of material line elements along $\mathbf{m}$ and $\mathbf{s}$ is unsheared. Note that equation (5.5) may also be obtained from a formula derived for finite strain by Joly in terms of quaternions [5].

There are, however, two exceptional cases:

(i) when equation (5.5) gives $\mathbf{m} = \mathbf{0}$ so that the direction of $\mathbf{m}$ is undetermined;

(ii) when the choice of $\mathbf{s}$ in (5.5) gives $\mathbf{m} = \mathbf{s}$ so that there is no companion element to $\mathbf{s}$ in π.

We now consider these two cases.

(i) Direction of $\mathbf{m}$ *undetermined.* From equation (5.5) it follows that $\mathbf{m} = \mathbf{0}$, so that the direction of $\mathbf{m}$ is undetermined, if and only if

$$(\mathbf{n}\times\mathbf{Es})\cdot\mathbf{s} = 0, \qquad E_{(\mathbf{s})} = E_{(\mathbf{n}\times\mathbf{s})}. \tag{5.9}$$

Now $(5.9)_1$ may be written as $(\mathbf{n}\times\mathbf{s})\cdot\mathbf{Es} = 0$, which expresses the condition that the directions of $\mathbf{n}\times\mathbf{s}$ and $\mathbf{s}$ are conjugate with respect to the ellipsoid $\mathcal{E}$. Also by $(5.9)_2$, the radii to $\mathcal{E}$ along $\mathbf{n}\times\mathbf{s}$ and $\mathbf{s}$ are equal in length. But $\mathbf{n}\times\mathbf{s}$ and $\mathbf{s}$ are orthogonal. Hence, $\mathbf{n}\times\mathbf{s}$ and $\mathbf{s}$ lie in the plane of $\mathcal{C}^+$ or $\mathcal{C}^-$.

Thus, we conclude that the case of undetermined direction for **m** arises only when the given plane π with normal **n** is a plane of central circular section of the ellipsoid $\mathcal{E}$.

(ii) **m** *parallel to* **s**. From equation (5.8), it follows that **m** is parallel to **s** if and only if equation $(5.9)_2$ holds. Now, however, equation $(5.9)_1$ need not be valid so that $\mathbf{n} \times \mathbf{s}$ and **s** need not be in conjugate directions with respect to the ellipsoid $\mathcal{E}$. However, the radii to $\mathcal{S}$ along the orthogonal vectors $\mathbf{n} \times \mathbf{s}$ and **s** are equal in length, and hence $\mathbf{n} \times \mathbf{s}$ and **s** are equally inclined at $\pi/4$ to **I** and **J**. Thus, in the present case when **m** and **s** are parallel, the orthogonal vectors $\mathbf{n} \times \mathbf{s}$ and **s** are along $(\mathbf{I} \pm \mathbf{J})/\sqrt{2}$, the bisectors of the angle between **I** and **J** in the plane π. These are the directions of those elements which have no conjugate elements in π which form unsheared pairs (Section 3).

To sum up, equation (5.5) gives the companion **m** to **s** such that the pair $(\mathbf{m}, \mathbf{s})$ in the plane π, with normal **n**, is unsheared. If the right side of (5.5) is zero so that the direction of **m** is undetermined, then $\mathbf{n} \times \mathbf{s}$ and **s** lie in a plane of central circular section of $\mathcal{E}$, whilst if **s** is so chosen that **m** is parallel to **s**, then **s** is along $\mathbf{I} + \mathbf{J}$ or $\mathbf{I} - \mathbf{J}$ in π, that is along a limiting direction.

References

1. Ph. Boulanger and M. Hayes, Shear, shear stress and shearing. *J. Mech. Phys. Solids* **40** (1992) 1449–1457.
2. Ph. Boulanger and M. Hayes, *Bivectors and Waves in Mechanics and Optics*. Chapman & Hall, London (1993).
3. Ph. Boulanger and M. Hayes, The common conjugate directions of plane sections of two concentric ellipsoids. In: J. Casey and M.J. Crochet (eds), *Theoretical, Experimental, and Numerical Contributions to the Mechanics of Fluids and Solids. Special Issue of Z. angew. Math. Phys.* **46** (1995) 356–371.
4. M. Hayes, On strain and straining. *Arch. Rational Mech. Anal.* **100** (1988) 265–273.
5. Ch. Joly, *A Manual of Quaternions*. MacMillan, New York (1905).
6. A.E.H. Love, *A Treatise on the mathematical Theory of Elasticity*, 4th edn. Cambridge Univ. Press, Cambridge (1927).
7. T.Y. Thomas, *Plastic Flow and Fracture in Solids*. Academic Press, New York (1961).

Journal of Elasticity **59:** 237–266, 2000.

Reflection and Refraction of Anti-Plane Shear Waves from a Moving Phase Boundary

THOMAS J. PENCE and HUNGYU TSAI
Department of Materials Science and Mechanics, Michigan State University, East Lansing, MI 48824, U.S.A. E-mail: pence@egr.msu.edu, hytsai@egr.msu.edu

Received 1 March 1999; in revised form 8 November 1999

Dedicated to Roger Fosdick on the occasion of his 60th birthday.

Abstract. The reflection and refraction of anti-plane shear waves from an interface separating half-spaces with different moduli is well understood in the linear theory of elasticity. Namely, an oblique incident wave gives rise to a reflected wave that departs at the same angle and to a refracted wave that, after transmission through the interface, departs at a possibly different angle. Here we study similar issues for a material that admits mobile elastic phase boundaries in anti-plane shear. We consider an energy minimal equilibrium state in anti-plane shear involving a planar phase boundary that is perturbed due to an incident wave of small magnitude. The phase boundary is allowed to move under this perturbation. As in the linear theory, the perturbation gives rise to a reflected and a refracted wave. The orientation of these waves is independent of the phase boundary motion and determined as in the linear theory. However, the phase boundary motion affects the amplitudes of the departing waves. Perturbation analysis gives these amplitudes for general small phase boundary motion, and also permits the specification of the phase boundary motion on the basis of additional criteria such as a kinetic relation. A standard kinetic relation is studied to quantify the subsequent energy partitioning and dissipation on the basis of the properties of the incident wave.

Mathematics Subject Classifications (2000): 73D25, 73C50.

Key words: anti-plane shear, wave propagation, wave reflection and refraction, phase transition.

1. Introduction

The continuum theory of deformable solids capable of stress-induced displacive transformations emphasizes the role of a nonconvex elastic energy density in the determination of multi-phase equilibrium states. These states, involving sharp interfaces that support discontinuities in the displacement gradient, retrieve the microstructures associated with the crystallographic theory of martensitic transformation provided that the strain energy density embodies the lattice symmetry of the participating phases [5]. In this context, anti-plane shear deformations have been widely studied in order to model material microstructures, such as crystallographic twins [13], in a mathematical setting that is simpler than that for other types of deformation. Due to the possible loss of ellipticity in the underlying governing equations, what might normally be a rather standard boundary value problem may instead exhibit multiple equilibrium solutions in the form of differing multi-phase microstructures. Consequently, abrupt changes may occur in the stability of various quasi-static solution paths associated with a particular microstructural evolution. The dynamical equations of motion for anti-plane shear are then central to treating the dynamical evolution of such microstructures. For example, Rosakis and Tsai [14] determine possible shapes of twin lamellae consistent with steady state growth. In addition, dynamical perturbations to a multi-phase equilibrium may trigger various kinds of microstructural change depending on the stability of the original equilibrium state.

Here we consider an energy minimal two-phase equilibrium state that is subject to a small dynamical perturbation in the form of a finite width plane wave under the setting of anti-plane shear deformations. At issue is the interaction of this wave with a preexisting phase boundary. Problems of this type have been the object of much study in the context of uniaxial deformation, in which case the interaction of the wave with the phase boundary can cause the phase boundary to change its position. This process gives rise to both a reflected and a transmitted wave [3, 10]. The motion of the phase boundary is typically not determined on the basis of the dynamical equations alone, although the second law of thermodynamics restricts the various motion possibilities [2, 11]. This is true either in an isothermal setting or when the theory is extended to include thermal effects [8]. Here we provide an isothermal treatment of a corresponding problem in anti-plane shear where the plane wave intersects a planar phase boundary at a general oblique angle. The assumed small amplitude of the dynamical perturbation permits a constitutive description that leads to linear equations of motion in each phase, which are formally equivalent to those for linear elasticity. However, unlike the classical theory of linear elasticity, the dynamical perturbation may trigger phase boundary movement. In the absence of any such movement, the problem reduces to a standard one for the reflection and refraction of anti-plane shear waves from an interface separating half-spaces with different moduli. This standard problem is well understood in the linear theory of elasticity. Namely, an oblique incident wave gives rise to a reflected wave that departs at the same angle, and to a refracted wave that, after transmission through the stationary interface, departs at a possibly different angle [4]. The properties of these waves are then dependent on the incident wave profile, the various material moduli, and the orientation of the incident wave with respect to the interface. In the problem considered here, it is also generally the case that a reflected and refracted wave are generated, and the orientation of these waves is independent of the phase boundary motion. However, the phase boundary motion does affect the wave amplitudes of the departing waves, the dissipation rate at the phase boundary, the energy partitioning between the various incoming and outgoing waves, and the energy change due to the accretion of one phase at the expense of the other.

This paper is organized as follows. The problem is formulated in Section 2 under a requirement of steady state phase boundary motion that correlates with the advance of the incident wave. The classical jump conditions then give formal equations for the reflected and refracted waves on the basis of the incident wave and the phase boundary motion. A perturbation analysis gives the first order solution to these equations for arbitrary (small) phase boundary motion and incident wave profile. The driving traction associated with such motion is given in Section 3, and two classes of kinetic relation are introduced for the purpose of determining the phase boundary motion. In the first class, there is a positive phase boundary mobility, which is the ratio of the phase boundary normal velocity to the driving traction in the small driving traction limit. In this case the leading order reflected and refracted waves are given by mobility dependent expressions that scale linearly with the amplitude of the incident wave. In contrast, for a second class of kinetic relation suggestive of frictional effects, the phase boundary may be stationary for finite regions of contact with the incident wave. The overall process energetics are explored in Section 4, where the energy rate associated with wave advance is defined for all three travelling waves. The sum of these rates is equal to the combined dissipation rate and energy accretion rate due to material conversion from one phase to the other. For the mobility mediated kinetic relation, the details of this partitioning are sensitive to the angle of incidence, the phase boundary mobility, and the moduli of the various phases. In particular, the energy rates associated with wave advance and the energy rate associated with material accretion are generally first order in the perturbation parameter, whereas the dissipation rate is second order. The mobility associated with maximum dissipation is obtained, as is the mobility dependent angle of incidence that suppresses the reflected wave.

2. Anti-Plane Shear Formulation

Consider the following time-dependent anti-plane shear deformation

$$\mathbf{y}(\mathbf{x}, t) = \mathbf{x} + u(x_1, x_2, t)\mathbf{e}_3 \tag{2.1}$$

in which the displacement is along the x_3-direction only and dependent on the coordinates within the infinite plane Π spanned by $\mathbf{e}_1$ and $\mathbf{e}_2$. The shear strain vector $\boldsymbol{\gamma}$ and shear stress vector $\boldsymbol{\tau}$ associated with this plane are defined by

$$\boldsymbol{\gamma} = \gamma_\alpha \mathbf{e}_\alpha, \quad \gamma_\alpha = \frac{\partial u}{\partial x_\alpha}, \qquad \boldsymbol{\tau} = \tau_\alpha \mathbf{e}_\alpha, \quad \tau_\alpha = \sigma_{3\alpha}. \tag{2.2}$$

The repeated Greek index α implies summation over $\alpha = 1$ and 2. The displacement u is assumed to be at least twice continuously differentiable with respect to x_α and t, except on a collection of piecewise smooth curves $\Sigma_t \subset \Pi$. These curves include the location of shock waves as well as phase boundaries. For a class of elastic materials, the balance of linear momentum leads to a single equation involving the out-of-plane displacement and the shear stress components:

$$\tau_{\alpha,\alpha} = \rho\ddot{u} \quad \text{on } \Pi - \Sigma_t, \tag{2.3}$$

where the referential mass density ρ is assumed to be constant. The other two (in-plane) equilibrium equations are automatically satisfied if certain restrictions on the material constitution are imposed. A discussion of this issue can be found in [16].

Along the curves Σ_t, the shear strains are allowed to have jump discontinuities while keeping the displacement continuous. In addition, the balance of linear momentum localized to the curves Σ_t imposes another condition on the stresses. These jump conditions can be written as

$$\begin{cases} [\![u]\!] = 0, \\ [\![\tau_\alpha]\!]n_\alpha + \rho[\![\dot{u}]\!]V_{\mathbf{n}} = 0 \end{cases} \quad \text{on } \Sigma_t, \tag{2.4}$$

where $\mathbf{n} = n_\alpha \mathbf{e}_\alpha$ is the unit normal to the curve Σ_t on Π and $V_{\mathbf{n}}$ is the normal velocity of Σ_t. The symbol $[\![\cdot]\!]$ denotes the difference of the limit values across Σ_t.

The material under consideration is assumed to be compressible and hyperelastic. For the anti-plane shear deformation (2.1), the stored energy density W can be reduced to a function of the shear strains $\boldsymbol{\gamma}$. Let this reduced stored energy function be denoted by $\mathcal{W}$, then the shear stresses τ_α are given by

$$\tau_\alpha = \frac{\partial \mathcal{W}}{\partial \gamma_\alpha}(\gamma_1, \gamma_2). \tag{2.5}$$

It is assumed that the material under consideration admits a particularly simple type of equilibrium two-phase deformation that is energy minimal when $\boldsymbol{\tau} = \sigma^*\mathbf{e}_2$. Here the constant σ^* is the special Maxwell stress. These energy minimal deformations involve alternating bands with shear strain states $\boldsymbol{\gamma}^A \equiv \gamma^A\mathbf{e}_2$ and $\boldsymbol{\gamma}^B \equiv \gamma^B\mathbf{e}_2 = (\gamma^A + \gamma^*)\mathbf{e}_2$, where γ^A, γ^B, γ^* are constant. The bands are parallel to the x_1-axis and, hence, have interface normal $\mathbf{e}_2$. If $\boldsymbol{\tau} = \sigma\mathbf{e}_2$ with σ slightly less than σ^* then energy minimal equilibrium configurations are states of simple shear, $\boldsymbol{\gamma} = \gamma\mathbf{e}_2$, with γ slightly less than γ^A. Conversely, if $\boldsymbol{\tau} = \sigma\mathbf{e}_2$ with σ slightly greater than σ^* then energy minimal equilibrium configurations are states of simple shear, $\boldsymbol{\gamma} = \gamma\mathbf{e}_2$, with γ slightly greater than γ^B. Let $\boldsymbol{\gamma}^* \equiv \gamma^*\mathbf{e}_2 = (\gamma^B - \gamma^A)\mathbf{e}_2$ be the difference between these two special Maxwell strains. The type of behavior just described can occur if the strain energy density has separate wells near $\boldsymbol{\gamma}^A = \gamma^A\mathbf{e}_2$ and $\boldsymbol{\gamma}^B = \gamma^B\mathbf{e}_2$ where γ^A and γ^B are determined on the basis of a well known tangent line construction (Figure 1(b)). The energy density at $\boldsymbol{\gamma}^B$ then exceeds the energy density at $\boldsymbol{\gamma}^A$ by the amount $\sigma^*\gamma^*$. This behavior is a simple anti-plane shear extension of a standard uniaxial model [6].

In the subsequent analysis, we will only consider deformations with strains close to the two Maxwell strains $\boldsymbol{\gamma}^A$ and $\boldsymbol{\gamma}^B$. Material points with strains near $\boldsymbol{\gamma}^A$

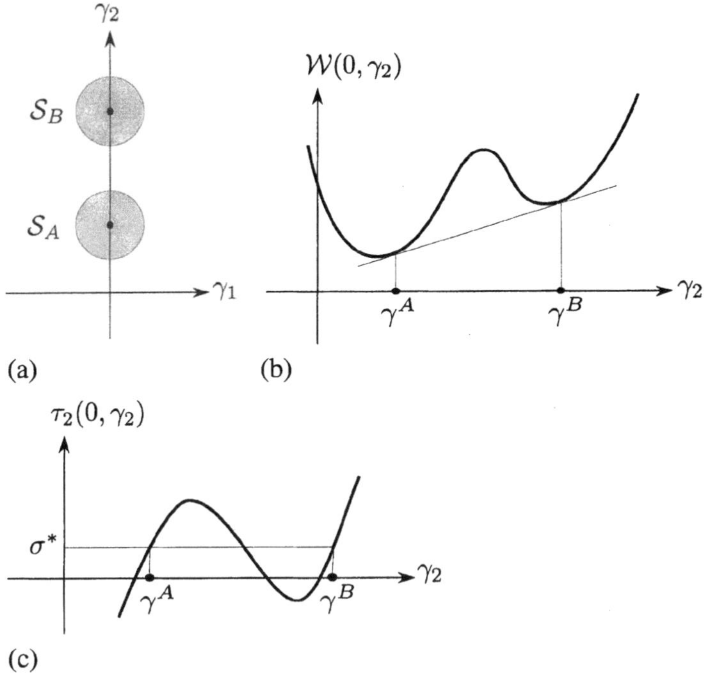

Figure 1. Constitutive model: (a) Two disjoint potential wells $\mathcal{S}_A$ and $\mathcal{S}_B$ in the shear strain space. (b) Dependence of the reduced stored energy function $\mathcal{W}$ on γ_2 for $\gamma_1 = 0$. (c) stress τ_2-strain γ_2 relation for $\gamma_1 = 0$.

(respectively, $\boldsymbol{\gamma}^B$) will be referred to as being in phase A (respectively, phase B). For these two phases, we define two regions in strain space:

$$\begin{cases} \mathcal{S}_A = \{\boldsymbol{\gamma} \mid |\boldsymbol{\gamma} - \gamma^A \mathbf{e}_2| < \delta\}, \\ \mathcal{S}_B = \{\boldsymbol{\gamma} \mid |\boldsymbol{\gamma} - \gamma^B \mathbf{e}_2| < \delta\}, \end{cases} \tag{2.6}$$

where $\delta < \gamma^*/2$ so that $\mathcal{S}_A$ and $\mathcal{S}_B$ are disjoint (Figure 1(a)). Inside these two regions, the material is assumed to be isotropic and linearly elastic with reduced stored energy of the form

$$\mathcal{W} = \begin{cases} \mathring{\mathcal{W}}_A + \sigma^* \mathbf{e}_2 \cdot (\boldsymbol{\gamma} - \gamma^A \mathbf{e}_2) + \dfrac{\mu^A}{2} |\boldsymbol{\gamma} - \gamma^A \mathbf{e}_2|^2 & \text{on } \mathcal{S}_A, \\ \mathring{\mathcal{W}}_B + \sigma^* \mathbf{e}_2 \cdot (\boldsymbol{\gamma} - \gamma^B \mathbf{e}_2) + \dfrac{\mu^B}{2} |\boldsymbol{\gamma} - \gamma^B \mathbf{e}_2|^2 & \text{on } \mathcal{S}_B, \end{cases} \tag{2.7}$$

where $\mathring{\mathcal{W}}_B = \mathring{\mathcal{W}}_A + \sigma^* \gamma^*$. Here μ^A and μ^B are the shear moduli for phase A and B, respectively. It follows that the associated constitutive relation governing anti-plane shear is

$$\tau_\alpha \mathbf{e}_\alpha = \begin{cases} \sigma^* \mathbf{e}_2 + \mu^A (\boldsymbol{\gamma} - \gamma^A \mathbf{e}_2) & \text{on } \mathcal{S}_A, \\ \sigma^* \mathbf{e}_2 + \mu^B (\boldsymbol{\gamma} - \gamma^B \mathbf{e}_2) & \text{on } \mathcal{S}_B. \end{cases} \tag{2.8}$$

With the constitutive relation (2.8), the field equation (2.3) reduces to the two-dimensional wave equation:

$$u_{,11} + u_{,22} = \begin{cases} \dfrac{1}{c_A^2}\ddot{u} & \text{for phase } A, \\ \dfrac{1}{c_B^2}\ddot{u} & \text{for phase } B \end{cases} \quad \text{on } \Pi - \Sigma_t, \tag{2.9}$$

where $c_A = \sqrt{\mu^A/\rho}$ and $c_B = \sqrt{\mu^B/\rho}$ are the shear wave speeds in phase A and B, respectively. The planar wave solution to (2.9) is given by

$$u = U(\mathbf{x} \cdot \mathbf{m} - ct), \tag{2.10}$$

where U gives the displacement profile, $\mathbf{m}$ is the unit vector along the direction of the propagation, and $c = c_A$ or c_B depending on the phase. On the curves of discontinuity Σ_t, the jump conditions (2.4) can be written as

$$\begin{cases} [\![u]\!] = 0, \\ [\![\tau_\alpha]\!] n_\alpha + \rho V_{\mathbf{n}}^2 [\![\gamma_\alpha]\!] n_\alpha = 0 \end{cases} \quad \text{on } \Sigma_t. \tag{2.11}$$

If the curve of discontinuity separates materials of the same phase, then the jump conditions (2.11) imply that the normal velocity $V_{\mathbf{n}}$ must be equal to the corresponding shear wave speed. Consider a two-phase energy minimal equilibrium state with $\boldsymbol{\tau} = \sigma^* \mathbf{e}_2$ consisting of two half-spaces separated by the x_1-axis such that the region $x_2 < 0$ involves $\boldsymbol{\gamma} = \gamma^A \mathbf{e}_2$ and $u = \gamma^A x_2$ whereas the region $x_2 > 0$ involves $\boldsymbol{\gamma} = \gamma^B \mathbf{e}_2$ and $u = \gamma^B x_2$. Our purpose is to determine the effect of a dynamic perturbation in the form of an oblique incident plane wave. Attention is restricted to a plane wave of finite thickness D that is incident through phase A and intersects the phase boundary at the angle θ_I. The incident wave region will be denoted by $\mathcal{P}_4$ (see Figure 2), and the incident wave is characterized by the displacement:

$$\begin{aligned} &u^I = \gamma^A x_2 + \varepsilon U_I(\xi \sin\theta_I + x_2 \cos\theta_I) \quad \text{in } \mathcal{P}_4, \\ &U_I(0) = 0, \qquad U_I'(0) = 0, \qquad \xi \equiv x_1 - \frac{c_A}{\sin\theta_I} t. \end{aligned} \tag{2.12}$$

The strains and the particle velocity in $\mathcal{P}_4$ are given by

$$\left. \begin{aligned} &\boldsymbol{\gamma}^I = \gamma^A \mathbf{e}_2 + \varepsilon U_I'(\xi \sin\theta_I + x_2 \cos\theta_I) \times (\sin\theta_I \mathbf{e}_1 + \cos\theta_I \mathbf{e}_2), \\ &v^I = -\varepsilon c_A U_I'(\xi \sin\theta_I + x_2 \cos\theta_I) \end{aligned} \right\} \text{ in } \mathcal{P}_4. \tag{2.13}$$

Here U_I is a continously differentiable function that gives the displacement profile of the incident wave, and ξ is an appropriate similarity variable incorporating the propagation velocity $\overset{\circ}{V} \equiv c_A / \sin\theta_I$ of the incident wave with respect to the

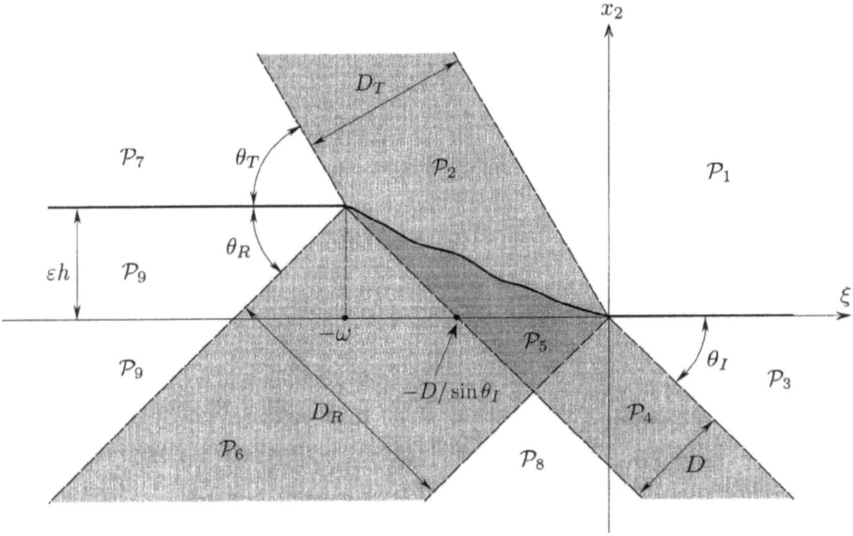

Figure 2. The interaction of the incident, reflected and transmitted waves with the phase boundary. The partitioned regions $\mathcal{P}_i$, $i = 1, 2, \ldots, 9$, are as shown. The system translates to the right with speed $\mathring{V}$.

x_1-axis. In particular, $\xi = 0$ gives the leading edge of the disturbance, and $\xi = -\omega$ gives the trailing edge. The parameter ε, which is assumed to be small with respect to γ^*, sets the scale of the perturbation.

The interaction of this plane wave with the phase boundary gives rise to a wave that is reflected back into phase A at the angle θ_R, and a wave that is transmitted into phase B at the angle θ_T. The properties of the reflected and transmitted waves depend both on the incident plane wave and on the accompanying motion of the phase boundary. Attention is restricted to steady state motion of the phase boundary that correlates with the propagation of the incident wave along the interface. In particular, this correlation is assumed to limit the phase boundary motion to an $O(\varepsilon)$ perturbation, such that the phase boundary is again stationary after wave passage (Figure 2). This state of affairs is consistent with the underlying energy minimal equilibrium as discussed, for example, in [9] for the simpler uniaxial setting. For the anti-plane shear motion under consideration, the reflected and transmitted wave are also steady state, and the overall wave geometry is as depicted in Figure 2. Namely, the displacement field is continuous throughout the $(x_1$–$x_2)$-plane, which, in turn, is partitioned into 9 regions $\mathcal{P}_i$, $i = 1, 2, \ldots, 9$, such that regions $\mathcal{P}_2$, $\mathcal{P}_4$, $\mathcal{P}_5$, $\mathcal{P}_6$ represent waves (nonzero particle velocity and hence kinetic energy), whereas regions $\mathcal{P}_1$, $\mathcal{P}_3$, $\mathcal{P}_7$, $\mathcal{P}_8$, $\mathcal{P}_9$ represent simple equilibrated states (constant strain and zero particle velocity). The steady-state propagation of the reflected and transmitted waves requires that

$$\mathring{V} \equiv \frac{c_A}{\sin\theta_I} = \frac{c_A}{\sin\theta_R} = \frac{c_B}{\sin\theta_T}. \tag{2.14}$$

Denote the ratio of the two wave speeds by $\lambda = c_B/c_A$, then

$$\theta_R = \theta_I, \qquad \sin\theta_T = \lambda \sin\theta_I, \qquad \lambda = \frac{c_B}{c_A} = \sqrt{\frac{\mu^B}{\mu^A}}. \tag{2.15}$$

Although the angle of the reflected wave is preserved, that of the transmitted wave is not. Following common usage, the transmitted wave is also referred to as the refracted wave. The value θ_T as defined by (2.15) will often appear explicitly in what follows. When $\lambda > 1$, there is a critical value $\theta_c = \arcsin(1/\lambda)$ for the incident angle θ_I. If $\theta_I > \theta_c$, then θ_T cannot be defined on the basis of (2.15) and the transmitted wave does not take the form of a plane wave (2.10) [4]. In all cases, a plane wave correlates with a steady state propagation speed $\mathring{V}$ that is supersonic with respect to the shear wave speed, while the non-plane wave form correlates with a steady state propagation speed $\mathring{V}$ that is subsonic with respect to the shear wave speed. Here we only consider the case involving plane waves in both phase A and phase B. Note that the aspects of the problem described above are formally identical to that of wave transmission from one elastic medium into another with differing elastic constants. The new features to the problem under consideration will result from the possible motion of the phase boundary.

The displacement perturbation in the reflected wave is described by the $O(\varepsilon)$ displacement profile U_R. Hence, the field associated with the reflected wave in $\mathcal{P}_6$, together with the strain and particle velocity, are given by

$$\left.\begin{aligned} u^R &= \gamma^A x_2 + \varepsilon U_R(\xi \sin\theta_I - x_2\cos\theta_I) + \varepsilon U_I(-D), \\ \boldsymbol{\gamma}^R &= \gamma^A \mathbf{e}_2 + \varepsilon U_R'(\xi\sin\theta_I - x_2\cos\theta_I) \times (\sin\theta_I \mathbf{e}_1 - \cos\theta_I \mathbf{e}_2), \\ v^R &= -\varepsilon c_A U_R'(\xi\sin\theta_I - x_2\cos\theta_I) \end{aligned}\right\} \text{ in } \mathcal{P}_6. \tag{2.16}$$

The displacement perturbation in the transmitted wave is described by $O(\varepsilon)$ displacement profile U_T. Hence, the transmitted wave in $\mathcal{P}_2$ is described as follows:

$$\left.\begin{aligned} u^T &= \gamma^B x_2 + \varepsilon U_T(\xi\sin\theta_T + x_2\cos\theta_T), \\ \boldsymbol{\gamma}^T &= \gamma^B \mathbf{e}_2 + \varepsilon U_T'(\xi\sin\theta_T + x_2\cos\theta_T) \times (\sin\theta_T \mathbf{e}_1 + \cos\theta_T \mathbf{e}_2), \\ v^T &= -\varepsilon c_B U_T'(\xi\sin\theta_T + x_2\cos\theta_T) \end{aligned}\right\} \text{ in } \mathcal{P}_2. \tag{2.17}$$

The region $\mathcal{P}_5$ is that in which the incident and reflected wave overlap, giving

$$\left.\begin{aligned} u^m &= \gamma^A x_2 + \varepsilon U_I(\xi\sin\theta_I + x_2\cos\theta_I) + \varepsilon U_R(\xi\sin\theta_I - x_2\cos\theta_I), \\ \boldsymbol{\gamma}^m &= \gamma^A \mathbf{e}_2 + \varepsilon U_I'(\xi\sin\theta_I + x_2\cos\theta_I) \times (\sin\theta_I\mathbf{e}_1 + \cos\theta_I\mathbf{e}_2) \\ &\quad + \varepsilon U_R'(\xi\sin\theta_I - x_2\cos\theta_I) \times (\sin\theta_I\mathbf{e}_1 - \cos\theta_I\mathbf{e}_2), \\ v^m &= -\varepsilon c_A U_I'(\xi\sin\theta_I + x_2\cos\theta_I) - \varepsilon c_A U_R'(\xi\sin\theta_I - x_2\cos\theta_I) \end{aligned}\right\} \text{ in } \mathcal{P}_5. \tag{2.18}$$

The fields in the remaining equilibrated regions are trivial as listed below:

$$\begin{array}{llll} u=\gamma^B x_2, & \boldsymbol{\gamma}=\gamma^B\mathbf{e}_2, & v=0 & \text{in } \mathcal{P}_1;\\ u=\gamma^A x_2, & \boldsymbol{\gamma}=\gamma^A\mathbf{e}_2, & v=0 & \text{in } \mathcal{P}_3;\\ u=\gamma^B x_2+\varepsilon U_T(-D_T), & \boldsymbol{\gamma}=\gamma^B\mathbf{e}_2, & v=0 & \text{in } \mathcal{P}_7;\\ u=\gamma^A x_2+\varepsilon U_I(-D), & \boldsymbol{\gamma}=\gamma^A\mathbf{e}_2, & v=0 & \text{in } \mathcal{P}_8;\\ u=\gamma^A x_2+\varepsilon U_I(-D)+\varepsilon U_R(-D_R), & \boldsymbol{\gamma}=\gamma^A\mathbf{e}_2, & v=0 & \text{in } \mathcal{P}_9. \end{array} \tag{2.19}$$

Notice that the fields in $\mathcal{P}_8$, $\mathcal{P}_7$, $\mathcal{P}_9$ are the formal continuations of the incident, transmitted and reflected wave fields $\mathcal{P}_4$, $\mathcal{P}_2$, $\mathcal{P}_6$ for their associated constant final displacements.

The phase boundary is capable of motion, and its location is given by $x_2 = s(x_1, t)$, with s being continuously differentiable and satisfying

$$s(x_1,t)=\begin{cases} 0, & \xi \geqslant 0,\\ \varepsilon\bar{s}(\xi) & -\omega\leqslant\xi\leqslant 0, \text{ with } \bar{s}_0=0\\ \varepsilon\bar{s}(-\omega) & \xi\leqslant-\omega \end{cases} \tag{2.20}$$

Here $-\omega \leqslant \xi \leqslant 0$ projects the intersection of the phase boundary with the incident wave onto the ξ-axis, and $\varepsilon\bar{s}(-\omega) = \varepsilon h$ is the possibly nonzero displacement of the phase boundary after wave passage. This displacement, along with D and θ_I, determine the trailing edge similarity value ω by $-\omega\sin\theta_I + \varepsilon h\cos\theta_I = -D$. Hence,

$$h=\bar{s}(-\omega), \qquad \omega=\frac{D+\varepsilon\cos\theta_I\bar{s}(-\omega)}{\sin\theta_I}. \tag{2.21}$$

In particular, if $h > 0$ then material transforms from phase B to phase A and $\omega > D/\sin\theta_I$. The thickness of the reflected and transmitted waves, D_R and D_T, depend on h and the geometry of Figure 2 establishes that

$$D_T=\lambda D+\varepsilon(\lambda\cos\theta_I-\cos\theta_T)h \quad\text{and}\quad D_R=D+2\varepsilon h\cos\theta_I. \tag{2.22}$$

The state of affairs described thus far is consistent with the previously stated continuity conditions and momentum balance principles provided that the functions U_I, U_R, U_T, and $\bar{s}$ satisfy the jump conditions (2.11) at the phase boundary. Indeed, if all of these functions vanish, then the jump conditions are satisfied trivially. In particular, it is only necessary to enforce these jump conditions on the active portion of the phase boundary, namely, the interface separating $\mathcal{P}_2$ and $\mathcal{P}_5$.

In the event that the incident wave U_I and the phase boundary motion $\bar{s}$ are known, then (2.11) provides two equations for the determination of U_R and U_T. Substituting from (2.18) and (2.17) into the continuity of displacement condition $(2.11)_1$ gives

$$\begin{aligned} &U_I(\xi\sin\theta_I+\varepsilon\bar{s}(\xi)\cos\theta_I)+U_R(\xi\sin\theta_I-\varepsilon\bar{s}(\xi)\cos\theta_I)\\ &\quad -U_T(\xi\sin\theta_T+\varepsilon\bar{s}(\xi)\cos\theta_T)=\gamma^*\bar{s}(\xi). \end{aligned} \tag{2.23}$$

We henceforth assume that $\varepsilon\bar{s}'(\xi)$ is sufficiently small so that the arguments of U_I, U_T and U_R are increasing in ξ. Namely,

$$\max\{-\tan\theta_I, -\tan\theta_T\} < \varepsilon\bar{s}'(\xi) < \tan\theta_I. \tag{2.24}$$

This gives a proper evolutionary status to the phase boundary advance as it assures that each wave characteristic terminates at this boundary with a single intersection.

An expression involving derivatives of the functions U_I, U_T and U_R results from the momentum balance condition $(2.11)_2$:

$$\begin{aligned}&\big(\sin^2\theta_I + \varepsilon^2(1+\sin^2\theta_I)\bar{s}'^2(\xi)\big)\big(\cos\theta_I - \varepsilon\bar{s}'(\xi)\sin\theta_I\big)\\ &\quad\times U_I{}'\big(\xi\sin\theta_I + \varepsilon\bar{s}(\xi)\cos\theta_I\big)\\ &\quad-\big(\sin^2\theta_I + \varepsilon^2(1+\sin^2\theta_I)\bar{s}'^2(\xi)\big)\big(\cos\theta_I + \varepsilon\bar{s}'(\xi)\sin\theta_I\big)\\ &\quad\times U_R{}'\big(\xi\sin\theta_I - \varepsilon\bar{s}(\xi)\cos\theta_I\big)\\ &\quad-\big(\sin^2\theta_T + \varepsilon^2(1+\sin^2\theta_T)\bar{s}'(\xi)\big)\big(\cos\theta_T - \varepsilon\bar{s}'(\xi)\sin\theta_T\big)\\ &\quad\times U_T{}'\big(\xi\sin\theta_T + \varepsilon\bar{s}(\xi)\cos\theta_T\big) = \gamma^*\varepsilon\bar{s}'^2(\xi).\end{aligned} \tag{2.25}$$

Here we have used

$$\mathbf{n} = \frac{-\varepsilon\bar{s}'\mathbf{e}_1 + \mathbf{e}_2}{\sqrt{1+(\varepsilon\bar{s}')^2}} \quad\text{and}\quad V_{\mathbf{n}} = -\frac{c_A}{\sin\theta_I}\frac{\varepsilon\bar{s}'}{\sqrt{1+(\varepsilon\bar{s}')^2}}. \tag{2.26}$$

The relations (2.23) and (2.25) are nontrivial on the projection of the active phase boundary portion onto the ξ-axis, namely the interval $-\omega \leqslant \xi \leqslant 0$. The equation obtained by differentiating (2.23) with respect to ξ, in conjunction with (2.25), provides a system of two linear algebraic equations for U_R' and U_T'. The solution of this set will be represented as

$$U_R{}'\big(\xi\sin\theta_I - \varepsilon\bar{s}(\xi)\cos\theta_I\big) = \Lambda_R\big(\xi,\bar{s}(\xi),\bar{s}'(\xi)\big), \tag{2.27}$$

$$U_T{}'\big(\xi\sin\theta_T + \varepsilon\bar{s}(\xi)\cos\theta_T\big) = \Lambda_T\big(\xi,\bar{s}(\xi),\bar{s}'(\xi)\big), \tag{2.28}$$

where expressions for Λ_R and Λ_T are given in the appendix. Formal integral expressions follow with the aid of the initial condition $\bar{s}(0) = 0$,

$$\begin{aligned}&U_R\big(\xi\sin\theta_I - \varepsilon\bar{s}(\xi)\cos\theta_I\big)\\ &\quad= \int_0^\xi \big(\sin\theta_I - \varepsilon\bar{s}'(\zeta)\cos\theta_I\big)\Lambda_R\big(\zeta,\bar{s}(\zeta),\bar{s}'(\zeta)\big)\,\mathrm{d}\zeta,\end{aligned} \tag{2.29}$$

$$\begin{aligned}&U_T\big(\xi\sin\theta_T + \varepsilon\bar{s}(\xi)\cos\theta_T\big)\\ &\quad= \int_0^\xi \big(\sin\theta_T + \varepsilon\bar{s}'(\zeta)\cos\theta_T\big)\Lambda_T\big(\zeta,\bar{s}(\zeta),\bar{s}'(\zeta)\big)\,\mathrm{d}\zeta.\end{aligned} \tag{2.30}$$

The preceding development gives U_T and U_R for a given U_I and $\bar{s}$. The prevailing smoothness assumptions on U_I and $\bar{s}$ give that U_T and U_R are also continuously differentiable. It is typically the case that the phase boundary location $\bar{s}(\xi)$ is itself an a priori unknown that must be determined within the context of any solution procedure. This is because the continuity condition (2.23) and momentum balance law (2.25) are generally not sufficient to determine the phase boundary motion in the present framework [7]. Second law considerations do serve to determine whether or not a particular $\bar{s}$ is consistent with a particular U_I. Specifically, as discussed further in Section 3, this requirement prohibits spontaneous phase boundary motion for the energy minimal equilibrium configuration under consideration, and also ensures that the phase boundary is again stationary after wave passage. However, for $U_I \neq 0$ the second law only narrows the possibilities for phase boundary motion without determining it uniquely. The issue of uniquely determining phase boundary motion has led to the posing of various additional criteria for the determination of phase boundary motion. In the present context, these criteria would then augment (2.23) and (2.25) so as to provide a well-posed problem for the simultaneous determination of U_T, U_R and $\bar{s}$ on the basis of the driving incident wave U_I. In the following, the concept of a phase boundary kinetic relation is utilized for this purpose [2].

Small amplitude strain disturbances would then generally correlate with low phase boundary velocities. In these situations, explicit forms for the leading order wave profiles in the reflected and transmitted waves can be obtained in terms of the incident wave and the as yet unknown phase boundary motion by evaluating (2.29) and (2.30) in the limit as ε tends to zero. This gives

$$\begin{cases} U_1^T(\xi \sin\theta_T) = \dfrac{2\cos\theta_I U_I(\xi \sin\theta_I) - \cos\theta_I \gamma^* s_1(\xi)}{\cos\theta_I + \lambda\cos\theta_T}, \\[2ex] U_1^R(\xi \sin\theta_I) = \dfrac{(\cos\theta_I - \lambda\cos\theta_T)U_I(\xi \sin\theta_I) + \lambda\cos\theta_T \gamma^* s_1(\xi)}{\cos\theta_I + \lambda\cos\theta_T} \end{cases} \tag{2.31}$$

for $-D/\sin\theta_I < \xi < 0$. To denote the result of this limit, we have formally replaced $U_R, U_T, \bar{s}$ with their leading order counterparts U_1^R, U_1^T, s_1. Note that the quantities U_1^R and U_1^T are no longer evaluated on the actual phase boundary location, but rather on the unperturbed phase boundary location $\omega_o = -D/\sin\theta_I < \xi < 0$. Here ω_o is the leading order counterpart of ω. The reason for acknowledging a leading order counterpart to $\bar{s}$, in addition to those for U_R and U_T, is explained in what follows.

More generally, U_1^R and U_1^T are leading order terms in a formal expansion of U_R and U_T,

$$U_R(\eta) = \sum_{n=1}^{\infty} \varepsilon^{n-1} U_n^R(\eta) \quad \text{and} \quad U_T(\eta) = \sum_{n=1}^{\infty} \varepsilon^{n-1} U_n^T(\eta). \tag{2.32}$$

Expressions for U_n^R and U_n^T may be obtained by a formal expansion of (2.31) Alternatively, one may proceed directly from expanded forms of (2.25) and the differentiated form of (2.23). This gives a sequence of $O(\varepsilon^n)$ equation pairs for the successive refinements $U_n^{T'}$ and $U_n^{R'}$. Each equation pair in this sequence involves the same linear operator, but with an increasingly complicated inhomogenous term that involves the effect of the lower order refinements. Such a direct treatment of (2.25) and the differentiated form of (2.23) is advantageous for the case in which $\bar{s}$ is also an a priori unknown and so to be determined by an additional requirement. One may then presume that the mathematical description of this additional requirement also admits an expansion which then generates a third $\mathrm{O}(\varepsilon^n)$ equation, augmenting the equation pair arising from (2.23) and (2.25). To consistently solve this sequence of equation triplets requires the additional expansion of the function $\bar{s}(\xi)$ via

$$\bar{s}(\xi) = \sum_{n=1}^{\infty} \varepsilon^{n-1} s_n(\xi), \tag{2.33}$$

where $s_n(0) = 0$ for $n = 1, 2, 3, \ldots$. The expansion (2.33) accounts for our earlier replacement of $\bar{s}(\xi)$ with $s_1(\xi)$ in (2.31). Since the treatment of $U_R{}'(\xi \sin\theta_I - \varepsilon\bar{s}(\xi)\cos\theta_I)$ and $U_T{}'(\xi \sin\theta_T + \varepsilon\bar{s}(\xi)\cos\theta_T)$ requires both the expansion of the function and its argument, the refinements $U_n^{T'}(\eta)$ and $U_n^{R'}(\eta)$ for $n > 1$ are sensitive to the scheme used for evaluating the arguments. A consistent procedure is obtained by evaluating the order functions U_n^R and U_n^T on a phase boundary location that involves terms through $\bar{s}_{n-1}(\xi)$.

In what follows we do not explicitly develop formal expressions for the perturbation expansions beyond the leading order. Our motivation is not only simplicity, but also stems from the sensitivity of our results with respect to the particular constitutive theory that we have employed. Namely, the expressions (2.31) are general for any material such that (2.7) represents the local expansions of the strain energy density function about the Maxwell strains $\boldsymbol{\gamma}^A$ and $\boldsymbol{\gamma}^B$. However, the expressions for U_n^R and U_n^T with $n > 1$ are also sensitive to the possible cubic and higher order terms in these energy well expansions.

3. Energy Balance and Kinetic Relation

We now begin to examine the energetics of these propagating waves with the aim of introducing a kinetic relation for the determination of the phase boundary motion. To this end, in the three-dimensional setting consider a bounded region $\mathcal{R}_t$ with the outward unit normal $\mathbf{n}$. The boundary $\partial\mathcal{R}_t$ is moving with the velocity $\mathbf{v}$. The normal velocity is $v_\mathbf{n} = \mathbf{v}\cdot\mathbf{n}$. Define the total mechanical energy of the region at time t by $E(t)$, then

$$E(t) = \int_{\mathcal{R}_t} \frac{\rho}{2}\dot{u}_i\dot{u}_i + W\,\mathrm{d}V. \tag{3.1}$$

The time rate of change for $E(t)$ is

$$\dot{E}(t) = \int_{\mathcal{R}_t} \rho\dot{u}_i\ddot{u}_i + \sigma_{ij}\dot{\gamma}_{ij}\,\mathrm{d}V + \int_{\partial\mathcal{R}_t}\left(\frac{\rho}{2}\dot{u}_i\dot{u}_i + W\right)v_\mathbf{n}\,\mathrm{d}A. \tag{3.2}$$

By using the divergence theorem and the local balance of linear momentum, it follows that

$$\dot{E}(t) = \int_{\partial \mathcal{R}_t} \sigma_{ij} n_j \dot{u}_i \, \mathrm{d}A + \int_{\partial \mathcal{R}_t} \left(\frac{\rho}{2} \dot{u}_i \dot{u}_i + W \right) v_{\mathbf{n}} \, \mathrm{d}A. \tag{3.3}$$

The first integral is the rate of work done on the boundary; the second is the energy flux into the region. For anti-plane shear, we define the power

$$P(t) = \int_{\partial \mathcal{R}_t} \tau_\alpha n_\alpha \dot{u} \, \mathrm{d}S \tag{3.4}$$

and the flux

$$F(t) = \int_{\partial \mathcal{R}_t} \left(\frac{\rho}{2} v^2 + \mathcal{W} \right) v_{\mathbf{n}} \, \mathrm{d}S. \tag{3.5}$$

Here, $\mathcal{R}_t$ is interpreted as a 2-dimensional region on the cross section Π, and its boundary $\partial \mathcal{R}_t$ is a closed piecewise smooth curve. For a smooth displacement field, the usual conservation of energy applies, namely $\dot{E} = P + F$. But if the region $\mathcal{R}_t$ intersects the phase boundary Σ_t, then there is an additional contribution to the energy balance due to the accretive motion of the phase boundary (motion with nonzero normal velocity $V_{\mathbf{n}}$). The second law of thermodynamics requires that this contribution is dissipative. The rate of energy loss due to moving phase boundary motion is characterized by the dissipation rate $D(t)$:

$$D(t) = \int_{\Sigma_t \cap \mathcal{R}_t} f V_{\mathbf{n}} \, \mathrm{d}S, \tag{3.6}$$

in which case the energy balance is amended to $\dot{E} = P + F - D$. Here f is the phase boundary driving traction [1] which is given by

$$f = [\![\mathcal{W}]\!] - \frac{1}{2} [\![\gamma_\alpha]\!] (\tau_\alpha^+ + \tau_\alpha^-), \tag{3.7}$$

where the superscripted "+" and "−" indicate the limiting values on the two sides of Σ_t, so that $[\![\mathcal{W}]\!] = \mathcal{W}^+ - \mathcal{W}^-$. For convenience, we will assign material in phase A to be the $-$ side; phase B the $+$ side. In this paper, the unit normal of Σ_t has been taken to be pointing into the $+$ side (2.26). As such, the normal velocity of the phase boundary $V_{\mathbf{n}}$ is positive if Σ_t is moving into the $+$ side, transforming material from phase B to A. The second law requirement $D(t) \geqslant 0$ gives $f V_{\mathbf{n}} \geqslant 0$ since the control region $\mathcal{R}_t$ is arbitrary.

For the constitutive description (2.7), the driving traction f is given by

$$\begin{aligned} f &= -\frac{1}{2} \gamma^* \mathbf{e}_2 \cdot \left[\mu^B (\boldsymbol{\gamma}^+ - \gamma^B \mathbf{e}_2) + \mu^A (\boldsymbol{\gamma} - \gamma^A \mathbf{e}_2) \right] \\ &\quad + \left(\frac{\mu^B - \mu^A}{2} \right) (\boldsymbol{\gamma}^+ - \gamma^B \mathbf{e}_2) \cdot (\boldsymbol{\gamma} - \gamma^A \mathbf{e}_2). \end{aligned} \tag{3.8}$$

This driving traction is zero both before and after the passage of the incident wave. On the potentially active portion of the phase boundary, use of the strain representations (2.17) and (2.18) gives that

$$\begin{aligned} f = & -\frac{\varepsilon\gamma^*}{2}\{\mu^B \cos\theta_T U_T'(\xi \sin\theta_T + \varepsilon\bar{s}(\xi)\cos\theta_T) \\ & + \mu^A \cos\theta_I [U_I'(\xi\sin\theta_I + \varepsilon\bar{s}(\xi)\cos\theta_I) - U_R'(\xi\sin\theta_I - \varepsilon\bar{s}(\xi)\cos\theta_I)]\} \\ & + \frac{\mu^B - \mu^A}{2}\varepsilon^2\{\cos(\theta_T - \theta_I)U_I'(\xi\sin\theta_I + \varepsilon\bar{s}(\xi)\cos\theta_I) \\ & \times U_T'(\xi\sin\theta_T + \varepsilon\bar{s}(\xi)\cos\theta_T) - \cos(\theta_T + \theta_I)U_R'(\xi\sin\theta_I - \varepsilon\bar{s}(\xi)\cos\theta_I) \\ & \times U_T'(\xi\sin\theta_T + \varepsilon\bar{s}(\xi)\cos\theta_T)\} \end{aligned} \tag{3.9}$$

for $-\omega \leqslant \xi \leqslant 0$. Note also that (3.9), (2.27), (2.28) permit the representation

$$f = \Xi(U_I'(\xi\sin\theta_I + \varepsilon\bar{s}(\xi)\cos\theta_I), \bar{s}'(\xi)), \tag{3.10}$$

where

$$\begin{aligned} \Xi(U_I', \bar{s}') = & \ \varepsilon\gamma^*\Xi_1(\bar{s}') + \varepsilon U_I'\Xi_2(\bar{s}') \\ & + \varepsilon^2\gamma^{*2}\Xi_3(\bar{s}') + \varepsilon^2\gamma^* U_I'\Xi_4(\bar{s}') + \varepsilon^2 U_I'^2\Xi_5(\bar{s}'), \end{aligned} \tag{3.11}$$

and expressions for $\Xi_i(\bar{s}')$, $i = 1, 2, \ldots, 5$, are given in the appendix. It is to be noted that $f = 0$ if both $\bar{s}' = 0$ and $U_I' = 0$, but that generally $f \neq 0$ if either $\bar{s}' \neq 0$ or $U_I' \neq 0$. Further, if $U_I' = 0$ then

$$f = \frac{\lambda\mu^A\gamma^{*2}\cos\theta_I\cos\theta_T}{\sin\theta_I\cos\theta_I + \sin\theta_T\cos\theta_T}\varepsilon\bar{s}' + O((\varepsilon\bar{s}')^2). \tag{3.12}$$

It thus follows from $(2.26)_2$ that if $U_I' = 0$ and if $\bar{s}' \neq 0$ with $|\bar{s}'|$ sufficiently small, then $fV_{\mathbf{n}} \neq 0$ with the sign of $fV_{\mathbf{n}}$ being the same as the sign of

$$-\frac{c_A}{\sin\theta_I}\frac{\lambda\mu^A\gamma^{*2}\cos\theta_I\cos\theta_T}{\sin\theta_I\cos\theta_I + \sin\theta_T\cos\theta_T}. \tag{3.13}$$

This expression is negative and consequently the second law requirement $fV_{\mathbf{n}} \geqslant 0$ is not met under all of these conditions. Since the conditions $\bar{s}' \neq 0$ with $|\bar{s}'|$ sufficiently small must occur for phase boundary motion under the prevailing continuity assumption on $\bar{s}'$, it thus follows that spontaneous phase boundary motion ($U_I' = 0$ and $\bar{s}' \neq 0$) is excluded by the second law (this would not be the case if the original equilibrium configuration were not energy minimal, as the form of (3.12) would be modified). However, for $U_I' \neq 0$ the second law alone is not sufficient to determine the phase boundary motion, which instead requires further consideration of the underlying mechanics and physics.

To capture the localized physical processes activating the phase boundary motion, it is common to posit an additional relation at the phase boundary. Such a kinetic relation invariably involves both f and $V_{\mathbf{n}}$. A standard form for such a relation is

$$V_{\mathbf{n}} = \phi(f), \tag{3.14}$$

where the second law requirement indicates that the function ϕ must obey $f\phi(f) \geqslant 0$. The effect of different forms for such a kinetic function ϕ upon the ensuing phase boundary motion has been investigated widely in the uniaxial context involving two-phase bars [12]. For the problem governed by (3.14) , the determination of the phase boundary motion requires the solution of the first order nonlinear ordinary differential equation

$$\Psi\big(U_I'(\xi \sin\theta_I + \varepsilon\bar{s}(\xi)\cos\theta_I), \bar{s}'(\xi)\big) = 0 \tag{3.15}$$

on $-\omega \leqslant \xi \leqslant 0$, subject to $\bar{s}(0) = 0$, where

$$\Psi\big(U_I', \bar{s}'\big) = \frac{c_A}{\sin\theta_I}\frac{\varepsilon\bar{s}'}{\sqrt{1+(\varepsilon\bar{s}')^2}} + \phi\big(\Xi(U_I', \bar{s}')\big). \tag{3.16}$$

The formulation (3.15), (3.16) permits both qualitative and quantitative analysis of phase boundary motion on the basis of the properties of the kinetic relation as embodied by ϕ and the form of the incident wave as embodied by U_I'. In particular, issues of solution existence, uniqueness, and smoothness are addressable via ordinary differential equation theory. Central to such a treatment is the extent to which (3.15) renders $\bar{s}'$ as a function of $\bar{s}$ and ξ. To this end we note that $\phi(0) = 0$ gives $\Psi(0, 0) = 0$. Evaluating the partial derivative $\partial\Psi/\partial\bar{s}'$ at $(U_I', \bar{s}') = (0, 0)$ now establishes that

$$\frac{\partial\Psi}{\partial\bar{s}'}(0,0) = \frac{\varepsilon c_A}{\sin\theta_I} + \varepsilon\phi'(0)\frac{\lambda\mu^A\gamma^{*2}\cos\theta_I\cos\theta_T}{\sin\theta_I\cos\theta_I + \sin\theta_T\cos\theta_T} > 0. \tag{3.17}$$

Consequently, the inverse function theorem ensures the existence of a neighborhood of $(U_I', \bar{s}') = (0, 0)$ upon which (3.15) gives rise to a relation of the form

$$\bar{s}' = \Gamma\big(U_I'(\xi \sin\theta_I + \varepsilon\bar{s}(\xi)\cos\theta_I)\big) \equiv \widehat{\Gamma}(\xi, \bar{s}), \tag{3.18}$$

where $\Gamma(0) = 0$ and

$$\Gamma'(0) = \frac{2\phi'(0)\lambda\mu^A\gamma^*\cos\theta_I\cos\theta_T\sin\theta_T}{c_A(\cos\theta_I + \lambda\cos\theta_T) + \phi'(0)\lambda\mu^A\gamma^{*2}\cos\theta_I\cos\theta_T}. \tag{3.19}$$

In particular, if $U_I''(0) \neq 0$ and $\phi'(0) \neq 0$ then a local analysis near $\xi = 0$ gives

$$\bar{s}(\xi) = \frac{1}{2}\sin\theta_I\Gamma'(0)U_I''(0)\xi^2 + o(\xi^2) \quad \text{as } \xi \to 0^-. \tag{3.20}$$

We defer more detailed study of the formulation (3.15), (3.16), other than to remark that a Lipschitz condition on $\widehat{\Gamma}$ would generally be required for existence and uniqueness, and so in the present context would generally require Lipschitz conditions on U_I' and ϕ.

We now turn to examine certain more immediate features of the phase boundary motion for two important classes of kinetic functions ϕ in the event that ε is small. The first class involves monotone increasing ϕ such that $\phi'(0) \equiv \kappa > 0$. This gives the simplest type of kinetic relation, in which case κ is the phase boundary mobility at small driving traction [15]. For this class of kinetic function, the phase boundary is generally set in motion by the leading edge of the incident wave, i.e., $\bar{s}'(0) \neq 0$ provided that $U_I'(0) \neq 0$. The second class of kinetic relations provides a frictional effect in that $\phi(f) = 0$ for $-f^* < f < f^*$. For this second class of kinetic relation, the phase boundary is set in motion by the leading edge of the incident wave only if $|U_I'(0)|$ is sufficiently large. More generally, this second class of kinetic relation may give rise to finite regions of phase boundary immobility between $\mathcal{P}_2$ and $\mathcal{P}_5$.

Regardless of the class of kinetic relation under consideration, determination of the phase boundary motion for small ε can proceed via a direct perturbation analysis of (3.15). Alternatively, one may proceed in the context of the previous perturbation procedure, provided that the kinetic relation is expanded in the same manner as that of the displacement and momentum jump conditions. The expression (3.9) will then generate a full power series expansion in ε upon introduction of (2.32). This formally gives

$$f(\xi) = \sum_{n=1}^{\infty} \varepsilon^n f_n(\xi). \tag{3.21}$$

The associated $O(\varepsilon^n)$ equation obtained via (3.14), in concert with (2.25) and the differentiated form of (2.23), provides a set of three equations for the determination of $U_n^{R\prime}$, $U_n^{T\prime}$ and s_n' in terms of the lower order functions $U_m^{R\prime}$, $U_m^{T\prime}$, s_m and s_m' $(m < n)$. Integration of these equations then gives U_n^R, U_n^T and s_n. Here, in accordance with our previous commentary, attention is subsequently restricted to the determination of the leading order behavior, namely, U_1^R, U_1^T and s_1. This merely requires the explicit expression for f_1,

$$\begin{aligned} f_1(\xi) &= -\frac{\mu^A\gamma^*}{2}\big\{\lambda^2\cos\theta_T U_1^{T\prime}(\xi\sin\theta_T) \\ &\quad +\cos\theta_I\big[U_I'(\xi\sin\theta_I) - U_1^{R\prime}(\xi\sin\theta_I)\big]\big\} \\ &= -\mu^A\gamma^*\lambda^2\cos\theta_I\frac{\cos\theta_T}{\sin\theta_T}\frac{2\sin\theta_I U_I'(\xi\sin\theta_I) - \gamma^* s_1'(\xi)}{\cos\theta_I + \lambda\cos\theta_T} \end{aligned} \tag{3.22}$$

for $-D/\sin\theta_I < \xi < 0$.

For the leading order analysis of the first class of kinetic relation, no further characterization of the kinetic relation beyond the relation $\phi'(0) \equiv \kappa > 0$ is relevant. Specifically, (3.14), (3.22) and (2.26) give

$$\begin{aligned} &-\frac{c_A}{\sin\theta_I}s_1'(\xi) \\ &= -\kappa\mu^A\gamma^*\lambda^2\frac{\cos\theta_I\cos\theta_T}{\sin\theta_T}\frac{2\sin\theta_I U_I'(\xi\sin\theta_I) - \gamma^* s_1'(\xi)}{\cos\theta_I + \lambda\cos\theta_T}, \end{aligned} \tag{3.23}$$

so that s_1 is given in terms of U_I as

$$\begin{aligned} s_1(\xi) &= \frac{1}{\gamma^*}\frac{2\hat{\kappa}\cos\theta_I\cos\theta_T}{(\cos\theta_I + \lambda\cos\theta_T) + \hat{\kappa}\cos\theta_I\cos\theta_T}U_I(\xi\sin\theta_I), \\ \hat{\kappa} &\equiv \frac{\kappa\mu^B\gamma^{*2}}{c_B}. \end{aligned} \tag{3.24}$$

In particular,

$$h = \frac{1}{\gamma^*}\frac{2\hat{\kappa}\cos\theta_I\cos\theta_T}{(\cos\theta_I + \lambda\cos\theta_T) + \hat{\kappa}\cos\theta_I\cos\theta_T}U_I(-D) + \mathrm{O}(\varepsilon). \tag{3.25}$$

The leading order displacement profile functions for the reflected and transmitted waves (2.31) in this case are

$$\begin{cases} U_1^T(\xi\sin\theta_T) = \dfrac{2\cos\theta_I}{(\cos\theta_I + \lambda\cos\theta_T) + \hat{\kappa}\cos\theta_I\cos\theta_T}U_I(\xi\sin\theta_I), \\[2ex] U_1^R(\xi\sin\theta_I) = \dfrac{(\cos\theta_I - \lambda\cos\theta_T) + \hat{\kappa}\cos\theta_I\cos\theta_T}{(\cos\theta_I + \lambda\cos\theta_T) + \hat{\kappa}\cos\theta_I\cos\theta_T}U_I(\xi\sin\theta_I) \end{cases} \tag{3.26}$$

for $-D/\sin\theta_I < \xi < 0$.

For the second class of kinetic relation, phase boundary motion is not triggered until the driving force reaches the critical value f^*. In view of (3.22), this correlates with a critical slope for the incident wave profile,

$$|U_I'| > \frac{\cos\theta_I + \lambda\cos\theta_T}{2\mu^A\gamma^*\lambda\cos\theta_I\cos\theta_T}f^* \equiv U_*'. \tag{3.27}$$

Phase boundary motion is possible where the above condition is satisfied. If the kinetic relation is specified by

$$\phi(f) = \begin{cases} \kappa(f - f^*) & \text{for } f > f^*, \\ 0 & \text{for } |f| \leqslant f^*, \\ \kappa(f + f^*) & \text{for } f < -f^*, \end{cases} \tag{3.28}$$

then the first order phase boundary location function s_1 is determined by integrating

$$s_1'(\xi) = \begin{cases} \dfrac{1}{\gamma^*}\dfrac{2\hat{\kappa}\cos\theta_I\sin\theta_I\cos\theta_T}{(\cos\theta_I + \lambda\cos\theta_T) + \hat{\kappa}\cos\theta_I\cos\theta_T}\left[U_I'(\xi\sin\theta_I) + U_*'\right], \\ \qquad\qquad\qquad\qquad\qquad\qquad \text{where } U_I' < -U_*'; \\ 0, \qquad\qquad\qquad\qquad\qquad\qquad \text{where } |U_I'| \leqslant U_*'; \\ \dfrac{1}{\gamma^*}\dfrac{2\hat{\kappa}\cos\theta_I\sin\theta_I\cos\theta_T}{(\cos\theta_I + \lambda\cos\theta_T) + \hat{\kappa}\cos\theta_I\cos\theta_T}\left[U_I'(\xi\sin\theta_I) - U_*'\right], \\ \qquad\qquad\qquad\qquad\qquad\qquad \text{where } U_I' > U_*; \end{cases} \tag{3.29}$$

using $s_1(0) = 0$.

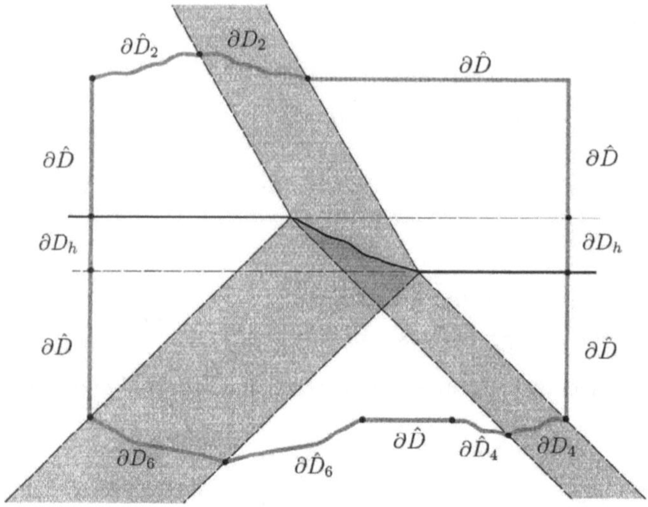

Figure 3. Generic control volume $\mathcal{D}$ with the boundary partitions as used for defining the energy rate balance.

4. Energy Partitioning and Dissipation

In this section, we turn to consider the energy partitioning associated with the incident, reflected and transmitted waves, together with the accretion of one phase at the expense of the other due to the phase boundary motion. In particular, consider a control volume $\mathcal{D}$ as depicted in Figure 3 that is rigidly translating with the steady state propagation velocity $\mathring{V}\mathbf{e}_1$. The total energy within $\mathcal{D}$ maintains a steady state value so that $\dot{E}(t) = 0$ where $E(t)$ is given by (3.1). For this particular control volume, the energy rate balance takes the form:

$$P(t) + F(t) - D(t) = 0. \tag{4.1}$$

According to Figure 3, the boundary of the control volume $\partial\mathcal{D}$ is partitioned into eight parts: the single component curves $\partial\mathcal{D}_2$, $\partial\mathcal{D}_4$, $\partial\mathcal{D}_6$ and their respective neighboring companion curves $\partial\widehat{\mathcal{D}}_2$, $\partial\widehat{\mathcal{D}}_4$, $\partial\widehat{\mathcal{D}}_6$, the two part curve $\partial\mathcal{D}_h$, and the five part curve $\partial\widehat{\mathcal{D}}$. The parts $\partial\mathcal{D}_2$, $\partial\mathcal{D}_4$, $\partial\mathcal{D}_6$ denote the intersection of $\partial\mathcal{D}$ with the transmitted, incident and reflected waves, respectively. They are accompanied by $\partial\widehat{\mathcal{D}}_2$, $\partial\widehat{\mathcal{D}}_4$, $\partial\widehat{\mathcal{D}}_6$, which are outside the waves. The boundary $\partial\mathcal{D}_h$ has two segments between $x_2 = 0$ and $x_2 = \varepsilon h$; the rest of the boundary is denoted by $\partial\widehat{\mathcal{D}}$. The shapes of $\partial\mathcal{D}_2$, $\partial\mathcal{D}_4$ and $\partial\mathcal{D}_6$ are arbitrary. The boundaries $\partial\widehat{\mathcal{D}}_2$, $\partial\widehat{\mathcal{D}}_4$, $\partial\widehat{\mathcal{D}}_6$ are chosen to have the same height in the x_2 direction as their counterparts; the shapes are otherwise not essential. Let the flux and the power (defined in (3.5) and (3.4)) associated with each segment be denoted by $\widehat{F}$, F_h, F_i, $\widehat{F}_i$ and $\widehat{P}$, P_h, P_i, $\widehat{P}_i$ correspondingly, with $i = 2, 4, 6$. Outside the waves, the particle velocity is zero, so $\widehat{P} = P_h = \widehat{P}_i = 0$. Also, since the flux through the leading edge of $\partial\widehat{\mathcal{D}}$ cancels the flux through the trailing edge, $\widehat{F} = 0$. Therefore, the boundary $\partial\widehat{\mathcal{D}}$ has no contribution to the energetics of the control volume. The same conclusion holds if the segments defining $\partial\widehat{\mathcal{D}}$ are altered from the simple collection of horizontal and vertical segments provided that the endpoints with the other parts of $\partial\mathcal{D}$ are maintained. The energy rate balance becomes

$$P_2 + P_4 + P_6 + F_2 + F_4 + F_6 + \widehat{F}_2 + \widehat{F}_4 + \widehat{F}_6 + F_h - D = 0. \tag{4.2}$$

Now observe on $\partial\widehat{\mathcal{D}}_i$ that the particle velocity is zero and the potential energy $\mathcal{W}$ is at its base value, $\mathring{\mathcal{W}}_A$ for phase A and $\mathring{\mathcal{W}}_B = \mathring{\mathcal{W}}_A + \sigma^*\gamma^*$ for phase B. One finds

$$\begin{aligned} F_i + \widehat{F}_i &= \int_{\partial\mathcal{D}_i} \left(\frac{\rho}{2}v^2 + \mathcal{W}\right) v_{\mathbf{n}}\,\mathrm{d}S + \int_{\partial\widehat{\mathcal{D}}_i} \mathring{\mathcal{W}} v_{\mathbf{n}}\,\mathrm{d}S \\ &= \int_{\partial\mathcal{D}_i} \left(\frac{\rho}{2}v^2 + \mathcal{W} - \mathring{\mathcal{W}}\right) v_{\mathbf{n}}\,\mathrm{d}S, \end{aligned} \tag{4.3}$$

where $\mathring{\mathcal{W}}$ denotes the base value for $\mathcal{W}$ at the corresponding phase. In view of (4.3), define the surplus flux associated with $\partial\mathcal{D}_i$ by

$$\mathbb{F}_i = F_i + \widehat{F}_i = \int_{\partial\mathcal{D}_i} \left[\frac{\rho}{2}v^2 + (\mathcal{W} - \mathring{\mathcal{W}})\right] v_{\mathbf{n}}\,\mathrm{d}S, \quad i = 2, 4, 6. \tag{4.4}$$

In addition, the energy rate associated with $\partial\mathcal{D}_i$ is the sum of power and surplus flux so that

$$\begin{aligned} \mathbb{E}_{inc.} &= P_4 + \mathbb{F}_4, \\ \mathbb{E}_{ref.} &= P_6 + \mathbb{F}_6, \quad \text{with } P_i = \int_{\partial\mathcal{D}_i} \tau_\alpha n_\alpha \dot{u}\,\mathrm{d}S, \quad i = 2, 4, 6. \\ \mathbb{E}_{tran.} &= P_2 + \mathbb{F}_2, \end{aligned} \tag{4.5}$$

The energy accretion rate due to phase transformation of the material between $x_2 = 0$ and $x_2 = \varepsilon h$ is given by

$$\mathbb{E}_{acc.} = -(P_h + F_h) = -F_h = -\gamma^*\sigma^*\mathring{V}\varepsilon h, \tag{4.6}$$

while the energy dissipation rate due to phase boundary motion is

$$\mathbb{E}_{diss.} = \int_{\Sigma_t \cap \mathcal{D}} f V_{\mathbf{n}}\,\mathrm{d}S. \tag{4.7}$$

With the above definitions, the energy rate balance becomes

$$\mathbb{E}_{inc.} + \mathbb{E}_{ref.} + \mathbb{E}_{tran.} = \mathbb{E}_{acc.} + \mathbb{E}_{diss.}. \tag{4.8}$$

Clearly, as long as the control volume encloses the whole mobile (curved) portion of the phase boundary, the dissipation rate $\mathbb{E}_{diss.}$ and the energy accretion rate $\mathbb{E}_{acc.}$ remain the same regardless of the shape of $\mathcal{D}$. Moreover, the values of $\mathbb{E}_{inc.}$, $\mathbb{E}_{ref.}$ and $\mathbb{E}_{tran.}$ are also independent of the shapes of $\partial\mathcal{D}_i$ and $\partial\widehat{\mathcal{D}}_i$. For example, consider a new control volume $\mathcal{D}^*$ which is constructed by replacing $\partial\mathcal{D}_4$ and $\partial\widehat{\mathcal{D}}_4$

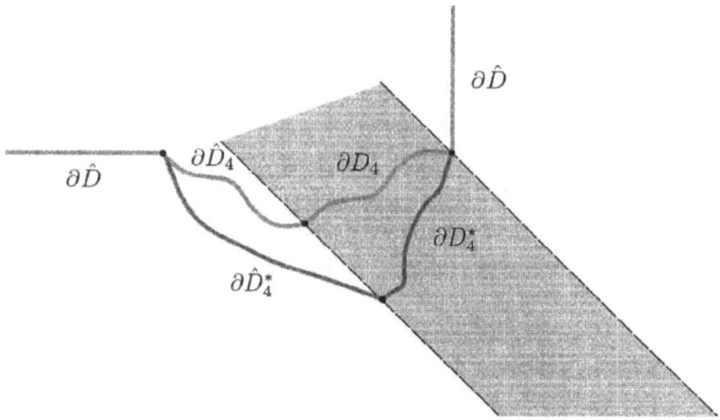

Figure 4. Variation of the shape of a typical control volume as used for demonstrating the well posed value of the energy rate $\mathbb{E}_{inc.}$.

in $\partial\mathcal{D}$ by new shapes $\partial\mathcal{D}_4^*$ and $\partial\widehat{\mathcal{D}}_4^*$ (Figure 4). Let the new incident energy rate be denoted by

$$\begin{aligned}\mathbb{E}^*_{inc.} &= P_4^* + \mathbb{F}_4^* \\ &= \int_{\partial\mathcal{D}_4^*} \tau_\alpha n_\alpha \dot{u}\,\mathrm{d}S + \int_{\partial\mathcal{D}_4^*} \left(\frac{\rho}{2}v^2 + \mathcal{W}\right) v_{\mathbf{n}}\,\mathrm{d}S + \int_{\partial\widehat{\mathcal{D}}_4^*} \mathring{\mathcal{W}} v_{\mathbf{n}}\,\mathrm{d}S. \qquad (4.9)\end{aligned}$$

Now, consider the control volume $\mathcal{Q}$ enclosed by $\partial\mathcal{D}_4$, $\partial\widehat{\mathcal{D}}_4$, $\partial\mathcal{D}_4^*$ and $\partial\widehat{\mathcal{D}}_4^*$. This control volume does not contain any phase boundary and hence is dissipation free. The energy balance leads to

$$\begin{aligned}&\int_{\partial\mathcal{D}_4} \tau_\alpha n^*_\alpha \dot{u}\,\mathrm{d}S + \int_{\partial\mathcal{D}_4} \left(\frac{\rho}{2}v^2 + \mathcal{W}\right) v^*_{\mathbf{n}}\,\mathrm{d}S + \int_{\partial\widehat{\mathcal{D}}_4} \mathring{\mathcal{W}} v^*_{\mathbf{n}}\,\mathrm{d}S \\ &\quad + \int_{\partial\mathcal{D}_4^*} \tau_\alpha n^*_\alpha \dot{u}\,\mathrm{d}S + \int_{\partial\mathcal{D}_4^*} \left(\frac{\rho}{2}v^2 + \mathcal{W}\right) v^*_{\mathbf{n}}\,\mathrm{d}S + \int_{\partial\widehat{\mathcal{D}}_4^*} \mathring{\mathcal{W}} v^*_{\mathbf{n}}\,\mathrm{d}S = 0, \qquad (4.10)\end{aligned}$$

where $\mathbf{n}^*$ is the unit outward normal to the control volume $\mathcal{Q}$. It is noticed that $\mathbf{n}^* = \mathbf{n}$ and $v^*_{\mathbf{n}} = v_{\mathbf{n}}$ on $\partial\mathcal{D}_4^*$ and $\partial\widehat{\mathcal{D}}_4^*$, while $\mathbf{n}^* = -\mathbf{n}$ and $v^*_{\mathbf{n}} = -v_{\mathbf{n}}$ on $\partial\mathcal{D}_4$ and $\partial\widehat{\mathcal{D}}_4$. On using (4.3), (4.4), (4.5) and (4.9), the energy balance for $\mathcal{Q}$ implies $\mathbb{E}^*_{inc.} = \mathbb{E}_{inc.}$. Similar arguments apply to $\mathbb{E}_{ref.}$ and $\mathbb{E}_{tran.}$. The quantities $\mathbb{E}_{inc.}$, $\mathbb{E}_{ref.}$ and $\mathbb{E}_{tran.}$, all independent of the shape of the control volume, represent the rate of energy contribution associated with the incident, reflected and transmitted waves. Note however that the power contributions P_i and surplus flux contributions $\mathbb{F}_i$ appearing in (4.5) are shape-dependent. In computing $\mathbb{E}_{inc.}$, $\mathbb{E}_{ref.}$ and $\mathbb{E}_{tran.}$, one can choose special shapes for $\partial\mathcal{D}_i$ in order to simplify calculation. For example:

(1) Horizontal segments: The surplus flux $\mathbb{F}_i$ then vanishes. The energy $\mathbb{E}$ comes solely from the power term P_i.

(2) Propagationally aligned segments: Here the normal $\mathbf{n}$ to $\partial\mathcal{D}_i$ is perpendicular to the direction of wave propagation $\mathbf{m}$. In general, both power and flux terms will then exist, however the power term reduces to

$$P_i = (\sigma^*\mathbf{e}_2 \cdot \mathbf{n}) \int_{\partial\mathcal{D}_i} \dot{u}\, \mathrm{d}S \quad \text{for propagationally aligned segments.}$$

(3) "Power free" segments: It is possible to construct a special $\partial\mathcal{D}_i$ so that the power term P_i vanishes. This is achieved by choosing the normal direction $\mathbf{n}$ to be perpendicular to the shear stress vector, rendering $\tau_\alpha n_\alpha = 0$, and thus requiring

$$\begin{cases} [\sigma^*\mathbf{e}_2 + \mu^A(\boldsymbol{\gamma} - \gamma^A\mathbf{e}_2)] \cdot \mathbf{n} = 0 & \text{in phase } A; \\ [\sigma^*\mathbf{e}_2 + \mu^B(\boldsymbol{\gamma} - \gamma^B\mathbf{e}_2)] \cdot \mathbf{n} = 0 & \text{in phase } B. \end{cases}$$

For the problem under consideration, the strain terms (such as $\boldsymbol{\gamma} - \gamma^A\mathbf{e}_2$) are of the order of $\mathrm{O}(\varepsilon)$. So the shape of a power-free segment is vertical to leading order.

In what follows, we calculate the energy contribution terms using a control volume with horizontal segments (Figure 5). In view of (2.13), (2.16) and (2.17), it is then found that

$$\begin{aligned} \mathbb{E}_{inc.} &= P_4 = \int_{\partial\mathcal{D}_4} \tau_2 \dot{u}\, \mathrm{d}S \\ &= -\varepsilon\sigma^*\mathring{V} U_I(-D) + \varepsilon^2\mu^A c_A \cos\theta_I \int_{-D/\sin\theta_I}^{0} \left[U_I'(\xi\sin\theta_I)\right]^2 \mathrm{d}\xi, \end{aligned} \tag{4.11}$$

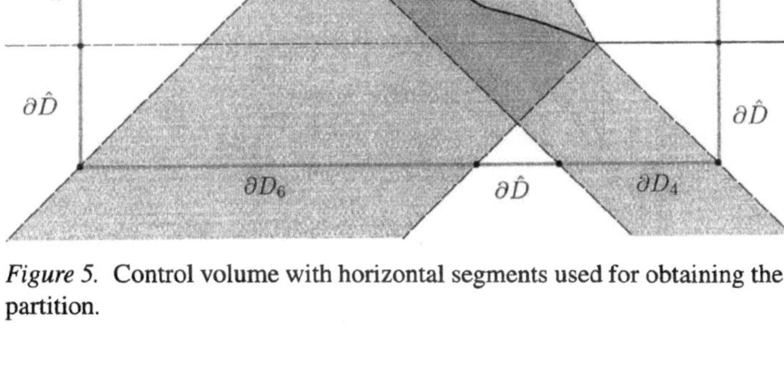

Figure 5. Control volume with horizontal segments used for obtaining the explicit energy rate partition.

$$\begin{aligned}\mathbb{E}_{ref.} &= P_6 = \int_{\partial\mathcal{D}_6} \tau_2 \dot{u}\, \mathrm{d}S \\ &= -\varepsilon\sigma^* \overset{\circ}{V} U_R(-D_R) - \varepsilon^2 \mu^A c_A \cos\theta_I \int_{-D_R/\sin\theta_I}^{0} \left[U_R'(\xi \sin\theta_I)\right]^2 \mathrm{d}\xi,\end{aligned} \tag{4.12}$$

$$\begin{aligned}\mathbb{E}_{tran.} &= P_2 = \int_{\partial\mathcal{D}_2} \tau_2 \dot{u}\, \mathrm{d}S \\ &= \varepsilon\sigma^* \overset{\circ}{V} U_T(-D_T) - \varepsilon^2 \mu^B c_B \cos\theta_T \int_{-D_T/\sin\theta_T}^{0} \left[U_T'(\xi \sin\theta_T)\right]^2 \mathrm{d}\xi.\end{aligned} \tag{4.13}$$

The following observations regarding the energy rates now follow from the above expressions in conjunction with (2.26), (3.9) and (4.7):

(1) $\mathbb{E}_{inc.}$, $\mathbb{E}_{ref.}$, $\mathbb{E}_{tran.}$ and $\mathbb{E}_{acc.}$ are each first order in ε.
(2) $\mathbb{E}_{diss.}$ is second order and independent of σ^*.
(3) $\mathbb{E}_{acc.}$ is linear in σ^*.
(4) The first order terms in the expansions of $\mathbb{E}_{inc.}$, $\mathbb{E}_{ref.}$ and $\mathbb{E}_{tran.}$ are also linear in σ^*.

Since the rate balance (4.8) is valid regardless of the value of the Maxwell stress σ^*, it follows that the portion of (4.8) that is linear in σ^* must individually balance, as must the portion of (4.8) that is independent of σ^*. The σ^* portion of the balance becomes (after eliminating ε, $\overset{\circ}{V}$ and σ^*):

$$-U_I(-D) - U_R(-D_R) + U_T(-D_T) = -\gamma^* h. \tag{4.14}$$

This relation is ensured by the displacement jump condition as can be verified by setting $\xi = -\omega$ in (2.23) and then using (2.21) and (2.22).

Explicit expressions to $\mathrm{O}(\varepsilon^2)$ for the expansions of $\mathbb{E}_{ref.}$, $\mathbb{E}_{tran.}$, $\mathbb{E}_{acc.}$ and $\mathbb{E}_{diss.}$ are determined with the aid of expansions (2.33), (2.32), (3.21) and (3.22) to be:

$$\begin{aligned}\mathbb{E}_{ref.} &= -\varepsilon\sigma^* \overset{\circ}{V} U_1^R(-D) + 2\varepsilon^2\sigma^* \overset{\circ}{V} \cos\theta_I s_1(-\omega_1) U_1^{R\prime}(-D) \\ &\quad - \varepsilon^2\sigma^* \overset{\circ}{V} U_2^R(-D_1^R) \\ &\quad - \varepsilon^2 \mu^A c_A \cos\theta_I \int_{-D/\sin\theta_I}^{0} \left[U_1^{R\prime}(\xi\sin\theta_I)\right]^2 \mathrm{d}\xi + \mathrm{O}(\varepsilon^3),\end{aligned} \tag{4.15}$$

$$\begin{aligned}\mathbb{E}_{tran.} &= \varepsilon\sigma^* \overset{\circ}{V} U_1^T(-\lambda D) \\ &\quad - \varepsilon^2\sigma^* \overset{\circ}{V} \cos\theta_T(\lambda\cos\theta_I - \cos\theta_T) s_1(-\omega_1) U_1^{T\prime}(-\lambda D) \\ &\quad + \varepsilon^2\sigma^* \overset{\circ}{V} U_2^T(-D_1^T) \\ &\quad - \varepsilon^2 \mu^B c_B \cos\theta_T \int_{-\lambda D/\sin\theta_T}^{0} \left[U_1^{T\prime}(\xi\sin\theta_T)\right]^2 \mathrm{d}\xi + \mathrm{O}(\varepsilon^3),\end{aligned} \tag{4.16}$$

$$\mathbb{E}_{acc.} = -\varepsilon\gamma^*\sigma^*\mathring{V}s_1(-\omega_0) + \varepsilon^2\gamma^*\sigma^*\mathring{V}\frac{\cos\theta_I}{\sin\theta_I}s_1'(-\omega_1)s_1(-\omega_1) - \varepsilon^2\gamma^*\sigma^*\mathring{V}s_2(-\omega_1) + \mathrm{O}(\varepsilon^3), \tag{4.17}$$

$$\mathbb{E}_{diss.} = \varepsilon^2\mu^A c_A\gamma^*\lambda^2\frac{\cos\theta_I\cos\theta_T}{\sin\theta_I\sin\theta_T} \times \int_{-D/\sin\theta_I}^{0}\frac{2\sin\theta_I U_I'(\xi\sin\theta_I) - \gamma^* s_1'(\xi)}{\cos\theta_I + \lambda\cos\theta_T}s_1'(\xi)\,\mathrm{d}\xi + \mathrm{O}(\varepsilon^3). \tag{4.18}$$

The rate balance (4.8) for the terms that are linear in σ^* is ensured at each order by virtue of (4.14). For the second order terms that are independent of σ^*, the rate balance (4.8) can be verified by showing the balance of the associated integrands in (4.11), (4.15), (4.16) and (4.18). The above results are general in the sense that the motion of the phase boundary is not specified.

Recall for $\phi'(0) \equiv \kappa > 0$, that the leading order motion of the phase boundary is given by (3.24) and that the leading order reflected and transmitted wave profiles are given by (3.26). Consequently, the leading order energy rates are easily expressed in terms of the incident wave profile U_I. It is instructive to examine the energy partitioning in this case for the separate scenarios in which $\sigma^* \neq 0$ and $\sigma^* = 0$.

1. Non-zero Maxwell stress. This will occur if and only if $\mathring{\mathcal{W}}_A \neq \mathring{\mathcal{W}}_B$. The leading order energy balance involves the $\mathrm{O}(\varepsilon)$ terms in $\mathbb{E}_{inc.}$, $\mathbb{E}_{ref.}$, $\mathbb{E}_{tran.}$ and $\mathbb{E}_{acc.}$, all of which are linear in σ^*. In view of (3.24) and (3.26), they can be expressed as

$$\begin{aligned}
\mathbb{E}_{inc.} &= -\varepsilon\sigma^*\mathring{V}U_I(-D) + \mathrm{O}(\varepsilon^2),\\
\mathbb{E}_{ref.} &= \frac{(\cos\theta_I - \lambda\cos\theta_T) + \hat{\kappa}\cos\theta_I\cos\theta_T}{(\cos\theta_I + \lambda\cos\theta_T) + \hat{\kappa}\cos\theta_I\cos\theta_T}\mathbb{E}_{inc.} + \mathrm{O}(\varepsilon^2),\\
\mathbb{E}_{tran.} &= -\frac{2\cos\theta_I}{(\cos\theta_I + \lambda\cos\theta_T) + \hat{\kappa}\cos\theta_I\cos\theta_T}\mathbb{E}_{inc.} + \mathrm{O}(\varepsilon^2),\\
\mathbb{E}_{acc.} &= \frac{2\hat{\kappa}\cos\theta_I\cos\theta_T}{(\cos\theta_I + \lambda\cos\theta_T) + \hat{\kappa}\cos\theta_I\cos\theta_T}\mathbb{E}_{inc.} + \mathrm{O}(\varepsilon^2),\\
\mathbb{E}_{diss.} &= \mathrm{O}(\varepsilon^2) = \varepsilon^2\frac{4\mu^A c_A\lambda\hat{\kappa}\cos^2\theta_I\cos^2\theta_T}{[(\cos\theta_I + \lambda\cos\theta_T) + \hat{\kappa}\cos\theta_I\cos\theta_T]^2}\\
&\quad\times\int_{-D/\sin\theta_I}^{0}\left[U_I'(\xi\sin\theta_I)\right]^2\mathrm{d}\xi + \mathrm{O}(\varepsilon^3),
\end{aligned} \tag{4.19}$$

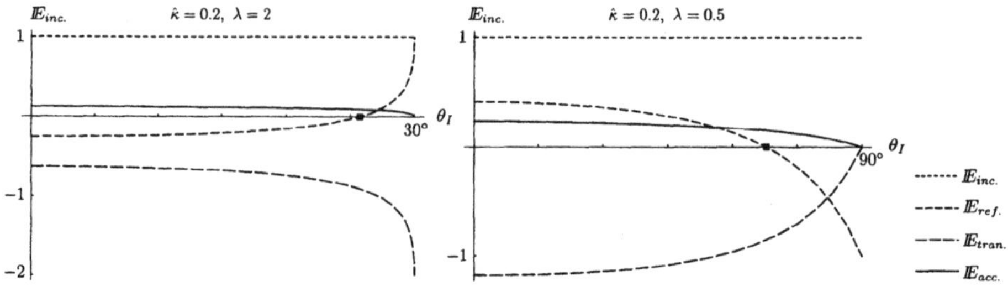

Figure 6. Leading order ($O(\varepsilon)$) energy rate balance for scenario 1 with fixed $\hat{\kappa}$ and λ. The incident angles at which the reflected wave disappears are shown with "■" at $\theta_I = 25.75°$ for $\lambda = 2$, and at $\theta_I = 67.89°$ for $\lambda = 0.5$.

where $\hat{\kappa}$ is again the normalized mobility, $\hat{\kappa} = \kappa\mu^B\gamma^{*2}/c_B$. For the purpose of comparing the dissipation rate in this scenario to that in the following scenario, we have also displayed the $O(\varepsilon^2)$ expression for $\mathbb{E}_{diss.}$ in (4.19). For the parameter values $\hat{\kappa} = 0.2$, $\lambda = 2$ and $\lambda = 0.5$, Figure 6 displays the individual contributions of the leading order parts of $\mathbb{E}_{inc.}$, $\mathbb{E}_{ref.}$ $\mathbb{E}_{tran.}$ and $\mathbb{E}_{acc.}$ to the rate balance (4.8) as a function of θ_I. Recall that $\lambda > 1$ gives rise to a critical angle θ_c ($\theta_c = 30°$ for $\lambda = 2$) such that $\theta_T \to 90°$ as $\theta_I \to \theta_c$. It then further follows, to leading order, that $\mathbb{E}_{acc.} \to 0$, $\mathbb{E}_{ref.} \to \mathbb{E}_{inc.}$, and $\mathbb{E}_{tran.} \to -2\mathbb{E}_{inc.}$ as $\theta_I \to \theta_c$ (Figure 6). In contrast, there is no critical angle for $\lambda < 1$, and the steady state waves under consideration are supported for all $0 \leqslant \theta_I \leqslant 90°$. In this case it follows, to leading order, that $\mathbb{E}_{acc.} \to 0$, $\mathbb{E}_{ref.} \to -\mathbb{E}_{inc.}$ and $\mathbb{E}_{tran.} \to 0$ as $\theta_I \to 90°$ (Figure 6). The limit $\theta_I \to 90°$ involves a reflected wave that overlaps with the incident wave, and, to leading order, these two waves cancel. Since the transmitted wave and the accretion zone also vanish to leading order in this limit, there is in fact no $O(\varepsilon)$ activity for $\lambda < 1$, $\theta_I = 90°$.

2. Zero Maxwell stress. This will occur if and only if $\overset{\circ}{W}_A = \overset{\circ}{W}_B$. In this scenario, $\mathbb{E}_{acc.} = 0$ and the leading order energy balance involves the $O(\varepsilon^2)$, σ^*-independent terms in $\mathbb{E}_{inc.}$, $\mathbb{E}_{ref.}$, $\mathbb{E}_{tran.}$ and $\mathbb{E}_{diss.}$. They are

$$\begin{aligned}
\mathbb{E}_{inc.} &= \varepsilon^2\mu^A c_A \cos\theta_I \int_{-D/\sin\theta_I}^{0} \left[U_I'(\xi\sin\theta_I)\right]^2 d\xi,\\
\mathbb{E}_{ref.} &= -\frac{[(\cos\theta_I - \lambda\cos\theta_T) + \hat{\kappa}\cos\theta_I\cos\theta_T]^2}{[(\cos\theta_I + \lambda\cos\theta_T) + \hat{\kappa}\cos\theta_I\cos\theta_T]^2}\mathbb{E}_{inc.} + O(\varepsilon^3),\\
\mathbb{E}_{tran.} &= -\frac{4\lambda\cos\theta_I\cos\theta_T}{[(\cos\theta_I + \lambda\cos\theta_T) + \hat{\kappa}\cos\theta_I\cos\theta_T]^2}\mathbb{E}_{inc.} + O(\varepsilon^3),\\
\mathbb{E}_{diss.} &= \frac{4\lambda\hat{\kappa}\cos\theta_I\cos^2\theta_T}{[(\cos\theta_I + \lambda\cos\theta_T) + \hat{\kappa}\cos\theta_I\cos\theta_T]^2}\mathbb{E}_{inc.} + O(\varepsilon^3).
\end{aligned} \tag{4.20}$$

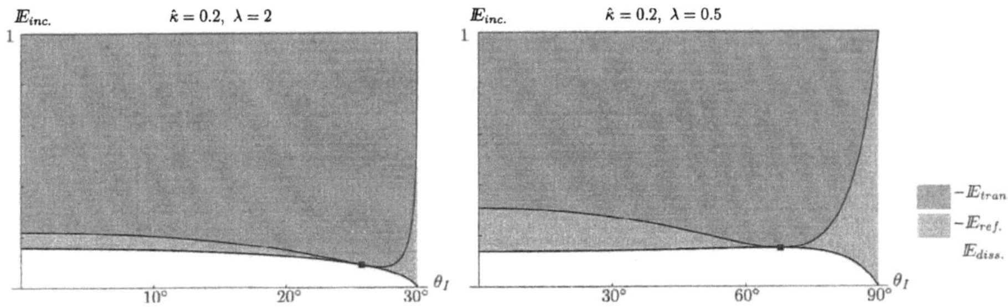

Figure 7. Leading order ($O(\varepsilon^2)$) energy rate balance for scenario 2 with fixed $\hat{\kappa}$ and λ. The incident angles at which the reflected wave disappears are shown with "■" at $\theta_I = 25.75°$ for $\lambda = 2$, and at $\theta_I = 67.89°$ for $\lambda = 0.5$.

For the parameter values $\hat{\kappa} = 0.2$, $\lambda = 2$ and $\lambda = 0.5$, Figure 7 displays the individual contributions of the leading order parts of $\mathbb{E}_{inc.}$, $\mathbb{E}_{ref.}$, $\mathbb{E}_{tran.}$ and $\mathbb{E}_{diss.}$ to the rate balance (4.8) as a function of θ_I. Unlike scenario 1, this scenario always involves leading order parts obeying the inequality restrictions $\mathbb{E}_{inc.} > 0$, $\mathbb{E}_{ref.} < 0$, $\mathbb{E}_{tran.} < 0$ and $\mathbb{E}_{diss.} > 0$ and so permits a graphical representation in which it is obvious that $\mathbb{E}_{inc.} = |\mathbb{E}_{ref.}| + |\mathbb{E}_{tran.}| + \mathbb{E}_{diss.}$. For this scenario it follows, to leading order, that $\mathbb{E}_{diss.} \to 0$, $\mathbb{E}_{ref.} \to -\mathbb{E}_{inc.}$ and $\mathbb{E}_{tran.} \to 0$ as θ_I approaches the maximum oblique angle, namely, as $\theta_I \to \theta_c$ when $\lambda > 1$ and as $\theta_I \to 90°$ when $\lambda < 1$ (Figure 7). Thus, in this limit there is no transmission through the interface and the reflected wave involves a displacement inversion. For the latter case ($\lambda < 1$, $\theta_I = 90°$) this again corresponds to a zero activity limit due to the overlap of reflected and incident waves.

Examination of (4.19) and (4.20) indicates that, to leading order, $\mathbb{E}_{ref.} = 0$ for both scenario 1 and scenario 2 if

$$(\cos\theta_I - \lambda\cos\theta_T) + \hat{\kappa}\cos\theta_I\cos\theta_T = 0. \tag{4.21}$$

If a plane wave is incident at an angle θ_I obeying (4.21) then there is no leading order reflected wave. The dependence of this angle on both wave speed ratio λ and the mobility $\hat{\kappa}$ is illustrated in Figure 8. For $\hat{\kappa} = 0$ the situation reduces to that of wave transmission from one conventional elastic medium into another with differing impedance. In this case there is exactly one angle θ_I obeying (4.21) for all $\lambda \neq 1$. In contrast, for $\hat{\kappa} > 0$ there is a range $1 < \lambda < 1 + \hat{\kappa}$ in which there is no incident angle θ_I that suppresses the reflected wave. If $\lambda < 1$ then the incident angle that suppresses the reflected wave is greater than 45°, whereas if $\lambda > 1$ (and $\hat{\kappa} < \lambda - 1$), then the incident angle that suppresses the reflected wave is less than 45°.

For both scenarios 1 and 2, examining the factor in front of the integral portion of $\mathbb{E}_{diss.}$ given by (4.19) and by (4.20) indicates that the leading order dissipation rate tends to zero as $\hat{\kappa}$ approaches either 0 or ∞. The former corresponds to an

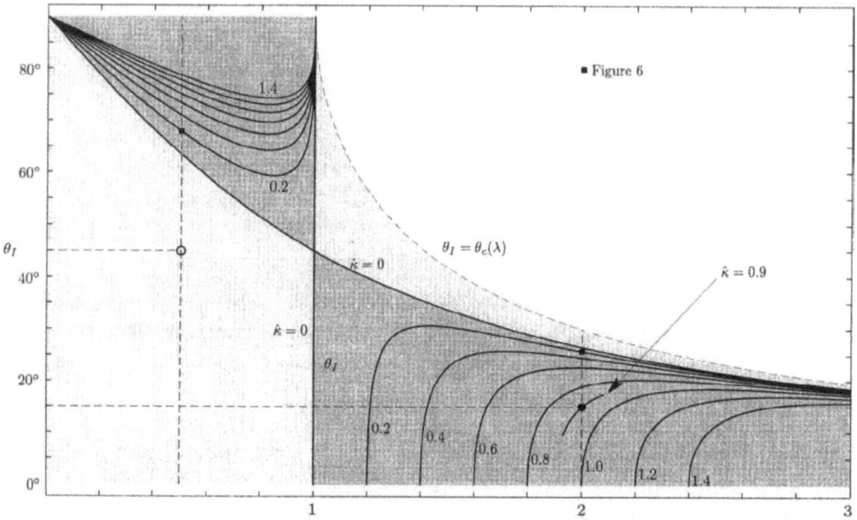

Figure 8. Incident angles that suppress the reflected wave as a function of λ (horizontal axis) and $\hat{\kappa}$ (separate curves).

immobile phase boundary; the latter enforces a zero driving traction condition. Between these two extremes there is a special mobility value:

$$\hat{\kappa}_{max} = \frac{\cos\theta_I + \lambda\cos\theta_T}{\cos\theta_I\cos\theta_T}, \tag{4.22}$$

which is common to both scenario 1 and scenario 2, that maximizes the dissipation rate as a function of incident angle θ_I. The corresponding maximal dissipation rate is given by

$$\mathbb{E}_{diss.} = \varepsilon^2 \frac{\mu^A c_A \lambda \cos\theta_I \cos\theta_T}{(\cos\theta_I + \lambda\cos\theta_T)} \int_{-D/\sin\theta_I}^{0} \left[U_I'(\xi \sin\theta_I)\right]^2 d\xi + O(\varepsilon^3) \quad \text{for scenario 1,} \tag{4.23}$$

$$\mathbb{E}_{diss.} = \frac{\lambda\cos\theta_T}{\cos\theta_I + \lambda\cos\theta_T}\mathbb{E}_{inc.} + O(\varepsilon^3) \quad \text{for scenario 2.}$$

The maximal dissipation result (4.22) is consistent with the uniaxial study of Lin and Pence [9] which would correspond to direct incidence ($\theta_I = 0$) in the present framework. Although the direct incidence case can no longer be interpreted in terms of steady state propagation (since $\overset{\circ}{V} \to \infty$), the maximal dissipation result (4.22) remains meaningful in the limit $\theta_I = 0$. In particular, the dissipation rate is maximized under direct incidence at $\hat{\kappa}_{max} = 1 + \lambda$, which further specializes to $\hat{\kappa}_{max} = 2$ in the event that the elastic constants characterizing the two phases are identical.

The effect of $\hat{\kappa}$ on the energy partitioning at fixed λ and θ_I is indicated in Figure 9 for scenario 1 and Figures 10 and 11 for scenario 2. The same values

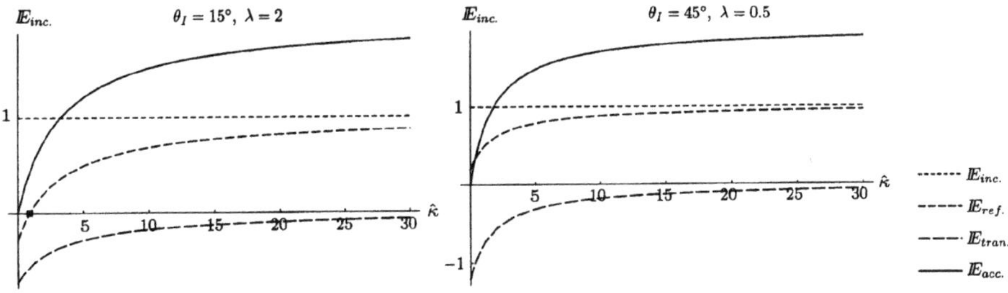

Figure 9. Leading order ($O(\varepsilon)$) energy rate balance for scenario 1 with fixed θ_I and λ. The mobility at which the reflected wave disappears is shown with "■" at $\hat{\kappa} = 0.9$ for the case $\theta_I = 15°$ and $\lambda = 2$.

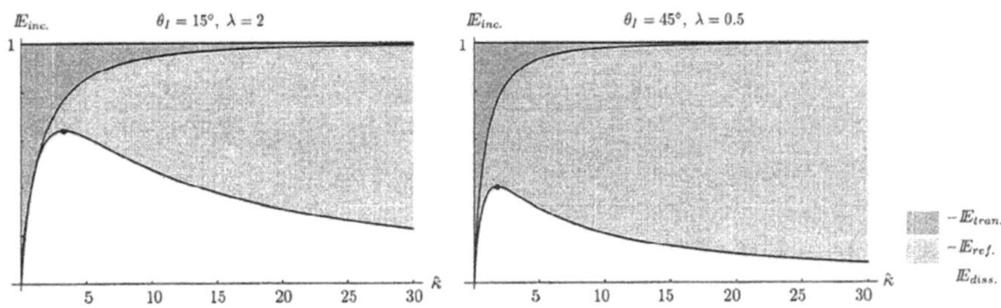

Figure 10. Leading order ($O(\varepsilon^2)$) rate balance for scenario 2 with fixed θ_I and λ. The values $\hat{\kappa}_{max}$ are marked.

of λ as used previously in Figures 6 and 7 are employed. As such, Figure 9 is a counterpart to Figure 6 whereas Figures 10 and 11 are counterparts to Figure 7. For scenario 1 with $\lambda = 2$, $\theta_I = \theta_C/2 = 15°$, and with $\lambda = 0.5$, $\theta_I = 45°$, Figure 9 shows that each rate component $\mathbb{E}_{acc.}$, $\mathbb{E}_{ref.}$ and $\mathbb{E}_{trans.}$ is increasing with $\hat{\kappa}$. In particular, the energy accretion rate increases from $\mathbb{E}_{acc.} = 0$ when $\hat{\kappa} = 0$ to the asymptotic value $2\mathbb{E}_{inc.}$ as $\hat{\kappa} \to \infty$. Conventional elastic mismatch values for $\mathbb{E}_{ref.}$ and $\mathbb{E}_{trans.}$ hold at $\hat{\kappa} = 0$, whereas $\mathbb{E}_{ref.} \to \mathbb{E}_{inc.}$ and $\mathbb{E}_{trans.} \to 0$ in the zero driving traction limit $\hat{\kappa} \to \infty$. In the case $\lambda = 2$, $\theta_I = \theta_C/2 = 15°$, it is evident from Figure 9 that the reflected wave is suppressed at a particular $\hat{\kappa}$ (essentially 0.9). In the case $\lambda = 0.5$, $\theta_I = 45°$ a nonzero reflected wave is generated for all $\hat{\kappa}$. This is consistent with Figure 8. In both cases, the maximally dissipative $\hat{\kappa}_{max}$ is not evident from Figure 9 since $\mathbb{E}_{diss.}$ is $O(\varepsilon^2)$, whereas the leading order energy balances depicted here are $O(\varepsilon)$.

For scenario 2, $\mathbb{E}_{acc.} = 0$ and the leading order energy balances for the remaining rates $\mathbb{E}_{inc.}$, $\mathbb{E}_{diss.}$, $\mathbb{E}_{ref.}$, $\mathbb{E}_{tran.}$ are $O(\varepsilon^2)$. Figure 10 is the direct counterpart to Figure 9 for this scenario (so that either $\lambda = 2$, $\theta_I = \theta_C/2 = 15°$ or $\lambda = 0.5$, $\theta_I = 45°$). Figure 11 magnifies Figure 10 near $\hat{\kappa} = 0$. As mentioned previously, $\mathbb{E}_{diss.}$ tends to zero both as $\hat{\kappa}$ tends to zero and as $\hat{\kappa}$ tends to ∞. The maximally dissipative mobility values are evident from Figure 10 ($\hat{\kappa}_{max} = 3.24$ for $\lambda = 2$, $\theta_I = \theta_C/2 = 15°$, and $\hat{\kappa}_{max} = 1.78$ for $\lambda = 0.5$, $\theta_I = 45°$). As the absence of

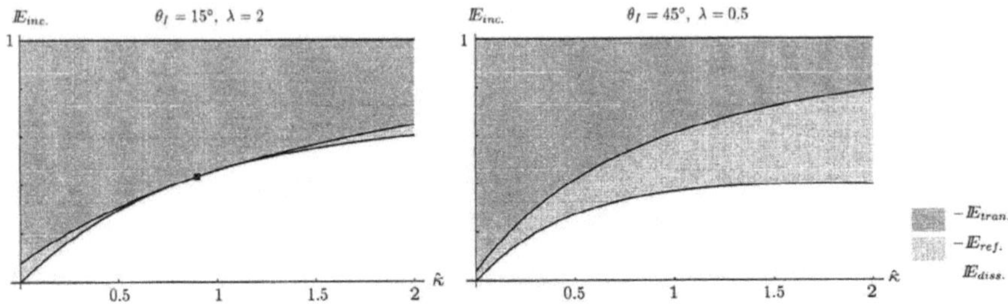

Figure 11. Detailed views of Figure 10 for the range $0 < \hat{\kappa} < 2$. The mobility at which the reflected wave disappears is shown with "■" at $\hat{\kappa} = 0.9$ for the case $\theta_I = 15°$ and $\lambda = 2$.

reflection is determined solely by relation (4.21) between λ, θ_I and $\hat{\kappa}$, it follows again that the reflected wave can be suppressed for $\lambda = 2$, $\theta_I = \theta_C/2 = 15°$ (at $\hat{\kappa} = 0.9$ as shown in Figure 11) but not for $\lambda = 0.5$, $\theta_I = 45°$. Note in both Figures 9 and 10 that the transmitted wave is suppressed in the limit as $\hat{\kappa}$ tends to ∞, in which case the response is one of pure reflection.

5. Acknowledgements

The authors thank the financial support of the Research Excellence Fund from the State of Michigan through the Composite Materials and Structure Center at Michigan State University.

6. Appendix

Expressions for $\Lambda_R(\xi, \bar{s}, \bar{s}')$ and $\Lambda_T(\xi, \bar{s}, \bar{s}')$ which appear in (2.27) and (2.28) are given by

$$\Lambda_R(\xi, \bar{s}, \bar{s}') = \frac{A_R(\bar{s}')}{\Delta(\bar{s}')} U_I(\xi \sin\theta_I + \varepsilon\bar{s}\cos\theta_I) + \frac{B_R(\bar{s}')}{\Delta(\bar{s}')}\gamma^* \tag{A.1}$$

and

$$\Lambda_T(\xi, \bar{s}, \bar{s}') = \frac{A_T(\bar{s}')}{\Delta(\bar{s}')} U_I(\xi \sin\theta_I + \varepsilon\bar{s}\cos\theta_I) + \frac{B_T(\bar{s}')}{\Delta(\bar{s}')}\gamma^*, \tag{A.2}$$

where the functions Δ, A_R, B_R, A_T and B_T are given by

$$\begin{aligned}\Delta(\bar{s}') = {} & \big(\sin^2\theta_T + \varepsilon^2(1+\sin^2\theta_T)\bar{s}'^2\big)(\sin\theta_I - \varepsilon\bar{s}'\cos\theta_I) \\ & \times (\cos\theta_T - \varepsilon\bar{s}'\sin\theta_T) + \big(\sin^2\theta_I + \varepsilon^2(1+\sin^2\theta_T)\bar{s}'^2\big) \\ & \times (\sin\theta_T + \varepsilon\bar{s}'\cos\theta_T)(\cos\theta_I + \varepsilon\bar{s}'\sin\theta_I),\end{aligned} \tag{A.3}$$

$$\begin{aligned}A_R(\bar{s}') = {} & -\big(\sin^2\theta_T + \varepsilon^2(1+\sin^2\theta_T)\bar{s}'^2\big)(\sin\theta_I + \varepsilon\bar{s}'\cos\theta_I) \\ & \times (\cos\theta_T - \varepsilon\bar{s}'\sin\theta_T) + \big(\sin^2\theta_I + \varepsilon^2(1+\sin^2\theta_I)\bar{s}'^2\big) \\ & \times (\sin\theta_T + \varepsilon\bar{s}'\cos\theta_T)(\cos\theta_I - \varepsilon\bar{s}'\sin\theta_I),\end{aligned} \tag{A.4}$$

$$B_R(\bar{s}') = \sin\theta_T \bar{s}'\left[\sin\theta_T\cos\theta_T - \varepsilon\bar{s}'\left(1+\sin^2\theta_T\right) - \varepsilon^3\bar{s}'^3\right], \tag{A.5}$$

$$A_T(\bar{s}') = 2\sin\theta_I\cos\theta_I\left(1+\varepsilon^2\bar{s}'^2\right)\left(\sin^2\theta_I + \varepsilon^2\left(1+\sin^2\theta_I\right)\bar{s}'^2\right), \tag{A.6}$$

$$B_T(\bar{s}') = -\sin\theta_I \bar{s}'\left[\sin\theta_I\cos\theta_I + \varepsilon\bar{s}'\left(1+\sin^2\theta_I\right) + \varepsilon^3\bar{s}'^3\right]. \tag{A.7}$$

Expressions for Ξ_i which appear in (3.10) and (3.11) are given by

$$\begin{aligned}
\Xi_1 &= -\frac{\gamma^*\mu^A}{2\Delta}\left(\lambda^2\cos\theta_T - \cos\theta_I B_R\right),\\
\Xi_2 &= -\frac{\gamma^*\mu^A}{2\Delta}\left[\lambda^2\cos\theta_T A_T + \cos\theta_I(\Delta - A_R)\right],\\
\Xi_3 &= -\frac{\mu^B-\mu^A}{2\Delta^2}\cos(\theta_T-\theta_I)B_R B_T,\\
\Xi_4 &= \frac{\mu^B-\mu^A}{2\Delta^2}\left[\cos(\theta_T-\theta_I)B_T\Delta - \cos(\theta_T+\theta_I)(A_R B_T + A_T B_R)\right],\\
\Xi_5 &= \frac{\mu^B-\mu^A}{2\Delta^2}A_T\left[\cos(\theta_T-\theta_I)\Delta - \cos(\theta_T+\theta_I)A_R\right].
\end{aligned} \tag{A.8}$$

References

1. R. Abeyaratne and J.K. Knowles, On the driving traction acting on a surface of strain discontinuity in a continuum. *J. Mech. Phys. Solids* **38** (1990) 345–360.
2. R. Abeyaratne and J.K. Knowles, Kinetic relations and the propagation of phase boundaries in solids. *Arch. Rational Mech. Anal.* **114** (1991) 119–154.
3. R. Abeyaratne and J.K. Knowles, Reflection and transmission of waves from an interface with a phase-transforming solid. *J. Intelligent Mater. Stuct.* **3** (1992) 345–360.
4. J.D. Achenbach, *Wave Propagation in Elastic Solids*. Elsevier/North-Holland, New York (1980).
5. K. Bhattacharya, Wedge-like microstructures in martensite. *Acta Metallica* **39** (1991) 2431–2444.
6. J.L. Ericksen, Equilibrium of bars. *J. Elasticity* **5** (1975) 191–201.
7. R.D. James, The propagation of phase boundaries in elastic bars. *Arch. Rational Mech. Anal.* **73** (1980) 125–158.
8. R.D. James, A relation between the jump in temperature across a propagating phase boundary and the stability of solid phases. *J. Elasticity* **13** (1983) 357–378.
9. J. Lin and T.J. Pence, Pulse attenuation by kinetically active phase boundary scattering during displacive phase transformations. *J. Mech. Phys. Solids* **46** (1998) 1183–1211.
10. T.J. Pence, On the encounter of an acoustic shear pulse with a phase boundary in an elastic material: reflection and transmission behavior. *J. Elasticity* **25** (1991) 31–74.
11. T.J. Pence, On the encounter of an acoustic shear pulse with a phase boundary in an elastic material: energy and dissipation. *J. Elasticity* **26** (1991) 95–146.
12. P. Rosakis and J.K. Knowles, Unstable kinetic relations and the dynamics of solid–solid phase transitions. *J. Mech. Phys. Solids* **45** (1997) 1183–1211.
13. P. Rosakis and H. Tsai, On the role of shear instability in the modelling of crystal twinning. *Mech. Mater.* **17** (1994) 245–259.

14. P. Rosakis and H. Tsai, Dynamic twinning processes in crystals. *Internat. J. Solids Struct.* **32** (1995) 2711–2723.
15. L.M. Truskinovsky, About the "normal growth" approximation in the dynamical theory of phase transitions. *Continuum Mech. Thermodyn.* **6** (1994) 185–208.
16. H. Tsai and P. Rosakis, On anisotropic compressible materials that can sustain dynamic anti-plane shear. *J. Elasticity* **35** (1994) 213–222.

Journal of Elasticity **59:** 267–318, 2000.

Microstructure in a Cubic to Orthorhombic Transition

KEVIN F. HANE* and THOMAS W. SHIELD
Department of Aerospace Engineering and Mechanics, University of Minnesota, Minneapolis, MN 55455, U.S.A. E-mail: hane@engin.brown.edu; shield@aem.umn.edu

Received 1 March 1999; in revised form 17 February 2000

Dedicated to Roger Fosdick on the occasion of his 60th birthday.

Abstract. Microstructures for a cubic to orthorhombic transition are constructed using a geometrically nonlinear, thermoelastic theory of martensitic transformations. Such microstructures are of interest because they provide low energy paths along which a specimen can transform. The particular microstructures considered are the twinned martensite, austenite–martensite, wedge, triangle, and diamond. More specifically, all possible twins are found along with the corresponding twinning elements and magnitude of the twin shear. Further, two kinds of austenite–martensite microstructures are studied: those with a single variant of martensite and those with twinned martensite. The regions in the space of transformation stretches in which each of these microstructures exist are determined, and the shape strains and habit plane normals are found as well. In addition, special microstructures, the wedge, triangle, and diamond, are constructed with both the austenite-single variant and austenite-twinned martensite microstructures. These special microstructures are of interest because they provide a mechanism through which the transformation may proceed more easily, and they are possible only in alloys with particular transformation stretches. Numerically computed level curves in the space of the stretches are presented on which the special microstructures are possible. These results may be useful in providing guidelines for alloy design.

1. Introduction

There is an ever-growing list of special alloys which are known to exhibit two unique and potentially useful thermo-mechanical properties: the shape memory effect and superelastic behavior. The shape memory effect is the ability of a specimen to be deformed at low temperatures and then to return to its original shape by heating; while, superelastic behavior is the ability of a specimen to be deformed at high temperatures and then to return to its original shape with unloading [22, 34, 36]. Both of these properties are the result of microstructure nucleating and growing in the body as it undergoes a thermoelastic martensitic transformation. This type of transformation is reversible, diffusionless, and structural, usually between a high temperature, high symmetry parent phase called austenite and a low temperature,

* Currently: Brown University, Division of Engineering, Box D, Providence, RI 02912, U.S.A.

low symmetry product phase called martensite, where the martensite has multiple variants: symmetry related configurations. Microstructures are geometric arrangements of the two phases, and they provide a means by which the transformation between the phases can occur. In particular, it is important to know the various microstructures which are possible in order to be able to successfully predict the response of these alloys under different loading programs.

The most popular shape memory material for applications and research is a nearly stoichiometric alloy of Ti–Ni. This material exhibits the martensitic transformation for a narrow range of compositions, and the transition is cubic to monoclinic. On the other hand, the ternary alloys with nominal compositions $Ti_{50}Ni_{50-x}Cu_x$, $0 \leqslant x \leqslant 20$, also exhibit a martensitic transformation. Depending upon the percentage of Cu, the alloy can undergo a myriad of different transformations – either cubic to orthorhombic, cubic to monoclinic or orthorhombic to monoclinic. For each alloy, the lattice parameters of the unit cells of the two phases are different, and consequently, the kinds of microstructures which are possible are different. An alloy with a large number of microstructures, providing low energy paths by which the material can transform, is expected to have good shape memory and superelastic response especially in polycrystalline specimens. Microstructures in several different transformations have been recently studied (see [25–29] and the references therein), and it is the purpose of this paper to consider microstructures in a cubic to orthorhombic transition with the goal of highlighting interesting combinations of lattice parameters for which many microstructures are possible.

A cubic to orthorhombic transition is described by three parameters: three transformation stretches, ratios of lattice parameters, along three mutually perpendicular directions. Several alloys undergo this transition such as Au–Cd, Cu–Al–Ni, Ti–Ta, and Ti–Ni–Cu to name a few. The Cu–Al–Ni alloy with approximately 14 weight percent Al and 4 weight percent Ni is one of the most widely studied shape memory alloys because of the relative ease of preparing single crystal specimens (see references in [5, 7, 8, 27]); while, a cubic to orthorhombic transition is observed in Ti–Ni–Cu alloys over a range of compositions [35, 45–47]. The latter alloys show great promise for applications.

The theory used to construct microstructures is a constrained formulation of a geometrically nonlinear, thermoelastic theory of martensite [2–7, 30, 31]. Microstructures in the constrained theory are continuous deformations with piecewise constant gradient taking values only on certain energy wells; thus, elastic strains are ignored. At the boundaries between the phases or between different variants of martensite, compatibility equations must be satisfied. If solutions exist to these equations with different variants, then it can be shown that such solutions are related by a symmetry transformation in the austenite Laue group – the group of rotations which map the austenite lattice into itself. Further, this symmetry amongst the solutions can be used to enumerate and organize the possible realizations of a particular microstructure and to help in the construction of several special microstructures. The latter are of interest because they provide a mechanism by which the transformation can proceed more easily.

In this article, it is shown that a rich collection of microstructures is possible in a cubic to orthorhombic transition depending upon the transformation stretches. Specifically, the twinned martensite microstructure between two martensite variants always exists for any choice of the stretches, and the twins are either compound, type I or type II. Also, two kinds of austenite–martensite microstructures are considered: either a single variant of martensite or twinned martensite coexisting with austenite. In particular, restrictions upon the stretches for which these various microstructures exists are given, and expressions for the shape strain and habit plane normal vectors are found when convenient. Further, several special microstructures, the wedge, triangle and diamond, are constructed. These microstructure are special in that they are possible only in materials with particular stretches. Specifically, the wedge can grow from a line on a free surface of a specimen or a grain boundary within the specimen; while, the triangle and diamond microstructures can grow from a point on a free surface completely surrounded by the austenite phase. It is found that these microstructures are possible both with and without twins. Another interesting result is that for very special transformation stretches an austenite-single of variant martensite, austenite-twinned martensite, and several different kinds of wedges, triangles and diamonds are all possible. In fact, the austenite-twinned martensite microstructure is possible for every volume fraction and any fineness of the twins. Such an alloy would be the ultimate shape memory material, because it would be able to form an infinite variety of microstructures at essentially zero energy.

Previously, Bhattacharya [5–8, 10] gave some analysis for the twinned martensite microstructure in a cubic to orthorhombic transition, as well as calculations of habit planes and the twinned wedge microstructure. Also, Chu [13], Chu and James [14, 15], and Abeyaratne et al. [1] investigated microstructures with the martensite variants in a Cu–Al–Ni alloy; while, Bhattacharya [5–8], Shield [44], and Hane and Shield [27] calculated all of the twins and habit planes in this alloy. The notation is the same as that used in [26], where specifically, a counterclockwise rotation $\mathbf{Q}$ of ψ degrees about the axis $\hat{\mathbf{p}}$ is written as $\mathbf{Q}[\psi, \hat{\mathbf{p}}]$, and the set of all rotations is denoted by SO(3). Also, no summation is implied by any of the repeated indices as they are for identification only.

2. Thermoelastic Martensitic Transformation

2.1. BACKGROUND

A thermoelastic martensitic transformation is a diffusionless structural transformation between a high temperature austenite parent phase and a low temperature martensite product phase (see for example [22, 24, 32, 34, 36, 42, 49]). Typically, during experiments on unstressed specimens, it is observed that there is a temperature interval in which the phases coexist and that the transformation is reversible.

The forward and reverse transformation sequences and the characteristic temperatures are as follows: upon cooling a specimen initially entirely in the austenite phase, there is a temperature M_s, the martensite start temperature, at which the martensite phase first forms. Surface relief appears on the specimen because the lattice parameters of the two phases are different. With further cooling, more and more of the body transforms from austenite to martensite until the temperature M_f, the martensite finish temperature, is reached below which the specimen is completely in the martensite phase. Heat is released during this transformation, and the specimen consists of many regions each containing a different variant of the martensite phase. These regions can grow or shrink under small applied loads. Upon heating the specimen, there is a temperature A_s, the austenite start temperature, at which the austenite phase first forms in the martensite. In general, the temperature A_s is not equal to the temperature M_s; so, there is thermal hysteresis between the forward and reverse transformations. With further heating, more and more of the body transforms from martensite to austenite until the temperature A_f, the austenite finish temperature, is reached above which the specimen is completely transformed back to the austenite phase. During this transformation, heat is absorbed, and the specimen recovers its initial shape because there is only one, high temperature crystal structure which is stable.

The crystal lattice of the austenite phase usually has higher symmetry than the crystal lattice of the martensite phase. On account of the difference in symmetry between the phases, the martensite exhibits multiple variants. The variants of martensite are symmetry related configurations of the product lattice, which arise because there are multiple, though distinct transformations which take the parent lattice to a product lattice. Furthermore, in experiments, it is observed that the deformation which takes one lattice into the other is an exact linear deformation. The components of this deformation depend upon the lattice parameters of both phases and the change in symmetry, and the strains produced by the transformation may be on the order of several percent. Thermal expansion of the crystal lattice is ignored in this paper because thermal strains are small when compared to the transformation strains [4]. On the other hand, the macroscopic deformation of the specimen is observed to be continuous across regions containing the different phases.

In most alloys, the arrangement of the atoms on a crystal lattice is not random, but rather specific atoms occupy specific sites on the lattice. Such a structure is called an ordered structure or a superlattice. An ordered structure may be described by multiple, interpenetrating Bravais lattices (see [38]). The analysis herein applies if a skeletal Bravais lattice is chosen so that only the position of some of the atoms is considered. Two specific examples for Cu–Al–Ni and Ti–Ni shape memory alloys are presented in [27, 28], respectively.

2.2. THEORY OF MARTENSITE

A recently developed geometrically nonlinear, thermoelastic theory of martensitic transformations essentially due to Ball and James [3, 4] (also [5–8, 30, 31]) is the main tool used to construct microstructures. The theory starts from a lattice viewpoint and then proceeds to a continuum viewpoint via the Cauchy–Born Rule. The constitutive assumptions for the free energy density give an energy well structure for the material. It is these energy wells which are used in the construction of microstructures herein. Below is a brief summary of the main ideas of the theory and the reader is directed to [4, Section 2] for a more complete discussion.

The starting point is an underlying atomic Bravais lattice of the austenite and martensite phases. Associated with each of the lattices is an invariance group: the group of transformations which map the lattice into itself. A traditional view is that the symmetry of a solid is the same through out the constitutive domain; however, for materials which undergo a martensitic transformation, this assumption is not valid. Ericksen [16–18] shows that a local theory of material symmetry is needed. More specifically, the local theory of material symmetry constructs a neighborhood about a stable equilibrium point. The invariance group of this neighborhood is a finite subgroup of an infinite invariance group, so that the equilibrium point is isolated, and this finite invariance group of the neighborhood corresponds to the point group of the solid. Pitteri [37], Fosdick and Hertog [21], and Ball and James [4] all prove theorems showing the existence of a neighborhood with the above properties. For thermoelastic martensitic transformations, a neighborhood is constructed about the lattice vectors of the parent phase, and it is assumed that this neighborhood also contains the lattice vectors of the product phase. The neighborhood becomes the domain of the free energy, and the invariance group of the neighborhood becomes the invariance group of the free energy.

For an alloy which undergoes a thermoelastic martensitic transformation, it is assumed that the austenite phase has a Bravais lattice with point group $\mathcal{P}^a$. From experiments, a positive-definite symmetric matrix $\mathbf{U}$ is found which takes the austenite lattice into the martensite lattice with point group $\mathcal{P}^m$, where $\mathcal{P}^m \subset \mathcal{P}^a$. There are other matrices related to the matrix $\mathbf{U}$ which also map the austenite lattice to a different configuration of the martensite lattice. These matrices are given by the unique elements in the set $\mathcal{S} = \{\mathbf{RUR}^{\mathrm{T}} \colon \mathbf{R} \in \mathcal{P}^a\}$, where the number of distinct elements μ in this set is found by dividing the order of the group $\mathcal{P}^a$ by the order of the group $\mathcal{P}^m$. The set $\{\mathbf{U}_1, \ldots, \mathbf{U}_\mu\}$ of distinct elements of the set $\mathcal{S}$ is called the variants of martensite (also [48]). From the definition of the set $\mathcal{S}$, it can be seen that only the proper rotations in the groups $\mathcal{P}^a$ and $\mathcal{P}^m$ need to be considered hereafter; that is, only the Laue groups, $\mathcal{P}^a \cap \mathrm{SO}(3)$ and $\mathcal{P}^m \cap \mathrm{SO}(3)$ for the parent and product lattices, respectively.

The atomic motion is related to the macroscopic motion by the Cauchy–Born rule [19, 50, 51]. The rule adopted herein assumes that the lattice vectors behave as material line elements under a deformation. If a bounded, open set $\Omega \subset \mathbb{R}^3$ is the reference configuration of the body, then let $\mathbf{y} \colon \Omega \to \mathbb{R}^3$ denote the deformation which the body undergoes. The gradient of the deformation is denoted as $\nabla \mathbf{y} = \mathbf{F}$, where $\det(\mathbf{F}) > 0$. On following [4], the Cauchy–Born rule states that a free energy density $W \colon \mathcal{D} \to \mathbb{R}^3$ is defined for each temperature $\theta \in I \subset \mathbb{R}^+$ by $W(\mathbf{F}, \theta)$, where $\mathcal{D}$ is the set of all linear transformations $\mathbf{K}$ with positive determinant, so that all lattice vectors are contained in some neighborhood. In addition, the free energy density W has the usual invariance properties: frame indifference and material symmetry. In particular, for each temperature $\theta \in I$, the free energy density W has the invariance given by

$$W(\mathbf{FQ}, \theta) = W(\mathbf{F}, \theta) \quad \text{for all } \mathbf{Q} \in \mathcal{P}^a \cap \mathrm{SO}(3), \tag{1}$$

and it is invariant under all rotations;

$$W(\mathbf{RF}, \theta) = W(\mathbf{F}, \theta) \quad \text{for all } \mathbf{R} \in \mathrm{SO}(3). \tag{2}$$

Moreover, the deformation gradient $\mathbf{F}$ has positive determinant; therefore, the polar decomposition theorem gives that there is a unique rotation $\overline{\mathbf{R}}$ and a unique, positive-definite symmetric matrix $\overline{\mathbf{U}}$ so that $\mathbf{F} = \overline{\mathbf{R}}\,\overline{\mathbf{U}}$. The invariance property (2) along with the polar decomposition of the deformation gradient $\mathbf{F}$ gives that only the symmetric part of this gradient needs to be considered.

A thermoelastic martensitic transformation is modeled by a free energy W with multiple energy wells. The energy wells are for the austenite phase and for each variant of the martensite phase. This type of free energy permits a change of stability between the two phases as either the temperature is changed or loads are applied. Further, it is assumed that there is a critical temperature $\theta_\mathrm{c} \in I$ at which both the forward and reverse transformations take place: thermal hysteresis is ignored. The critical temperature θ_c is often taken to be the midpoint between M_s and A_s. The assumptions for the free energy are such that for temperatures $\theta > \theta_\mathrm{c}$, the austenite phase is the energy minimizer, and therefore, it is the stable phase; while, the martensite phase is a relative minimizer, and thus, it is only a metastable phase. Upon cooling to the temperature $\theta = \theta_\mathrm{c}$, both phases are energy minimizers, and therefore, both phases are stable. For temperatures $\theta < \theta_\mathrm{c}$, the martensite phase is the energy minimizer, and thus, it is the stable phase; while, the austenite phase is a relative minimizer, and therefore, it is only a metastable phase.

More specifically, taking a specimen in the undistorted, high temperature austenite phase as the reference configuration, then from equation (1), for each temperature $\theta > \theta_\mathrm{c}$, the absolute minimizer of the free energy density W is the set of rotations SO(3). This set is called the austenite well and is denoted as $\mathcal{A}$. From equations (1) and (2), for each temperature $\theta < \theta_\mathrm{c}$, the absolute minimizers of the free energy density W are the sets $\mathrm{SO}(3)\mathbf{U}_1, \ldots, \mathrm{SO}(3)\mathbf{U}_\mu$. Each of these sets is called a martensite well, and their union is denoted as $\mathcal{M}$. At the critical temperature θ_c, the austenite and martensite wells are both energy minimizers. Thus, the constitutive assumptions for the free energy density W are: for each temperature $\theta \in I$ such that $\theta > \theta_\mathrm{c}$, the minimizers of the energy are

$$W(\mathbf{F}, \theta)\big|_{\mathbf{F}\in\mathcal{A}} \leqslant W(\mathbf{F}, \theta)\big|_{\mathbf{F}\in\mathcal{D}/\mathcal{A}};$$

while, for each temperature $\theta \in I$ such that $\theta < \theta_\mathrm{c}$, the minimizers of the energy are

$$W(\mathbf{F}, \theta)\big|_{\mathbf{F}\in\mathcal{M}} \leqslant W(\mathbf{F}, \theta)\big|_{\mathbf{F}\in\mathcal{D}/\mathcal{M}}.$$

At the transition temperature θ_c, deformations with gradients in $\mathcal{A} \cup \mathcal{M}$ are minimizers.

The construction of energy minimizing microstructures reduces to considering various continuous deformations with piecewise constant gradient on the austenite and martensite energy wells. In particular, consider a body Ω which undergoes a homogeneous deformation with piecewise constant gradient of the form

$$\mathbf{y} = \begin{cases} \mathbf{R}\mathbf{F}_1\mathbf{x} + \mathbf{c}_1 & \text{for } \mathbf{x} \cdot \hat{\mathbf{n}} \leqslant 0, \ \mathbf{x} \in \Omega, \\ \mathbf{F}_2\mathbf{x} + \mathbf{c}_2 & \text{for } \mathbf{x} \cdot \hat{\mathbf{n}} > 0, \ \mathbf{x} \in \Omega, \end{cases} \tag{3}$$

where $\mathbf{c}_1$ and $\mathbf{c}_2$ are constant vectors, $\mathbf{F}_1 \neq \mathbf{F}_2$, $\mathbf{R} \in SO(3)$, and $\hat{\mathbf{n}}$ is the normal to the surface dividing the body into regions with either deformation gradient. Necessary and sufficient conditions that the deformation given in equation (3) is continuous are: $\mathbf{c}_1 = \mathbf{c}_2$; $\hat{\mathbf{n}}$ is a constant vector; and

$$\mathbf{R}\mathbf{F}_1 - \mathbf{F}_2 = \mathbf{a} \otimes \hat{\mathbf{n}}. \tag{4}$$

Therefore, the regions undergo a common translation, the dividing surface is a plane, and the deformation gradients differ by a rank-1 matrix. The Hadamard compatibility condition (4) is the most important equation to consider in the construction of microstructures, and solutions to this equation can be found as follows: if the deformation gradients $\mathbf{F}_1$ and $\mathbf{F}_2$ are known, then define the symmetric stretch $\mathbf{C}$ to be $(\mathbf{F}_1\mathbf{F}_2^{-1})^T(\mathbf{F}_1\mathbf{F}_2^{-1})$, which from equation (4) can also be written as $\mathbf{C} = (\mathbf{I} + \mathbf{a} \otimes \mathbf{F}_2^{-T}\hat{\mathbf{n}})^T(\mathbf{I} + \mathbf{a} \otimes \mathbf{F}_2^{-T}\hat{\mathbf{n}})$. Necessary and sufficient conditions that solutions to equation (4) for the vectors $\mathbf{a}$ and $\hat{\mathbf{n}}$ exist are given by:

PROPOSITION 1 [3, 4]. *Necessary and sufficient conditions for a symmetric* 3×3 *matrix* $\mathbf{C} \neq \mathbf{I}$ *with eigenvalues* $\lambda_1 \leqslant \lambda_2 \leqslant \lambda_3$ *to be expressible in the form*

$$\mathbf{C} = (\mathbf{I} + \hat{\mathbf{m}} \otimes \mathbf{b})(\mathbf{I} + \mathbf{b} \otimes \hat{\mathbf{m}}),$$

with $1 + \mathbf{b} \cdot \hat{\mathbf{m}} > 0$ *and* $\mathbf{b} \neq \mathbf{0}$, $\hat{\mathbf{m}} \neq \mathbf{0}$ *are that* $\lambda_1 > 0$ *and* $\lambda_2 = 1$. *The solutions are given by*

$$\mathbf{b} = \frac{\rho}{\sqrt{\lambda_3 - \lambda_1}}\left(\sqrt{\lambda_3(1-\lambda_1)}\hat{\mathbf{e}}_1 + \kappa\sqrt{\lambda_1(\lambda_3 - 1)}\hat{\mathbf{e}}_3\right), \tag{5}$$

$$\hat{\mathbf{m}} = \frac{1}{\rho}\left(\frac{\sqrt{\lambda_3} - \sqrt{\lambda_1}}{\sqrt{\lambda_3 - \lambda_1}}\right)\left(-\sqrt{1-\lambda_1}\hat{\mathbf{e}}_1 + \kappa\sqrt{\lambda_3 - 1}\hat{\mathbf{e}}_3\right) \tag{6}$$

with ρ *a nonzero constant, and* $\hat{\mathbf{e}}_1$ *and* $\hat{\mathbf{e}}_3$ *are the eigenvectors of* $\mathbf{C}$ *corresponding to the eigenvalues* λ_1 *and* λ_3, *respectively, and* $\kappa = \pm 1$.

From Proposition 1, there are at most two solutions to equation (4), and for each solution, the rotation $\mathbf{R}$ can be found by direct substitution of the vectors $\mathbf{a}$ and $\hat{\mathbf{n}}$ back into equation (4).

3. Transformation Stretch Matrices

A transformation between a cubic austenite Bravais lattice and an orthorhombic martensite Bravais lattice can occur in two distinct ways. In particular, there are three cubic Bravais lattices, simple, body-centered, and face-centered cubic; while, there are four orthorhombic Bravais lattices, simple, body-centered, face-centered, and base-centered. The cubic to orthorhombic transition can be divided into several categories: the simple cubic lattice may transform into either the simple or the base-centered orthorhombic lattices. The body-centered and face-centered cubic lattices are both able to transform into either the body-centered or the face-centered orthorhombic lattices. In all of these cases, the homogeneous deformation that takes a cubic lattice into an orthorhombic lattice has gradient denoted as $\mathbf{U}$ which is called herein a transformation stretch matrix, and, as shown in [4, Theorem 2.10] (also [40, 41, 52]), there are two kinds of such matrices for this transition. Following [40, 41], the transformation stretch matrix either has dilatation components exclusively (a diagonal matrix) called orthorhombic "cubic axes" variants or has both dilatation and shear components called orthorhombic "mixed axes" variants, where the components of these matrices are in a basis parallel to the edges of a cubic unit cell. Despite the fact that there are two possible transformation stretch matrices, experimentalists find that the second kind is most commonly observed. In fact, the only material which we know of exhibiting the "cubic axes" variants is a Ni_2–Mn–Ga alloy [33]. Therefore, all of the calculations hereafter are with the "mixed axes" variants.

The cubic and orthorhombic Laue groups have 24 and 4 elements, respectively; thus, there are six orthorhombic variants. Further, it is assumed that the variants have transformation stretch matrices denoted as $\mathbf{U}_1$, $\mathbf{U}_2$, $\mathbf{U}_3$, $\mathbf{U}_4$, $\mathbf{U}_5$, and $\mathbf{U}_6$ with components

$$
U_1 = \begin{pmatrix} \beta & 0 & 0 \\ 0 & \dfrac{\alpha+\gamma}{2} & \dfrac{\alpha-\gamma}{2} \\ 0 & \dfrac{\alpha-\gamma}{2} & \dfrac{\alpha+\gamma}{2} \end{pmatrix}, \qquad U_2 = \begin{pmatrix} \beta & 0 & 0 \\ 0 & \dfrac{\alpha+\gamma}{2} & \dfrac{\gamma-\alpha}{2} \\ 0 & \dfrac{\gamma-\alpha}{2} & \dfrac{\alpha+\gamma}{2} \end{pmatrix},
$$

$$
U_3 = \begin{pmatrix} \dfrac{\alpha+\gamma}{2} & 0 & \dfrac{\alpha-\gamma}{2} \\ 0 & \beta & 0 \\ \dfrac{\alpha-\gamma}{2} & 0 & \dfrac{\alpha+\gamma}{2} \end{pmatrix}, \qquad U_4 = \begin{pmatrix} \dfrac{\alpha+\gamma}{2} & 0 & \dfrac{\gamma-\alpha}{2} \\ 0 & \beta & 0 \\ \dfrac{\gamma-\alpha}{2} & 0 & \dfrac{\alpha+\gamma}{2} \end{pmatrix}, \tag{7}
$$

$$
U_5 = \begin{pmatrix} \dfrac{\alpha+\gamma}{2} & \dfrac{\alpha-\gamma}{2} & 0 \\ \dfrac{\alpha-\gamma}{2} & \dfrac{\alpha+\gamma}{2} & 0 \\ 0 & 0 & \beta \end{pmatrix}, \qquad U_6 = \begin{pmatrix} \dfrac{\alpha+\gamma}{2} & \dfrac{\gamma-\alpha}{2} & 0 \\ \dfrac{\gamma-\alpha}{2} & \dfrac{\alpha+\gamma}{2} & 0 \\ 0 & 0 & \beta \end{pmatrix},
$$

respectively, where α, β, and γ are the transformation stretches. The stretches α, β, and γ are assumed to be all positive, and α and γ are assumed to be unequal. If the

Table I. Symmetry transformation amongst the orthorhombic variants. The numbers i are the variants $\mathbf{U}_i$ which are given by $\mathbf{Q}\mathbf{U}_j\mathbf{Q}^T$ for j from 1 to 6 and $\mathbf{Q} \in \mathcal{L}^c$.

Rotations in cubic Laue group $\mathcal{L}^c$	$\mathbf{U}_1$	$\mathbf{U}_2$	$\mathbf{U}_3$	$\mathbf{U}_4$	$\mathbf{U}_5$	$\mathbf{U}_6$
$\mathbf{I} = \hat{\imath}_1 \otimes \hat{\imath}_1 + \hat{\imath}_2 \otimes \hat{\imath}_2 + \hat{\imath}_3 \otimes \hat{\imath}_3$	1	2	3	4	5	6
$\mathbf{Q}[120°, \hat{\imath}_1 + \hat{\imath}_2 + \hat{\imath}_3] = \hat{\imath}_1 \otimes \hat{\imath}_3 + \hat{\imath}_2 \otimes \hat{\imath}_1 + \hat{\imath}_3 \otimes \hat{\imath}_2$	3	4	5	6	1	2
$\mathbf{Q}[-120°, \hat{\imath}_1 + \hat{\imath}_2 + \hat{\imath}_3] = \hat{\imath}_1 \otimes \hat{\imath}_2 + \hat{\imath}_2 \otimes \hat{\imath}_3 + \hat{\imath}_3 \otimes \hat{\imath}_1$	5	6	1	2	3	4
$\mathbf{Q}[120°, -\hat{\imath}_1 + \hat{\imath}_2 + \hat{\imath}_3] = -\hat{\imath}_1 \otimes \hat{\imath}_2 + \hat{\imath}_2 \otimes \hat{\imath}_3 - \hat{\imath}_3 \otimes \hat{\imath}_1$	6	5	2	1	3	4
$\mathbf{Q}[-120°, -\hat{\imath}_1 + \hat{\imath}_2 + \hat{\imath}_3] = -\hat{\imath}_1 \otimes \hat{\imath}_3 - \hat{\imath}_2 \otimes \hat{\imath}_1 + \hat{\imath}_3 \otimes \hat{\imath}_2$	4	3	5	6	2	1
$\mathbf{Q}[120°, \hat{\imath}_1 - \hat{\imath}_2 + \hat{\imath}_3] = -\hat{\imath}_1 \otimes \hat{\imath}_2 - \hat{\imath}_2 \otimes \hat{\imath}_3 + \hat{\imath}_3 \otimes \hat{\imath}_1$	5	6	2	1	4	3
$\mathbf{Q}[-120°, \hat{\imath}_1 - \hat{\imath}_2 + \hat{\imath}_3] = \hat{\imath}_1 \otimes \hat{\imath}_3 - \hat{\imath}_2 \otimes \hat{\imath}_1 - \hat{\imath}_3 \otimes \hat{\imath}_2$	4	3	6	5	1	2
$\mathbf{Q}[120°, \hat{\imath}_1 + \hat{\imath}_2 - \hat{\imath}_3] = \hat{\imath}_1 \otimes \hat{\imath}_2 - \hat{\imath}_2 \otimes \hat{\imath}_3 - \hat{\imath}_3 \otimes \hat{\imath}_1$	6	5	1	2	4	3
$\mathbf{Q}[-120°, \hat{\imath}_1 + \hat{\imath}_2 - \hat{\imath}_3] = -\hat{\imath}_1 \otimes \hat{\imath}_3 + \hat{\imath}_2 \otimes \hat{\imath}_1 - \hat{\imath}_3 \otimes \hat{\imath}_2$	3	4	6	5	2	1
$\mathbf{Q}[180°, \hat{\imath}_1 + \hat{\imath}_2] = \hat{\imath}_1 \otimes \hat{\imath}_2 + \hat{\imath}_2 \otimes \hat{\imath}_1 - \hat{\imath}_3 \otimes \hat{\imath}_3$	4	3	2	1	5	6
$\mathbf{Q}[180°, \hat{\imath}_1 - \hat{\imath}_2] = -\hat{\imath}_1 \otimes \hat{\imath}_2 - \hat{\imath}_2 \otimes \hat{\imath}_1 - \hat{\imath}_3 \otimes \hat{\imath}_3$	3	4	1	2	5	6
$\mathbf{Q}[180°, \hat{\imath}_1 + \hat{\imath}_3] = \hat{\imath}_1 \otimes \hat{\imath}_3 - \hat{\imath}_2 \otimes \hat{\imath}_2 + \hat{\imath}_3 \otimes \hat{\imath}_1$	6	5	3	4	2	1
$\mathbf{Q}[180°, \hat{\imath}_1 - \hat{\imath}_3] = -\hat{\imath}_1 \otimes \hat{\imath}_3 - \hat{\imath}_2 \otimes \hat{\imath}_2 - \hat{\imath}_3 \otimes \hat{\imath}_1$	5	6	3	4	1	2
$\mathbf{Q}[180°, \hat{\imath}_2 + \hat{\imath}_3] = -\hat{\imath}_1 \otimes \hat{\imath}_1 + \hat{\imath}_2 \otimes \hat{\imath}_3 + \hat{\imath}_3 \otimes \hat{\imath}_2$	1	2	6	5	4	3
$\mathbf{Q}[180°, \hat{\imath}_2 - \hat{\imath}_3] = -\hat{\imath}_1 \otimes \hat{\imath}_1 - \hat{\imath}_2 \otimes \hat{\imath}_3 - \hat{\imath}_3 \otimes \hat{\imath}_2$	1	2	5	6	3	4
$\mathbf{Q}[180°, \hat{\imath}_1] = \hat{\imath}_1 \otimes \hat{\imath}_1 - \hat{\imath}_2 \otimes \hat{\imath}_2 - \hat{\imath}_3 \otimes \hat{\imath}_3$	1	2	4	3	6	5
$\mathbf{Q}[180°, \hat{\imath}_2] = -\hat{\imath}_1 \otimes \hat{\imath}_1 + \hat{\imath}_2 \otimes \hat{\imath}_2 - \hat{\imath}_3 \otimes \hat{\imath}_3$	2	1	3	4	6	5
$\mathbf{Q}[180°, \hat{\imath}_3] = -\hat{\imath}_1 \otimes \hat{\imath}_1 - \hat{\imath}_2 \otimes \hat{\imath}_2 + \hat{\imath}_3 \otimes \hat{\imath}_3$	2	1	4	3	5	6
$\mathbf{Q}[90°, \hat{\imath}_1] = \hat{\imath}_1 \otimes \hat{\imath}_1 - \hat{\imath}_2 \otimes \hat{\imath}_3 + \hat{\imath}_3 \otimes \hat{\imath}_2$	2	1	6	5	3	4
$\mathbf{Q}[-90°, \hat{\imath}_1] = \hat{\imath}_1 \otimes \hat{\imath}_1 + \hat{\imath}_2 \otimes \hat{\imath}_3 - \hat{\imath}_3 \otimes \hat{\imath}_2$	2	1	5	6	4	3
$\mathbf{Q}[90°, \hat{\imath}_2] = \hat{\imath}_1 \otimes \hat{\imath}_3 + \hat{\imath}_2 \otimes \hat{\imath}_2 - \hat{\imath}_3 \otimes \hat{\imath}_1$	5	6	4	3	2	1
$\mathbf{Q}[-90°, \hat{\imath}_2] = -\hat{\imath}_1 \otimes \hat{\imath}_3 + \hat{\imath}_2 \otimes \hat{\imath}_2 + \hat{\imath}_3 \otimes \hat{\imath}_1$	6	5	4	3	1	2
$\mathbf{Q}[90°, \hat{\imath}_3] = -\hat{\imath}_1 \otimes \hat{\imath}_2 + \hat{\imath}_2 \otimes \hat{\imath}_1 + \hat{\imath}_3 \otimes \hat{\imath}_3$	4	3	1	2	6	5
$\mathbf{Q}[-90°, \hat{\imath}_3] = \hat{\imath}_1 \otimes \hat{\imath}_2 - \hat{\imath}_2 \otimes \hat{\imath}_1 + \hat{\imath}_3 \otimes \hat{\imath}_3$	3	4	2	1	6	5

latter two stretches are equal, then the orthorhombic variants become the tetragonal variants considered in [3–5, 26, 43]. The symmetry transformation amongst the variants is given in Table I with the rotations from the cubic Laue group $\mathcal{L}^c$.

The microstructures for this cubic to orthorhombic transition are continuous deformations with piecewise constant gradient on the austenite and martensite energy wells. In particular, the austenite energy well $\mathcal{A}$ is SO(3), and the martensite energy wells $\mathcal{M}$ are

$$\mathcal{M} = \bigcup_{i=1}^{6} \bigl\{\mathbf{R}\mathbf{U}_i\colon \mathbf{R} \in SO(3) \quad \text{and} \quad \mathbf{U}_i \text{ variants of martensite (7)}\bigr\}. \tag{8}$$

The remainder of this article constructs various microstructures which are either commonly observed or have the potential to be observed in shape memory materials. In particular, Section 4 considers the twinned martensite microstructure between two variants of martensite. Section 5 studies the austenite–martensite microstructures, where three kinds are considered: those with a single variant of martensite; those with compound twins; and those with either type I or type II twins. Sections 6–8 consider the special microstructures, the wedge, triangle, and diamond, respectively, with all three of the possible austenite–martensite microstructures.

4. Twinned Martensite Microstructure

At or below the critical temperature θ_c, a specimen of shape memory alloy may consist of many regions each containing a different variant of the martensite phase. Further, it is found from experimental observations that any two adjacent regions form a twin. A twin is neighboring regions of the same crystal lattice separated by a plane and related by a simple shear, where the surface of separation is called the twin plane (for example, [20, 23, 24, 32, 39, 50]). Macroscopically, a twinned martensite microstructure is a continuous deformation with gradient in two distinct martensite wells.

Any two distinct martensite variants $\mathbf{U}_i$ and $\mathbf{U}_j$ form a variant pair denoted as $(i : j)$, and the compatibility equation (4) between them, called the twinning equation, can be written as

$$\mathbf{R}_{ij}\mathbf{U}_i - \mathbf{U}_j = \mathbf{a} \otimes \hat{\mathbf{n}}, \tag{9}$$

where $\mathbf{R}_{ij}$ is the twin rotation, $\mathbf{a}$ is parallel to the twin shear, and $\mathbf{U}_j^{-1}\hat{\mathbf{n}}$ is parallel to the twin plane normal; also, the magnitude of the twin shear s is given by $|\mathbf{U}_j^{-1}\hat{\mathbf{n}}||\mathbf{a}|$. In particular, there are thirty possible variant pairs between the six orthorhombic variants, and these variant pairs can be segregated into two sets:

$$S_1 = \begin{pmatrix} (1:2) \\ (3:4) \\ (5:6) \end{pmatrix} \quad \text{and} \quad S_2 = \begin{pmatrix} (1:3) & (1:4) & (1:5) & (1:6) \\ (2:3) & (2:4) & (2:5) & (2:6) \\ (3:5) & (3:6) & (4:5) & (4:6) \end{pmatrix},$$

where it is to be understood that if the variant pair $(i : j)$ is contained in a particular set, then so is the variant pair $(j : i)$. In either set S_1 or S_2, the variant pairs are all related to each other by symmetry transformations using the rotations in $\mathcal{L}^c$, and there is no symmetry transformation in $\mathcal{L}^c$ relating a variant pair in one of the sets to any variant pair in the other.

Solutions to the twinning equation (9) with any variant pair in sets S_1 and S_2 can be found from Proposition 1. However, for every variant pair $(i : j)$, there is at least one 180 degree rotation $\mathbf{Q}$ in $\mathcal{L}^c$ such that $\mathbf{U}_j = \mathbf{Q}\mathbf{U}_i\mathbf{Q}$ as given in Table I. Thus, all of the variant pairs with these orthorhombic variants have solutions to the twinning equation (9) from a proposition of Bhattacharya [5] (also [20, 23]). This latter proposition provides explicit expressions for the unknown vectors and rotation in the twinning equation.

4.1. COMPOUND TWINS

For the first variant pair $(i : j) = (1 : 2)$ in the set S_1, the twinning equation (9) has unknowns $\mathbf{R}_{12}$, $\mathbf{a}$, and $\hat{\mathbf{n}}$. In particular, it can be shown that one solution to equation (9) has twin shear $\mathbf{a}^{\mathrm{I}}$ and twin plane normal $\hat{\mathbf{n}}^{\mathrm{I}}$ with components

$$a^{\mathrm{I}} = \frac{\alpha^2 - \gamma^2}{\alpha^2 + \gamma^2}\begin{pmatrix} 0 \\ \gamma - \alpha \\ \alpha + \gamma \end{pmatrix} \quad \text{and} \quad \hat{n}^{\mathrm{I}} = \begin{pmatrix} 0 \\ 1 \\ 0 \end{pmatrix}, \tag{10a}$$

respectively, and twin rotation $\mathbf{R}_{12}^{\mathrm{I}}$ with components

$$R_{12}^{\mathrm{I}} = \begin{pmatrix} 1 & 0 & 0 \\ 0 & \dfrac{2\alpha\gamma}{\alpha^2+\gamma^2} & \dfrac{\gamma^2-\alpha^2}{\alpha^2+\gamma^2} \\ 0 & \dfrac{\alpha^2-\gamma^2}{\alpha^2+\gamma^2} & \dfrac{2\alpha\gamma}{\alpha^2+\gamma^2} \end{pmatrix}. \tag{10b}$$

The other solution to the twinning equation (9) has twin shear $\mathbf{a}^{\mathrm{II}}$ and twin plane normal $\hat{\mathbf{n}}^{\mathrm{II}}$ with components

$$a^{\mathrm{II}} = \frac{\alpha^2 - \gamma^2}{\alpha^2 + \gamma^2}\begin{pmatrix} 0 \\ \alpha + \gamma \\ \gamma - \alpha \end{pmatrix} \quad \text{and} \quad \hat{n}^{\mathrm{II}} = \begin{pmatrix} 0 \\ 0 \\ 1 \end{pmatrix}, \tag{11}$$

respectively; while $\mathbf{R}_{12}^{\mathrm{II}} = (\mathbf{R}_{12}^{\mathrm{I}})^{\mathrm{T}}$. The magnitude of the twin shear s is the same for both twin solutions and is given by $s = |\alpha^2 - \gamma^2|/(\alpha\gamma)$, which is a function of the orthorhombic lattice parameters only.

A twin is commonly defined by its twinning elements (see for example [24, 32, 42]). The specific elements are: the twin plane K_1; the direction of the twin shear η_1; the second undistorted plane K_2; and an undistorted direction η_2 in the plane K_2. On following [26, 50], the twin solution $(\mathbf{R}_{12}^{\mathrm{I}}, \mathbf{a}^{\mathrm{I}}, \hat{\mathbf{n}}^{\mathrm{I}})$ in equation (10) has twinning elements, relative to the cubic basis, $(K_1^{\mathrm{I}}, \eta_1^{\mathrm{I}}, K_2^{\mathrm{I}}, \eta_2^{\mathrm{I}}) = ((010), [001], (001), [010])$, respectively; while, the solution $(\mathbf{R}_{12}^{\mathrm{II}}, \mathbf{a}^{\mathrm{II}}, \hat{\mathbf{n}}^{\mathrm{II}})$ in equation (11) has twinning elements, relative to the cubic basis, $(K_1^{\mathrm{II}}, \eta_1^{\mathrm{II}}, K_2^{\mathrm{II}}, \eta_2^{\mathrm{II}}) = ((001), [010], (010), [001])$, respectively. Because all of the twinning elements are rational, that is, along crystallographic directions, both of these twin solutions are called compound twins. Further, the twin solutions in equations (10) and (11) are related to each other by a rotation in $\mathcal{L}^{c}$; and the twin solutions for any other variant pair in set S_1 can be found from those for the variant pair $(1:2)$ by a symmetry transformation using the rotations in $\mathcal{L}^{c}$ (see [26]). There are a total of twelve compound twins.

4.2. TYPE I AND TYPE II TWINS

For the first variant pair $(i : j) = (1 : 3)$ in the set S_2, the twinning equation (9) has the unknowns $\mathbf{R}_{13}$, $\mathbf{a}$, and $\hat{\mathbf{n}}$. In this case, it can be shown that one solution to equation (9) has twin shear $\mathbf{a}^{\mathrm{I}}$ with components

$$a^{\mathrm{I}} = \frac{\sqrt{2}}{4\alpha^2\gamma^2 + 2\beta^2(\alpha^2+\gamma^2)} \begin{pmatrix} (\alpha+\gamma)(4\alpha\gamma\beta^2 - 2\alpha^2\gamma^2 - \beta^2(\alpha^2+\gamma^2)) \\ 2\beta(\beta^2(\alpha^2+\gamma^2) - 2\alpha^2\gamma^2) \\ (\gamma-\alpha)(4\alpha\gamma\beta^2 + 2\alpha^2\gamma^2 + \beta^2(\alpha^2+\gamma^2)) \end{pmatrix}, \quad (12)$$

while the twin plane normal $\hat{\mathbf{n}}^{\mathrm{I}}$ has components

$$\hat{n}^{\mathrm{I}} = \frac{1}{\sqrt{2}} \begin{pmatrix} 1 \\ -1 \\ 0 \end{pmatrix}. \quad (13)$$

The twin rotation is $\mathbf{R}^{\mathrm{I}}_{13} = -\mathbf{I} + (2/|\mathbf{U}_3^{-1}\hat{\mathbf{n}}^{\mathrm{I}}|^2)(\mathbf{U}_3^{-1}\hat{\mathbf{n}}^{\mathrm{I}} \otimes \mathbf{U}_3^{-1}\hat{\mathbf{n}}^{\mathrm{I}})\mathbf{Q}$, where $\mathbf{Q} = \mathbf{Q}[180^\circ, \hat{\mathbf{p}}] \in \mathcal{L}^c$ such that $\mathbf{U}_3 = \mathbf{Q}\mathbf{U}_1\mathbf{Q}$ (see [10, 23, 26]). Another solution to the twinning equation (9) has the twin shear $\mathbf{a}^{\mathrm{II}}$ with components

$$a^{\mathrm{II}} = \frac{\sqrt{8\beta^2(\beta^2-\alpha^2-\gamma^2) + 6\alpha^4 - 4\alpha^2\gamma^2 + 6\gamma^4}}{2(\alpha^2+\gamma^2+2\beta^2)} \begin{pmatrix} \alpha+\gamma \\ -2\beta \\ \alpha-\gamma \end{pmatrix} \quad (14)$$

and the twin plane normal $\hat{\mathbf{n}}^{\mathrm{II}}$ with components

$$\hat{n}^{\mathrm{II}} = \frac{1}{\sqrt{8\beta^2(\beta^2-\alpha^2-\gamma^2) + 6\alpha^4 - 4\alpha^2\gamma^2 + 6\gamma^4}} \begin{pmatrix} 2\beta^2-\alpha^2-\gamma^2 \\ 2\beta^2-\alpha^2-\gamma^2 \\ 2(\gamma^2-\alpha^2) \end{pmatrix}. \quad (15)$$

Here, the twin rotation is $\mathbf{R}^{\mathrm{II}}_{13} = -\mathbf{I} + (2/|\mathbf{U}_3\hat{\mathbf{n}}^{\mathrm{I}}|^2)(\mathbf{U}_3\hat{\mathbf{n}}^{\mathrm{I}} \otimes \mathbf{U}_3\hat{\mathbf{n}}^{\mathrm{I}})\mathbf{Q}$ with the rotation $\mathbf{Q}$ as above. Further, the magnitude of the twin shear s is the same for both twin solutions and can be written as

$$s = \frac{\sqrt{(2\alpha^2\gamma^2 + \beta^2(\alpha^2+\gamma^2))(\alpha^2+2\beta^2+\gamma^2) - 16\alpha^2\beta^2\gamma^2}}{2\alpha\beta\gamma}.$$

The first twin solution $(\mathbf{R}^{\mathrm{I}}_{13}, \mathbf{a}^{\mathrm{I}}, \hat{\mathbf{n}}^{\mathrm{I}})$ has twinning elements, relative to the cubic basis, $(K^{\mathrm{I}}_1, \eta^{\mathrm{I}}_1, K^{\mathrm{I}}_2, \eta^{\mathrm{I}}_2)$ given by $(\hat{\mathbf{n}}^{\mathrm{I}}, (\mathbf{U}_3^{-1}\mathbf{a}^{\mathrm{I}})/|\mathbf{U}_3^{-1}\mathbf{a}^{\mathrm{I}}|, \hat{\mathbf{n}}^{\mathrm{II}}, \hat{\mathbf{n}}^{\mathrm{I}})$, respectively. This twin solution is a type I twin because the twinning elements K^{I}_1 and η^{I}_2 are rational (crystallographic directions) and the twinning elements K^{I}_2 and η^{I}_1 are irrational (non-crystallographic directions). For the second twin solution $(\mathbf{R}^{\mathrm{II}}_{13}, \mathbf{a}^{\mathrm{II}}, \hat{\mathbf{n}}^{\mathrm{II}})$, the twinning elements, also relative to the cubic basis, $(K^{\mathrm{II}}_1, \eta^{\mathrm{II}}_1, K^{\mathrm{II}}_2, \eta^{\mathrm{II}}_2)$ are given by $(\hat{\mathbf{n}}^{\mathrm{II}}, \hat{\mathbf{n}}^{\mathrm{I}}, \hat{\mathbf{n}}^{\mathrm{I}}, (\mathbf{U}_3^{-1}\mathbf{a}^{\mathrm{I}})/|\mathbf{U}_3^{-1}\mathbf{a}^{\mathrm{I}}|)$, respectively. This twin solution is a type II twin because the twinning elements K^{II}_2 and η^{II}_1 are rational and the twinning elements K^{II}_1 and η^{II}_2 are irrational. It is straight forward to show that if the transformation stretches α and γ are equal, then the above solutions degenerate into the twins for the cubic to tetragonal transition [3, 5, 8, 26, 43]. Moreover, the type I (II) twin solution for any other variant pair in set S_2 is related to the type I (II) solution for variant pair (1 : 3) by a symmetry transformation in $\mathcal{L}^c$; and, the type I and type II solutions for any variant pair in S_2 are not related by a symmetry transformation in $\mathcal{L}^c$. There are a total of twenty-four type I and twenty-four type II twins.

More complicated microstructures with more than two variants of martensite may be considered. For example, Bhattacharya [10] studies a microstructure which forms when twins cross. In particular, it is shown that several kinds of twin crossings are possible for this transition, and a number of these microstructures have been observed in the experiments of Chu and James [14, 15]; while, Chu [13] (also [1]) has studied other complicated microstructures using only two variants of martensite; while, general microstructures with the orthorhombic variants are considered in [11].

Finally, it is interesting to contrast the twins found above with the orthorhombic "mixed axes" variants to the twins which are possible using the orthorhombic "cubic axes" variants. It is straight-forward to show that not all "cubic axes" variants are rank-1 connected with every other variant unless the symmetry is reduced further to tetragonal; that is, $\alpha = \gamma$. Also, all twins are compound.

5. Austenite–Martensite Microstructure

The austenite and martensite phases may coexist in a body when the specimen is either at the critical temperature θ_c or above it, but under applied stress. The austenite–martensite microstructures are the mechanism through which the transformation between the phases occurs. Four kinds of austenite–martensite microstructures are considered in this section: an austenite–martensite microstructure with a single variant of martensite and compound, type I, and type II twinned martensites.

5.1. AUSTENITE-SINGLE VARIANT OF MARTENSITE MICROSTRUCTURE

If a single variant of martensite and austenite are to form a continuous deformation, then from Proposition 1, the martensite variant must have an eigenvalue greater than one, an eigenvalue less than one, and an eigenvalue equal to one. In particular, the variant $\mathbf{U}_1$ from equation (7) has eigenvalue–eigenvector pairs $(\lambda_1, \hat{\mathbf{e}}_1)$, $(\lambda_2, \hat{\mathbf{e}}_2)$, and $(\lambda_3, \hat{\mathbf{e}}_3)$, where the specific eigenvalues are $\lambda_1 = \beta$, $\lambda_2 = \alpha$, and $\lambda_3 = \gamma$ and the components of the eigenvectors are

$$\hat{e}_1 = \begin{pmatrix} 1 \\ 0 \\ 0 \end{pmatrix}, \quad \hat{e}_2 = \frac{1}{\sqrt{2}} \begin{pmatrix} 0 \\ 1 \\ 1 \end{pmatrix}, \quad \hat{e}_3 = \frac{1}{\sqrt{2}} \begin{pmatrix} 0 \\ -1 \\ 1 \end{pmatrix}, \tag{16}$$

respectively. Two cases arise which are considered separately below.

In the first case, let $\alpha = 1$ and choose $\beta < 1$ and $\gamma > 1$. The compatibility equation (4) for this microstructure* with the variant $\mathbf{U}_1$ is written as

$$\mathbf{R}\mathbf{U}_1 - \mathbf{I} = \mathbf{b} \otimes \widehat{\mathbf{m}}, \tag{17}$$

where the unknowns are the rotation $\mathbf{R}$, the shape strain $\mathbf{b}$, and the habit plane normal $\widehat{\mathbf{m}}$. The habit plane is the plane between the untransformed austenite region and the transformed martensite region; while, the shape strain gives the magnitude and sense of the transformation. Equation (17) is called the habit plane equation. From Proposition 1, it is found that at most two solutions exist to this habit plane equation, where the components of the shape strain $\mathbf{b}$ and the habit plane normal $\widehat{\mathbf{m}}$ are

$$b = \sqrt{\frac{\gamma - \beta}{\gamma + \beta}} \begin{pmatrix} \gamma\sqrt{1-\beta^2} \\ -\kappa\beta\sqrt{\gamma^2-1}/\sqrt{2} \\ \kappa\beta\sqrt{\gamma^2-1}/\sqrt{2} \end{pmatrix} \tag{18a}$$

and

$$\widehat{m} = \frac{1}{\sqrt{\gamma^2 - \beta^2}} \begin{pmatrix} -\sqrt{1-\beta^2} \\ -\kappa\sqrt{\gamma^2-1}/\sqrt{2} \\ \kappa\sqrt{\gamma^2-1}/\sqrt{2} \end{pmatrix}, \tag{18b}$$

respectively, where $\kappa = \pm 1$. The shape strain $\mathbf{b}$ and the habit plane $\widehat{\mathbf{m}}$ are both perpendicular to the eigenvector $\hat{\mathbf{e}}_2$ in equation (16), and thus, this vector is also parallel to the axis of the rotation $\mathbf{R}$. The angle Θ of this rotation is found from $\cos(\Theta) = (1 + \beta\gamma)/(\beta + \gamma)$. Similarly, solutions with the stretches $\beta > 1$ and $\gamma < 1$ can be found as well; and one unique solution exists to the habit plane equation (17) if either of the stretches β or γ equals one. On the other hand, with the stretch $\gamma = 1$, solutions to the habit plane equation (17) exist for $\alpha < 1, \beta > 1$ and $\alpha > 1, \beta < 1$. The shape strain and habit plane normal vectors are very similar to the ones given in equations (18).

For all of the cases above, it can be shown that the two solutions to equation (17) with $\kappa = +1$ and $\kappa = -1$ for any orthorhombic variant are related by a symmetry transformation using a rotation in $\mathcal{L}^c$. Thus, there are six orthorhombic variants and each variant has two solutions to the habit plane equation (17) for a total of twelve solutions. All of these solutions are related by a symmetry transformation using the rotations in $\mathcal{L}^c$.

In the second case, let $\beta = 1$ and choose $\alpha < 1$ and $\gamma > 1$. The habit plane equation is the same as equation (17) with the variant $\mathbf{U}_1$ and the same unknowns.

* Strictly speaking this is not a microstructure since there is no internal length scale, but we abuse the terminology for convenience. It would be more appropriate to call it an interface.

Solutions to this equation are the shape strain **b** and the habit plane normal $\widehat{\mathbf{m}}$ with components

$$b = \frac{1}{\sqrt{2}}\sqrt{\frac{\gamma-\alpha}{\gamma+\alpha}}\begin{pmatrix} 0 \\ \gamma\sqrt{1-\alpha^2}-\kappa\alpha\sqrt{\gamma^2-1} \\ \gamma\sqrt{1-\alpha^2}+\kappa\alpha\sqrt{\gamma^2-1} \end{pmatrix} \tag{19a}$$

and

$$\widehat{m} = \frac{1}{\sqrt{2(\gamma^2-\alpha^2)}}\begin{pmatrix} 0 \\ -\sqrt{1-\alpha^2}-\kappa\sqrt{\gamma^2-1} \\ -\sqrt{1-\alpha^2}+\kappa\sqrt{\gamma^2-1} \end{pmatrix}, \tag{19b}$$

respectively, where $\kappa = \pm 1$. The axis of the rotation **R** is parallel to the eigenvector $\hat{\mathbf{e}}_1$ in equation (16), and the angle Θ of rotation is found from $\cos(\Theta) = (1+\alpha\gamma)/(\alpha+\gamma)$. Alternatively, if the transformation stretches are chosen such that $\alpha > 1$ and $\gamma < 1$, then the shape strain and the habit plane normal of the solution to equation (17) are found by interchanging α and γ in equations (19) and by multiplying the second component in each vector by -1. Also, one unique solution exists to the habit plane equation (17) even with a stretch α or γ equal to one. Furthermore, the two solutions from equations (19) to the habit plane equation (17) with $\kappa = +1$ and $\kappa = -1$ are related by a symmetry transformation using a rotation in $\mathcal{L}^c$. Therefore, a total of twelve solutions exist to the habit plane equation (17) using the six orthorhombic variants, and as above, all of the solutions are related by a symmetry transformation in $\mathcal{L}^c$.

In order for an alloy to exhibit the austenite-single variant of martensite microstructure, it must have special transformation stretches. In fact, two alloys Ti–Ta [12] and Ti–Ni–Cu [35, 45–47] transform via this microstructure. In both cases, the composition of the alloys was changed until the required special stretches were reached.

5.2. AUSTENITE-TWINNED MARTENSITE MICROSTRUCTURE

The austenite-twinned martensite microstructure consists of a region of undeformed austenite adjacent to a region of twinned martensite. The twinned martensite region consists of fine layers containing alternately two variants of martensite. In [3], it is shown that in order for such a deformation to be continuous, a transition layer is needed between the austenite and twinned martensite regions. This transition layer necessarily involves deformations with gradient away from the energy wells, and, therefore, it is not an energy minimizer. If, however, the width of the twins and the width of the transition layer are both scaled by $1/k$, then, in the limit as $k \to \infty$, the width of the twins and transition layer both go to zero, and the deformation is continuous and uses gradients only on the energy wells. Following [3], the compatibility equations for the austenite-twinned martensite microstructure for the variant pair $(i : j)$ are the twinning equation (9) and the habit plane equation

$$\overline{\mathbf{R}}_{ij}\big(\lambda \mathbf{R}_{ij}\mathbf{U}_i + (1-\lambda)\mathbf{U}_j\big) - \mathbf{I} = \mathbf{b} \otimes \widehat{\mathbf{m}}, \tag{20}$$

which gives compatibility between austenite and the weighed average of two martensite variants which satisfy the twinning equation. The solutions to the twinning equation (9) are given in Section 4, and substituting this equation into equation (20) gives

$$\overline{\mathbf{R}}_{ij}\big(\mathbf{U}_j + \lambda \mathbf{a} \otimes \hat{\mathbf{n}}\big) - \mathbf{I} = \mathbf{b} \otimes \widehat{\mathbf{m}}, \tag{21}$$

where the unknowns are the rotation $\overline{\mathbf{R}}_{ij}$, the volume fraction λ of variant $\mathbf{U}_i$, the shape strain $\mathbf{b}$, and the habit plane normal $\widehat{\mathbf{m}}$. From [3], necessary and sufficient conditions that equation (21) has solutions are;

$$\delta = \mathbf{a} \cdot \mathbf{U}_j\big(\mathbf{U}_j^2 - \mathbf{I}\big)^{-1}\hat{\mathbf{n}} \leqslant -2 \tag{22}$$

and

$$\eta = \mathrm{tr}\big(\mathbf{U}_j^2\big) - \det\big(\mathbf{U}_j^2\big) - 2 + \frac{|\mathbf{a}|^2}{2\delta} \geqslant 0, \tag{23}$$

where, in equation (23), the functions $\mathrm{tr}(\cdot)$ and $\det(\cdot)$ are the trace and the determinant, respectively. Equation (22) insures that an eigenvalue of the symmetric stretch $\mathbf{C}_{ij}(\lambda) = (\mathbf{U}_j + \lambda \mathbf{a} \otimes \hat{\mathbf{n}})^{\mathrm{T}}(\mathbf{U}_j + \lambda \mathbf{a} \otimes \hat{\mathbf{n}})$ equals one; while, equation (23) gives that of the remaining eigenvalues, one is less than one, and the other is greater than one. Furthermore, the volume fraction λ of variant $\mathbf{U}_i$ is given by

$$\lambda = \frac{1}{2}\left(1 - \sqrt{1 + \frac{2}{\delta}}\right), \tag{24}$$

where the function δ is found from equation (22). There are three possible austenite-twinned martensite microstructures: the martensite region may contain either compound, type I or type II twins. The case where the martensite region contains compound twins is considered separately from the case where the region contains either type I or type II twins, which are considered together.

5.2.1. *Austenite-compound twinned martensite microstructure*

For the variant pair $(i : j) = (1 : 2)$ in set S_1, the compatibility equations for the austenite-twinned martensite microstructure are the twinning equation (9) and the habit plane equation (21). The solutions to the twinning equation (9) are given in Section 4.1, where it is shown that compound twins exist for all choices of the transformation stretches. With these twin solutions, it can be shown that solutions exist to equation (21), that is, conditions (22) and (23) are satisfied, if and only if, either

$$\beta < 1, \qquad (\alpha - 1)(\gamma - 1) < 0, \quad \text{and} \quad \frac{1}{\alpha^2} + \frac{1}{\gamma^2} \leqslant 2 \tag{25}$$

or

$$\beta > 1, \qquad (\alpha - 1)(\gamma - 1) < 0, \quad \text{and} \quad \alpha^2 + \gamma^2 \leqslant 2, \tag{26}$$

(also [5]). So, there are two regions in the space of transformation stretches α, β, and γ in which the austenite-compound twinned martensite microstructure is possible. These regions are shown below in Figures 1 and 2 as the shaded region between the dashed curve and the lines $\alpha = 1$ and $\gamma = 1$. Furthermore, if the transformation stretches fall within either region (25) or region (26), then the volume fraction λ of variant $\mathbf{U}_i$ is given by

$$\lambda = \frac{1}{2}\left(1 - \sqrt{\frac{(\alpha^2 + \gamma^2 - 2)(2\alpha^2\gamma^2 - \alpha^2 - \gamma^2)}{(\alpha^2 - \gamma^2)^2}}\right). \tag{27}$$

The volume fraction λ of variant $\mathbf{U}_i$ is a half if equality holds in either case.

In order to find the solutions to equation (21), the symmetric stretch $\mathbf{C}_{12}(\lambda) = (\mathbf{U}_2 + \lambda\mathbf{a} \otimes \hat{\mathbf{n}})^{\mathrm{T}}(\mathbf{U}_2 + \lambda\mathbf{a} \otimes \hat{\mathbf{n}})$ with the volume fraction λ of variant $\mathbf{U}_1$ given by equation (27) and the vectors $\mathbf{a}$ and $\hat{\mathbf{n}}$ given by equations (10) is composed, and then its eigenvalues and eigenvectors are found in order to be able to use Proposition 1. For case (25), the components of the shape strain $\mathbf{b}$ and habit plane normal $\widehat{\mathbf{m}}$ are

$$b = \sqrt{\frac{\alpha\gamma - \beta}{\alpha\gamma + \beta}}\begin{pmatrix} \alpha\gamma\sqrt{1-\beta^2} \\ \kappa\beta\sqrt{2\alpha^2\gamma^2 - \alpha^2 - \gamma^2}/\sqrt{2} \\ -\chi\kappa\beta\sqrt{\alpha^2 + \gamma^2 - 2}/\sqrt{2} \end{pmatrix} \tag{28a}$$

and

$$\widehat{m} = \frac{1}{\sqrt{\alpha^2\gamma^2 - \beta^2}}\begin{pmatrix} -\sqrt{1-\beta^2} \\ \kappa\sqrt{2\alpha^2\gamma^2 - \alpha^2 - \gamma^2}/\sqrt{2} \\ -\chi\kappa\sqrt{\alpha^2 + \gamma^2 - 2}/\sqrt{2} \end{pmatrix}, \tag{28b}$$

respectively, where $\chi = \mathrm{sign}(\alpha - \gamma)$ and $\kappa = \pm 1$. The rotation $\overline{\mathbf{R}}_{12}$ is found by direct substitution of the shape strain and habit plane normal from equations (28) into equation (21). On the other hand, for case (26), the components of the shape strain $\mathbf{b}$ and habit plane normal $\widehat{\mathbf{m}}$ are

$$b = \sqrt{\frac{\beta - \alpha\gamma}{\beta + \alpha\gamma}}\begin{pmatrix} \kappa\alpha\gamma\sqrt{\beta^2 - 1} \\ \beta\sqrt{\alpha^2 + \gamma^2 - 2\alpha^2\gamma^2}/\sqrt{2} \\ \chi\beta\sqrt{2 - \alpha^2 - \gamma^2}/\sqrt{2} \end{pmatrix} \tag{29a}$$

and

$$\widehat{m} = \frac{1}{\sqrt{\beta^2 - \alpha^2\gamma^2}}\begin{pmatrix} \kappa\sqrt{\beta^2 - 1} \\ -\sqrt{\alpha^2 + \gamma^2 - 2\alpha^2\gamma^2}/\sqrt{2} \\ -\chi\sqrt{2 - \alpha^2 - \gamma^2}/\sqrt{2} \end{pmatrix}, \tag{29b}$$

respectively, where $\chi = \text{sign}(\alpha - \gamma)$ and $\kappa = \pm 1$. The rotation $\overline{\mathbf{R}}_{12}$ is found as above.

Each of the variant pairs $(1 : 2)$, $(3 : 4)$, and $(5 : 6)$ can have up to eight solutions to the habit plane equation (21): two for each twin solution with volume fraction λ of variant $\mathbf{U}_i$ and two for each twin solution with volume fraction $1 - \lambda$ of variant $\mathbf{U}_i$ replacing λ in equation (21). Also, variant pairs $(2 : 1)$, $(4 : 3)$, and $(6 : 5)$ have the same habit plane solutions, vectors $\mathbf{b}$ and $\widehat{\mathbf{m}}$, as the variant pairs $(1 : 2)$, $(3 : 4)$, and $(5 : 6)$, respectively. Further, the two solutions from either equations (28) or (29) to the habit plane equation (21) with $\kappa = +1$ and $\kappa = -1$ are related by a symmetry transformation using a rotation in $\mathcal{L}^c$. Therefore, twenty-four solutions to the habit plane equation (21) with compound twins exists, and each solution is related to every other solution by a symmetry transformation using the rotations in $\mathcal{L}^c$ (see [26]).

For most choices of the transformation stretches, the austenite-compound twinned martensite microstructure exists for only specific values of the volume fraction λ of variant $\mathbf{U}_i$ as given in equation (27). On the other hand, this microstructure may be possible for all volume fractions, but with special stretches. In particular, it can be shown that the symmetric stretch $\mathbf{C}_{12}(\lambda) = (\mathbf{U}_2 + \lambda \mathbf{a} \otimes \hat{\mathbf{n}})^{\mathrm{T}}(\mathbf{U}_2 + \lambda \mathbf{a} \otimes \hat{\mathbf{n}})$ has an eigenvalue $\lambda_2 = 1$ if, and only if, either equation (27) holds or the transformation stretch $\beta = 1$. Assume that the latter is true and, further, define the following function:

$$\begin{aligned} \eta \stackrel{\text{def}}{=} (1 - \lambda_1)(\lambda_3 - 1) &= \operatorname{tr}\big(\mathbf{C}_{12}(\lambda)\big) - \det\big(\mathbf{C}_{12}(\lambda)\big) - 2, \\ &= \big(\alpha^2 - 1\big)\big(1 - \gamma^2\big) - 2\lambda(1 - \lambda)\frac{(\alpha^2 - \gamma^2)^2}{(\alpha^2 + \gamma^2)}, \end{aligned} \tag{30}$$

in particular. The remaining two eigenvalues of $\mathbf{C}_{12}(\lambda)$ are ordered as $\lambda_1 \leqslant 1$ and $\lambda_3 \geqslant 1$ if, and only if, the function η in equation (30) is greater than or equal to zero; in fact, from equation (30), the transformation stretches α and γ must lie in either the sector $\alpha \geqslant 1$ and $\gamma \leqslant 1$ or the sector $\alpha \leqslant 1$ and $\gamma \geqslant 1$ for all volume fractions λ. Clearly, there are many possible habit planes with the stretch $\beta = 1$ and compound twins, and the shape strains and habit plane normals can be found from Proposition 2.2. All such vectors lie in the $\{100\}$ family of planes.

5.2.2. *Austenite type I and II twinned martensite microstructure*

As above, the compatibility equations for the austenite-twinned martensite microstructure with any variant pair $(i : j)$ in set S_2 are the twinning equation (9) and the habit plane equation (21). For any such variant pair $(i : j)$ with either the type I twin solution in equations (12) and (13) or the type II twin solution in equations (14) and (15), the analytic expressions for the functions δ from equation (22) and η from equation (23) are too long and difficult to analyze algebraically. In order to estimate the region in the space of transformation stretches α, β, and γ in which conditions (22) and (23) are satisfied, a numerical computation is performed.

For fixed stretch β, the numerical program searches through values of the other two stretches α and γ noting those points where the conditions (22) and (23) are satisfied. It is, however, found that part of the boundary determined from the numerical computation is given by functions which can be determined analytically. For the variant pair $(i : j) = (1 : 3)$, the surfaces are found by setting the function δ from equation (22) identically equal to -2. For each twin type, there are two functions for which equality holds in equation (22) (see [25] for details). The results of all of these numerical computations are presented in Figure 1 with type I twins and Figure 2 with type II twins. Also included in these figures is the domain in which an austenite-compound twinned martensite microstructure is possible; this is shown as a shaded region bounded by the dashed curve and the lines $\alpha = 1$ and $\gamma = 1$. For each fixed stretch β in Figures 1 and 2, the admissible domain in which an austenite-twinned martensite microstructure is possible is the region between the solid curve for that fixed β and the lines $\alpha = 1$ and $\gamma = 1$. Another feature of some curves in these figures are curves with corners. For example, the solid curve in Figure 1 for the transformation stretch $\beta = 0.95$ is made up of two parts from the two curves along which the function δ is identically -2 with type I twins, and the intersection of these two curves gives rise to the corners in this figure. For Figure 2, it is the intersection of the corresponding curves with type II twins which creates the corners.

Moreover, if the symmetric stretch $\mathbf{C}_{13}(\lambda)$ is written as $(\mathbf{U}_3 + \lambda\mathbf{a} \otimes \hat{\mathbf{n}})^T(\mathbf{U}_3 + \lambda\mathbf{a} \otimes \hat{\mathbf{n}})$, then it can be shown that habit plane solutions exist even if a transformation stretch is one. In particular, with the type I twins from equations (12) and (13), the symmetric stretch $\mathbf{C}_{13}(\lambda)$ has an eigenvalue less than one, greater than one, and equal to one for all volume fractions λ if, and only if, either $\alpha = 1$ and $\gamma = \beta/\sqrt{3\beta^2 - 2}$ or $\beta = 1$ and $\gamma = \alpha/\sqrt{2\alpha^2 - 1}$. Also, the same is true with the type II twins from equations (14) and (15) if, and only if, either $\alpha = 1$ and $\gamma = \sqrt{3 - 2\beta^2}$ or $\beta = 1$ and $\gamma = \sqrt{2 - \alpha^2}$. It is the points along these curves which are shown in Figures 1 and 2.

The algebra involved in determining solutions to the habit plane equation (21) with either type I or type II twins is prohibitively complex; so, solutions must be determined numerically. An algorithm to compute the solutions to the twinning equation (9) and the habit plane equation (21) is outlined in [28]. On the other hand, there are up to four possible categories of twinned habit plane solutions for the variant pairs in set S_2 : (THa) type I twins and $\kappa = +1$; (THb) type I twins and $\kappa = -1$; (THc) type II twins and $\kappa = +1$; and (THd) type II twins and $\kappa = -1$. There can be at most 96 solutions to the habit plane equation with the variant pairs in set S_2 : 48 with type I twins and 48 with type II twins. Each of the categories above, (THa), (THb), (THc), and (THd), has at most 24 solutions. All twenty-four habit plane solutions in category (THa) are related by a symmetry transformation using the rotations in $\mathcal{L}^c$. Similarly, the solutions in categories (THb), (THc), and (THd) are also related by using the rotations in $\mathcal{L}^c$ (see [26]). In general, there is no symmetry transformation in the group of rotations $\mathcal{L}^c$ which relates the habit

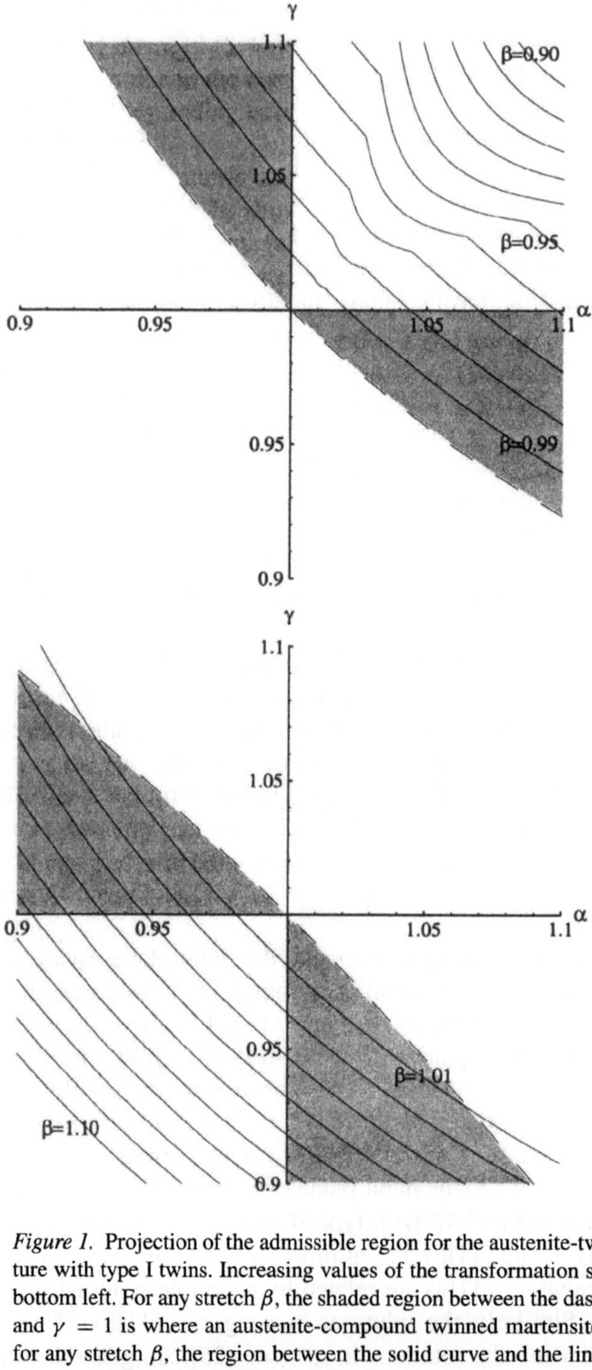

Figure 1. Projection of the admissible region for the austenite-twinned martensite microstructure with type I twins. Increasing values of the transformation stretch β are from top right to bottom left. For any stretch β, the shaded region between the dashed curve and the lines $\alpha = 1$ and $\gamma = 1$ is where an austenite-compound twinned martensite microstructure exists. Also, for any stretch β, the region between the solid curve and the lines $\alpha = 1$ and $\gamma = 1$ is where an austenite-twinned martensite microstructure exists with type I twins.

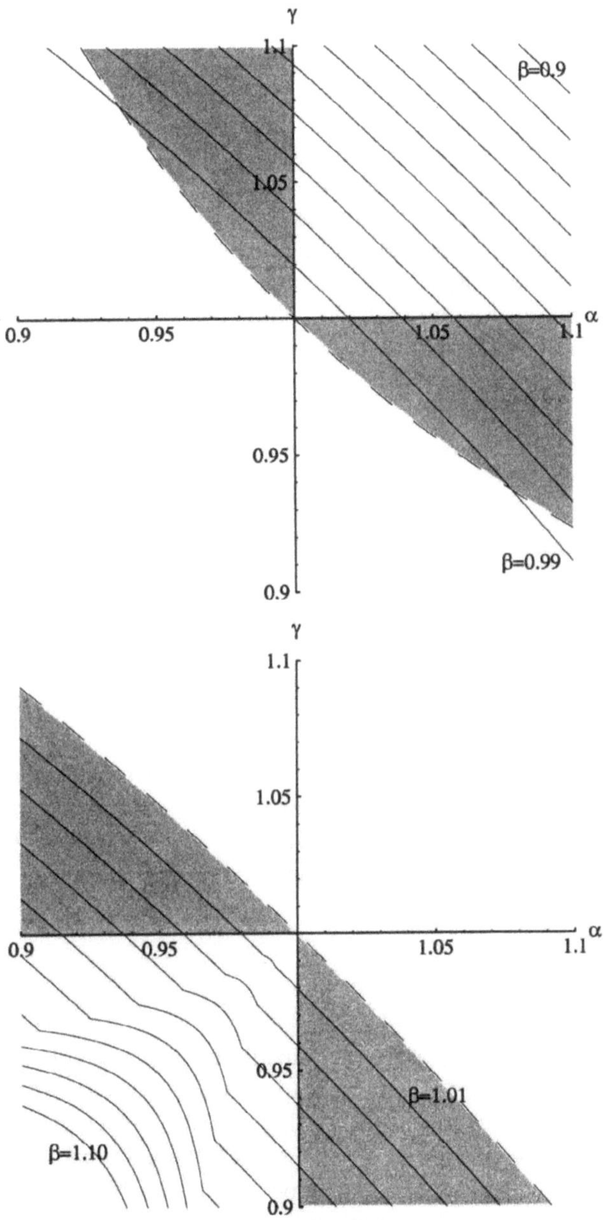

Figure 2. Projection of the admissible region for the austenite-twinned martensite microstructure with type II twins. See caption to Figure 1 for details.

plane solutions in either categories (THa) and (THb) or (THc) and (THd); and a habit plane solution with type I twins is not related to a habit plane solution with type II twins by any rotation.

6. Wedge Microstructure

The remaining microstructures considered in this article are the so-called special microstructures, the wedge, triangle, and diamond. All of these microstructures involve fitting multiple austenite–martensite microstructures together, and they are of interest because they provide a mechanism through which a specimen can easily undergo the martensitic transformation. In particular, the remainder of this section considers the wedge microstructure in detail; while, the triangle and diamond microstructures are discussed in Sections 7 and 8, respectively.

A wedge microstructure consists of two austenite–martensite microstructures with habit planes which intersect along a line and two martensite regions which meet along a midrib plane. The wedge can grow from a line on either a free surface or a grain boundary. For the cubic to orthorhombic transition, three possible wedge microstructures are considered corresponding to the three types of austenite–martensite microstructures: single variant wedge; compound twinned wedge; and type I and II twinned wedge. As discussed in [5, 7, 8, 26], the wedge microstructures are possible only in materials with special transformation stretches, and it is the purpose of this section to find these stretches for this transition.

6.1. SINGLE VARIANT WEDGE MICROSTRUCTURE

The wedge microstructure with single variants of martensite is formed from two austenite-single variant of martensite microstructures with intersecting habit planes and martensite variants which are mutually compatible. For distinct variants $\mathbf{U}_i$ and $\mathbf{U}_j$, the compatibility equations for the single variant wedge are

$$\mathbf{R}_i\mathbf{U}_i - \mathbf{I} = \mathbf{b}' \otimes \widehat{\mathbf{m}}', \tag{31}$$

$$\mathbf{R}_j\mathbf{U}_j - \mathbf{I} = \mathbf{b}'' \otimes \widehat{\mathbf{m}}'', \tag{32}$$

$$\mathbf{R}_i\mathbf{U}_i - \mathbf{R}_j\mathbf{U}_j = \mathbf{b} \otimes \widehat{\mathbf{m}}, \tag{33}$$

respectively. Equations (31) and (32) are the compatibility equations at the two austenite-single variant of the martensite interfaces forming the wedge, and equation (33) is the compatibility equation at the wedge midrib; in fact, it is just a twinning equation. By following [5], necessary and sufficient conditions that equations (31) to (33) are all satisfied are that the shape strain vectors $\mathbf{b}'$ and $\mathbf{b}''$ are parallel and the habit plane normals $\widehat{\mathbf{m}}'$ and $\widehat{\mathbf{m}}''$ are not parallel. From Section 5.1, there are two cases to be considered, one with $\alpha = 1$ and the other with $\beta = 1$. For either of these cases, the compatibility equation (31) with variant $\mathbf{U}_i = \mathbf{U}_1$ has solutions as given in either equations (18) or (19); moreover, as remarked in Section 5.1, the compatibility equation (32) with variant $\mathbf{U}_j$ has solutions related by a symmetry transformation in $\mathcal{L}^c$ to the aforementioned solutions of equation (31). Thus, by following [26], a single variant wedge microstructure is possible if and only if the shape strain $\mathbf{b}'$ is either parallel or perpendicular to the axis of a rotation in $\mathcal{L}^c$ and the habit plane normal $\widehat{\mathbf{m}}'$ is neither parallel nor perpendicular to this axis.

For the case with the transformation stretches $\alpha = 1$, $\beta \leqslant 1$, and $\gamma \geqslant 1$, it can be shown that the single variant wedge exists if, and only if, either the first and the second components or the first and the third components of the shape strain $\mathbf{b}'$ in equation (18) have the same absolute value. This is true when

$$\gamma = \frac{\beta}{\sqrt{3\beta^2 - 2}}, \tag{34}$$

where the stretch γ is greater than one for $\sqrt{2/3} < \beta < 1$. If equation (34) is substituted into equations (18) for the shape strain vector $\mathbf{b}'$ and habit plane normal $\widehat{\mathbf{m}}'$, then their components are

$$b' = \frac{\beta}{\sqrt{3}} \frac{1 - \sqrt{3\beta^2 - 2}}{\sqrt{3\beta^2 - 2}} \begin{pmatrix} 1 \\ -\kappa \\ \kappa \end{pmatrix} \tag{35a}$$

and

$$\widehat{m}' = \frac{1}{\sqrt{3}\beta} \begin{pmatrix} -\sqrt{3\beta^2 - 2} \\ -\kappa \\ \kappa \end{pmatrix}, \tag{35b}$$

respectively, where $\kappa = \pm 1$. All of the possible single variant wedge microstructures with the transformation stretch $\alpha = 1$ are listed in Table II. There are twelve unique single variant wedges, and each habit plane solution can form a wedge with two other habit plane solutions with two distinct variants. All midrib planes are parallel to a type I twin plane normal.

For the case with the stretches $\alpha = 1$, $\beta \geqslant 1$, and $\gamma \leqslant 1$, a short calculation shows that the conditions for the single variant wedge microstructure are the same as above. Moreover, along the curve given by equation (34) in this region, the shape strain vector is the negative of the vector with components given in equation (35), while the habit plane normal has the same components as given in equation (35). Therefore, the single variant wedges given in Table II are possible for all transformation stretches $\beta > \sqrt{2/3}$. Even with $\beta = 1$, all variants are the identity, and thus, all microstructure is trivially possible. On the other hand, interchanging the stretches α and γ is equivalent to a relabeling of the numbering of the variants; so, no new wedges are found.

For the second case with the stretches $\beta = 1$, $\alpha \leqslant 1$ and $\gamma \geqslant 1$, it can be shown that the single variant wedge exists if, and only if, either of the last two components of the shape strain $\mathbf{b}'$ from equation (19) are zero. This holds when

$$\gamma = \frac{\alpha}{\sqrt{2\alpha^2 - 1}}, \tag{36}$$

Table II. Single variant wedge microstructures with the transformation stretches $\alpha = 1$ and $\gamma = \beta/\sqrt{3\beta^2 - 2}$. The component μ for the shape strains and the components μ_1 and μ_2 for the habit plane normals are given in equations (35). The components of the midrib plane normal $\widehat{\mathbf{m}}$ are also listed.

$\mathbf{U}_i$	$\mathbf{U}_j$	b' and b''	$\widehat{m}'$	$\widehat{m}''$	$\sqrt{2}\,\widehat{m}$
1	3	$\mu(1,1,-1)$	$(-\mu_1,\mu_2,-\mu_2)$	$(\mu_2,-\mu_1,-\mu_2)$	$(1,-1,0)$
1	6	$\mu(1,1,-1)$	$(-\mu_1,\mu_2,-\mu_2)$	(μ_2,μ_2,μ_1)	$(1,0,1)$
3	6	$\mu(1,1,-1)$	$(\mu_2,-\mu_1,-\mu_2)$	(μ_2,μ_2,μ_1)	$(0,1,1)$
1	4	$\mu(1,-1,1)$	$(-\mu_1,-\mu_2,\mu_2)$	(μ_2,μ_1,μ_2)	$(1,1,0)$
1	5	$\mu(1,-1,1)$	$(-\mu_1,-\mu_2,\mu_2)$	$(\mu_2,-\mu_2,-\mu_1)$	$(1,0,-1)$
4	5	$\mu(1,-1,1)$	(μ_2,μ_1,μ_2)	$(\mu_2,-\mu_2,-\mu_1)$	$(0,1,1)$
2	3	$\mu(-1,1,1)$	(μ_1,μ_2,μ_2)	$(-\mu_2,-\mu_1,\mu_2)$	$(1,1,0)$
2	5	$\mu(-1,1,1)$	(μ_1,μ_2,μ_2)	$(-\mu_2,\mu_2,\mu_1)$	$(1,0,1)$
3	5	$\mu(-1,1,1)$	$(-\mu_2,-\mu_1,\mu_2)$	$(-\mu_2,\mu_2,-\mu_1)$	$(0,1,-1)$
2	4	$\mu(1,1,1)$	$(-\mu_1,\mu_2,\mu_2)$	$(\mu_2,-\mu_1,\mu_2)$	$(1,-1,0)$
2	6	$\mu(1,1,1)$	$(-\mu_1,\mu_2,\mu_2)$	$(\mu_2,\mu_2,-\mu_1)$	$(1,0,-1)$
4	6	$\mu(1,1,1)$	$(\mu_2,-\mu_1,\mu_2)$	$(\mu_2,\mu_2,-\mu_1)$	$(0,1,-1)$

where the stretch γ is greater than one for $\sqrt{1/2} < \alpha < 1$. If equation (36) is substituted into equations (19) for the shape strain $\mathbf{b}'$ and habit plane normal $\widehat{\mathbf{m}}'$, then their components are

$$b' = \frac{\alpha}{2}\frac{1-\sqrt{2\alpha^2-1}}{\sqrt{2\alpha^2-1}}\begin{pmatrix}0\\1-\kappa\\1+\kappa\end{pmatrix} \tag{37a}$$

and

$$\widehat{m}' = \frac{1}{2\alpha}\begin{pmatrix}0\\-1-\kappa\sqrt{2\alpha^2-1}\\-1+\kappa\sqrt{2\alpha^2-1}\end{pmatrix}, \tag{37b}$$

respectively, where $\kappa = \pm 1$. All of the possible single variant wedge microstructures with the stretch $\beta = 1$ are listed in Table III. There are eighteen unique wedges, and each habit plane solution can form a wedge with three other habit plane solutions with three distinct variants. Depending upon the variant pair making up the single variant wedge, the midrib planes are parallel to either the compound or the type I twin plane normals.

For the case with the stretches $\beta = 1$, $\alpha \geqslant 1$, and $\gamma \leqslant 1$, another short calculation shows that the conditions for the single variant wedge microstructure are the same as above. Moreover, along the curve given by equation (36) in this

Table III. Single variant wedge microstructures with the transformation stretches $\beta = 1$ and $\gamma = \alpha/\sqrt{2\alpha^2 - 1}$. The component μ for the shape strains and the components μ_1 and μ_2 for the habit plane normals are given in equations (37). The components of the midrib plane normal $\widehat{\mathbf{m}}$ are listed, but some are not normalized.

$\mathbf{U}_i$	$\mathbf{U}_j$	b' and b''	$\widehat{m}'$	$\widehat{m}''$	$\widehat{m}$
3	4	$(\mu, 0, 0)$	$(\mu_2, 0, -\mu_1)$	$(\mu_2, 0, \mu_1)$	$(0, 0, 1)$
3	5	$(\mu, 0, 0)$	$(\mu_2, 0, -\mu_1)$	$(\mu_2, -\mu_1, 0)$	$(0, 1, -1)$
3	6	$(\mu, 0, 0)$	$(\mu_2, 0, -\mu_1)$	$(\mu_2, \mu_1, 0)$	$(0, 1, 1)$
4	5	$(\mu, 0, 0)$	$(\mu_2, 0, \mu_1)$	$(\mu_2, -\mu_1, 0)$	$(0, 1, 1)$
4	6	$(\mu, 0, 0)$	$(\mu_2, 0, \mu_1)$	$(\mu_2, \mu_1, 0)$	$(0, 1, -1)$
5	6	$(\mu, 0, 0)$	$(\mu_2, -\mu_1, 0)$	$(\mu_2, \mu_1, 0)$	$(0, 1, 0)$
1	2	$(0, \mu, 0)$	$(0, \mu_2, -\mu_1)$	$(0, \mu_2, \mu_1)$	$(0, 0, 1)$
1	5	$(0, \mu, 0)$	$(0, \mu_2, -\mu_1)$	$(-\mu_1, \mu_2, 0)$	$(1, 0, -1)$
1	6	$(0, \mu, 0)$	$(0, \mu_2, -\mu_1)$	$(\mu_1, \mu_2, 0)$	$(1, 0, 1)$
2	5	$(0, \mu, 0)$	$(0, \mu_2, \mu_1)$	$(-\mu_1, \mu_2, 0)$	$(1, 0, 1)$
2	6	$(0, \mu, 0)$	$(0, \mu_2, \mu_1)$	$(\mu_1, \mu_2, 0)$	$(1, 0, -1)$
5	6	$(0, \mu, 0)$	$(-\mu_1, \mu_2, 0)$	$(\mu_1, \mu_2, 0)$	$(1, 0, 0)$
1	2	$(0, 0, \mu)$	$(0, -\mu_1, \mu_2)$	$(0, \mu_1, \mu_2)$	$(0, 1, 0)$
1	3	$(0, 0, \mu)$	$(0, -\mu_1, \mu_2)$	$(-\mu_1, 0, \mu_2)$	$(1, -1, 0)$
1	4	$(0, 0, \mu)$	$(0, -\mu_1, \mu_2)$	$(\mu_1, 0, \mu_2)$	$(1, 1, 0)$
2	3	$(0, 0, \mu)$	$(0, \mu_1, \mu_2)$	$(-\mu_1, 0, \mu_2)$	$(1, 1, 0)$
2	4	$(0, 0, \mu)$	$(0, \mu_1, \mu_2)$	$(\mu_1, 0, \mu_2)$	$(1, -1, 0)$
3	4	$(0, 0, \mu)$	$(-\mu_1, 0, \mu_2)$	$(\mu_1, 0, \mu_2)$	$(1, 0, 0)$

region, the shape strain vector is the negative of the vector with components given in equation (37), while the habit plane normal has the same components as given in equation (37). Therefore, the single variant wedges given in Table III are possible for all stretches $\alpha > \sqrt{1/2}$, including $\alpha = 1$ as all microstructures are trivial. These are the only single variant wedges that can be constructed.

6.2. TWINNED WEDGE

The twinned wedge microstructure consists of two regions of twinned martensite separated individually from the austenite phase by a twinned habit plane and separated from each other by a midrib plane. For variant pairs $(i : j)$ and $(k : l)$, the compatibility equations for the twinned wedge are

$$\mathbf{R}_{ij}\mathbf{U}_i - \mathbf{U}_j = \mathbf{a}' \otimes \hat{\mathbf{n}}', \tag{38}$$

$$\overline{\mathbf{R}}_{ij}\big(\lambda'\mathbf{R}_{ij}\mathbf{U}_i + (1-\lambda')\mathbf{U}_j\big) - \mathbf{I} = \mathbf{b}' \otimes \widehat{\mathbf{m}}', \tag{39}$$

$$\mathbf{R}_{kl}\mathbf{U}_k - \mathbf{U}_l = \mathbf{a}'' \otimes \hat{\mathbf{n}}'', \tag{40}$$

$$\overline{\mathbf{R}}_{kl}\big(\lambda''\mathbf{R}_{kl}\mathbf{U}_k + (1-\lambda'')\mathbf{U}_l\big) - \mathbf{I} = \mathbf{b}'' \otimes \widehat{\mathbf{m}}'', \tag{41}$$

$$\overline{\mathbf{R}}_{ij}\big(\lambda'\mathbf{R}_{ij}\mathbf{U}_i + (1-\lambda')\mathbf{U}_j\big) - \overline{\mathbf{R}}_{kl}\big(\lambda''\mathbf{R}_{kl}\mathbf{U}_k + (1-\lambda'')\mathbf{U}_l\big) = \mathbf{b} \otimes \widehat{\mathbf{m}}. \tag{42}$$

Equations (38) and (40) are the twinning equations discussed in Sections 4.1 and 4.2 which give the compatibility between the martensite variants. Equations (39) and (41) are the habit plane equations discussed in Section 5.2 which give the compatibility between the twinned martensite and austenite regions. Finally, equation (42) is the compatibility equation between the two twinned martensite regions and gives the wedge midrib plane normal $\widehat{\mathbf{m}}$. On following [5], two austenite-twinned martensite microstructures form a wedge if, and only if, the shape strains $\mathbf{b}'$ and $\mathbf{b}''$ are parallel and the habit plane normals $\widehat{\mathbf{m}}'$ and $\widehat{\mathbf{m}}''$ are not parallel. Only four of the possible twinned wedge microstructures are considered in this section; the first is a twinned wedge with compound twins solely; the next two are twinned wedges with either type I or type II twins only; and the last twinned wedges contain both type I and type II twins.

6.2.1. *Compound Twinned Wedge*

For the variant pairs $(i : j)$ and $(k : l)$ both from set S_1, the compatibility equations for the compound twinned wedge are the equations (38)–(42). As discussed in Section 5.2.1, all of the habit plane solutions with variant pairs in set S_1 are related by a symmetry transformation using the rotations in $\mathcal{L}^c$; thus, the compound twinned wedge microstructure is possible if, and only if, the shape strain $\mathbf{b}'$ is either parallel or perpendicular to an axis of a rotation in $\mathcal{L}^c$ and the habit plane normal $\widehat{\mathbf{m}}'$ is neither parallel nor perpendicular to this axis [26]. Several cases arise in which it is assumed that none of the transformation stretches are one.

For the case (25), the solutions from Section 5.2.1 to equations (38) and (39) with variant pair $(i : j) = (1 : 2)$ are given in equations (28). For the case (26), the solutions from Section 5.2.1 to equations (38) and (39) with variant pair $(i : j) = (1 : 2)$ are given in equations (29). The possible compound twinned wedges are

1. For the shape strain $\mathbf{b}'$ with components from either equation (28) or equation (29), the first and the second components of $\mathbf{b}'$ have equal magnitude when

$$\gamma = \frac{\alpha\beta}{\sqrt{4\alpha^2\beta^2 - 2\alpha^2 - \beta^2}}, \tag{43}$$

a result reported in [5]. If equation (43) is substituted into either equation (28) or equation (29) for the shape strain $\mathbf{b}'$, then this vector has components of the form $(\mu_1, \pm\mu_1, \mu_2)$ with nonzero μ_1 and μ_2; while, the habit plane normal $\widehat{\mathbf{m}}'$ from either equations (28) or (29) has components of the form (ξ_1, ξ_2, ξ_3) with nonzero ξ_1, ξ_2, and ξ_3 and the first two components never have the same absolute value. Furthermore, the shape strain $\mathbf{b}'$ from equations (28) and (29) has all three components with the same absolute value if the transformation stretch $\alpha = 1$, which is in agreement with the calculation for the single variant wedge from Section 6.1 (compare with equation (34)). By using the symmetry amongst the habit plane solutions, then it can be shown that there are a total of twelve compound twinned wedges, each wedge has a midrib plane in the {110} family of planes, and each deformation forms a wedge with exactly one other deformation.

A special case of the previous twinned wedge occurs when the first two components of the shape strain $\mathbf{b}'$ have the same absolute value and the third component vanishes. The last condition is true when

$$\alpha = \sqrt{2 - \gamma^2}, \tag{44}$$

where the stretch α from equation (44) is greater than one for $\gamma < 1$ and is less than one for $1 < \gamma < \sqrt{2}$. On assuming that the condition (44) holds, the remaining two components of the shape strain have equal absolute value when

$$\beta = \gamma \sqrt{\frac{\gamma^2 - 2}{4\gamma^2 - 2\gamma^4 - 1}}. \tag{45}$$

The transformation stretch β from equation (45) is greater than one for all $\gamma < \sqrt{2}$; therefore, the habit plane solution with components from equations (29) exists. If equations (44) and (45) are substituted into equations (29) for the shape strain $\mathbf{b}'$ and the habit plane normal $\widehat{\mathbf{m}}'$, then these vectors have components

$$b' = \frac{\gamma}{\sqrt{2}} \sqrt{\frac{2 - \gamma^2}{4\gamma^2 - 2\gamma^4 - 1}} \begin{pmatrix} \kappa \\ 1 \\ 0 \end{pmatrix} \tag{46a}$$

and

$$\widehat{m}' = \frac{1}{\sqrt{2\gamma^2(2 - \gamma^2)}} \begin{pmatrix} \kappa \\ -\sqrt{4\gamma^2 - 2\gamma^4 - 1} \\ 0 \end{pmatrix}, \tag{46b}$$

respectively, where $\kappa = \pm 1$. The possible compound twinned wedge microstructures with transformation stretches on the curve given by equations (44) and (45) are a special case of the stretches on the surface (43). For the wedges here, the volume fraction λ is one half, and, thus, there are six unique compound twinned wedge microstructures, and each deformation forms a wedge with only one other deformation.

2. For the shape strain $\mathbf{b}'$ with components from either equations (28) or (29), the first and the third components of the shape strain $\mathbf{b}'$ have the same magnitude when

$$\gamma = \beta \sqrt{\frac{\alpha^2 - 2}{2\alpha^2 - 2\alpha^2\beta^2 - \beta^2}}; \tag{47}$$

a result also given in [5]. If equation (47) is substituted into either equation (28) or equation (29) for the shape strain $\mathbf{b}'$, then its components are of the form $(\mu_1, \mu_2, \pm\mu_1)$ with nonzero μ_1 and μ_2; while, from either equation (28) or (29), the habit plane normal $\widehat{\mathbf{m}}'$ has components of the form (ξ_1, ξ_2, ξ_3) for nonzero ξ_1, ξ_2, and ξ_3 and the first and third components never have the same absolute value. Likewise, it can be shown that the shape strain $\mathbf{b}'$ from equations (28) and (29) has all three components with the same absolute value if the transformation stretch $\alpha = 1$, which is in agreement with the calculation for the single variant wedge from Section 6.1 (compare with equation (36)). As for the case (1), there are twelve possible twinned wedges and each deformation here forms a wedge with only one other deformation.

A special case of the previous twinned wedge occurs when the first and last components of the shape strain $\mathbf{b}'$ have the same absolute value and the second component vanishes. The last condition is true when

$$\alpha = \frac{\gamma}{\sqrt{2\gamma^2 - 1}}. \tag{48}$$

The stretch α from equation (48) is greater than one for $\sqrt{1/2} < \gamma < 1$ and is less than one for $\gamma > 1$. On assuming that the condition (48) holds, the remaining two components of the shape strain have equal absolute value when

$$\beta = \frac{\gamma^2}{\sqrt{2\gamma^4 - 2\gamma^2 + 1}}. \tag{49}$$

The transformation stretch β from equation (49) is less than one for all $\gamma > 0$; therefore, the habit plane solution with components from equations (28) exists. If equations (48) and (49) are substituted into equations (28) for the shape strain $\mathbf{b}'$ and habit plane normal $\widehat{\mathbf{m}}'$, then they have components

$$b' = \gamma^2 \left(\frac{\sqrt{2\gamma^4 - 2\gamma^2 + 1} - \sqrt{2\gamma^2 - 1}}{\sqrt{(2\gamma^4 - 2\gamma^2 + 1)(2\gamma^2 - 1)}} \right) \begin{pmatrix} -1 \\ 0 \\ \chi\kappa \end{pmatrix} \tag{50a}$$

and

$$\widehat{m}' = \frac{1}{\sqrt{2}\gamma^2} \begin{pmatrix} \sqrt{2\gamma^2 - 1} \\ 0 \\ \chi\kappa\sqrt{2\gamma^4 - 2\gamma^2 + 1} \end{pmatrix}, \tag{50b}$$

respectively, where $\chi = \operatorname{sign}(\alpha - \gamma)$ and $\kappa = \pm 1$. The possible compound twinned wedge microstructures for the transformation stretches on the curve given by equations (48) and (49) are a special case of the stretches on the surface (47). For the wedges here, the volume fraction λ is one half; thus, there are only six unique compound twinned wedge microstructures, and each deformation forms a wedge with only one other deformation.

6.2.2. *Compound Twinned Wedge with a Transformation Stretch Equal to One*

As discussed in Section 5.2, the austenite-twinned martensite microstructure is possible for a large set of transformation stretches, but with only specific values of the volume fraction. On the other hand, this microstructure may be possible for all volume fractions, but with special transformation stretches, and, consequently, the twinned wedge microstructure may exist at particular values of the volume fraction λ. In this section, several cases are explored which are suggested by the previous analysis of the single variant wedge.

The first case is: from equation (47) with $\beta = 1$, then $\gamma = \sqrt{2-\alpha^2}$, and equation (30) becomes

$$\eta = (1-2\lambda)^2\left(1-\gamma^2\right)^2,$$

which is greater than or equal to zero for all values of the stretch γ and for all volume fractions $0 \leqslant \lambda \leqslant 1$; therefore, solutions exist to the habit plane equation (39). The second case is: from equation (43) also with $\beta = 1$, then $\gamma = \alpha/\sqrt{2\alpha^2-1}$, and equation (30) becomes

$$\eta = \left(1-\gamma^2\right)^2\frac{(2\lambda-1)^2}{2\gamma^2-1},$$

which is greater than or equal to zero for all stretches $\gamma > 1/\sqrt{2}$ and for all volume fractions $0 \leqslant \lambda \leqslant 1$; thus, solutions exist to the habit plane equation (39). Notice that, in both cases, the relation between the stretches are exactly those for which a single variant wedge is possible.

In both of the cases above, the volume fraction λ is a free parameter, and it can be adjusted in order to find a compound twinned wedge microstructure. The algebra is too complex to perform by hand; so, a numerical calculation is performed to find the wedges. The algorithm is as follows: with the transformation stretch $\beta = 1$ and for each value of the stretch γ with the stretch $\alpha = \sqrt{2-\gamma^2}$, the volume fraction λ is found so that the shape strain $\mathbf{b}'$ is parallel to a vector with components of the form $(0, \mu, \mu)$ with nonzero μ and the habit plane normal $\widehat{\mathbf{m}}$ is parallel to a vector with components $(0, 1, 0)$. The results of the numerical computation are shown as the solid curve in Figure 3. Similarly, along the curve $\alpha = \gamma/\sqrt{2\gamma^2-1}$, the volume fraction λ is found so that the shape strain $\mathbf{b}'$ is parallel to a vector with components of the form $(0, \tilde{\mu}, \tilde{\mu})$ with nonzero $\tilde{\mu}$ and the habit plane normal $\widehat{\mathbf{m}}'$ is parallel to a vector with components of the form $(0, \bar{\mu}, \bar{\nu})$ with unequal and nonzero $\bar{\mu}$ and $\bar{\nu}$. There are six possible compound twinned wedges with transformation stretches $\beta = 1$ and either $\alpha = \sqrt{2-\gamma^2}$ or $\alpha = \gamma/\sqrt{2\gamma^2-1}$ as listed in Table IV. On the curve $\alpha = \sqrt{2-\gamma^2}$, the components of the habit plane normals $\widehat{\mathbf{m}}'$ and $\widehat{\mathbf{m}}''$ are $\bar{\mu} = 1$ and $\bar{\nu} = 0$; therefore, the habit plane normals $\widehat{\mathbf{m}}'$ and $\widehat{\mathbf{m}}''$ are parallel to the twin plane normals $\hat{\mathbf{n}}'$ and $\hat{\mathbf{n}}''$, respectively, and the two habit planes forming this compound twinned wedge are perpendicular. On the curve $\alpha = \gamma/\sqrt{2\gamma^2-1}$, the components $\bar{\mu}$ and $\bar{\nu}$ are nonzero and unequal.

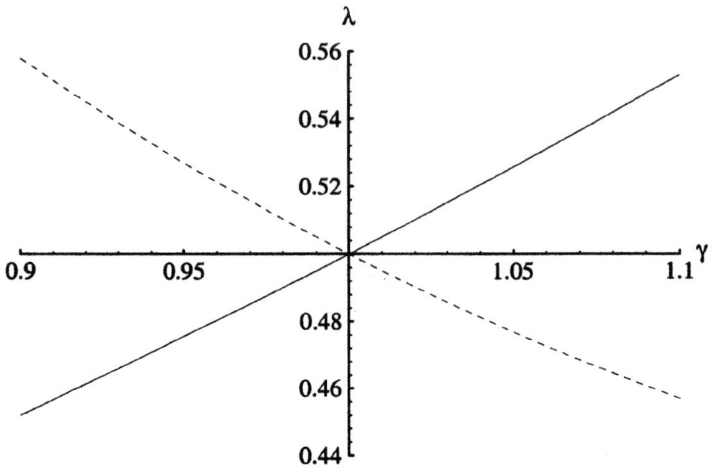

Figure 3. Plot of volume fraction λ verses transformation stretch γ to form the compound twinned wedge microstructure. The stretch $\beta = 1$, and along the dashed curve, the stretch $\alpha = \gamma/\sqrt{2\gamma^2 - 1}$; while, along the solid curve, the stretch $\alpha = \sqrt{2 - \gamma^2}$.

Table IV. Compound twinned wedge microstructures with the transformation stretch $\beta = 1$. The components of both of the twin shears $\mathbf{a}'$ and $\mathbf{a}''$ can be found by a symmetry transformation in $\mathcal{L}^c$ of the twin solutions for the variant pair $(1:2)$, and specific components for the shape strains and habit plane normals can be found as detailed in the text; while, the volume fraction λ is given in Figure 3.

$(i:j)$	$(1:2)$	$(1:2)$	$(3:4)$	$(3:4)$	$(5:6)$	$(5:6)$
$(k:l)$	$(1:2)$	$(1:2)$	$(3:4)$	$(3:4)$	$(5:6)$	$(5:6)$
$\hat{n}'$	$(0,1,0)$	$(0,1,0)$	$(1,0,0)$	$(1,0,0)$	$(0,1,0)$	$(0,1,0)$
$\hat{n}''$	$(0,0,1)$	$(0,0,1)$	$(0,0,1)$	$(0,0,1)$	$(1,0,0)$	$(1,0,0)$
λ' and λ''	λ	$1-\lambda$	λ	$1-\lambda$	λ	$1-\lambda$
b' and b''	$(0,\mu,\mu)$	$(0,\mu,-\mu)$	$(\mu,0,\mu)$	$(\mu,0,-\mu)$	$(\mu,\mu,0)$	$(\mu,-\mu,0)$
$\widehat{m}'$	$(0,\bar{\mu},\bar{\nu})$	$(0,\bar{\mu},-\bar{\nu})$	$(\bar{\mu},0,\bar{\nu})$	$(\bar{\mu},0,-\bar{\nu})$	$(\bar{\mu},\bar{\nu},0)$	$(\bar{\mu},-\bar{\nu},0)$
$\widehat{m}''$	$(0,\bar{\nu},\bar{\mu})$	$(0,-\bar{\nu},\bar{\mu})$	$(\bar{\nu},0,\bar{\mu})$	$(-\bar{\nu},0,\bar{\mu})$	$(\bar{\nu},\bar{\mu},0)$	$(-\bar{\nu},\bar{\mu},0)$
$\sqrt{2}\,\widehat{m}$	$(0,1,-1)$	$(0,1,1)$	$(1,0,-1)$	$(1,0,1)$	$(1,-1,0)$	$(1,1,0)$

For the stretches either $\alpha = 1$ or $\gamma = 1$, equations (43) and (47) both become equation (34) discussed in Section 6.1 for the single variant wedge. So, equations (43) and (47) extend continuously to the lines $\alpha = 1$ and $\gamma = 1$. One last item of import is to note that each austenite-compound twinned martensite microstructure can form a twinned wedge with only one other austenite-compound twinned martensite microstructure, unless the microstructures degenerate to the single variant microstructures. Finally, there may be other values of the transformation stretches α and γ and volume fractions λ at which a compound twinned wedge with transformation stretch $\beta = 1$ is possible, but these are not pursued here.

6.2.3. *Type I and Type II Twinned Wedge*

In this section, twinned wedge microstructures are considered in which the twinned martensite regions contain either type I twins, type II twins or both. Consequently, all variant pairs are in set S_2. For distinct variant pairs $(i : j)$ and $(k : l)$, the compatibility equations for the twinned wedge microstructure are the equations (38) to (42). As above, the necessary and sufficient conditions that all of these equations are satisfied are that the shape strains $\mathbf{b}'$ and $\mathbf{b}''$ are parallel and the habit plane normals $\widehat{\mathbf{m}}'$ and $\widehat{\mathbf{m}}''$ are not parallel. Solutions to the habit plane equations (39) and (41) are discussed in Section 5.2.2, where the habit plane solutions are divided into four categories, (THa), (THb), (THc), and (THd), depending upon the twin type, either type I twins or type II twins, and whether $\kappa = +1$ or $\kappa = -1$. From these four categories of habit plane solutions, seven categories of twinned wedge microstructures are possible: (TWa) variant pairs $(i : j)$ and $(k : l)$ both with habit plane solutions from category (THa); (TWb) variant pairs $(i : j)$ and $(k : l)$ both with habit plane solutions from category (THb); (TWc) variant pairs $(i : j)$ and $(k : l)$ both with habit plane solutions from category (THc); (TWd) variant pairs $(i : j)$ and $(k : l)$ both with habit plane solutions from category (THd); (TWe) variant pair $(i : j)$ with habit plane solution from category (THa) and variant pair $(k : l)$ with habit plane solution from category (THb) or vice versa; (TWf) variant pair $(i : j)$ with habit plane solution from category (THc) and variant pair $(k : l)$ with habit plane solution from category (THd) or vice versa; (TWg) variant pair $(i : j)$ with habit plane solution from either category (THa) or (THb) and variant pair $(k : l)$ with habit plane solution from either category (THc) or (THd). The habit plane solutions within each of the four categories (THa) to (THd) from Section 5.2 are related by a symmetry transformation using the rotations in $\mathcal{L}^c$. Thus, the first four categories of twinned wedges (TWa) to (TWd) are possible if and only if the shape strain $\mathbf{b}'$ is either parallel or perpendicular to a vector which is an axis of one of the rotations in $\mathcal{L}^c$ and the habit plane normal $\widehat{\mathbf{m}}'$ is neither parallel nor perpendicular to that vector [26]. For the categories of twinned wedges (TWe) to (TWf), no twinned wedge can exist if the two variant pairs are related by an 180 degree rotation in $\mathcal{L}^c$ as shown in [26]. For the other rotations in $\mathcal{L}^c$, no general restriction on the shape strains $\mathbf{b}'$ and $\mathbf{b}''$ and the habit plane normals $\widehat{\mathbf{m}}'$ and $\widehat{\mathbf{m}}''$ can be found. For the last category of twinned wedges (TWg), no symmetry exists amongst the habit plane solutions in general, and, similarly to the previous case, no general restriction on the shape strains and habit plane normals can be found.

In order to find the special transformation stretches at which the twinned wedge microstructure is possible, a numerical computation is performed for the cubic to orthorhombic transition. The numerical program exploits the symmetry amongst the habit plane solutions for categories (TWa) to (TWd) of the twinned wedges to limit the number of computations performed, and it uses the domain calculation discussed in Section 5.2.2 to the limit the area searched. For the categories (TWe) to (TWg) of the twinned wedges, a separate calculation is performed at each point within the search domain. The numerical program performs the habit plane calculations for the variant pair $(i : j) = (1 : 3)$ only on account of the symmetry amongst the habit plane solutions. Details are given in [25]. On the other hand, Bhattacharya [5, 7] gives numerically computed level curves in the space of the transformation stretches α, β, and γ for the same cubic to orthorhombic transition on which the twinned wedge microstructure is possible; however, the accuracy of his level curves is believed to be low [9]. In addition, it is found herein that some of the curves were missed.

6.2.4. *Type I Twinned Wedges for Categories (TWa) and (TWb)*

The numerical computation finds the level curves of three surfaces on which twinned wedge microstructures with type I twins only are possible. Figure 4 shows the level curves of one of these surfaces in the space of the transformation stretches. In this figure, increasing values of the stretch β are from top to bottom in each half of the plane. For any stretch $\beta < 1$, the level curve starts at the point $(\alpha, \gamma) = (1, 1)$ and proceeds into the $\alpha > 1$ half plane. For any stretch $\beta > 1$, the level curve starts at the point $(\alpha, \gamma) = (1, 1)$ and proceeds into the $\alpha < 1$ half plane. On the solid curves, the shape strain for the variant pair $(i : j) = (1 : 3)$ with volume fraction λ of variant $\mathbf{U}_i$ is perpendicular to a vector with components $(1, 0, 0)$; while, on the dashed curves, the shape strain for the variant pair $(i : j) = (1 : 3)$ with volume fraction λ of variant $\mathbf{U}_i$ is perpendicular to the vector with components $(0, 1, 0)$. The open circles between the solid and dashed curves are inadmissible points, because at these points, the austenite-twinned martensite microstructure with type I twins does not exist. The level curves of another surface are obtained by interchanging the transformation stretches α and γ in Figure 4; however, the properties of the twinned wedges on these level curves are the same as the twinned wedges on the level curves in Figure 4. The level curves are continuous through the lines $\alpha = 1$ and $\gamma = 1$, and these twinned wedge microstructures are discussed below. Portions of the level curves are densely spaced, but closer inspection shows that the level curves are distinct for each value of the stretch β. The curves as shown in Figure 4 are not reported in [5].

The variant pair $(k : l)$ with which the variant pair $(i : j) = (1 : 3)$ with volume fraction λ of variant $\mathbf{U}_i$ can form a wedge is found by a symmetry transformation using the 180 degree rotation in $\mathcal{L}^c$ that has axis parallel to $(1, 0, 0)$. This variant pair is $(k : l) = (1 : 4)$, and the twinned wedge is denoted as $((i : j), (k : l))$. From [26], the variant pair $(j : i)$ with volume fraction λ of variant $\mathbf{U}_j$ has the same shape strain and habit plane normal vectors as the variant pair $(i : j)$ with volume fraction $1-\lambda$ of variant $\mathbf{U}_i$. Thus, for the twinned wedge microstructure, all variant pairs $(j : i)$ with volume fraction λ $\mathbf{U}_j$ obtained by a symmetry transformation are replaced by the variant pair $(i : j)$ with volume fraction $1 - \lambda$ of variant $\mathbf{U}_i$. This replacement is denoted as $[i : j]$. For example, the twinned wedge microstructure denoted as $([i : j], (k : l))$ means a twinned wedge with an austenite-twinned

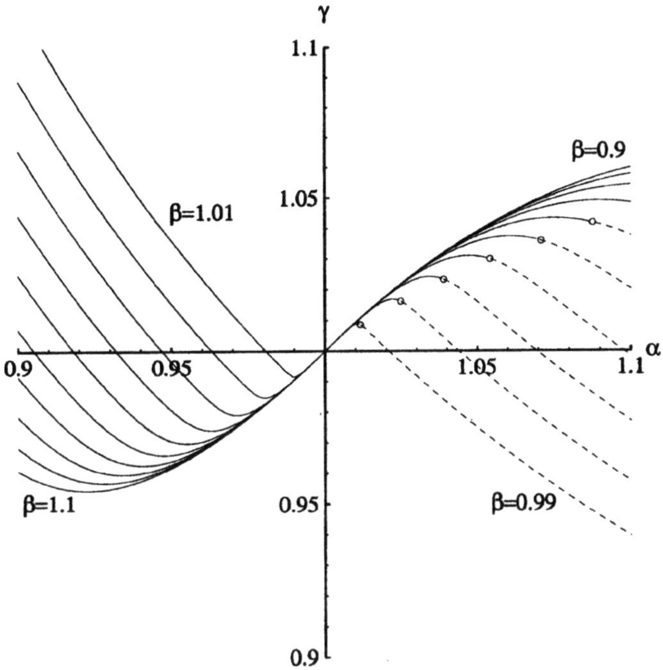

Figure 4. First level curves on which the twinned wedge microstructure with type I twins is possible. See text for details.

martensite microstructure for variant pair $(i : j)$ with volume fraction $1 - \lambda$ of variant $\mathbf{U}_i$ and an austenite-twinned martensite microstructure for variant pair $(k : l)$ with volume fraction λ of variant $\mathbf{U}_k$. For materials with stretches that lie on the solid curves in Figure 4, the results of the symmetry transformation are that the following twinned wedges are possible:

$$\begin{pmatrix} ((1:3),(1:4)) & ([1:3],[2:3]) & ([1:4],[2:4]) \\ ((1:5),(1:6)) & ([1:5],[2:5]) & ([1:6],[2:6]) \\ ((2:3),(2:4)) & ((2:5),(2:6)) & ([3:5],[4:5]) \\ ((3:5),(3:6)) & ([3:6],[4:6]) & ((4:5),(4:6)) \end{pmatrix}. \tag{51}$$

The other case is the dashed curves in Figure 4 on which the shape strain is perpendicular to the vector with components (0, 1, 0). Repeating the above calculation to find all of the possible twinned wedges gives the set

$$\begin{pmatrix} ((1:3),(2:3)) & ([1:3],[1:4]) & ((1:4),(2:4)) \\ ((1:5),(2:5)) & ([1:5],[1:6]) & ((1:6),(2:6)) \\ ([2:3],[2:4]) & ([2:5],[2:6]) & ((3:5),(4:5)) \\ ([3:5],[3:6]) & ((3:6),(4:6)) & ([4:5],[4:6]) \end{pmatrix}. \tag{52}$$

The wedges in both sets (51) and (52) have midrib planes contained in the {100} family of planes, and there are twelve distinct twinned wedges. The wedges in

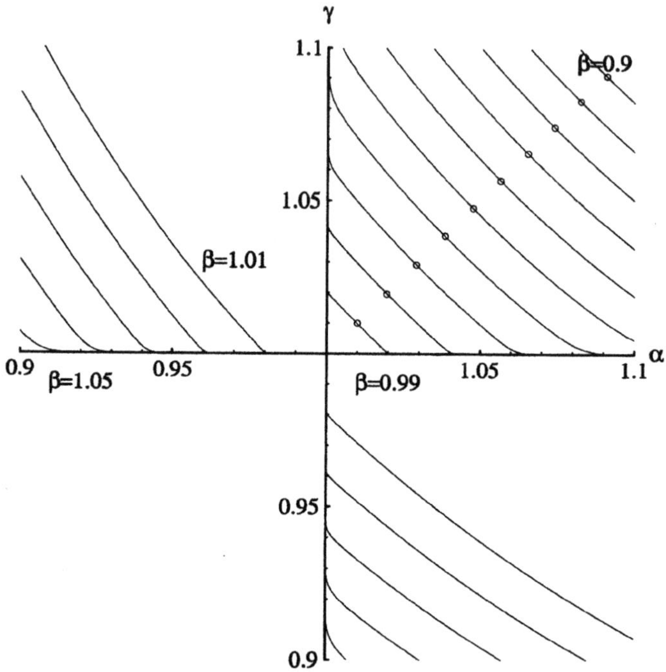

Figure 5. Second level curves on which the twinned wedge microstructure with type I twins is possible. See text for details.

these two sets contain only three distinct variants, and each variant pair forms a twinned wedge microstructure with exactly one other.

Figure 5 shows the level curves of another of the surfaces in the space of the transformation stretches on which the twinned wedge microstructure with type I twins only is possible. Here, increasing values of the stretch β are from top right to lower left. On the level curves, the shape strain for the variant pair $(i : j) = (1 : 3)$ with volume fraction λ of variant $\mathbf{U}_i$ is perpendicular to a vector with components $(1, 1, 0)$; the points where each level curve intersects the line $\alpha = \gamma$ as marked by circles are inadmissible, because, at these points, the shape strain and habit plane normal vectors are both perpendicular to a vector with components $(1, 1, 0)$. These level curves are symmetric about the line $\alpha = \gamma$. For stretches $\beta > 1.05$, level curves are not found by the numerical computation in the region of stretches $\alpha \in [0.90,\ 1.1]$ and $\gamma \in [0.90,\ 1.1]$. The level curves extend continuously to the lines $\alpha = 1$ and $\gamma = 1$. Along the line $\alpha = 1$ with $\gamma = \beta/\sqrt{3\beta^2 - 2}$, a single variant wedge is possible with the shape strain vector perpendicular to a vector with components $(1, 1, 0)$ from Table II in Section 6.1. Thus, the volume fraction λ of variant $\mathbf{U}_i$ is zero for those points in Figure 5 along the lines $\alpha = 1$ and $\gamma = 1$. The level curves as shown in Figure 5 are reported in [5]; however, it is believed that the

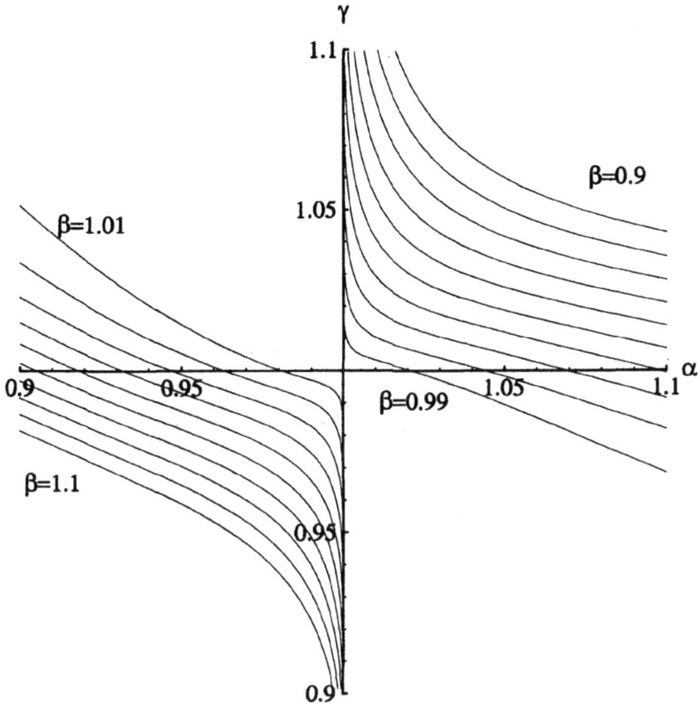

Figure 6. Third level curves on which the twinned wedge microstructure with type I twins is possible. See text for details.

level curves presented here are more accurate due to the simpler calculation which exploits symmetry.

The other variant pair $(k : l)$ with which the variant pair $(i : j) = (1 : 3)$ can form a twinned wedge microstructure is $(k : l) = [2 : 4]$, which is the variant pair $(2 : 4)$ with volume fraction $1 - \lambda$ of variant $\mathbf{U}_2$. For materials with stretches that lie on the curves in Figure 5, the results of the symmetry transformation are that the following twinned wedges are possible:

$$\begin{pmatrix} ((1:3),[2:4]) & ([1:3],(2:4)) & ((1:4),[2:3]) \\ ([1:4],(2:3)) & ((1:5),[2:6]) & ([1:5],(2:6)) \\ ((1:6),[2:5]) & ([1:6],(2:5)) & ((3:5),[4:6]) \\ ([3:5],(4:6)) & ((3:6),[4:5]) & ([3:6],(4:5)) \end{pmatrix}. \tag{53}$$

All of the midrib planes belong to the $\{110\}$ family of planes, and again, twelve distinct twinned wedges are possible. Unlike the twinned wedges in the sets (51) and (52), the wedges given in the set (53) contain four distinct variants, but each variant pair forms a twinned wedge microstructure with exactly one other.

Figure 6 gives the level curves of the last surface in the space of the transformation stretches on which the twinned wedge microstructure with type I twins only is possible. Increasing values of the stretch β are from top right to bottom left. On the level curves, the shape strain for the variant pair $(i : j) = (1 : 3)$ with volume fraction λ of variant $\mathbf{U}_i$ is perpendicular to a vector with components $(0, 1, -1)$; while, interchanging the stretches α and γ gives other level curves on which the shape strain for the variant pair $(i : j) = (1 : 3)$ with volume fraction λ of variant $\mathbf{U}_i$ is perpendicular to a vector with components $(0, 1, 1)$. The level curves are continuous through the lines $\alpha = 1$ and $\gamma = 1$, and these twinned wedge microstructures are discussed below. The level curves as shown in Figure 6 are reported [5], but it is believed that the level curves presented here are more accurate for the reasons previously given.

The other variant pair $(k : l)$ with which the variant pair $(i : j) = (1 : 3)$ can form a twinned wedge microstructure is $(k : l) = (1 : 5)$. For materials with stretches that lie on the curves in Figure 6, the results of the symmetry transformation are that the following twinned wedges are possible:

$$\begin{pmatrix} ((1:3),(1:5)) & ([1:3],(3:5)) & ((1:4),(1:6)) \\ ([1:4],(4:6)) & ([1:5],[3:5]) & ([1:6],[4:6]) \\ ((2:3),(2:6)) & ([2:3],(3:6)) & ((2:4),(2:5)) \\ ([2:4],(4:5)) & ([2:5],[4:5]) & ([2:6],[3:6]) \end{pmatrix}. \tag{54}$$

The other case is interchanging the stretches α and γ in Figure 6 on which the shape strain is perpendicular to a vector with components (0,1,1). Repeating the above calculation to find all of the twinned wedge microstructures gives a set which can be found from set (54) by interchanging the labels of the variants: $1 \leftrightarrow 2$, $3 \leftrightarrow 4$, and $5 \leftrightarrow 6$. For the wedges in these two sets, the midrib planes belong to the $\{110\}$ family of planes; twelve distinct twinned wedges are possible; each wedge contains only three distinct variants; and each variant pair once again forms a twinned wedge microstructure with exactly one other variant pair.

The level curves in Figures 4–6 are the only curves found by the numerical computation on which the twinned wedge microstructures with type I twins only can form. At almost all of the points on the level curves, twelve twinned wedge microstructures are possible. There are several special points discussed below at which more than twelve twinned wedges can be constructed.

6.2.5. *Type I Twinned Wedge with a Transformation Stretch Equal to One*

As discussed in Section 5.2.2, solutions to the habit plane equation with type I twins are possible even if a transformation stretch is one. Therefore, for certain volume fractions, the twinned wedge may exist as well.

With stretches $\alpha = 1$ and $\gamma = \beta/\sqrt{3\beta^2 - 2}$, the austenite-twinned martensite microstructure with type I twins is possible for all volume fractions λ. Recall that along this curve in the space of the stretches a single variant wedge is possible as discussed in Section 6.1. The twinned wedge microstructure with type I twins and transformation stretches restricted to the above curve is found numerically. For each fixed stretch β, the other two stretches are known, and so, the free parameter is the volume fraction λ. Volume fractions are found iteratively for which the twinned wedge microstructure is possible. The results of the numerical computation are given as the solid line in Figure 7. For stretches $\beta < 1$, the shape strain for the variant pair $(i : j) = (1 : 3)$ with volume fraction λ of variant $\mathbf{U}_i$ is perpendicular to a vector with components $(0, 1, 0)$; while, for stretches $\beta > 1$, the shape strain for the variant pair $(i : j) = (1 : 3)$ with volume fraction λ of variant $\mathbf{U}_i$ is perpendicular to a vector with components $(1, 0, 0)$. This is in agreement with Figure 4, where the level curves on which the twinned wedge microstructures exist are continuous through the lines $\alpha = 1$ and $\gamma = 1$.

Other twinned wedge microstructures with the stretches $\alpha = 1$ and $\gamma = \beta/\sqrt{3\beta^2 - 2}$ exist. The same numerical computation as used above finds for a volume fraction λ equal to one half that the shape strain for the variant pair $(i : j) = (1 : 3)$ is parallel to a vector with components $(1, 1, -1)$, while the habit plane normal is parallel to a vector with components (μ_1, μ_1, μ_2), where the components μ_1 and μ_2 are nonzero and have unequal absolute value. If, on the other hand, the stretches are $\gamma = 1$ and $\alpha = \beta/\sqrt{3\beta^2 - 2}$, then the shape strain is parallel to a vector with components $(1, 1, 1)$, while the habit plane normal is parallel to a vector with components (μ_1, μ_1, μ_2), where again the components μ_1 and μ_2 are nonzero and have unequal absolute value. This is in agreement with Figure 6, where the level curves on which the twinned wedge microstructures exist are continuous through the lines $\alpha = 1$ and $\gamma = 1$.

The twinned wedge microstructure with type I twins and the transformation stretch $\beta = 1$ is not investigated in this article; however, the single variant wedge

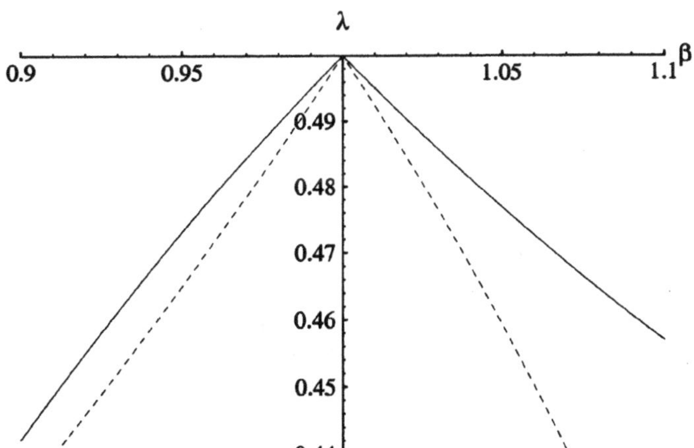

Figure 7. Volume fraction λ verses transformation stretch β. The solid curve is the type I twinned wedge, where the stretches $\alpha = 1$ and $\gamma = \beta/\sqrt{3\beta^2 - 2}$. The dashed curve is the type II twinned wedge, where the stretches $\alpha = 1$ and $\gamma = \sqrt{3 - 2\beta^2}$.

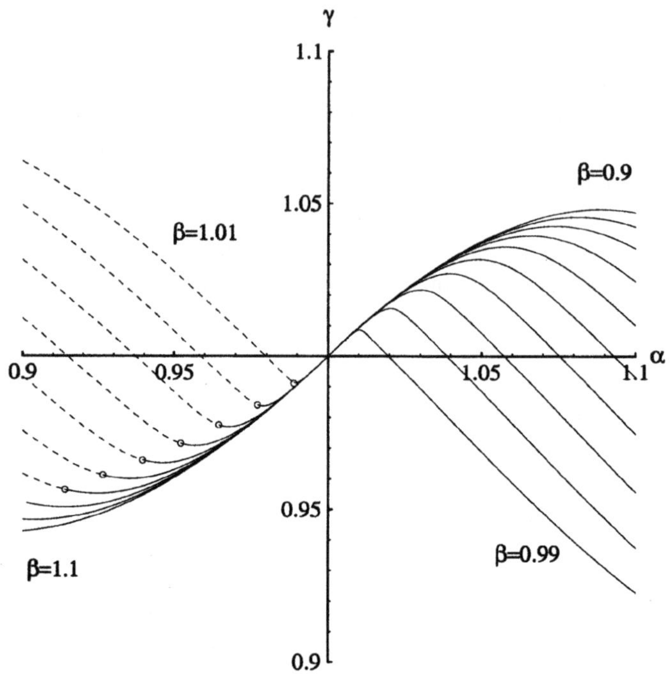

Figure 8. First level curves on which the twinned wedge microstructure with type II twins is possible. See text for details.

microstructure with the stretch $\beta = 1$ is possible as discussed in Section 6.1, but more wedges with twins may exist.

6.2.6. *Type II Twinned Wedges for Categories (TWc) and (TWd)*

The numerical computation finds the level curves of six surfaces on which the twinned wedge microstructures with type II twins only are possible. Figures 8–11 give the level curves of five of these surfaces in the space of the transformation stretches. The first three figures are very similar to Figures 4–6, respectively, with type I twins solely. In fact, the exact same discussion as given for the type I twinned wedges in Figure 4 applies for the type II twinned wedges in Figure 8. Also, the type II wedges in Figure 9 are analogous to the type I wedges in Figure 5. And the type II wedges in Figure 10 are similar to the type I wedges in Figure 6; the difference being that the level curves in Figure 10 are not continuous through the lines $\alpha = 1$ and $\gamma = 1$. For the level curves in the quadrant $\alpha > 1$, $\gamma < 1$, the stretches β are 0.96 to 0.99 from right to left.

The solid lines in Figure 11 indicate another series of level curves of a surface in the space of the transformation stretches on which the twinned wedge microstructure with type II twins only is possible. Increasing values of the stretch β are from right to left. On the level curves, the shape strain for the variant pair

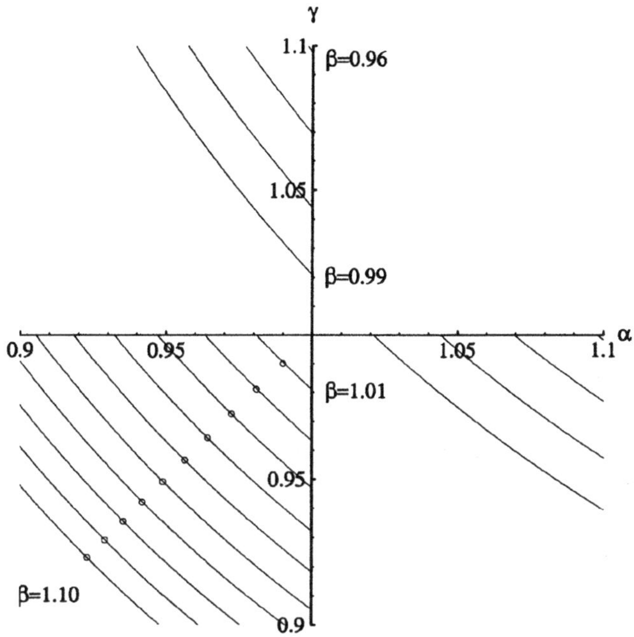

Figure 9. Second level curves on which the twinned wedge microstructure with type II twins is possible. See text for details.

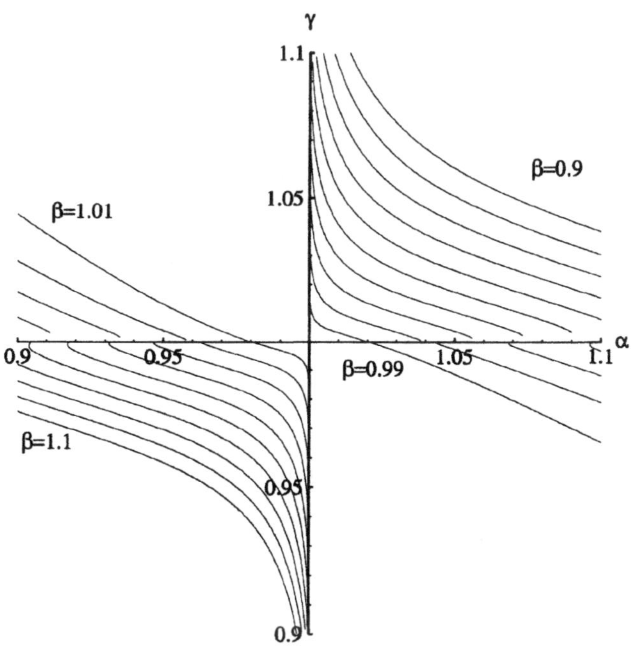

Figure 10. Third level curves on which the twinned wedge microstructure with type II twins is possible. See text for details.

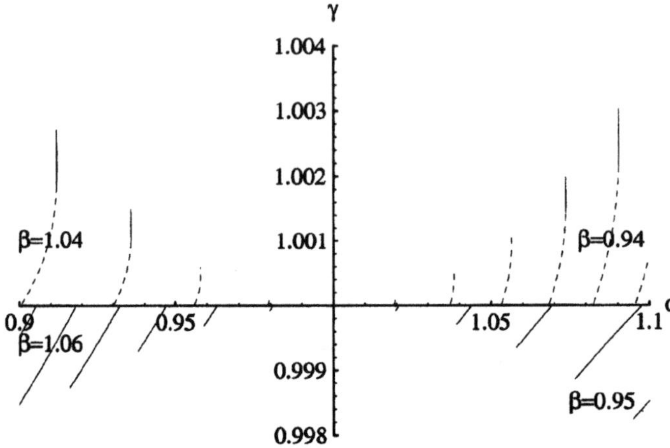

Figure 11. Fourth and fifth level curves on which the twinned wedge microstructure with type II twins is possible. See text for details.

$(i : j) = (1 : 3)$ with volume fraction λ of variant $\mathbf{U}_i$ is perpendicular to a vector with components $(1, -1, 0)$; other level curves are obtained by switching the stretches α and γ, but the shape strain for the same variant pair and volume fraction λ of variant $\mathbf{U}_i$ is also perpendicular to a vector with components $(1, -1, 0)$. The curves as shown in Figure 11 are not reported in [5]. The level curves end so abruptly because they are restricted by the boundary of the admissible domain for the austenite-twinned martensite microstructures as shown in Figures 2. The reason that there are no analogous curves for the type I twin wedges is unknown.

The other variant pair $(k : l)$ with which the variant pair $(i : j) = (1 : 3)$ with volume fraction λ of variant $\mathbf{U}_i$ can form a twinned wedge microstructure is $(k : l) = [1 : 3]$. For the twinned wedges in Figure 11, the results of the symmetry transformation are that the following twinned wedges are possible:

$$\begin{pmatrix} ((1:3),[1:3]) & ((1:4),[1:4]) & ((1:5),[1:5]) \\ ((1:6),[1:6]) & ((2:3),[2:3]) & ((2:4),[2:4]) \\ ((2:5),[2:5]) & ((2:6),[2:6]) & ((3:5),[3:5]) \\ ((3:6),[3:6]) & ((4:5),[4:5]) & ((4:6),[4:6]) \end{pmatrix} \tag{55}$$

These are interesting twinned wedges because they can form with only two variants, where all of the other cases need three or four variants. In set (55), it is the same variants, but with different volume fractions on each side and none of which are found to one half. All of the midrib planes belong to the $\{110\}$ family of planes for the twelve distinct twinned wedges, and each variant pair forms a twinned wedge with exactly one other.

Figure 11 also shows the level curves of a fifth surface in the space of the transformation stretches on which the twinned wedge microstructure with type II twins only is possible. They are indicated in this figure by the dashed lines.

Increasing values of the stretch β are from right to left. On the level curves, the shape strain for the variant pair $(i : j) = (1 : 3)$ with volume fraction λ of variant $\mathbf{U}_i$ is perpendicular to the vector with components $(1, 0, -1)$; other level curves are obtained by switching the stretches α and γ, and here the shape strain for the variant pair $(i : j) = (1 : 3)$ with volume fraction λ of variant $\mathbf{U}_i$ is perpendicular to the vector with components $(1, 0, 1)$. These level curves are also not reported in [5], and they end so abruptly because they are restricted by the boundary of the admissible domain for the austenite-twinned martensite microstructures as shown in Figures 2. As for the other curves in Figure 11, the reason that there are no analogous curves for the type I twin wedges is unknown.

The other variant pair $(k : l)$ with which the variant pair $(i : j) = (1 : 3)$ with volume fraction λ of variant $\mathbf{U}_i$ can form a twinned wedge microstructure is $(k : l) = [3 : 5]$. For materials with stretches which lie on one of these curves, the results of the symmetry transformation are that the following twinned wedges are possible:

$$\begin{pmatrix} ((1:3),[3:5]) & ([1:3],[1:5]) & ((1:4),[4:6]) \\ ([1:4],[1:6]) & ((1:5),(3:5)) & ((1:6),(4:6)) \\ ((2:3),[3:6]) & ([2:3],[2:6]) & ((2:4),[4:5]) \\ ([2:4],[2:5]) & ((2:5),(4:5)) & ((2:6),(3:6) \end{pmatrix}. \tag{56}$$

The other possibility is on the curves in Figure 11 where the stretches α and γ are switched. On these curves, the shape strain is perpendicular to the vector with components (1, 0, 1). Repeating the above calculation to find all of the possible twinned wedges microstructures gives a set which can be found from set (56) by interchanging the labels of the variants: $1 \leftrightarrow 2$, $3 \leftrightarrow 4$, and $5 \leftrightarrow 6$ (see discussion after set (54)). In these sets, all of the midrib planes belong to the $\{110\}$ family of planes; there are twelve distinct twinned wedges; and each variant pair forms a twinned wedge with exactly one other variant pair.

The last numerical result is that in the domain $\alpha \in [0.90,\ 1.1]$ and $\gamma \in [0.90,\ 1.1]$ with the transformation stretch $\beta = 0.99$, a curve exists on which the shape strain for the variant pair $(i : j) = (1 : 3)$ with volume fraction λ of variant $\mathbf{U}_i$ is perpendicular to the vector with components $(0, 0, 1)$. It happens that this curve is exactly given by $\gamma = \alpha/\sqrt{2\alpha^2 - 1}$, which forms part of the boundary of the austenite-compound twinned martensite microstructure domain. From Figure 2, it is only the $\beta = 0.99$ curve which intersects the austenite-compound twinned martensite microstructure boundary (dashed line), and it is that small segment along the dashed line which is found by the numerical computation. With other stretches β, these wedges exist, but outside of the aforementioned domain. The other variant pair $(k : l)$ with which the variant pair $(i : j) = (1 : 3)$ with volume fraction λ of variant $\mathbf{U}_i$ can form a twinned wedge microstructure is

$(k : l) = (2 : 4)$. For these wedges, the results of the symmetry transformation are that the following twinned wedges are possible:

$$\begin{pmatrix} ((1:3),(2:4)) & ([1:3],[2:4]) & ((1:4),(2:3)) \\ ([1:4],[2:3]) & ((1:5),(2:6)) & ([1:5],[2:6]) \\ ((1:6),(2:5)) & ([1:6],[2:5]) & ((3:5),(4:6)) \\ ([3:5],[4:6]) & ((3:6),(4:5)) & ([3:6],[4:5]) \end{pmatrix}. \tag{57}$$

These twinned wedges have midrib planes contained in the {100} family of planes, and there are twelve distinct twinned wedges. Like the set from equation (53), the twinned wedges in the set from equation (57) contain four distinct orthorhombic variants. Each variant pair forms a twinned wedge microstructure with exactly one other variant pair. It is not clear why there are no analogous curves with type I twins.

The level curves in the six cases above are the only curves found by the numerical computation on which the twinned wedge microstructures with type II twins only can form. At almost all of the points on the level curves, twelve twinned wedge microstructures are possible. As for the type I twin wedges, there are several special points discussed below at which more than twelve twinned wedge microstructures can be constructed.

6.2.7. *Type II Twinned Wedge with a Transformation Stretch Equal to One*

As for the twinned wedges with type I twins, wedges with type II twins and a transformation stretch equal to one may exist. From Section 5.2.2 with stretches $\alpha = 1$ and $\gamma = \sqrt{3 - 2\beta^2}$, the austenite-twinned martensite microstructure with type II twins is possible for all volume fractions λ. Notice that these conditions are those needed for a single variant wedge as discussed in Section 6.1.

The twinned wedge microstructure with stretches that lie on the above curve is found numerically by the same approach as for the type I twinned wedge, and the results of the numerical computation are given as the dashed line in Figure 7. For stretches $\beta < 1$, the shape strain for the variant pair $(i : j) = (1 : 3)$ with volume fraction λ of variant $\mathbf{U}_i$ is perpendicular to the vector with components $(1, 0, 0)$; and the habit plane normal is parallel to the vector with components $(1, 1, -1)$. For stretches $\beta > 1$, the shape strain for the variant pair $(i : j) = (1 : 3)$ with volume fraction λ of variant $\mathbf{U}_i$ is perpendicular to the vector with components $(0, 1, 0)$; and the habit plane normal is parallel to the vector with components $(1, 1, -1)$. This is in agreement with Figures 8 and 10. In Figure 8, the level curves on which the twinned wedge microstructure exist are continuous through the lines $\alpha = 1$ and $\gamma = 1$. On the contrary, in Figure 10, the shape strain and habit plane normal are both perpendicular to the same vector, which is the reason for the discontinuity in the level curves at the lines $\alpha = 1$ and $\gamma = 1$.

As for the type I twinned wedge, the twinned wedge microstructure with type II twins and the transformation stretch $\beta = 1$ is not investigated in this article;

however, the single variant wedge microstructure with the stretch $\beta = 1$ is possible as discussed in Section 6.1.

6.2.8. *Twinned Wedge for Categories (TWe) and (TWf)*

The numerical computation finds no twinned wedge microstructures such that the shape strain $\mathbf{b}'$ with $\kappa = +1$ for the variant pair $(i : j)$ is parallel to the shape strain $\mathbf{b}''$ with $\kappa = -1$ for the variant pair $(k : l)$. It may be that this kind of wedge occurs at isolated points which are missed by the numerical algorithm or these twinned wedges may not be possible.

6.2.9. *Twinned Wedge for Category (TWg) – a Mixed Twinned Wedge*

A mixed twinned wedge is a wedge microstructure such that one of the martensite regions contains type I twins and the other martensite region contains type II twins. The numerical calculation finds that there is a curve in the space of three transformation stretches α, β and γ on which this microstructure is possible. A plot of line segments joining the computed points is shown in Figure 12. In this figure, for each value of the stretch β, the stretch α can be taken from the solid curve, while the stretch γ is given on the dashed curve. There are twenty-four distinct variant pairs which can form a mixed twinned wedge as listed in the set

$$\begin{pmatrix} ((1:3)_{\mathrm{I}}, (2:5)_{\mathrm{II}}) & ([1:3]_{\mathrm{I}}, (4:5)_{\mathrm{II}}) & ((1:4)_{\mathrm{I}}, (2:6)_{\mathrm{II}}) \\ ([1:4]_{\mathrm{I}}, (3:6)_{\mathrm{II}}) & ((1:5)_{\mathrm{I}}, (2:3)_{\mathrm{II}}) & ([1:5]_{\mathrm{I}}, [3:6]_{\mathrm{II}}) \\ ((1:6)_{\mathrm{I}}, (2:4)_{\mathrm{II}}) & ([1:6]_{\mathrm{I}}, [4:5]_{\mathrm{II}}) & ((2:3)_{\mathrm{I}}, (1:6)_{\mathrm{II}}) \\ ([2:3]_{\mathrm{I}}, (4:6)_{\mathrm{II}}) & ((2:4)_{\mathrm{I}}, (1:5)_{\mathrm{II}}) & ([2:4]_{\mathrm{I}}, (3:5)_{\mathrm{II}}) \\ ((2:5)_{\mathrm{I}}, (1:4)_{\mathrm{II}}) & ([2:5]_{\mathrm{I}}, [4:6]_{\mathrm{II}}) & ((2:6)_{I}, (1:3)_{\mathrm{II}}) \\ ([2:6]_{\mathrm{I}}, [3:5]_{\mathrm{II}}) & ((3:5)_{\mathrm{I}}, [1:4]_{\mathrm{II}}) & ([3:5]_{\mathrm{I}}, [1:6]_{\mathrm{II}}) \\ ((3:6)_{\mathrm{I}}, [2:4]_{\mathrm{II}}) & ([3:6]_{\mathrm{I}}, [2:5]_{\mathrm{II}}) & ((4:5)_{\mathrm{I}}, [2:3]_{\mathrm{II}}) \\ ([4:5]_{\mathrm{I}}, [2:6]_{\mathrm{II}}) & ((4:6)_{\mathrm{I}}, [1:3]_{\mathrm{II}}) & ([4:6]_{\mathrm{I}}, [1:5]_{\mathrm{II}}) \end{pmatrix}. \tag{58}$$

In the set (58), the notation is that variant pairs $(i : j)_{\mathrm{I}}$ are with type I twins, and variant pairs $(k : l)_{\mathrm{II}}$ are with type II twins. Also, each variant pair forms a wedge with exactly one other variant pair, and the set is the same for all stretches in Figure 12. Of course, interchanging the stretches α and γ gives other mixed twinned wedges which can be found from those in set (58) by just relabeling the variants. As an example, the numerical computation shows that for the stretch $\beta = 0.9$, the stretches to give the mixed twinned wedge are $\alpha = 1.0374$ and $\gamma = 1.0791$. The components of the vector solutions to the twinning equation (38) and the habit plane equation (39) with the variant pair $(i : j) = (1 : 3)_I$ and type I twins are

$$a' = (-0.2404, -0.2037, 0.0543), \tag{59}$$

$$\hat{n}' = (0.7071, -0.7071, 0), \tag{60}$$

$$b' = 0.1142(-0.2346, 0.6475, 0.7251), \tag{61}$$

$$\widehat{m}' = (-0.1212, -0.7141, 0.6895), \tag{62}$$

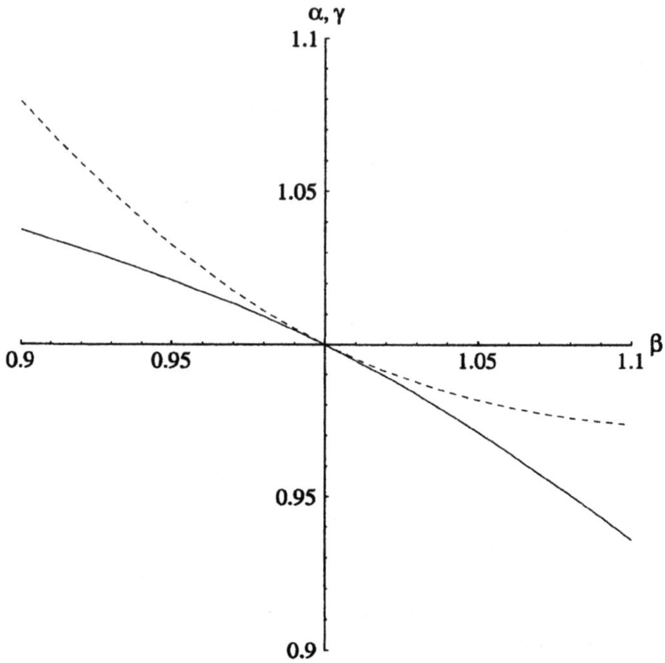

Figure 12. Transformation stretches α, β, and γ for which a mixed twinned wedge is possible. For each stretch β, the stretch α is given on the solid curve, and and the stretch γ is given on the dashed curve.

where the volume fraction of variant $\mathbf{U}_i$ is $\lambda' = 0.3298$. The variant pair $(k : l)$ with which $(i : j) = (1 : 3)_{\mathrm{I}}$ can form a mixed twinned wedge is $(k : l) = (2 : 5)_{\mathrm{II}}$. For this case, the components of the vector solutions to the twinning equation (40) and the habit plane equation (41) with the variant pair $(k : l) = (2 : 5)_{\mathrm{II}}$ and type II twins are

$$a'' = (-0.2454, 0.00484, -0.2087), \tag{63}$$

$$\hat{n}'' = (0.6932, -0.1972, -0.6932), \tag{64}$$

$$b'' = 0.1111(-0.2346, 0.6475, 0.7251), \tag{65}$$

$$\widehat{m}'' = (-0.0696, 0.7783, -0.6240), \tag{66}$$

where the volume fraction of variant $\mathbf{U}_k$ is $\lambda'' = 0.3398$. Clearly, the conditions for the twinned wedge are satisfied. Moreover, it can be shown that the midrib shape strain $\mathbf{b}$ and the midrib plane normal $\widehat{\mathbf{m}}$ have components $b = 0.2240$ $(-0.2346,\ 0.6475,\ 0.7251)$ and $\widehat{m} = (-0.0273,\ -0.7500,\ 0.6609)$, respectively.

6.2.10. *Parallel Twinned Wedge*

So far, the wedge microstructures considered have had shape strains which are perpendicular to an axis of a rotation in $\mathcal{L}^c$. In contrast, the last kind of wedge

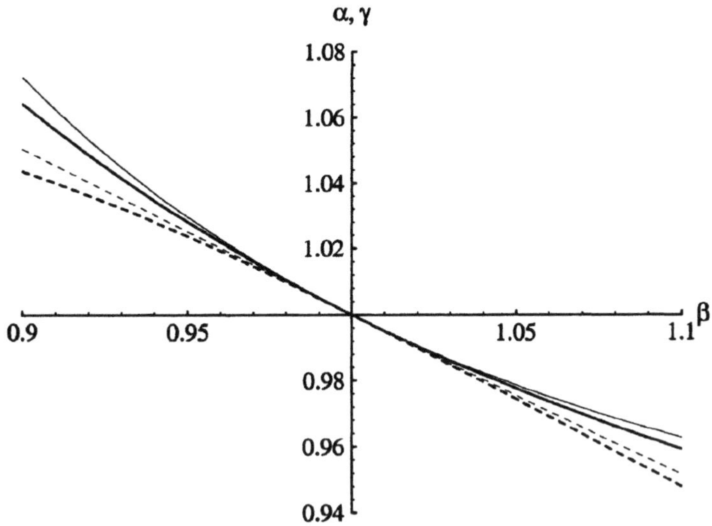

Figure 13. Transformation stretches for which the parallel twinned wedge with type I or type II twins is possible. The thin (thick) lines are with type I (II) twins. See text for more details.

microstructure constructed in this article is a twinned wedge with shape strains parallel to the axis of a rotation in $\mathcal{L}^c$. This type of twinned wedge is called a parallel twinned wedge, and as discussed in [26], there are two condition which must be satisfied by the transformation stretches in order for the shape strains to be parallel to the axis of a rotation in $\mathcal{L}^c$. Thus, for the cubic to orthorhombic transition, the parallel twinned wedge is possible only on curves in the space of the stretches. To find these curves, the procedure is to search through all of the data points as plotted in Figures 4 and 11 looking for habit plane solutions with the shape strain parallel to a vector with components in the $\langle 100 \rangle$, $\langle 110 \rangle$ or $\langle 111 \rangle$ families of directions and the habit plane normal not parallel to the same vector. This search gives, for both the type I twinned wedges and the type II twinned wedges, a curve in the space of transformation stretches as shown in Figure 13, where the solid curve is the stretch α and the dashed curve is the stretch γ. Along this curve for the variant pair $(i : j) = (1 : 3)$ with volume fraction λ of variant $\mathbf{U}_i$, the shape strain is parallel to the vector with components (0, 1, 1); and the habit plane normal is not parallel to this vector. On the other hand, if α and γ are switched, then the variant pair $(i : j) = (1 : 3)$ with volume fraction λ of variant $\mathbf{U}_i$ has shape strain parallel to the vector with components $(0, 1, -1)$; and habit plane normal is not parallel to this vector. For the twinned wedge with either type I or type II twins, the variant pairs $((i : j), (k : l))$ forming a parallel twinned wedge are listed in Table V, where the components of the shape strains and habit plane normals can be found numerically. Moreover, it is found that the components μ_1, μ_2, and μ_3 are nonzero and have different absolute values. So, for this case, 36 twinned wedges are possible; the midrib planes are in the $\{100\}$ and $\{110\}$ families of planes; and each variant pair forms a wedge with three other distinct variant pairs.

The transformation stretches in Figure 13 for type I twins are contained in Figures 4 and 6. On the other hand, fixing the stretch β, there are stretches which are common to the curves in Figures 4 and 5. At these stretches, the habit plane equation (39) has two solutions, one with $\kappa = +1$ and one with $\kappa = -1$; and both of these solutions satisfy different conditions to form a wedge: the shape strains are perpendicular to different vectors. A similar result holds at any stretches common to the curves Figures 5 and 6 and for all of the figures with type II twinned wedges.

7. Triangle Microstructure

Another special microstructure is the triangle microstructure. This microstructure is a triangular pyramid made from three deformations which are mutually compatible and are each individually compatible with the austenite phase. Each deformation has gradient either as a single martensite variant or an average of two twinned martensite variants. The three habit planes intersect at a point and through this point passes a line which is the intersection of the midrib planes between the three deformations forming the triangle microstructure. The necessary compatibility equations for the triangle are given in [26]; therein, it is shown that these equations are all satisfied when the three shape strain vectors are parallel; the three habit plane normals are not coplanar; and the three midrib planes are coplanar. Candidates to form the triangle have the property that any deformation must be able to form a wedge microstructure with at least two other distinct deformations and the midrib planes of the three wedges forming the triangle microstructure must be coplanar. There are two categories of wedges for the cubic to orthorhombic transition, single variant and twinned, that may form a triangle microstructure.

The single variant wedges are discussed in Section 6.1, where two kinds are considered. The first kind is with transformation stretches $\alpha = 1$ and $\gamma = \beta/\sqrt{3\beta^2 - 2}$; from Table II, each deformation is able to form a wedge with two distinct deformations. There are four distinct triangle microstructures which can be constructed, and they are listed in the aforementioned table segregated by vertical space. The second kind is a single variant wedge with stretches $\beta = 1$ and $\gamma = \alpha/\sqrt{2\alpha^2 - 1}$; from Table III, each deformation is able to form a wedge with three distinct deformations. From each of the three sets of wedges as segregated by vertical space within Table III, it can be shown that four distinct triangle microstructures are possible. Thus, twelve distinct triangle microstructures are possible. On the other hand, for the twinned wedges, the only deformations which can possibly form the triangle microstructures are the parallel twinned wedges which are listed in the set in Table V. For the variant pairs with parallel shape strains, the triangle microstructure is not possible, because the condition that the midrib planes are non-coplanar can not be satisfied.

Table V. Parallel twinned wedge microstructures. See text for details

$((i:j),(k:l))$	b' and b''	$\widehat{m}'$	$\widehat{m}''$	$\widehat{m}$
$((1:3),(1:4))$	$\mu(0,1,1)$	(μ_1,μ_2,μ_3)	$(-\mu_1,\mu_2,\mu_3)$	$(1,0,0)$
$((1:3),(1:5))$	$\mu(0,1,1)$	(μ_1,μ_2,μ_3)	(μ_1,μ_3,μ_2)	$(0,1,-1)$
$((1:3),(1:6))$	$\mu(0,1,1)$	(μ_1,μ_2,μ_3)	$(-\mu_1,\mu_3,\mu_2)$	$(0,1,1)$
$((1:4),(1:5))$	$\mu(0,1,1)$	$(-\mu_1,\mu_2,\mu_3)$	(μ_1,μ_3,μ_2)	$(0,1,1)$
$((1:4),(1:6))$	$\mu(0,1,1)$	$(-\mu_1,\mu_2,\mu_3)$	$(-\mu_1,\mu_3,\mu_2)$	$(0,1,1)$
$((1:5),(1:6))$	$\mu(0,1,1)$	(μ_1,μ_3,μ_2)	$(-\mu_1,\mu_3,\mu_2)$	$(1,0,0)$
$((2:3),(2:4))$	$\mu(0,1,-1)$	$(\mu_1,-\mu_2,\mu_3)$	$(-\mu_1,-\mu_2,\mu_3)$	$(1,0,0)$
$((2:3),(2:5))$	$\mu(0,1,-1)$	$(\mu_1,-\mu_2,\mu_3)$	$(\mu_1,\mu_3,-\mu_2)$	$(0,1,-1)$
$((2:3),(2:6))$	$\mu(0,1,-1)$	$(\mu_1,-\mu_2,\mu_3)$	$(-\mu_1,\mu_3,-\mu_2)$	$(0,1,1)$
$((2:4),(2:5))$	$\mu(0,1,-1)$	$(-\mu_1,-\mu_2,\mu_3)$	$(\mu_1,\mu_3,-\mu_2)$	$(0,1,1)$
$((2:4),(2:6))$	$\mu(0,1,-1)$	$(-\mu_1,-\mu_2,\mu_3)$	$(-\mu_1,\mu_3,-\mu_2)$	$(0,1,1)$
$((2:5),(2:6))$	$\mu(0,1,-1)$	$(\mu_1,\mu_3,-\mu_2)$	$(-\mu_1,\mu_3,-\mu_2)$	$(1,0,0)$
$([1:3],(3:5))$	$\mu(1,0,1)$	(μ_2,μ_1,μ_3)	(μ_3,μ_1,μ_2)	$(1,0,-1)$
$([1:3],(3:6))$	$\mu(1,0,1)$	(μ_2,μ_1,μ_3)	$(\mu_3,-\mu_1,\mu_2)$	$(1,0,1)$
$((3:5),(3:6))$	$\mu(1,0,1)$	(μ_3,μ_1,μ_2)	$(\mu_3,-\mu_1,\mu_2)$	$(0,1,0)$
$([1:3],[2:3])$	$\mu(1,0,1)$	(μ_2,μ_1,μ_3)	$(\mu_2,-\mu_1,\mu_3)$	$(0,1,0)$
$([2:3],(3:5))$	$\mu(1,0,1)$	$(\mu_2,-\mu_1,\mu_3)$	(μ_3,μ_1,μ_2)	$(1,0,1)$
$([2:3],(3:6))$	$\mu(1,0,1)$	$(\mu_2,-\mu_1,\mu_3)$	$(\mu_3,-\mu_1,\mu_2)$	$(1,0,-1)$
$([1:4],(4:5))$	$\mu(1,0,-1)$	$(\mu_2,-\mu_1,\mu_3)$	$(\mu_3,\mu_1,-\mu_2)$	$(1,0,-1)$
$([1:4],(4:6))$	$\mu(1,0,-1)$	$(\mu_2,-\mu_1,\mu_3)$	$(\mu_3,-\mu_1,-\mu_2)$	$(1,0,1)$
$((4:5),(4:6))$	$\mu(1,0,-1)$	$(\mu_3,\mu_1,-\mu_2)$	$(\mu_3,-\mu_1,-\mu_2)$	$(0,1,0)$
$([1:4],[2:4])$	$\mu(1,0,-1)$	$(\mu_2,-\mu_1,\mu_3)$	$(\mu_2,\mu_1,-\mu_3)$	$(0,1,0)$
$([2:4],(4:5))$	$\mu(1,0,-1)$	$(\mu_2,\mu_1,-\mu_3)$	$(\mu_3,\mu_1,-\mu_2)$	$(1,0,1)$
$([2:4],(4:6))$	$\mu(1,0,-1)$	$(\mu_2,\mu_1,-\mu_3)$	$(\mu_3,-\mu_1,-\mu_2)$	$(1,0,-1)$
$([1:5],[2:5])$	$\mu(1,1,0)$	(μ_2,μ_3,μ_1)	$(\mu_2,\mu_3,-\mu_1)$	$(0,0,1)$
$([1:5],[3:5])$	$\mu(1,1,0)$	(μ_2,μ_3,μ_1)	(μ_3,μ_2,μ_1)	$(1,-1,0)$
$([1:5],[4:5])$	$\mu(1,1,0)$	(μ_2,μ_3,μ_1)	$(\mu_3,\mu_2,-\mu_1)$	$(1,1,0)$
$([2:5],[3:5])$	$\mu(1,1,0)$	$(\mu_2,\mu_3,-\mu_1)$	(μ_3,μ_2,μ_1)	$(1,1,0)$
$([2:5],[4:5])$	$\mu(1,1,0)$	$(\mu_2,\mu_3,-\mu_1)$	$(\mu_3,\mu_2,-\mu_1)$	$(1,-1,0)$
$([3:5],[4:5])$	$\mu(1,1,0)$	(μ_3,μ_2,μ_1)	$(\mu_3,\mu_2,-\mu_1)$	$(0,0,1)$
$([1:6],[2:6])$	$\mu(1,-1,0)$	$(\mu_2,-\mu_3,-\mu_1)$	$(\mu_2,-\mu_3,\mu_1)$	$(0,0,1)$
$([1:6],[3:6])$	$\mu(1,-1,0)$	$(\mu_2,-\mu_3,-\mu_1)$	$(\mu_3,-\mu_2,\mu_1)$	$(1,-1,0)$
$([1:6],[4:6])$	$\mu(1,-1,0)$	$(\mu_2,-\mu_3,-\mu_1)$	$(\mu_3,-\mu_2,-\mu_1)$	$(1,1,0)$
$([2:6],[3:6])$	$\mu(1,-1,0)$	$(\mu_2,-\mu_3,\mu_1)$	$(\mu_3,-\mu_2,\mu_1)$	$(1,1,0)$
$([2:6],[4:6])$	$\mu(1,-1,0)$	$(\mu_2,-\mu_3,\mu_1)$	$(\mu_3,-\mu_2,-\mu_1)$	$(1,-1,0)$
$([3:6],[4:6])$	$\mu(1,-1,0)$	$(\mu_3,-\mu_2,\mu_1)$	$(\mu_3,-\mu_2,-\mu_1)$	$(0,0,1)$

8. Diamond Microstructure

The third and final special microstructure is the diamond microstructure. This microstructure is a diamond pyramid made from four deformations which are mutually compatible and are each individually compatible with the austenite phase. The four habit planes intersect at a point, and through this point passes a line which is the intersection of the midrib planes between the four deformations forming the diamond microstructure. In [26], the necessary compatibility equations for the diamond to be possible are discussed. These are satisfied if all of the shape strain vectors are parallel; all of the habit plane normals are not coplanar; and all of the midrib planes are coplanar. This means that any deformation in the diamond microstructure must be able to form a wedge microstructure with at least two other distinct deformations and the midrib planes of the four wedges forming the diamond microstructure must be coplanar. A diamond may be thought of as two wedges placed back-to-back, and it may form with only two unique midrib planes.

On following the discussion of the triangle microstructure in Section 7, there are three possible deformations which can form the diamond microstructure, two single variant wedge deformations and one twinned wedge deformation. From Section 6.1, the first kind of single variant wedge is with transformation stretches $\alpha = 1$ and $\gamma = \beta/\sqrt{3\beta^2 - 2}$; and, from Table II, each deformation is able to form a wedge with two distinct deformations. The diamond microstructure is not possible for these transformation stretches, because at least four distinct deformations must have parallel shape strain, which is not the case as shown in Table II. The second kind is the single variant wedge with stretches $\beta = 1$ and $\gamma = \alpha/\sqrt{2\alpha^2 - 1}$. From Table III, each deformation is able to form a wedge with three distinct deformations. For this case, there are three diamond microstructures which are possible, one from each of the sets of six wedges listed in Table III. Finally, for the parallel twinned wedge listed in Table V, there are a total of eighteen distinct diamond microstructures. From each group of six wedges segregated by vertical space within Table V, three distinct diamond microstructures are possible: in fact, two of these three have the line of intersection of the midrib planes in the $\langle 110 \rangle$ family of directions; while, the remaining one has the line of intersection in the $\langle 100 \rangle$ family. Also, each of these diamond microstructures has only two distinct midrib planes which are perpendicular to each other.

9. Summary and Conclusions

A summary of some of the microstructures for a cubic to orthorhombic transition is given in Table VI. The table lists each of the microstructures considered in this paper, along with the number of unique, possible realizations of a particular microstructure using all of the orthorhombic variants. The specific references for the Cu–Al–Ni alloy are given in [27], where a detailed comparison between theoretical predictions and experimental data is given; and some references for the Ti–Ni–Cu alloy can be found in [25]. A large variety of different microstructures

Table VI. Summary of microstructures for a cubic to orthorhombic transition

Microstructure	Number	Observed	Comments
twinned martensite	12	Cu–Al–Ni	compound twins
	24	Cu–Al–Ni	type I twins
	24	Cu–Al–Ni	type II twins
austenite-single	12	Ti–Ni–Cu	with $\alpha = 1$
variant of martensite	12		with $\beta = 1$
austenite-twinned	24	Ti–Ni–Cu	with compound twins
martensite	48	Cu–Al–Ni	with type I twins
	48	Cu–Al–Ni	with type II twins
single variant wedge	12		$\alpha = 1,\ \gamma = \beta/\sqrt{3\beta^2 - 2}$
	18		$\beta = 1,\ \gamma = \alpha/\sqrt{2 - \alpha^2}$
twinned wedge	12		compound twins, on curve in equation (43)
	12		compound twins, on curve in equation (47)
	12	Cu–Al–Ni	type I twins, on surfaces
	12	Cu–Al–Ni	type II twins, on surfaces
	24		both type I and type II twins, mixed twinned wedge
single variant triangle	4		$\alpha = 1,\ \gamma = \beta/\sqrt{3\beta^2 - 2}$
	12		$\beta = 1,\ \gamma = \alpha/\sqrt{2 - \alpha^2}$
twinned triangle			not possible
single variant diamond	3		$\beta = 1,\ \gamma = \alpha/\sqrt{2 - \alpha^2}$
twinned diamond	18		type I twins on curve
	18		type II twins on curve

is possible for this transition, and this can be contrasted to the number of possible microstructures for the cubic to tetragonal and cubic to trigonal transitions as discussed in [26, 29], respectively.

In this paper, it is shown that a material undergoing a cubic to orthorhombic transition may exhibit a rich collection of microstructures depending on the specific lattice parameters or transformation stretches. While twins are possible for all values of the stretches, continuous microstructures between the austenite and martensite phases are possible in a subset contained in the set of stretches. Some of the more intriguing microstructures, which involve fitting multiple austenite–martensite microstructures together, are possible along surfaces in the space of transformation stretches, and these surfaces contain special curves of particular interest. Materials with stretches which lie on one of these curves are able to form the austenite-single variant of martensite, the single variant wedge, and the triangle microstructures. Further, the austenite-twinned martensite microstructures are possible for which the volume fraction and the fineness of the twins are free parameters. In particular, it can be imagined that a specimen forms a series of parallel single variant wedges, and while the interface between austenite and martensite is very jagged for a small number of bands, it becomes more planar as the number of bands (fineness) tends to infinity. Such materials could be the ultimate shape memory alloys, because they could form an infinite variety of different microstructures at essentially zero energy in order to accommodate internal grain boundary constraints or external boundary conditions.

Acknowledgements

This work was supported by the National Science Foundation (grant # NSF MSS-9257945) and the Office of Naval Research (grant # ONR N/N00014-91-J-4034).

References

1. R. Abeyaratne, C. Chu and R.D. James, Kinetics of materials with wiggly energies: Theory and application to the evolution of twinning microstructures in a Cu–Al–Ni shape memory alloy. *Philos. Mag. A* **73** (1996) 457–497.
2. J.M. Ball, C. Chu and R.D. James, Hysteresis during stress-induced variant rearrangement. In: R. Gotthardt and J. van Humbeeck (eds), *Proc. of the Internat. Conf. on Martensitic Transformations*, Lausanne, Switzerland, (1995); appear in *J. Phys. IV*, supplment to J. Phys. III **5**, colloque C8 (1995) 245–251.
3. J.M. Ball and R.D. James, Fine phase mixtures as minimizers of energy. *Arch. Rational Mech. Anal.* **100** (1987) 13–52.
4. J.M. Ball and R.D. James, Proposed experimental tests of a theory of fine microstructure and the two-well problem. *Philos. Trans. Roy. Soc. London A* **338** (1992) 389–450.
5. K. Bhattacharya, Wedge-like microstructure in martensites. *Acta Metallurgica Materialia* **39** (1991) 2431–2444.
6. K. Bhattacharya, Self-accommodation in martensite. *Arch. Rational Mech. Anal.* **120** (1992) 201–244.
7. K. Bhattacharya, Comparison of the geometrically nonlinear and linear theories of martensitic transformation. *Continuum Mech. Thermodyn.* **5** (1993) 205–242.
8. K. Bhattacharya, The microstructure of martensite and its implications for the shape-memory effect. In: D. Kinderlehrer, R. James, M. Luskin and J.L. Ericksen (eds), *Microstructure and Phase Transition*, IMA Vol. 54. Springer, New York (1994) pp. 1–25.
9. K. Bhattacharya, Private communication.
10. K. Bhattacharya, Kinematics of crossing twins. In: R.C. Batra and M.F. Beatty (eds), *Contemporary Research in the Mechanics and Mathematics of Materials*. CIMNE, Barcelona (1996) pp. 251–262.
11. K. Bhattacharya, B. Li and M. Luskin, The simply laminated microstructure in martensite crystals that undergo a cubic-to-orthorhombic phase transformation. *Arch. Rational Mech. Anal.* **149** (1999) 123–154.

12. K.A. Bywater and J.W. Christian, Martensitic transformations in titanium–tantalum alloys. *Philos. Mag.* **25** (1972) 1249–1273.
13. C. Chu, Hysteresis and microstructures: a study of biaxial loading on compound twins of copper–aluminum–nickel single crystals. PhD Dissertation, University of Minnesota (1993).
14. C. Chu and R.D. James, Biaxial-loading experiments on Cu–Al–Ni single crystals. In: K.S. Kim (ed.), *Experiments in Smart Materials and Structures*, AMD, Vol. 181, ASME, New York (1993) pp. 61–69.
15. C. Chu and R.D. James, Analysis of microstructures in Cu-14.0% Al-3.9% Ni by energy minimization. In: R. Gotthardt and J. van Humbeeck (eds), *Proc. of the Internat. Conf. on Martensitic Transformations*, Lausanne, Switzerland, (1995); appears in *J. Phys. IV*, Suppl. to J. Phys. III **5**, colloque C8 (1995) 143–149.
16. J.L. Ericksen, On the symmetry and stability of thermoelastic solids. *J. Appl. Mech.* **45** (1978) 740–744.
17. J.L. Ericksen, On the symmetry of deformable crystals. *Arch. Rational Mech. Anal.* **72** (1979) 1–13.
18. J.L. Ericksen, Some phase transitions in crystals. *Arch. Rational Mech. Anal.* **73** (1980) 99–124.
19. J.L. Ericksen, The Cauchy and Born hypotheses for crystals. In: M.E. Gurtin (ed.), *Phase Transformations and Material Instabilities in Solids*. Academic Press, New York (1984) pp. 61–77.
20. J.L. Ericksen, Twinning of crystals (I). In: S. Antman, J.L. Ericksen, D. Kinderlehrer and I. Müller (eds), IMA Vol. 3. Springer, New York (1987) pp. 77–93.
21. R.L. Fosdick and B. Hertog, Material symmetry and crystals. *Arch. Rational Mech. Anal.* **110** (1990) 43–72.
22. H. Funakubo (ed.), *Shape Memory Alloys*. Gordon and Breach, New York (1987).
23. M.E. Gurtin, Two-phase deformations of elastic solids. *Arch. Rational Mech. Anal.* **84** (1984) 1–29.
24. E.O. Hall, *Twinning and Diffusionless Transformations in Metals*. Butterworths, London (1954).
25. K.F. Hane, Microstructures in thermoelastic martensites. PhD Thesis, University of Minnesota (1998).
26. K.F. Hane and T.W. Shield, Symmetry and microstructure in martensite. *Philos. Mag. A* **78** (1998) 1215–1252.
27. K.F. Hane and T.W. Shield, Microstructure in a copper–aluminum–nickel shape memory alloy. *Proc. Roy. Soc. London A* **455** (1999) 3901–3915.
28. K.F. Hane and T.W. Shield, Microstructure in the cubic to monoclinic transition in titanium–nickel shape memory alloys. *Acta Materialia* **47** (1999) 2603–2617.
29. K.F. Hane and T.W. Shield, Microstructure in the cubic to trigonal transition. *Materials Sci. Engrg.* **A 291** (2000) 147–159.
30. R.D. James and K.F. Hane, Martensitic transformations and shape memory materials. *Acta Materialia* **48** (2000) 197–222.
31. R.D. James and D. Kinderlehrer, Theory of diffusionless phase transformations. In: M. Rascle, D. Serre and M. Slemrod (eds), *Partial Differential Equations and Continuum Models of Phase Transitions*, Lecture Notes in Physics, Vol. 344. Springer, New York (1989) pp. 51–84.
32. A. Kelly and G.W. Groves, *Crystallography and Crystal Defects*. Addison-Wesley, Reading, MA (1970).
33. V.V. Martynov and V.V. Kokorin, The crystal structure of thermally- and stress-induced martensites in Ni_2MnGa single crystals. *J. Phys. III* **2** (1992) 739–749.
34. S. Miyazaki and K. Otsuka, Development of shape memory alloys. *ISIJ Internat.* **29** (1989) 353–377.
35. W.J. Moberly, J.L. Proft, T.W. Duerig and R. Sinclair, Twinless martensite in TiNiCu shape memory alloys. *Materials Sci. Forum* **56–58** (1990) 605–610.

36. K. Otsuka and C.M. Wayman (eds), *Shape Memory Materials*. Cambridge Univ. Press, Cambridge (1998).
37. M. Pitteri, Reconciliation of local and global symmetries of crystals. *J. Elasticity* **14** (1984) 175–190.
38. M. Pitteri, On $\nu + 1$-lattices. *J. Elasticity* **15** (1985) 3–25.
39. M. Pitteri, On the kinematics of mechanical twinning in crystals. *Arch. Rational Mech. Anal.* **88** (1985) 25–57.
40. M. Pitteri, and G. Zanzotto, On the definition and classification of Bravais lattices. *Acta Crystallographica A* **A52** (1996) 830–838.
41. M. Pitteri and G. Zanzotto, Symmetry-breaking and transformation twinning. Preprint.
42. D.A. Porter and K.E. Easterling, *Phase Transformations in Metals and Alloys*, 2nd edn. Chapman and Hall, London (1993).
43. G. Ruddock, A microstructure of martensite which is not a minimizer of energy: The X-interface. *Arch. Rational Mech. Anal.* **127** (1994) 1–39.
44. T.W. Shield, Orientation dependence of the pseudoelastic behavior of single crystals of Cu–Al–Ni in tension. *J. Mech. Phys. Solids* **43** (1995) 869–895.
45. T. Saburi, Y. Watanabe and S. Nenno, Morphological characteristics of the orthorhombic martensite in a shape memory Ti–Ni–Cu alloy. ISIJ Internat. **29** (1989) 405–411.
46. T. Tadaki and C.M. Wayman, Electron microscopy studies of martensitic transformations in $Ti_{50}Ni_{50-x}Cu_x$ alloys. Part I. Compositional dependence of one-third reflections from the matrix phase. *Metallography* **15** (1982) 233–245.
47. T. Tadaki and C.M. Wayman, Electron microscopy studies of martensitic transformations in $Ti_{50}Ni_{50-x}Cu_x$ alloys. Part II. Morphology and crystal structure of martensites. *Metallography* **15** (1982) 247–258.
48. G. van Tendeloo and S. Amelinckx, Group-theoretical considerations concerning domain formation in ordered alloys. *Acta Crystallography A* **30** (1974) 431–440.
49. C.M. Wayman, *Introduction to the Crystallography of Martensitic Transformations*. Macmillan, New York (1964).
50. G. Zanzotto, On the material symmetry group of elastic crystals and the Born rule. *Arch. Rational Mech. Anal.* **121** (1992) 1–36.
51. G. Zanzotto, The Cauchy–Born hypothesis, nonlinear elasticity and mechanical twinning in crystals. *Acta Crystallographica A* **52** (1996) 839–849.
52. G. Zanzotto, Weak and symmetry-breaking transitions in simple lattices. Preprint.

Journal of Elasticity **59:** 319–355, 2000.

Finite Scale Microstructures in Nonlocal Elasticity

XIAOFENG REN[1,*] and LEV TRUSKINOVSKY[2,**]
[1]*Department of Mathematics and Statistics, Utah State University, Logan, UT 84322, U.S.A. E-mail: ren@sunfs.math.usa.edu*
[2]*Department of Aerospace Engineering and Mechanics, University of Minnesota, Minneapolis, MN 55455, U.S.A. E-mail: trusk@aem.umn.edn*

Received 22 March 1999; in revised form 21 January 2000

Dedicated to Roger Fosdick.

Abstract. In this paper we develop a simple one-dimensional model accounting for the formation and growth of globally stable finite scale microstructures. We extend Ericksen's model [9] of an elastic "bar" with nonconvex energy by including both oscillation-inhibiting and oscillation-forcing terms in the energy functional. The surface energy is modeled by a conventional strain gradient term. The main new ingredient in the model is a nonlocal term which is quadratic in strains and has a negative definite kernel. This term can be interpreted as an energy associated with the long-range elastic interaction of the system with the constraining loading device. We propose a scaling of the problem allowing one to represent the global minimizer as a collection of localized interfaces with explicitly known long-range interaction. In this limit the augmented Ericksen's problem can be analyzed completely and the equilibrium spacing of the periodic microstructure can be expressed as a function of the prescribed average displacement. We then study the inertial dynamics of the system and demonstrate how the nucleation and growth of the microstructures result in the predicted stable pattern. Our results are particularly relevant for the modeling of twined martensite inside the austenitic matrix.

1. Introduction

Self-accommodating coherent mixtures of solid phases are observed in shape memory alloys and other "active" materials in a variety of stable equilibrium configurations. Typically these microstructures exhibit almost periodic patterns of piecewise homogeneous elastic domains. Some of the most important properties of "active" materials are due to the fact that the microstructures are mobile so that large reversible deformation can be accomplished by the motion of the domain boundaries.

Intensive research in recent years has led to well defined constitutive theories which explained some general features of the equilibrium microstructures. It is now well established that it is the non(quasi)-convexity of the elastic energy which is

* Supported in part by NSF grant DMS-9703727.
** Supported in part by NSF grants DMS-9501433, DMS-9803572.

responsible for the formation of elastic domains. The adaptation of the thermoelasticity theory was successful in predicting the orientation of the microstructures and in quantifying the overall response of the phase mixtures (see the reviews of Roytburd [28], Khachaturian [17], Luskin [20], Müller [23], Ball and James [1], Pitteri and Zanzotto [25]). The classical nonlinear elasticity, however, was unable to predict the finite scale of the equilibrium domains, neither was it sufficient to describe such important dynamic phenomena as the microstructure nucleation and growth. A considerable challenge is still presented by a self-consistent description of the complex structure of the martensite-austenite interfaces separating the twinned regions from the homogeneous phases.

It has long been clear that an adequate theory should contain at least one intrinsic length scale. Recent attempts to bring a length scale into the theory were focused on gradient models which represent high order singular perturbation of the classical elasticity and are capable of describing "thick" phase boundaries (e.g., Kohn and Müller [19]). The most complete understanding was achieved in a one-dimensional setting where the analysis can be carried out explicitly. Unfortunately, the one-dimensional gradient models, having the merit of simplicity, fail to describe some significantly non-one-dimensional effects, in particular the strain compatibility constraint and, what is more important for the subject of this paper, the constraint provided by the fixation of the boundary displacements. This last limitation precludes the refinement of the microstructure in principle (e.g., Carr et al. [6]). The problem does not disappear if the gradient term describing the short range interaction is substituted by a fully nonlocal term with a positive definite kernel (e.g., Fosdick and Mason [12–14], Brandon et al. [5], Rogers and Truskinovsky [27], Bates et al. [4], Ren and Winter [26]).

The fact that the observed microstructures (at least away from the boundaries) are grossly one-dimensional suggests that the one-dimensional framework may still be sufficient for modeling the main features of the domain layering. To obtain a description of the stable finite scale microstructures in terms of a one-dimensional ansatz, it is necessary to include both oscillation-inhibiting and oscillation-forcing terms in the governing equations. The corresponding models must not only penalize sharp interfaces but also account for the incompatibility of the fixed displacements on the boundary with the coarse spacing of elastic domains. The two competing interactions operate at drastically different length scales: atomic distances in the case of surface effects and the size of the sample in the case of the boundary constraint. In order to preserve the long range component of the interaction in the one-dimensional setting one needs to introduce an additional non-constitutive compensating term into the energy functional; formally this term has the appearance of a nonlocal coupling. The most well known example of such a quasi-one-dimensional ansatz is the Pierls's model in the theory of dislocations [24].

Sometimes, as in the case of a shape-memory epitaxial thin film on an elastic substratum, the one-dimensional model is applicable directly. In this case, the elastic foundation plays the role of an oscillation-forcing agent (Ball et al. [2]). The "effective" one-dimensional setting was adopted in a series of recent papers (Müller [23], Truskinovsky and Zanzotto [33, 34], Vainchten et al. [35]), where the Ericksen's bar model with nonconvex energy was augmented by the introduction of the strain gradient surface energy and where the displacements of the bar were coupled quadratically with a rigid substratum. The "foundation" term, which depends on displacements (rather than strains), can also be rewritten in terms of strains as a nonlocal term with a nonpositive definite interaction kernel. The model was shown to present a rich variety of stable and metastable finite scale microstructures; it is quite clear that the refinement tendency is due to the negative definite ("antiferromagnetic") component of the kernel.

In this paper we adopt the idea of a negative definite kernel as a general framework in which the nonlocal oscillation-forcing interactions can be modeled in a one-dimensional setting. Exactly like in the Peierls's model, one can understand the corresponding contributions to the energy as a result of minimizing out the "non-one-dimensional" components of the elastic fields and expressing the energy of the body in terms of the relevant "order parameters" only. The reason why the nonlocal terms are quadratic can be traced to the fact that the elastic problem can usually be considered linear with respect to all variables other than "order parameters". Explicit examples of the equilibrium relaxation of the "irrelevant" components of the strain field leading to the nonpositive definite contribution to the nonlocal interaction kernel, can be found in [16, 18].

Although the power law kernel would be more appropriate, for technical reasons we focus in what follows on a model with a (negative definite) *exponential* kernel. The same model with a positive definite exponential kernel was previously considered by Rogers and Truskinovsky [27] who showed that the equivalent local problem reduces to the classical Timoshenko model in the theory of elastic beams (see also [5, 7, 12–14]). Here we perform an analogous reduction to the local model and obtain a static FitzHugh–Nagumo theory [11] which is known to describe stable periodic patterns. We study in some detail a particular asymptotic limit of the theory presuming a physically natural separation of scales: in our approximation the thickness of the phase boundaries (controlled by the strain gradient term in the energy) is much smaller than the range of nonlocal interaction. As a result one can neglect the nonlocal term inside the transition layers and drop the gradient term in the description of the interaction between the phase boundaries. We perform a rigorous Γ-limit analysis leading to the discrete problem describing N interacting phase boundaries. In the one-dimensional setting the phase boundaries are represented by point "particles"; the resulting effective "interatomic potential" can be calculated explicitly. We then show that the distribution of interfaces is necessarily periodic and compute equilibrium spacing of the corresponding lattice.

An obvious drawback of our analytic approach is the focus on the absolute minimizer only. It is a serious limitation, since the previous experience with analogous models (e.g., [33, 34]) strongly suggests a rich variety of local minimizers (metastable states). To check the accessibility of the calculated microstructures we study dynamics. Our dynamical model, which includes both inertia and viscosity, describes spontaneous nucleation and propagation of the microstructures; the pattern apparently settles down to a time independent state. As we show, the limiting microstructure is periodic and the period is in close agreement with the one calculated for the global minimizer. Effectively, our numerical results present a first example of self-consistent simulation of the kinetics of austenite–martensite interface.

The remaining sections of the paper are organized as follows. In Section 2 we revisit the Ericksen's bar model on an elastic foundation and provide a motivation for the introduction of a nonlocal energy with negative kernel. In Section 3 we specify the exponential kernel and prove the equivalence of our nonlocal theory to the local FitzHugh–Nagumo model in the theory of excitable reaction–diffusion systems. In Section 4 we nondimensionalize the governing equations and suggest a scaling leading to the desired separation of scales. The Γ-limit and the reduction to a finite-dimensional problem are performed in Section 5; the technical results are collected in the Appendix. In Section 6 we show that in the equilibrium configuration the effective "interface-particles" form a periodic pattern and calculate explicitly the energy of the corresponding lattice. In Section 7 we find, among all admissible lattices, the one which corresponds to the absolute minimum of the energy and give a simple approximate formula for the number of interfaces per unit length. In Section 8 we formulate our dynamical model. Computational results describing nucleation, propagation and stabilization of the microstructures are presented in Section 9. In Section 10 we briefly discuss an extension of the model capable of simulating sharp (rather than diffuse) phase transition fronts. In the last section we summarize the results and formulate the unsolved problems.

2. The Motivation

We begin with the problem studied by Ericksen [9]. Consider an elastic bar with nonconvex energy density. The bar is placed in a hard loading device which means that the total displacement is fixed. The mathematical problem reduces to constrained minimization of the energy functional

$$F(w) = \int_0^L W(w)\,\mathrm{d}x. \tag{2.1}$$

Here $w(x) = Du(x)$ is the strain field, $u(x)$ is the displacement field and D is the operator of differentiation in x. Without loss of generality, we may assume that the energy density $W(w)$ is a non-negative double-well function with zeros at -1 and $+1$; for example, $W(w) = (w^2 - 1)^2$ (see Figure 1). Physically, the wells correspond to different crystallographic phases of the material or different symmetry-related variants of the same phase (twins). The double-well energy function gives rise to a characteristic nonmonotone stress–strain relation $\sigma(w) = W(w)$

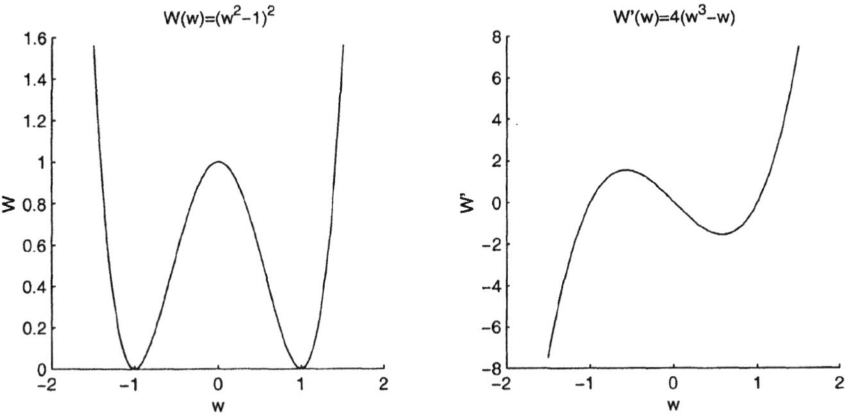

Figure 1. A double well energy $W(w)$ and its derivative $W'(w)$.

(see Figure 1). The constraint on the total displacement (or average strain) can be formulated in the form

$$\int_0^L w \, \mathrm{d}x = d. \tag{2.2}$$

Suppose that the trivial (homogeneous) solution of the variational problem (2.1), (2.2) is unstable, which takes place, for example, at $d = 0$ (see [9]). The double-well structure of the energy clearly favors configurations where the displacement gradient takes values at the bottoms of the wells, and the energy minimizing configuration represents a mixture of homogeneous states (phases) with w equal $+1$ or -1. The phase distribution must be compatible with the condition (2.2) specifying the corresponding "volume" fractions. This one-dimensional constraint is too weak, however, to affect the fineness of the microstructure which remains unspecified in this classical setting.

To penalize sharp interfaces between the phases one can include a surface energy into the model, by introducing a strain gradient term into the energy functional

$$F(w) = \int_0^L \left[W(w) + a|Dw|^2\right] \mathrm{d}x. \tag{2.3}$$

Here $a > 0$ is a constant parameter representing an internal length scale; throughout this paper we denote by $D, D^2, \ldots$ the corresponding derivatives of a function with respect to the space variable.

The strain gradient penalization turns sharp interfaces into diffuse ones and produces minimizers with one interface at most [6]. To obtain a fine scale microstructure with many interfaces, one needs to include an additional interaction into the system which acts against the strain gradient coarsening. For example, one can place the bar on an elastic foundation and consider the following functional [34]:

$$F(w) = \int_0^L \left[W(w) + a|Dw|^2 + bu^2\right] \mathrm{d}x. \tag{2.4}$$

Here $b > 0$ is a constant parameter which brings another length scale into the model. The "foundation" term is of non-constitutive nature which is clear from its dependence on displacements rather than displacement gradients. The introduction of this term guarantees the refinement of the microstructure: the strain gradients can now stay near the bottoms of the wells only if the displacements $u(x)$ are small and that requires fine oscillations of the displacement gradient. A competition between the coarsening tendency due to the surface energy and the refinement tendency due to the foundation term stabilizes the finite scale microstructures. In the limit $a \to 0$, $b \to 0$ the number of interfaces is finite and depends on the limit of the non-dimensional combination $c = ab^{-2}$ [34].

We observe that the " foundation" term in (2.4) can be rewritten in terms of strain gradients as well. Assume for determinacy that $u(0) = -d/2$ and $u(L) = d/2$ and let $\xi_x(y)$ be the characteristic function of the set $(0, x)$. Then

$$\begin{aligned}\int_0^L u^2\,\mathrm{d}x &= \int_0^L \left[\int_0^L \xi_x(y)w(y)\,\mathrm{d}y - \frac{d}{2}\right]^2 \mathrm{d}x \\ &= \int_0^L \left[\int_0^L \left(\xi_x(y)w(y) - \frac{1}{2}w(y)\right)\mathrm{d}y\right]^2 \mathrm{d}x \\ &= \int_0^L \int_0^L w(x)w(y)\left\{\int_0^L \left[\xi_z(x) - \frac{1}{2}\right]\left[\xi_z(y) - \frac{1}{2}\right]\mathrm{d}z\right\}\mathrm{d}x\,\mathrm{d}y.\end{aligned}$$

Now, since

$$\int_0^L \left[\xi_z(x) - \frac{1}{2}\right]\left[\xi_z(y) - \frac{1}{2}\right]\mathrm{d}z = \frac{L}{4} - \max\{x, y\} + \frac{1}{2}(x + y),$$

we obtain

$$F(w) = \int_0^L [W(w) + a|Dw|^2]\mathrm{d}x + b\int_0^L \int_0^L K(x, y)w(x)w(y)\,\mathrm{d}x\,\mathrm{d}y, \tag{2.5}$$

where

$$K(x, y) = \frac{L}{4} - \max\{x, y\} + \frac{1}{2}(x + y). \tag{2.6}$$

Here K is the interaction kernel defined in the square domain $[0, L]\times[0, L]$ (see Figure 2); notice that the scale of interaction is of the order of the size of the domain. The energy functional (2.5) can now be rewritten as

$$\begin{aligned}F(w) &= \int_0^L [W(w) + bk(x)w^2 + a|Dw|^2]\mathrm{d}x \\ &\quad -\frac{b}{2}\int_0^L \int_0^L K(x, y)\big[w(x) - w(y)\big]^2 \mathrm{d}x\,\mathrm{d}y,\end{aligned} \tag{2.7}$$

where

$$k(x) = \int_0^L K(x, y)\,\mathrm{d}y.$$

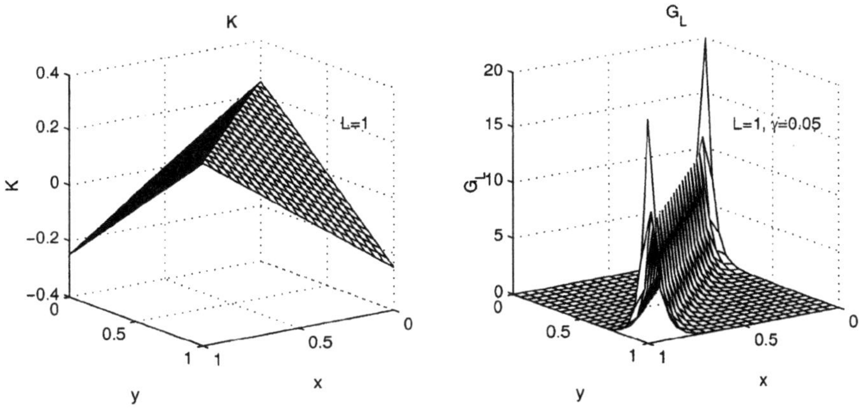

Figure 2. The kernels $K(x, y) = L/4 - \max\{x, y\} + (1/2)(x + y)$ and $G_L(x, y) = 1/(\gamma(e^{L/\gamma} - e^{-L/\gamma}))[\cosh((x + y - L)/\gamma) + \cosh((|x - y| - L)/\gamma)]$.

One can see from Figure 2 that the kernel of the *essentially* nonlocal interaction in (2.7) is negative exactly in the domain where the interaction is the strongest (near the diagonal $x = y$). This observation suggests a general framework for bringing a (non-one-dimensional) microstructure-stabilizing constraint into the semi-quantitative one-dimensional models.

3. The Model

Motivated by the example from the previous section, we consider the following functional:

$$F(w) = \int_0^L \left[W(w) + \frac{\epsilon^2}{2}|Dw|^2 \right] dx + \frac{1}{4}\int_0^L \int_0^L J(x, y)\big(w(x) - w(y)\big)^2 dx\, dy, \tag{3.1}$$

where $J \leqslant 0$. The Euler–Lagrange equation and the natural boundary conditions in this model are

$$\begin{cases} W'(w) - \epsilon^2 D^2 w + \big(jw - J[w]\big) = \sigma, \\ Dw(0) = Dw(L) = 0. \end{cases} \tag{3.2}$$

Here we introduced

$$j(x) = \int_0^L J(x, y)\, dy \quad \text{and} \quad J[w](x) = \int_0^L J(x, y) w(y)\, dy.$$

The constant parameter σ, which has a meaning of the overall stress, is to be determined together with the strain field $w(x)$ from the condition (2.2).

Consider a class of exponential kernels of the type $J(x, y) = -\delta G_\infty(|x-y|)$, where

$$G_\infty(|x-y|) = \frac{1}{2\gamma} e^{-|x-y|/\gamma} \tag{3.3}$$

and $\gamma > 0$, $\delta > 0$ are constant parameters characterizing the range and the scale of the nonlocal interaction. We notice that $G_\infty(|x-y|)$ is the Green function for the screened Poisson's equation

$$-\gamma^2 D^2 y + y = f$$

defined in the infinite domain $(-\infty, \infty)$. The fact that the kernel $G_\infty(|x-y|)$ is special allows one to reformulate the integro–differential equation (3.2) as a pair of second order differential equations. The corresponding local problem will contain explicit dependence on x (see, for instance, [13, 14]). To make the local problem spatially independent, one has to modify accordingly the kernel $J(x, y)$; it is clear that this modification will only affect the regions near the boundary of the interval $(0, L)$.

In fact, consider a new class of kernels

$$J(x, y) = -\delta G_L(x, y), \tag{3.4}$$

where $G_L(x, y)$ is the Green function for the following boundary value problem on $(0, L)$:

$$\begin{aligned} -\gamma^2 D^2 y + y &= f, \\ Dy(0) = Dy(L) &= 0. \end{aligned}$$

The expression of $G_L(x, y)$ can be given explicitly:

$$G_L(x, y) = \frac{1}{\gamma(e^{L/\gamma} - e^{-L/\gamma})} \left[\cosh\left(\frac{x+y-L}{\gamma}\right) + \cosh\left(\frac{|x-y|-L}{\gamma}\right)\right] \tag{3.5}$$

(see Figure 2). The above class of interaction kernels with a plus sign in (3.4) was previously considered in [27].

To illustrate the relation between $G_L(x, y)$ and $G_\infty(|x-y|)$, we rewrite the kernel $G_L(x, y)$ in the form

$$\begin{aligned} G_L\left(x + \frac{L}{2}, y + \frac{L}{2}\right) = \frac{1}{2\gamma} e^{-|x-y|/\gamma} + \frac{1}{2\gamma \sinh(L/\gamma)} \left[\cosh\left(\frac{x+y}{\gamma}\right)\right. \\ \left. + e^{-L/\gamma} \cosh\left(\frac{x-y}{\gamma}\right)\right]. \end{aligned}$$

If we fix points x and y and stretch L, we obtain

$$G_L\left(x + \frac{L}{2}, y + \frac{L}{2}\right) \to G_\infty(x, y) = \frac{1}{2\gamma} e^{-|x-y|/\gamma}$$

which means that in the limit $L \to \infty$ the nonlocal interaction is effectively described by the kernel $G_\infty(|x-y|)$; at finite L the two interactions differ near the boundaries of the domain only.

In what follows we fix the kernel $J(x,y) = -\delta G_L(x,y)$ and formulate the mathematical problem:

MODEL 1. Minimize the functional

$$F^1(w) = \int_0^L \left[W(w) + \frac{\epsilon^2}{2}|Dw|^2 \right] dx - \frac{\delta}{4} \int_0^L \int_0^L G_L(x,y)\big(w(x) - w(y)\big)^2 dx\, dy, \tag{3.6}$$

where w is taken from the set $W^{1,2}(0,L) \cap X$ and

$$X = \left\{ w \in L^2(0,L) \colon \int_0^L w(x)\, dx = d \right\}. \tag{3.7}$$

The Euler–Lagrange equation and the natural boundary conditions for this problem are

$$\begin{cases} W'(w) - \epsilon^2 D^2 w - \delta\big(w - G_L[w]\big) = \sigma, \\ Dw(0) = Dw(L) = 0, \end{cases} \tag{3.8}$$

where $G_L[w](x) = \int_0^L G_L(x,y) w(y)\, dy$.

In addition to Model 1 we consider an auxiliary problem:

MODEL 2. Minimize the functional

$$F^2(w) = \int_0^L \left\{ W(w) + \frac{\epsilon^2}{2}|Dw|^2 - \frac{\delta}{2}w^2 + \frac{\delta}{2}\left[\left(-\gamma^2 D^2 + 1\right)^{-1/2} w\right]^2 \right\} dx, \tag{3.9}$$

where $w \in W^{1,2}(0,L) \cap X$.

The operator $(-\gamma^2 D^2 + 1)^{-1/2}$ is defined as follows. Suppose v is a (unique) solution of the following linear boundary value problem:

$$\left(-\gamma^2 D^2 + 1\right) v = w, \qquad Dv(0) = Dv(L) = 0.$$

Then the action of the operator $(-\gamma^2 D^2 + 1)^{-1/2}$ is defined by

$$\int_0^L \left[\left(-\gamma^2 D^2 + 1\right)^{-1/2} w\right]^2 dx = \int_0^L vw\, dx.$$

The Euler–Lagrange equation and the natural boundary conditions in Model 2 can be written as

$$\begin{cases} W'(w) - \epsilon^2 D^2 w - \delta\big[w - \big(-\gamma^2 D^2 + 1\big)^{-1} w\big] = \sigma, \\ Dw(0) = Dw(L) = 0. \end{cases} \tag{3.10}$$

Now introduce the second auxiliary problem:

MODEL 3. Find critical points of

$$F^3(w, v) = \int_0^L \left\{ W(w) + \frac{\epsilon^2}{2}|Dw|^2 - \frac{\delta}{2}\big[(w - v)^2 + \gamma^2 |Dv|^2\big]\right\} \mathrm{d}x, \tag{3.11}$$

where $(w, v) \in (W^{1,2}(0, L) \cap X) \oplus W^{1,2}(0, L)$.

The Euler–Lagrange equations and the natural boundary conditions in this model take the form

$$\begin{cases} W'(w) - \epsilon^2 D^2 w - \delta(w - v) = \sigma, \\ \gamma^2 D^2 v - v = -w, \\ Dw(0) = Dw(L) = 0, \\ Dv(0) = Dv(L) = 0. \end{cases} \tag{3.12}$$

The proposition below establishes the relation among the three models. Let us select a particular critical point (w, v) of F^3 and set

$$\begin{aligned} \mathcal{M}_1 &= \big\{(\tilde{w}, \tilde{v}) \colon \tilde{w} \in W^{1,2}(0, L) \cap X,\ \tilde{v} = \big(-\gamma^2 D^2 + 1\big)^{-1} \tilde{w}\big\}, \\ \mathcal{M}_2 &= \big\{(w, \tilde{v}) \colon \tilde{v} \in W^{1,2}(0, L)\big\}. \end{aligned}$$

PROPOSITION 3.1. *The models 1, 2 and 3 are equivalent in the following sense:*

1. *For each* $w \in W^{1,2}(0, L) \cap X$, $F^1(w) = F^2(w)$.
2. *The pair* (w, v) *is a critical point of* F^3 *iff* $w \in W^{1,2}(0, L) \cap X$ *is a critical point of* F^1 *(and* F^2*) and* $v = (-\gamma^2 D^2 + 1)^{-1} w \in W^{1,2}(0, L)$. *Moreover, in this case* $F^1(w) = F^3(w, v)$.
3. *If* w *is a local or global minimum of* F^1 *in* $W^{1,2}(0, L) \cap X$, *then* (w, v), *where* $v = (-\gamma^2 D^2 + 1)^{-1} w$, *is a saddle point of* F^3 *in* $(W^{1,2}(0, L) \cap X) \oplus W^{1,2}(0, L)$. *Moreover,* (w, v) *is a minimum of* F^3 *on the manifold* $\mathcal{M}_1$, *and a maximum of* F^3 *on the manifold* $\mathcal{M}_2$.

Proof. To prove part 1 of the proposition, we note that

$$\begin{aligned} F^1(w) &= \int_0^L \left[W(w) + \frac{\epsilon^2}{2}|Dw|^2 - \frac{\delta}{2}w^2\right] \mathrm{d}x \\ &\quad + \frac{\delta}{2}\int_0^L \int_0^L w(x) G_L(x, y) w(y)\, \mathrm{d}x\, \mathrm{d}y. \end{aligned}$$

Here we have used the fact that

$$\int_0^L G_L(x, y)\,dy = 1$$

for all $x \in (0, 1)$. If we write

$$\int_0^L G_L(x, y)w(y)\,dy = \big[\big(-\gamma^2 D^2 + 1\big)^{-1} w\big](x),$$

we find

$$\int_0^L \int_0^L w(x)G_L(x, y)w(y)\,dx\,dy = \int_0^L w(x)\big[\big(-\gamma^2 D^2 + 1\big)^{-1} w\big](x)\,dx,$$

and since $(-\gamma^2 D^2 + 1)^{-1}$ is a self-adjoint positive operator, we obtain

$$\int_0^L w(x)\big[\big(-\gamma^2 D^2 + 1\big)^{-1} w\big](x)\,dx = \int_0^L \big[\big(-\gamma^2 D^2 + 1\big)^{-1/2} w\big]^2 dx.$$

Here $(-\gamma^2 D^2 + 1)^{-1/2}$ is the positive square root of $(-\gamma^2 D^2 + 1)^{-1}$. Now the functional $F^1(w)$ becomes

$$F^1(w) = \int_0^1 \left\{W(w) + \frac{\epsilon^2}{2}|Dw|^2 - \frac{\delta}{2}w^2 + \frac{\delta}{2}\big[\big(-\gamma^2 D^2 + 1\big)^{-1/2} w\big]^2\right\}dx,$$

which proves that $F^1(w) = F^2(w)$.

To see part 2 of the proposition, we first compare the Euler–Lagrange equations for the three models and find that they are identical. Now, suppose that for every $w \in W^{1,2}(0, L) \cap X$, v is such that

$$-\gamma^2 D^2 v + v = w, \qquad Dv(0) = Dv(L) = 0.$$

Then

$$F^2(w) = \int_0^L \left\{W(w) + \frac{\epsilon^2}{2}|Dw|^2 - \frac{\delta}{2}w^2 + \frac{\delta}{2}wv\right\}dx.$$

However

$$\int_0^L [wv]\,dx = \int_0^L \big[\gamma^2|Dv|^2 + v^2\big]dx,$$

which can also be rewritten as

$$\int_0^L [wv]\,dx = \int_0^L \big[-\big(\gamma^2|Dv|^2 + v^2\big) + 2wv\big]dx.$$

The functional $F^2(w)$ becomes

$$F^2(w) = \int_0^L \left\{W(w) + \frac{\epsilon^2}{2}|Dw|^2 - \frac{\delta}{2}\big[(w - v)^2 + \gamma^2|Dv|^2\big]\right\}dx,$$

which proves that $F^1(w) = F^2(w) = F^3(w, v)$

To see part 3 of the proposition, we note from part 2 that when $(w, v) \in \mathcal{M}_1$, $F^3(w, v) = F^1(w)$. Therefore, a minimum w of F^1 gives a minimum $(w, v = (-\gamma^2 D^2 + 1)^{-1} w)$ in $\mathcal{M}_1$. In $\mathcal{M}_2$, F^3 is quadratic and concave in $\tilde{v}$, so along this manifold v is the point of global maximum. □

REMARK 3.2. Model 3 describes the equilibrium states in the FitzHugh–Nagumo theory of activator–inhibitor reaction–diffusion systems (e.g, [11]). The internal parameter v can also be considered as an analog of the internal shear in a Timoshenko-like model of a beam with negative (!) bending stiffness (e.g., [27, 30]).

4. Nondimensionalization and Scaling

Our energy functional depends on 4-dimensional parameters ϵ, γ, δ and L. By rescaling the independent variable x, we can reduce this number to three. In fact, introduce $\bar{x} = x/L$ and $\overline{w}(\bar{x}) = w(x)$. Then the functional $\overline{F^2}(\overline{w}) = F^2(w)/L$ can be written as

$$\overline{F^2}(\overline{w}) = \int_0^1 \left\{ W(\overline{w}) + \frac{\bar{\epsilon}^2}{2}|\overline{D}\overline{w}|^2 - \frac{\bar{\delta}}{2}\overline{w}^2 + \frac{\bar{\delta}}{2}[(-\bar{\gamma}^2\overline{D}^2 + 1)^{-1/2}\overline{w}]^2 \right\} \mathrm{d}\bar{x}, \quad (4.1)$$

where the differential operator $\overline{D}$ is the derivative with respect to $\bar{x}$ and

$$\bar{\epsilon} = \frac{\epsilon}{L}, \qquad \bar{\delta} = \delta, \qquad \bar{\gamma} = \frac{\gamma}{L}$$

are the main non-dimensional parameters of the problem. Two of them, $\bar{\epsilon}$ and $\bar{\delta}$, are direct analogs of the parameters a and b in the model of a bar on an elastic foundation (2.4). Since our nonlocal interaction is a long-range one, we assume that the third parameter $\bar{\gamma} \sim 1$.

Notice, that in the limit $\bar{\epsilon} \to 0$, $\bar{\delta} \to 0$ the energy (4.1) formally converges to (2.1), and in the naive limit we obtain microstructures with unspecified number of interfaces. More careful analysis shows that, as in the case of a bar on an elastic foundation, the asymptotics of the number of interfaces depends on the joint limit of the non-dimensional parameters $\bar{\delta}$ and $\bar{\epsilon}$. Following the arguments presented in [34], one can show that the number of interfaces in the limit $\bar{\epsilon} \to 0$, $\bar{\delta} \to 0$ depends on the ratio $\bar{c} = \bar{\epsilon}^{-1}\bar{\delta}$. If $\bar{c} \sim 1$ the limiting solution will contain a finite number of interacting interfaces, each of width $\sim \bar{\epsilon}$. The fine structure of the phase boundaries is then completely controlled by the gradient term in the energy functional while their long range interaction is determined by the negative definite nonlocal contribution to the energy. With this limit in view, we assume that $\bar{\delta} = \bar{\epsilon}\bar{c}$ and rewrite our functional in the form

$$\begin{aligned} \overline{F}_{\bar{\epsilon}}(\overline{w}) = \int_0^1 \Big\{ & W(\overline{w}) + \frac{\bar{\epsilon}^2}{2}|\overline{D}\overline{w}|^2 - \frac{\bar{\epsilon}\bar{c}}{2}\overline{w}^2 \\ & + \frac{\bar{\epsilon}\bar{c}}{2}[(-\bar{\gamma}^2\overline{D}^2 + 1)^{-1/2}\overline{w}]^2 \Big\} \mathrm{d}\bar{x}. \end{aligned} \quad (4.2)$$

The constraint (2.2) becomes

$$\int_0^1 \overline{w}(\bar{x})\,\mathrm{d}\bar{x} = \bar{d}, \tag{4.3}$$

where $\bar{d} = d/L$. In the present formulation the only small parameter is $\bar{\epsilon}$ and we focus our analysis on the asymptotics as $\bar{\epsilon} \to 0$. In what follows, we omit the unnecessary overbars.

5. The $\epsilon \to 0$ Limit

As ϵ approaches 0, the minimizer w_ϵ of F_ϵ tends to a minimizer of another variational problem, F_0, the Γ-limit of $\epsilon^{-1}F_\epsilon$ (see [8] for the definition of Γ-convergence). In this section we explicitly describe the limiting problem; the rigorous mathematical proofs are given in the Appendix.

We first observe that because of the continuity of the nonlocal term, the behavior of our system in the limit of small ϵ is similar to the behavior of the corresponding gradient model (2.3). In particular, at $\epsilon \to 0$ the phase boundaries become infinitely sharp, while the displacement gradients inside the phases assume constant values in the bottoms of the energy wells. As a result, the problem reduces to finding the *geometry* of the domain occupied by one of the phases. This, in turn, means finding the corresponding characteristic function. Notice that the strain gradient term drives the system towards the lower "surface area", which means the minimization of the number of interfaces. In terms of the interface interaction this means repulsion and leads to the overall coarsening. Alternatively, the nonlocal term in the energy causes the interfaces to attract each other and produces a tendency towards the refinement.

We begin with the formal introduction of an adequate space for the limiting strain field. To make our analysis slightly more general, we assume that the energy density $W(w)$ has two nonsymmetric wells located at $w = \alpha$ and $w = \beta$. We take $\epsilon^{-1}F_\epsilon$ as a family of functionals defined on $W^{1,2}(0,1) \cap X$, where F_ϵ is given in (4.2).

We say that a function χ is in $BV((0,1),\{\alpha,\beta\}) \cap X$ if for every $x \in (0,1)$, $\chi(x) = \alpha$ or β, and there are finitely many points $x_1, x_2, \ldots, x_N$ in (0,1), where χ jumps between α and β. The jumps describe the phase boundaries whose locations, given by the numbers $x_1, x_2, \ldots, x_N$, remain unknown.

As we show in the Appendix, the family of strain fields w_ϵ in $W^{1,2}(0,1) \cap X$ with the property that $\epsilon^{-1}F_\epsilon(w_\epsilon)$ is bounded uniformly in ϵ converges to a function χ in $BV((0,1),\{\alpha,\beta\}) \cap X$. Moreover, in the original functional

$$\epsilon^{-1}F_\epsilon(w_\epsilon) = \int_0^1 \left\{ \frac{1}{\epsilon} W(w_\epsilon) + \frac{\epsilon}{2}|Dw_\epsilon|^2 - \frac{c}{2}w_\epsilon^2 + \frac{c}{2}\left[\left(-\gamma^2 D^2 + 1\right)^{-1/2} w_\epsilon\right]^2 \right\} \mathrm{d}x,$$

the "gradient" part

$$\int_0^1 \left\{ \frac{1}{\epsilon} W(w_\epsilon) + \frac{\epsilon}{2} |Dw_\epsilon|^2 \right\} \mathrm{d}x$$

Γ-converges to $c_0 N$, where N is the number of the points where χ jumps, and c_0 is defined by

$$c_0 = \sqrt{2} \int_\alpha^\beta [W(t)]^{1/2} \, \mathrm{d}t, \tag{5.1}$$

(e.g., [21]). The Γ-limit of the "nonlocal" part of the energy,

$$\int_0^1 \left\{ -\frac{c}{2} w_\epsilon^2 + \frac{c}{2} [(-\gamma^2 D^2 + 1)^{-1/2} w_\epsilon]^2 \right\} \mathrm{d}x,$$

turns out to be trivially

$$\int_0^1 \left\{ -\frac{c}{2} \chi^2 + \frac{c}{2} [(-\gamma^2 D^2 + 1)^{-1/2} \chi]^2 \right\} \mathrm{d}x.$$

Although the above integral represents an implicit function of the location of the phase boundaries $x_1, x_2, \ldots, x_N$, the first term can be calculated exactly. In fact, the constraint $\int_0^1 \chi \, \mathrm{d}x = \int_0^1 w_\epsilon \, \mathrm{d}x = d$ implies that the size of the set where $\chi = \alpha$ is $(\beta - d)/(\beta - \alpha)$ and the size of the set where $\chi = \beta$ is $(d - \alpha)/(\beta - \alpha)$. Then

$$\int_0^1 \chi^2 \, \mathrm{d}x = \alpha^2 \frac{\beta - d}{\beta - \alpha} + \beta^2 \frac{d - \alpha}{\beta - \alpha} = (\alpha + \beta) d - \alpha\beta.$$

To summarize, the limiting functional F_0 is defined by

$$F_0(\chi) = c_0 \mathcal{S}(\chi) + \frac{c}{2} \left\{ -(\alpha + \beta) d + \alpha\beta + \int_0^1 [(-\gamma^2 D^2 + 1)^{-1/2} \chi]^2 \mathrm{d}x \right\}. \tag{5.2}$$

The function $\mathcal{S}(\chi) = N$ denotes the number of points where χ jumps. The class of all admissible fields χ can be decomposed into

$$BV\big((0, 1), \{\alpha, \beta\}\big) \cap X = \bigcup_{N=1}^{\infty} A_N,$$

where $A_N = \{\chi \in BV((0, 1), \{\alpha, \beta\}) \cap X \colon \mathcal{S}(\chi) = N\}$. In other words A_N is the set of functions in $BV((0, 1), \{\alpha, \beta\})$ which have N jumps. The spaces A_N can be further decomposed into

$$A_N = A_N^\alpha \cup A_N^\beta,$$

where $A_N^\alpha = \{\chi \in A_N\colon \chi(x) = \alpha, \text{for } x \in (0, x_1)\}$ and $A_N^\beta = \{\chi \in A_N\colon \chi(x) = \beta,$ for $x \in (0, x_1)\}$. A_N^α is the part of A_N where functions equal α before their jump point x_1, and A_N^β is the rest part of A_N where functions equal β before their first jump point x_1. More specifically, the subspace A_N^α can be identified with

$$\big\{(x_1, x_2, \ldots, x_N) \in \mathbb{R}^N\colon x_1 < x_2 < \cdots < x_N,\ \alpha x_1 + \beta(x_2 - x_1) + \cdots = d\big\},$$

and the subspace A_N^β with

$$\big\{(x_1, x_2, \ldots, x_N) \in \mathbb{R}^N\colon x_1 < x_2 < \cdots < x_N,\ \beta x_1 + \alpha(x_2 - x_1) + \cdots = d\big\}.$$

The minimization of F_0, which turns out to be a problem in a *finite-dimensional* space of an unknown dimension, can be subdivided into two steps. First, we solve a N-dimensional problem in each A_N^α and A_N^β. The minimizer turns out to be unique in each of the subspaces: χ_N^α in A_N^α and χ_N^β in A_N^β. The symmetry between A_N^α and A_N^β is manifested through the fact that $F_0(\chi_N^\alpha) = F_0(\chi_N^\beta)$ (see Proposition 6.3). This allows us to define

$$F_0^*(N) = F_0\big(\chi_N^{\alpha,\beta}\big). \tag{5.3}$$

The second step of our procedure is the *discrete* minimization of $F_0^*(N)$ among all positive integers N.

We present our approach in the form of the following schematic diagram:

$$\begin{aligned}
&\min_w F_\epsilon(w), \quad w \in W^{1,2}(0,1) \cap X,\\
&\Gamma\text{-converge} \overset{\epsilon\to 0}{\longrightarrow} \min_\chi F_0(\chi), \quad \chi \in BV\big((0,1), \{\alpha,\beta\}\big) \cap X\\
&\qquad\Longrightarrow \left\langle \begin{array}{l} \min\limits_{(x_1,\ldots,x_N)} F_0(x_1,\ldots,x_N),\ (x_1,\ldots,x_N) \in A_N^\alpha \rightsquigarrow \chi_N^\alpha \\ \min\limits_{(x_1,\ldots,x_N)} F_0(x_1,\ldots,x_N),\ (x_1,\ldots,x_N) \in A_N^\beta \rightsquigarrow \chi_N^\beta \end{array} \right\rangle\\
&\qquad\Longrightarrow \min_N F_0^*(N), \quad N \in \{1, 2, \ldots\} \rightsquigarrow \chi_{N_*}^{\alpha,\beta}.
\end{aligned}$$

The Γ-limit of the functional (5.2) is computed in the Appendix. The first step of the minimization procedure is discussed in Section 7, the second step is detailed in Section 8.

We close this section with the computation of the partial derivatives of the implicit function

$$\begin{aligned} F_0(x_1, \ldots, x_N) &= c_0 N + \frac{c}{2}\bigg\{ -(\alpha+\beta)d + \alpha\beta \\ &\qquad + \int_0^1 \big[(-\gamma^2 D^2 + 1)^{-1/2}\chi\big]^2 \mathrm{d}x \bigg\}. \end{aligned} \tag{5.4}$$

Let χ be an element in A_N^α. Then the nonlocal part of (5.4) can be rewritten as

$$\begin{aligned} H(\chi) &= H(x_1,\ldots,x_N) = \int_0^1 \left[(-\gamma^2 D^2+1)^{-1/2}\chi\right]^2 \mathrm{d}x \\ &= \int_0^1\int_0^1 G_L(x,y)\chi(x)\chi(y)\,\mathrm{d}x\,\mathrm{d}y. \end{aligned}$$

Assuming $v = (-\gamma^2 D^2+1)^{-1}\chi$, we compute

$$\begin{aligned} \frac{\partial H}{\partial x_1} &= \frac{\partial}{\partial x_1}\left[\int_0^{x_1}\alpha v\,\mathrm{d}x + \int_{x_1}^{x_2}\beta v\,\mathrm{d}x + \cdots + \int_{x_N}^{L}\begin{pmatrix}\alpha\\ \text{or}\\ \beta\end{pmatrix} v\,\mathrm{d}x\right] \\ &= (\alpha-\beta)v(x_1) + \int_0^1 \chi(x)\frac{\partial v}{\partial x_1}(x)\,\mathrm{d}x. \end{aligned}$$

To find $(\partial v/\partial x_1)(x)$, we note that

$$\begin{aligned} \frac{\partial v}{\partial x_1}(x) &= \frac{\partial}{\partial x_1}\left[\int_0^{x_1}\alpha G_L(x,y)\,\mathrm{d}y + \int_{x_1}^{x_2}\beta G_L(x,y)\,\mathrm{d}y \right.\\ &\qquad \left. + \cdots + \int_{x_N}^{1}\begin{pmatrix}\alpha\\ \text{or}\\ \beta\end{pmatrix} G_L(x,y)\,\mathrm{d}y\right] \\ &= (\alpha-\beta)G_L(x,x_1). \end{aligned}$$

Therefore,

$$\frac{\partial H}{\partial x_1} = (\alpha-\beta)v(x_1) + (\alpha-\beta)\int_0^1 \chi(x)G_L(x,x_1)\,\mathrm{d}x = 2(\alpha-\beta)v(x_1).$$

Repeating a similar argument for the other partial derivatives of H, we deduce

$$\nabla H(x_1,x_2,\ldots,x_N) = 2(\alpha-\beta)\begin{pmatrix} v(x_1)\\ -v(x_2)\\ v(x_3)\\ \vdots\\ (-1)^{N+1}v(x_N)\end{pmatrix}. \tag{5.5}$$

If $\chi \in A_N^\beta$, then

$$\nabla H(x_1,x_2,\ldots,x_N) = 2(\beta-\alpha)\begin{pmatrix} v(x_1)\\ -v(x_2)\\ v(x_3)\\ \vdots\\ (-1)^{N+1}v(x_N)\end{pmatrix}. \tag{5.6}$$

This representation of the derivatives of the energy in the N-dimensional problem will be used in the next section which details the first step of our minimization procedure.

6. Minimizing over $x_1, x_2, \ldots, x_N$

We minimize F_0 in A_N^α and A_N^β separately. Our first result is:

PROPOSITION 6.1. *If χ_N^α minimizes F_0 in A_N^α, then χ_N^α is necessary periodic. More specifically,*

$$\chi_N^\alpha(x) = \begin{cases} \alpha, & x \in \left(0, \dfrac{-d+\beta}{(\beta-\alpha)N}\right), \\ \beta, & x \in \left(\dfrac{-d+\beta}{(\beta-\alpha)N}, \dfrac{d+\beta-2\alpha}{(\beta-\alpha)N}\right), \\ \alpha, & x \in \left(\dfrac{d+\beta-2\alpha}{(\beta-\alpha)N}, \dfrac{-d+3\beta-2\alpha}{(\beta-\alpha)N}\right), \\ \beta, & x \in \left(\dfrac{-d+3\beta-2\alpha}{(\beta-\alpha)N}, \dfrac{d+3\beta-4\alpha}{(\beta-\alpha)N}\right), \\ \alpha, & x \in \left(\dfrac{d+3\beta-4\alpha}{(\beta-\alpha)N}, \dfrac{-d+5\beta-4\alpha}{(\beta-\alpha)N}\right), \\ \beta, & x \in \left(\dfrac{-d+5\beta-4\alpha}{(\beta-\alpha)N}, \dfrac{d+5\beta-6\alpha}{(\beta-\alpha)N}\right), \\ \vdots \end{cases}$$

Also, if χ_N^β minimizes F_0 in A_N^β, then χ_N^β is periodic and

$$\chi_N^\beta(x) = \begin{cases} \beta, & x \in \left(0, \dfrac{d-\alpha}{(\beta-\alpha)N}\right), \\ \alpha, & x \in \left(\dfrac{d-\alpha}{(\beta-\alpha)N}, \dfrac{-d+2\beta-\alpha}{(\beta-\alpha)N}\right), \\ \beta, & x \in \left(\dfrac{-d+2\beta-\alpha}{(\beta-\alpha)N}, \dfrac{d+2\beta-3\alpha}{(\beta-\alpha)N}\right), \\ \alpha, & x \in \left(\dfrac{d+2\beta-3\alpha}{(\beta-\alpha)N}, \dfrac{-d+4\beta-3\alpha}{(\beta-\alpha)N}\right), \\ \beta, & x \in \left(\dfrac{-d+4\beta-3\alpha}{(\beta-\alpha)N}, \dfrac{d+4\beta-5\alpha}{(\beta-\alpha)N}\right), \\ \alpha, & x \in \left(\dfrac{d+4\beta-5\alpha}{(\beta-\alpha)N}, \dfrac{-d+6\beta-5\alpha}{(\beta-\alpha)N}\right), \\ \vdots \end{cases}$$

Proof. First consider A_N^α. We use the method of Lagrange multipliers to accommodate the constraint $\alpha x_1 + \beta(x_2 - x_1) + \cdots = d$. For the characteristic function χ to be a minimizer of (5.4), it is necessary that for some constant σ

$$2(\alpha-\beta)\begin{pmatrix} v(x_1) \\ -v(x_2) \\ v(x_3) \\ \vdots \\ (-1)^{N+1}v(x_N) \end{pmatrix} - \sigma(\alpha-\beta)\begin{pmatrix} 1 \\ -1 \\ 1 \\ \vdots \\ (-1)^{N+1} \end{pmatrix} = \begin{pmatrix} 0 \\ 0 \\ 0 \\ \vdots \\ 0 \end{pmatrix}.$$

This obviously implies

$$v(x_1) = v(x_2) = \cdots = v(x_N). \tag{6.1}$$

In particular, $v(x_1) = v(x_2)$ which means that v is symmetric on (x_1, x_2). The function $v(x)$ in this interval can be calculated explicitly:

$$v(x) = \beta + C\cosh\left(\frac{x - (x_1 + x_2)/2}{\gamma}\right),$$

where C is a constant of integration. Notice that $Dv(x_1) = -Dv(x_2)$. Now, by uniqueness and existence theorems for ordinary differential equations,we obtain that v on $(0, x_1)$ is a reflection of v on $(x_2, (x_2 + x_3)/2)$. Also because $Dv(0) = 0$, the length of $(0, x_1)$ is half of that of (x_2, x_3). By repeating this argument one can show that the intervals where $\chi = \alpha$ all have the same length with the exception of $(0, x_1)$ (and $(x_N, 1)$), whose length is only a half of the above length. The same can be concluded for the intervals where $\chi = \beta$, say $(x_1, x_2), (x_3, x_4), \ldots$: the length of $(0, x_1)$ is $\beta - d/((\beta - \alpha)N)$; the length of (x_1, x_2) is $2(d - \alpha)/((\beta - \alpha)N)$; and the length of (x_2, x_3) is $2(\beta - d)/((\beta - \alpha)N)$. This gives the desired expression for χ_N^α. A similar argument for A_N^β determines χ_N^β. □

REMARK 6.2. For each $\chi_N^{\alpha,\beta}$, by separating variables, the corresponding $v = (-\gamma^2 D^2 + 1)^{-1}\chi_N^{\alpha,\beta}$ can be calculated from the explicit series

$$v(x) = d + \sum_{k=1}^{\infty}\left[\frac{2\int_0^1 \chi_N^{\alpha,\beta}(x)\cos(k\pi x)\,\mathrm{d}x}{k^2\gamma^2\pi^2 + 1}\cos(k\pi x)\right].$$

Using Proposition 6.1, we can simplify this to

$$\left(-\gamma^2 D^2 + 1\right)^{-1}\chi_N^\alpha = d + 2\sum_{l=1}^{\infty}\frac{(\beta-\alpha)\sin(((\beta-d)/(\beta-\alpha))l\pi)}{l\pi(l^2N^2\gamma^2\pi^2 + 1)}\cos(lN\pi x),$$

$$\left(-\gamma^2 D^2 + 1\right)^{-1}\chi_N^\beta = d + 2\sum_{l=1}^{\infty}\frac{(\beta-\alpha)\sin(((d-\alpha)/(\beta-\alpha))l\pi)}{l\pi(l^2N^2\gamma^2\pi^2 + 1)}\cos(lN\pi x).$$

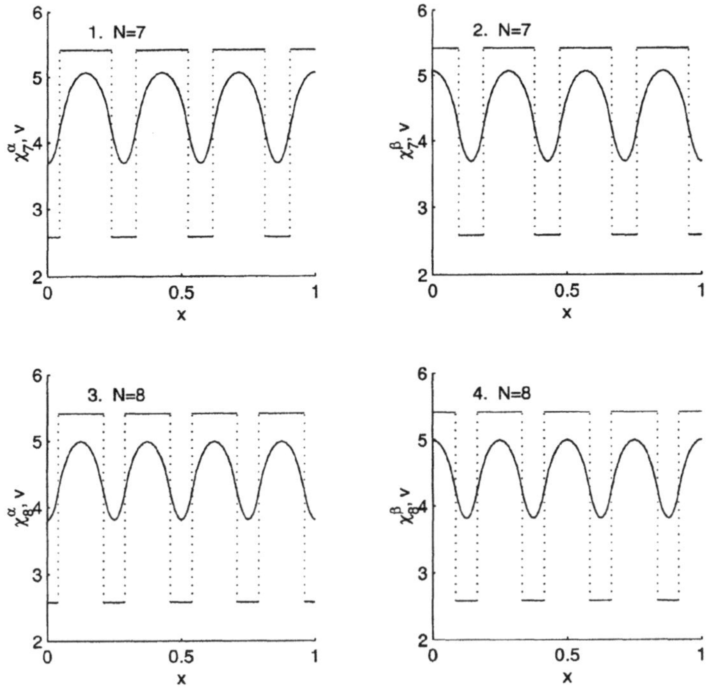

Figure 3. Metastable configurations with $N = 7$ and 8. Shown are the graphs of $\chi_N^{\alpha,\beta}$ and the corresponding $v = (-\gamma^2 D^2 + 1)^{-1}\chi_N^{\alpha,\beta}$. Here $\alpha = 2.59$, $\beta = 5.41$, $d = 4.5$.

Two examples of the functions χ_N^α and χ_N^β and the corresponding functions $v(x)$ are shown in Figure 3. In our numerical calculations we specified

$$W(w) = \left[\frac{1}{4}(w-4)^2 - \frac{1}{2}\right]^2, \tag{6.2}$$

which gives $\alpha = 4 - \sqrt{2}$, $\beta = 4 + \sqrt{2}$. We also choose $d = 4.5$ to be between α and β. In the first two pictures (Figure 3 (1,2)) $N = 7$ (odd), while in the last two pictures (Figure 3 (3,4)) $N = 8$ (even), which illustrate two different morphologies near the boundaries of the domain. Notice that the strains χ_N^α and χ_N^β are discontinuous even when the internal fields v's are smooth.

The next proposition provides an explicit expression for the energy of the minimizer at a given N. The corresponding states are only metastable in the sense of the original problem.

PROPOSITION 6.3.

$$\begin{aligned} F_0^*(N) &= F_0(\chi_N^{\alpha,\beta}) \\ &= c_0 N - \frac{c(\beta-\alpha)^2\gamma N \sinh(\frac{\beta-d}{(\beta-\alpha)\gamma N})\sinh(\frac{d-\alpha}{(\beta-\alpha)\gamma N})}{2\sinh(\frac{1}{\gamma N})}. \end{aligned}$$

Proof. We will only compute $F_0(\chi_N^\alpha)$ since the computation of $F_0(\chi_N^\beta)$ is identical. We start with

$$\int_0^1 \left[\left(-\gamma^2 D^2 + 1\right)^{-1/2} \chi_N^\alpha\right]^2 \mathrm{d}x = \int_0^1 \chi_N^\alpha v \,\mathrm{d}x.$$

Here we have used that $v = (-\gamma^2 D^2 + 1)^{-1} \chi_N^\alpha$. On a subinterval of (0,1), say (a, b), where $\chi_N^\alpha = \alpha$,

$$\begin{aligned} \int_a^b \chi_N^\alpha v &= \alpha \int_a^b v = \alpha \int_a^b (\gamma^2 D^2 v + \alpha) \\ &= \alpha\gamma^2\left[Dv(b) - Dv(a)\right] + \alpha^2(b-a). \end{aligned}$$

From $\int_0^1 \chi_N^\alpha(x)\,\mathrm{d}x = d$ we find that the measure of the set on which $\chi_N^\alpha = \alpha$ is $(\beta - d)/(\beta - \alpha)$ and the measure of the set on which $\chi_N^\alpha = \beta$ is $(d-\alpha)/(\beta-\alpha)$. Then

$$\begin{aligned} &\int_0^1 \left[\left(-\gamma^2 D^2 + 1\right)^{-1/2} \chi_N^\alpha\right]^2 \mathrm{d}x \\ &= \alpha^2 \frac{\beta - d}{\beta - \alpha} + \beta^2 \frac{d-\alpha}{\beta-\alpha} + (\alpha-\beta)\gamma^2\left[Dv(x_1) - Dv(x_2) + Dv(x_3) - \cdots\right] \\ &= (\alpha+\beta)d - \alpha\beta L + (\alpha-\beta)\gamma^2 N Dv(x_1). \end{aligned}$$

Now we need to calculate $Dv(x_1)$. First, observe that on $(0, x_1)$ we have $v(x) = \alpha + C' \cosh(x/\gamma)$ and on (x_1, x_2) we have $v(x) = \beta + C'' \cosh((x - (x_1+x_2)/2)/\gamma)$ for some appropriate C' and C''. Both functions as well as their derivatives must match at x_1. Therefore,

$$\begin{cases} \alpha + C' \cosh\left(\dfrac{x_1}{\gamma}\right) = \beta + C'' \cosh\left(\dfrac{x_1 - x_2}{2\gamma}\right), \\ \dfrac{C'}{\gamma} \sinh\left(\dfrac{x_1}{\gamma}\right) = \dfrac{C''}{\gamma} \sinh\left(\dfrac{x_1 - x_2}{2\gamma}\right). \end{cases}$$

Solving this system we find

$$\begin{cases} C' = \dfrac{(\alpha-\beta)\sinh((x_1 - x_2)/(2\gamma))}{\sinh(1/(\gamma N))}, \\ C'' = \dfrac{(\alpha-\beta)\sinh(x_1/\gamma)}{\sinh(1/(\gamma N))}. \end{cases}$$

Recall that

$$x_1 = \frac{\beta - d}{(\beta-\alpha)N}, \qquad x_2 - x_1 = \frac{2(d-\alpha)}{(\beta-\alpha)N}, \qquad \text{and} \qquad \frac{x_1 + x_2}{2} = \frac{L}{N}.$$

Then

$$Dv(x_1) = -\frac{(\alpha-\beta)\sinh((\beta-d)/((\beta-\alpha)\gamma N))\sinh((d-\alpha)/((\beta-\alpha)\gamma N))}{\gamma\sinh(1/(\gamma N))}$$

from which we define

$$\begin{aligned}&\int_0^1 [(-\gamma^2 D^2+1)^{-1/2}\chi_N^\alpha]^2 \mathrm{d}x\\ &\quad = (\alpha+\beta)d-\alpha\beta\\ &\qquad -\frac{(\beta-\alpha)^2\gamma N\sinh((\beta-d)/((\beta-\alpha)\gamma N))\sinh((d-\alpha)/((\beta-\alpha)\gamma N))}{\sinh(1/(\gamma N))}.\end{aligned}$$

Now the proposition follows from (5.2). □

PROPOSITION 6.4. *χ_N^α minimizes F_0 in A_N^α and χ_N^β minimizes F_0 in A_N^β.*

Proof. Suppose that χ_N^α is not a minimizer of F_0 in A_N^α. Then there exists a sequence $(x_{n,1}, x_{n,2}, \ldots, x_{n,N})$ in $\{(x_1, x_2, \ldots, x_N) \in R^N\colon x_1 < x_2 < \cdots < x_N,\ \alpha x_1 + \beta(x_2-x_1)+\cdots = d\}$ converging to a boundary point and such that

$$\lim_{n\to\infty} H(x_{n,1}, x_{n,2}, \ldots, x_{n,N}) \leqslant H(\chi_N^\alpha).$$

A boundary point of $\{(x_1, x_2, \ldots, x_N) \in \mathbb{R}^N\colon x_1 < x_2 < \cdots < x_N,\ \alpha x_1 + \beta(x_2-x_1)+\cdots = d\}$ can be considered as a point in $A_{N'}^\alpha$ (or $A_{N'}^\beta$) with $N' < N$. Let us call it $(y_1, y_2, \ldots, y_{N'})$. Now, since $H(\chi_N^{\alpha,\beta})$ decreases in N (see below Proposition 7.1, part 3), we find that $H(y_1, y_2, \ldots, y_{N'}) < H(\chi_{N'}^{\alpha,\beta})$. This implies that either $\chi_{N'}^\alpha$ is not a minimizer in $A_{N'}^\alpha$, or $\chi_{N'}^\beta$ is not a minimizer in $A_{N'}^\beta$. This recursive argument continues until we reach A_1^α (or A_1^β). A contradiction occurs since there is only one element in A_1^α (or A_1^β). □

7. Minimizing over N

The properties of the function $F_0^*(N)$, introduced in Proposition 6.3, are summarized in the following proposition.

PROPOSITION 7.1. *Assume that r takes real values in $(0,\infty)$.*

1. *$F_0^*(r)$ is convex in r.*
2. *$F_0^*(r)$ has the Laurent expansion*

$$F_0^*(r) = c_0 r - \frac{c(\beta-d)(d-\alpha)}{2} + \frac{c(\beta-d)^2(d-\alpha)^2}{6(\beta-\alpha)^2\gamma^2}\frac{1}{r^2} + \cdots.$$

3. *The function*

$$F_0^*(r) - c_0 r = -c(\beta-\alpha)^2\gamma r \sinh\left(\frac{\beta-d}{(\beta-\alpha)\gamma r}\right)\sinh\left(\frac{d-\alpha}{(\beta-\alpha)\gamma r}\right) \Big/ 2\sinh\left(\frac{1}{\gamma r}\right),$$

decreases with r.

4. *For large r the curve $F_0^*(r)$ has an asymptote*

$$l(r) = c_0 r - \frac{c(\beta - d)(d - \alpha)}{2}.$$

Proof. Introduce

$$p_1 = \frac{\beta - d}{(\beta - \alpha)\gamma}, \qquad p_2 = \frac{d - \alpha}{(\beta - \alpha)\gamma}.$$

Then

$$\begin{aligned}\frac{\mathrm{d}F_0^*(r)}{\mathrm{d}r} = c_0 - \frac{c(\beta-\alpha)^2\gamma}{2r(\sinh((p_1+p_2)/r))^2}\Big\{ & r \sinh\frac{p_1}{r}\sinh\frac{p_2}{r}\sinh\frac{p_1+p_2}{r} \\ & -p_1\cosh\frac{p_1}{r}\sinh\frac{p_2}{r}\sinh\frac{p_1+p_2}{r} \\ & -p_2\sinh\frac{p_1 r}{\cosh r}\frac{p_2}{r}\sinh\frac{p_1+p_2}{r} \\ & +(p_1+p_2)\sinh\frac{p_1}{r}\sinh\frac{p_2}{r}\cosh\frac{p_1+p_2}{r}\Big\},\end{aligned}$$

and

$$\begin{aligned}\frac{\mathrm{d}^2F_0^*(r)}{\mathrm{d}r^2} = \frac{c(\beta-\alpha)^2\gamma}{r^3(\sinh 1/(\gamma r))^3}\Big\{ & \cosh\frac{L}{\gamma r}\Big[p_1^2\Big(\sinh\frac{p_2}{r}\Big)^2 + p_2^2\Big(\sinh\frac{p_1}{r}\Big)^2\Big] \\ & -2p_1p_2\sinh\frac{p_1}{r}\sinh\frac{p_2}{r}\Big\}.\end{aligned}$$

Now, since $\cosh(1/(\gamma r)) > 1$, we obtain

$$\begin{aligned}\frac{\mathrm{d}^2F_0^*(r)}{\mathrm{d}r^2} > \frac{c(\beta-\alpha)^2\gamma}{r^3(\sinh 1/(\gamma r))^3}\Big\{ & p_1^2\Big(\sinh\frac{p_2}{r}\Big)^2 + p_2^2\Big(\sinh\frac{p_1}{r}\Big)^2 \\ & -2p_1p_2\sinh\frac{p_1}{r}\sinh\frac{p_2}{r}\Big\} \geqslant 0.\end{aligned}$$

The last inequality implies that the graph of $F_0^*(r)$ versus r is a convex curve. A straightforward calculation gives the Laurent expansion mentioned in the proposition. In particular, it implies

$$\lim_{r\to\infty}\frac{\mathrm{d}F_0^*(r)}{\mathrm{d}r} = c_0.$$

Then the inequality

$$\frac{\mathrm{d}F_0^*(r)}{\mathrm{d}r} < c_0$$

follows from $\mathrm{d}^2F_0^*(r)/\mathrm{d}r^2 > 0$. □

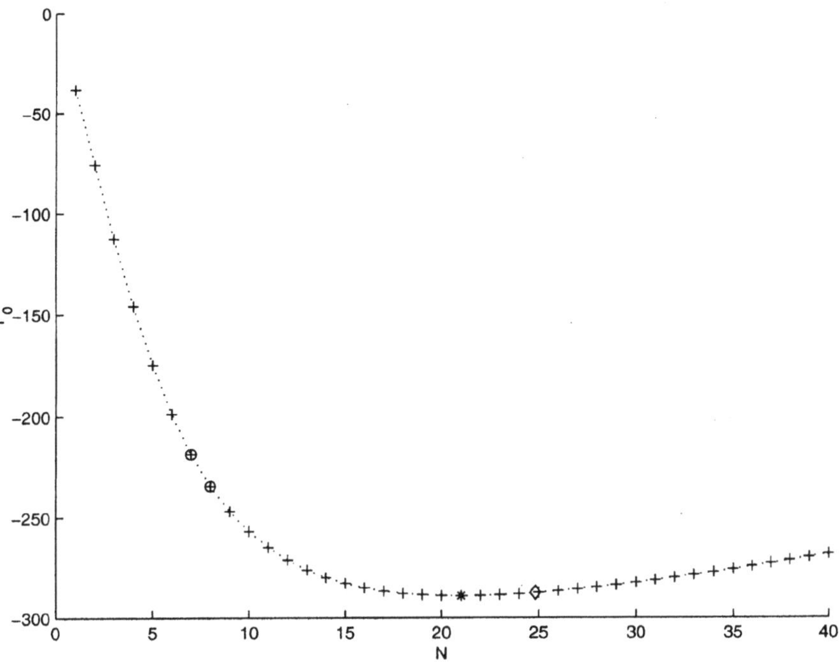

Figure 4. The graph of $F_0^*(N)$ versus N. Here $\alpha = 2.59$, $\beta = 5.41$, $\gamma = 0.05$, $c_0 = 1.33$, $c = 400$ and $d = 4.5$. The markers are explained in the text.

Figure 4 illustrates the behavior of the function $F_0^*(r)$. Here, in addition to the parameters of the energy selected for Figure 3, we choose $\gamma = 0.05$, $c_0 = 4/3$, and $c = 400$. The energies of the four metastable states with $N = 7$ and $N = 8$ (Figure 3) are indicated in Figure 4 by the "cross" markers.

In general, the curve $F_0^*(r)$ versus r, restricted to the interval $[1, \infty)$, has a unique global minimum, say $r = \nu$, where ν is usually not an integer. An integer N_* adjacent to ν minimizes F_0^* among all positive integers. Generically this N_* is unique and there is exactly one pair of global minimizers, $\chi_{N_*}^{\alpha}$ and $\chi_{N_*}^{\beta}$, corresponding to it. It may also happen that for the specific choice of the nondimensional constants there are two consecutive positive integers, N_* and $N_* + 1$, that both minimize $F_0^*(N)$. Then we have two pairs of global minimizers, $\chi_{N_*}^{\alpha}$, $\chi_{N_*}^{\beta}$, $\chi_{N_*+1}^{\alpha}$ and $\chi_{N_*+1}^{\beta}$.

The number N_* can be estimated with the help of the Laurent expansion calculated in part 2 of Proposition 6.3. It follows that

$$N_* \approx \left(\frac{c(\beta - d)^2 (d - \alpha)^2}{3c_0(\beta - \alpha)^2 \gamma^2} \right)^{1/3}. \tag{7.1}$$

In the example shown in Figure 4, this gives an estimate $N_* \approx 24.8\ldots$, shown as the "diamond" marker, while the exact minimum in the space of integers is achieved at $N_* = 21$ (indicated with the "star" marker). The agreement obviously improves with increasing c.

Finally, it is instructive to present a dimensional form of (7.1):

$$\frac{N_*}{L} \approx \left(\frac{c(\beta - d/L)^2(d/L - \alpha)^2}{3c_0(\beta - \alpha)^2\gamma^2}\right)^{1/3}. \tag{7.2}$$

This expression provides an estimate of the actual number of interfaces per unit length as a function of the average strain and other parameters of the model. Notice that the density of the interfaces does not depend on the size of the body.

8. Dynamics

In the previous sections we studied the structure of the global minimizer in the the variational problem (4.2), (4.3) in the limit $\epsilon \to 0$. To check whether our mechanical system can actually reach the corresponding configuration we need to consider a dynamical extension of the variational problem.

There are two major requirements. The dynamical model should be capable of describing fast processes such as nucleation and (explosive) growth of the microstructure. This requires consideration of the kinetic energy and inclusion of inertial terms into the governing equations. On the other hand, the model must be compatible with the final stabilization of the equilibrium microstructure through the much slower processes of coarsening or refinement of the elastic domains. The description of these phenomena requires the introduction of an adequate dissipative mechanism.

In this section we consider a particular dynamic counterpart of our variational model which introduces both inertia and dissipation in a simple way. To avoid integral equations, we consider a dynamical extension of the local model (Model 3, see (3.11) and (3.12)). In this case we need a system of partial differential equations governing the evolution of the strain field $w(x,t)$ and of an additional field of internal variable $v(x,t)$. We assume that the strain field $w(x,t)$ satisfies the equations of classical inertial visco-elastodynamics. For simplicity, the adjustment of the internal variable $v(x,t)$ is considered instantaneous and the corresponding field inertia-free: in the framework of FitzHugh–Nagumo dynamics this case is known as the fast-inhibitor limit. As we show below, the resulting model represents a nonlocal generalization of the viscosity–capillarity model of Truskinovsky [31] and Slemrod [29].

We begin with an introduction of a displacement field

$$u(x,t) = \int_0^x w(y,t)\,dy. \tag{8.1}$$

By adding into the Lagrangian a standard kinetic energy associated with the displacement field $u(x,t)$ and assuming the reference mass density to be equal to unity we obtain a system of (dimensional) equations:

$$\begin{cases} u_{tt} = D\big[-\epsilon^2 D^3 u - \delta(Du - v) + W'(Du)\big] + \eta D^2 u_t \\ -\gamma^2 D^2 v + v = Du. \end{cases} \tag{8.2}$$

Here $\eta > 0$ is a constant (Kelvin) viscosity. The initial and boundary conditions for the system (8.2) take the form

$$\begin{cases} u(x,0) = u_0(x), & x \in (0,L), \\ u_t(x,0) = u_1(x), & x \in (0,L), \\ u(0,t) = 0, \quad u(L,t) = d, & t \in (0,T), \\ D^2u(0,t) = D^2u(0,L) = 0, & t \in (0,T), \\ Dv(0,t) = Dv(L,t) = 0, & t \in (0,T). \end{cases} \tag{8.3}$$

When all small parameters δ, γ, η, and ϵ are equal to zero we obtain a mixed type nonlinear wave equation and the problem is ill-posed, which results in a dramatic lack of uniqueness. Partial regularization can be achieved if either $\delta = 0$ or $\gamma = 0$. In this case the (well-posed) model (8.2), (8.3) reduces to the viscosity–capillarity model which is rich enough to describe propagation of isolated shock waves and phase boundaries (e.g., [32]), but is incompatible with nucleation and spreading of the finite scale microstructures. In this sense the full system (8.2) is expected to have a completely new set of solutions comparing to straightforward dispersive and dissipative regularizations of the nonlinear wave equation.

The detailed analysis of the dynamical equations (8.2), (8.3) is beyond the scope of the present paper. In the next section we present some numerical results pertinent to the accessibility of the periodic minimizer.

In our numerical solution of equation (8.2), we choose an *explicit* scheme. Namely, we replace (8.2) with the iteration

$$\begin{aligned} u(x,t+\Delta t) \approx\ & u(x,t) + u_t(x,t)\Delta t + \frac{1}{2}\big\{D\big[-\epsilon^2 D^3 u(x,t) \\ & - \delta\big(Du(x,t) - v(x,t)\big) + W'\big(Du(x,t)\big)\big] \\ & + \eta D^2 u_t(x,t)\big\}\Delta t^2, \end{aligned}$$

where $v(x,t)$ is obtained at each time step by solving the boundary value problem

$$-\gamma^2 D^2 v(x,t) + v(x,t) = Du(x,t), \qquad Dv(0,t) = Dv(1,t) = 0$$

numerically.

9. Computational Results

The goal of this section is to illustrate numerically the generation and stabilization of the microstructures. Specifically, we shall follow the evolution of two sets

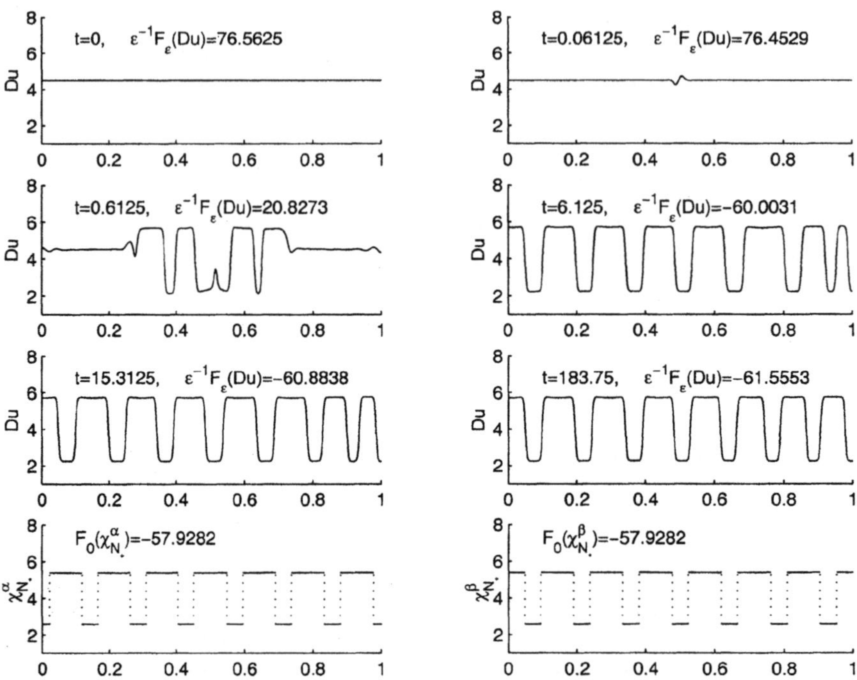

Figure 5. The first six pictures show the evolution of the strain field Du under equations (8.2) and (8.3) at six different times. Here $L = 1$, $d = 4.5$, $\gamma = 0.05$, $\eta = 0.01$, $\epsilon = 0.0025$, $\delta = 0.25$ and $c = 100$. The last two pictures present the minima $\chi^{\alpha}_{N_*}$ and $\chi^{\beta}_{N_*}$ of F_0. The energy density $W$is given by (6.2) with $\alpha = 2.59$, $\beta = 5.41$. In this case $N_* = 14$.

of initial data representing the breakdown of a metastable and an unstable state. The main question is whether the global minimizer is dynamically accessible or whether the system, governed by (8.2), (8.3) get, stuck in a metastable configuration. We remark that the minimizer is known only in the limiting case $\epsilon = 0$ while the calculations are performed at finite ϵ.

In the first set of computations we consider the case when ϵ is sufficiently small so that the functional F_ϵ can be adequately approximated by F_0. Thus, choose $L = 1$, $d = 4.5$, $\gamma = 0.05$, $\eta = 0.01$, $\epsilon = 0.0025$, $\delta = \epsilon c = 0.25$ and $c = 100$. The double well potential W is taken from (6.2).

Suppose initially $u_0(x) = 4.5x$, which corresponds to a uniform metastable strain field, and assume that the localized perturbation of the velocity $u_1(x)$ is small. The initial perturbation must be sufficiently strong to break the metastability. Figure 5 traces the evolution of the strain u_x at six different times t. We observe the nucleation of the microstructure which is followed by its propagation into the undisturbed area. When the growing microstructure reaches the boundary of the domain the process of coarsening begins which eventually leads to an almost periodic pattern. The observed stabilization of the microstructure at large t may mean

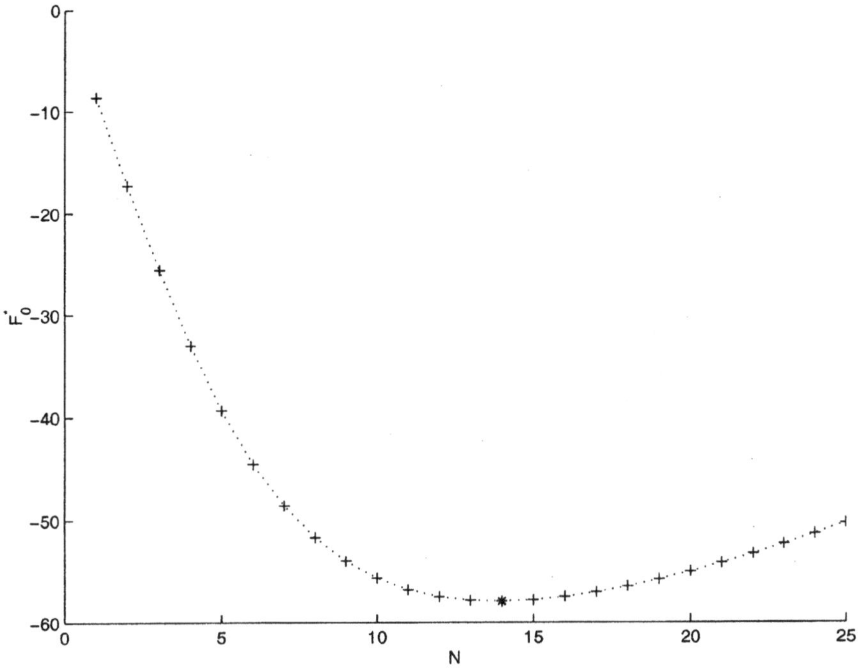

Figure 6. The dependence of F_0^* on N. Here $\alpha = 2.59$, $\beta = 5.41$, $\gamma = 0.05$, $c_0 = 1.33$, $c = 100$ and $d = 4.5$. Note $N_* = 14$.

that the system has reached a state of local/global minimum of the energy. It may also mean significant slowing down in the vicinity of a saddle point.

In order to check different possibilities we can compare the large t configuration with the microstructure delivering the global minimum to the functional F_0. In our case

$$c_0 = \sqrt{2}\int_\alpha^\beta \left[W(t)\right]^{1/2}\mathrm{d}t = \sqrt{2}\int_{4-\sqrt{2}}^{4+\sqrt{2}} \left[\frac{1}{2} - \frac{1}{4}(t-4)^2\right]\mathrm{d}t = \frac{4}{3}.$$

The dependence of $F_0^*(N)$ on N is shown in Figure 6. It is not hard to see that the minimum value of $F_0^*(N)$ is achieved at $N_* = 14$, indicated with the "star" marker. This number agrees quite well with the final picture in Figure 5 where at large t, we observe 15 jumps of u_x. The last two pictures in Figure 5 depict the minimizers $\chi_{N_*}^\alpha$ and $\chi_{N_*}^\beta$ of F_0. In spite of the discrepancy near the boundary, the general agreement between these pictures confirms that the numerical solution is quite close to the global minimizer.

We have also compared the energy $\epsilon^{-1}F_\epsilon(Du)$, calculated from (4.2), with the minimum energy $F_0(\chi_{N_*}^\alpha)$ $(= F_0(\chi_{N_*}^\beta))$, explicitly calculated in Proposition 6.3. These numbers are given together with the corresponding Du, $\chi_{N_*}^\alpha$ and $\chi_{N_*}^\beta$ in Figure 5. We find that they match quite well at large t.

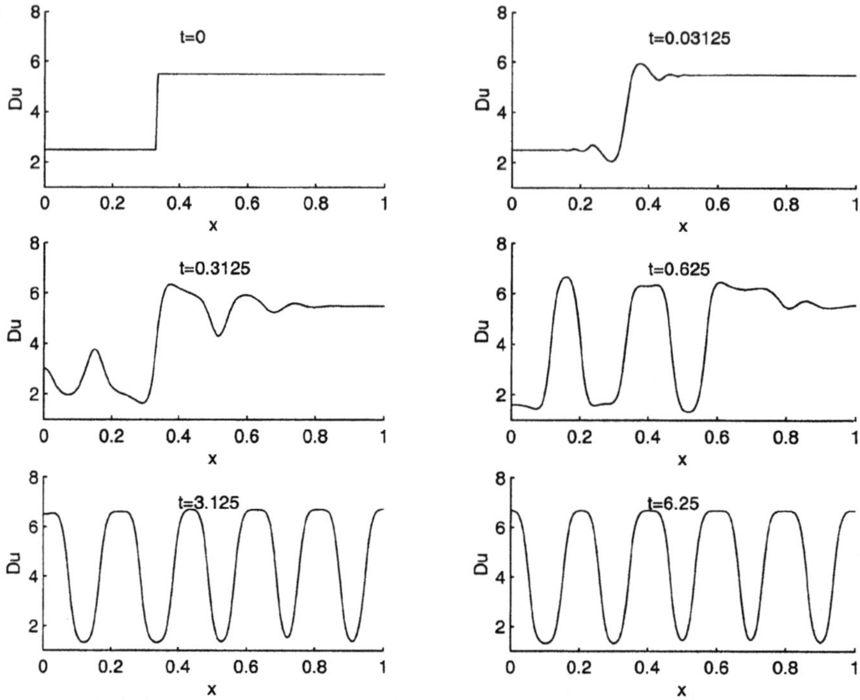

Figure 7. The evolution of the strain field Du at $\epsilon = 0.02$. The rest of the parameters are the same as in Figure 5.

In the second set of computations based on (8.2), (8.3) we use the same parameters except for the higher value of $\epsilon = 0.02$; here we expect a somewhat larger discrepancy with the results of the $\epsilon = 0$ theory. In addition, we change the initial data for $u(x, 0)$, making the configuration unstable, and assuming $u_t(x, 0) = 0$. Specifically we consider a special Riemann problem, when two homogeneous configurations are separated by a sharp interface:

$$u(x, 0) = \begin{cases} 2.5x & \text{if } 0 < x \leqslant 1/3, \\ 5.5(x-1) + 4.5 & \text{if } 1/3 \leqslant x < 1. \end{cases}$$

Notice that we preserve the same total elongation d as in the previous set of computations. The corresponding Du is depicted in the first plot of Figure 7.

From the static problem it is quite clear that the above Riemann data are unstable. As a result, the dynamical process begins without any initial perturbation of velocity. First, the shock waves are emitted and the single layer quickly develops into multiple layers. The oscillatory state then propagates in both directions. After the microstructure hits the boundaries, the process of stabilization and coarsening begins. The final pattern is again almost periodic. The interfaces are more diffuse now because of the larger value of ϵ. The count of the number of interfaces suggests that the achieved microstructure is at most metastable in the sense of the $\epsilon = 0$ limit.

10. Model Allowing for Sharp Interfaces

Although our simple one-dimensional model encompasses all necessary physical ingredients and produces qualitatively reasonable results, it has the problem common to all gradient models which use strain as an order parameter: misrepresentation of the dispersion relations for lattices which leads to over-regularized, unrealistically diffuse interfaces. In fact the typical TEM observations show that the phase or twin boundaries are often close to being atomically sharp. In order to overcome this difficulty, one can try to modify the "surface energy" part of our model and substitute the strain gradient term by a nonlocal term with a *positive* definite kernel. It is known that depending on the parameters of the kernel and the degree of the nonconvexity of the elastic energy, the "positive" nonlocal model can produce both sharp and diffuse interfaces (see, for instance, [3, 12–14]).

In this section we exploit this idea and briefly sketch a variant of the theory which can produce finite scale microstructures with sharp interfaces. Essentially, we combine positive and negative definite nonlocal terms into one sign-indefinite kernel $J(x, y)$. We assume that when x is close to y, the kernel $J(x, y)$ is positive, and when x is away from y, the kernel $J(x, y)$ is negative. Our new energy functional takes the form

$$F(w) = \int_0^1 W(w)\,\mathrm{d}x + \frac{1}{4}\int_0^1\int_0^1 J(x, y)\big(w(x) - w(y)\big)^2\,\mathrm{d}x\,\mathrm{d}y, \tag{10.1}$$

where

$$J(x, y) = \delta_1 G_1(x, y) - \delta_2 G_2(x, y). \tag{10.2}$$

The functions G_i are selected as Green functions of the two auxiliary problems

$$-\gamma_i^2 D^2 y + y = f \quad \text{in } (0, 1), \qquad Dy(0) = Dy(1) = 0$$

(see the explicit expression in (3.5)). Notice that $G_1(x, y)$ and $G_2(x, y)$ enter the kernel $J(x, y)$ with different signs; the representative graph of $J(x, y)$ is shown in Figure 8.

By choosing again the kernel to be exponential we are aiming at the equivalent local theory: a blend of the Timoshenko model considered by Rogers and Truskinovsky [27] and the FitzHugh–Nagumo model considered in this paper. In addition to the conventional strain field, the present theory contains two (!) new physical fields.

In order to rewrite the problem as local we introduce two internal variables,

$$p(x) = \int_0^1 G_1(x, y) w(y)\,\mathrm{d}y, \qquad q(x) = \int_0^1 G_2(x, y) w(y)\,\mathrm{d}y. \tag{10.3}$$

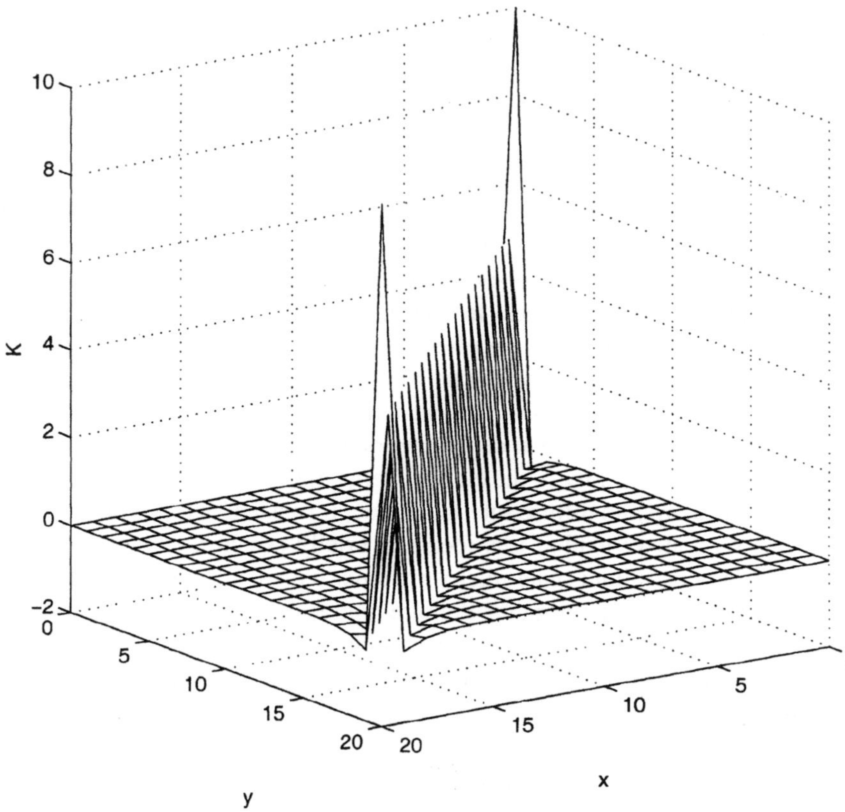

Figure 8. The representative graph of the sign-indefinite kernel $J = \delta_1 G_1 - \delta_2 G_2$ from (10.2). Here $\delta_1 = 1.0$, $\delta_2 = 1.0$, $\gamma_1 = 0.005$ and $\gamma_2 = 0.05$.

In terms of these new variables, the energy functional (3.11) takes the form

$$F^3(w, p, q) = \int_0^1 \left\{ W(w) + \frac{\delta_1 \gamma_1^2}{2}|Dp|^2 + \frac{\delta_1}{2}(p-w)^2 - \frac{\delta_2 \gamma_2^2}{2}|Dq|^2 - \frac{\delta_2}{2}(q-w)^2 \right\} \mathrm{d}x. \tag{10.4}$$

The derivation is identical to the one given in Section 3. Notice that the classical Ericksen's double well energy density is the only nonlinear term in the energy functional. The conventional strain field is linearly coupled with two additional fields, $p(x)$ and $q(x)$, which satisfy linear "field" equations. The Euler–Lagrange equations for this model and the corresponding natural boundary conditions take the forms

$$\begin{cases} \delta_1(w-p) - \delta_2(w-q) + W'(w) = \sigma, \\ -\gamma_1^2 D^2 p + p = w, \qquad Dp(0) = Dp(1) = 0, \\ -\gamma_2^2 D^2 q + q = w, \qquad Dq(0) = Dq(1) = 0. \end{cases} \tag{10.5}$$

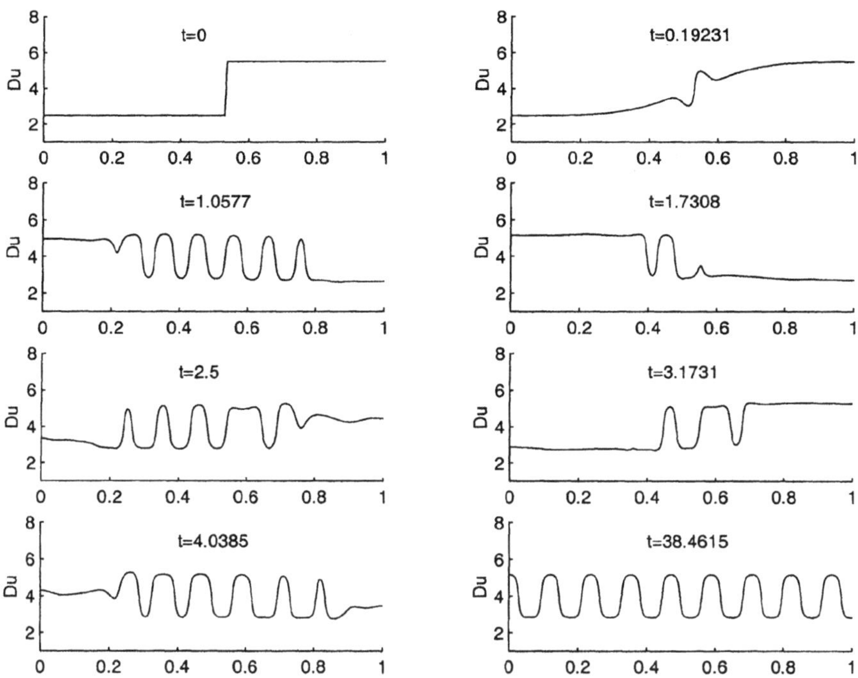

Figure 9. The evolution of the strain field Du according to (10.6). Here $\eta = 0.015$, $d = 3.9$, $\delta_1 = 1.0$, $\delta_2 = 1.0$, $\gamma_1 = 0.005$ and $\gamma_2 = 0.05$.

The detailed analysis of this model will be given elsewhere. Our preliminary results indicate that the new model exhibits periodic minimizers with sharp interfaces.

In dynamics the above model shows interesting reversible transformations between the twinned "martensite" and the surrounding "austenite".

To illustrate this point, consider the simplest dynamic counterpart of (10.1):

$$u_{tt} = D\left[W'(Du) + \int_0^1 J(x,y)\Big(Du(x,t) - Du(y,t)\Big)\mathrm{d}y\right] + \eta D^2 u_t, \qquad (10.6)$$

where again $u(x,t) = \int_0^x w(y,t)\,\mathrm{d}y$ is the displacement field. Equation (10.6) is subject to the initial and boundary conditions

$$\begin{cases} u(x,0) = u_0(x), \\ u_t(x,0) = u_1(x), \\ u(0,t) = 0, \qquad u(1,t) = d. \end{cases} \qquad (10.7)$$

The dynamical problem (10.6) can also be rewritten in the local form

$$\begin{cases} u_{tt} = D\big[\delta_1(Du - p) - \delta_2(Du - q) + W'(Du)\big] + \eta D^2 u_t, \\ -\gamma_1^2 D^2 p + p = Du, \\ -\gamma_2^2 D^2 q + q = Du. \end{cases} \qquad (10.8)$$

The additional boundary conditions for p and q are

$$\begin{cases} Dp(0,t) = Dp(1,t) = 0, \\ Dq(0,t) = Dq(1,t) = 0. \end{cases} \tag{10.9}$$

The results of the numerical simulation of equation (10.1) are shown in Figure 9. We took $\eta = 0.015$, $d = 3.9$ and chose the kernel $J(x, y)$ as in (10.2) with $\delta_1 = 1.0$, $\delta_2 = 1.0$, $\gamma_1 = 0.005$ and $\gamma_2 = 0.05$. In the simulation we start with the unstable Riemann data for Du and zero velocity. We observe that the single interface again develops into multiple layers. The oscillatory state propagates in both direction with clearly indicated fronts; these fronts simulate martensite–austenite interfaces. We observe that the "martensitic" domains behind the moving transition zone are almost stationary so the attractor is not a travelling wave. At some point the layered configuration fills the whole domain. Then the domains near the boundaries start to disappear and the uniform states propagate back but in a reversed order. This process, however, does not go all the way back to the configuration with a single interface. Instead, long before the number of domains reduces to one, the oscillations of the strain field Du start to grow again, propagating towards the boundary. After several repetitions, each time with more "leftover" oscillations, the configuration finally stabilizes, forming a periodic pattern. In the interim period the oscillatory state clearly behaves like an elastic "phase".

11. Conclusions

In this paper we introduced an easily tractable one-dimensional model describing stable finite scale microstructures. This model can be viewed as a one-dimensional ansatz, which in the simplest qualitative form reproduces most of the relevant effects of the full 2- and 3-dimensional formulations. To make a model amenable to analysis we specified the kernel of a nonlocal part of the interaction to be exponential; the more realistic power law kernels can presently be studied numerically at most. We investigated in some detail the important special case when the internal length scales associated with the coarsening and the refinement mechanisms can be separated. By sacrificing some of the metastable configurations, we focused on the asymptotic description of the global minimum. The corresponding limiting problem turned out to be finite-dimensional. We showed that this new problem has a periodic minimizer and explicitly calculated the optimal period. We then looked at the accessibility of the corresponding configuration from the perspective of a realistic dynamics. In particular, we demonstrated how the periodic microstructure nucleates, spreads into the metastable areas and stabilizes. Among other things, we examined the process of final coarsening leading to the microstructure with the predicted period. Our solution presents a new self-consistent description of a moving interface between a homogenous state and a two-phase mixture.

The main open problems are associated with the direct derivation of the nonlocal kernel from the 3-dimensional theory, with the careful analysis of the local minimizers outside the range of the Γ-limit technique, and with the homogenized description of the kinetics of the propagating microstructures.

Acknowledgments

We thank the reviewer for helpful comments. This work was initiated during the stay of XR as a postdoctoral fellow in the IMA, University of Minnesota. Both authors acknowledge the hospitality of the University of Padova, Italy, in the Fall of 1997.

Appendix

We begin with several technical definitions. Define the set

$$X = \left\{ w \in L^2(0,1)\colon \int_0^1 w(x)\,\mathrm{d}x = d \right\}.$$

The distance between two elements w_1 and w_2 is assumed to be $\|w_1 - w_2\|_{L^2(0,1)}$. We extend the domain of F_ϵ to X by setting $F_\epsilon(w) = \infty$ if $w \in X\backslash W^{1,2}(0,1)$, and the domain of F_0 to X by setting $F_0(\chi) = \infty$ if $\chi \in X\backslash BV((0,1),\{\alpha,\beta\})$. We introduce $BV((0,1),\{\alpha,\beta\})$ as a subset of the space of functions of bounded variation, $BV(0,1)$ (see, for instance, [10, Chapter 5]). A function χ of bounded variation is in $BV((0,1),\{\alpha,\beta\})$ if, for every $x \in (0,1)$, $\chi(x) = \alpha$ or β.

Suppose that the energy density W has two global minima at α and β, i.e., $W(t) \geqslant 0$ and $W(t) = 0$ if and only if $t = \alpha$ or β. We also assume for the technical reasons that W is continuous and there exist $k \geqslant 2$, $K_1 > 0$, $K_2 > 0$, and $\bar{t} > 1$ such that, for all t, $|t| > \bar{t}$,

$$K_1|t|^k \leqslant W(t) \leqslant K_2|t|^k. \tag{A.1}$$

We first observe that $\epsilon^{-1}F_\epsilon$ is uniformly bounded below for small ϵ. In fact, for $C > \max\{|\alpha|,|\beta|\}$, and $|w| > C$, we have $(1/\epsilon)W(w) - (c/2)w^2 > 0$ for small ϵ. Then

$$\begin{aligned}
\epsilon^{-1}F_\epsilon(w) &\geqslant \int_0^1 \left[\frac{1}{\epsilon}W(w) - \frac{c}{2}w^2\right]\mathrm{d}x \\
&\geqslant \int_{\{x:|w(x)|\leqslant C\}} \left[\frac{1}{\epsilon}W(w) - \frac{c}{2}w^2\right]\mathrm{d}x \\
&\geqslant \int_{\{x:|w(x)|\leqslant C\}} -\frac{c}{2}w^2\,\mathrm{d}x \geqslant -\frac{cC^2}{2}.
\end{aligned}$$

The relation between $\epsilon^{-1}F_\epsilon$ and F_0 is described here in the language of Γ-convergence, defined as follows (see [8]).

DEFINITION A.1. $\epsilon^{-1}F_\epsilon$ Γ-converges to F_0 if the following two statements hold:

1. For every family $\{w_\epsilon\} \subset X$ with $\lim_{\epsilon\to 0} w_\epsilon = w$,

$$\liminf_{\epsilon\to 0} \epsilon^{-1}F_\epsilon(w_\epsilon) \geqslant F_0(w).$$

2. For every $w \in X \cap BV((0,1),\{\alpha,\beta\})$, there exists a family $\{w_\epsilon\} \subset X$ such that $\lim_{\epsilon\to 0} w_\epsilon = w$ and

$$\limsup_{\epsilon\to 0} \epsilon^{-1}F_\epsilon(w_\epsilon) \leqslant F_0(w).$$

Our first key result is the following proposition.

PROPOSITION A.2. *$\epsilon^{-1}F_\epsilon$ Γ-converges to F_0.*

Proof. For every $w \in X$, we define

$$E_\epsilon(w) = \begin{cases} \displaystyle\int_0^L \left\{\frac{1}{\epsilon}W(w) + \frac{\epsilon}{2}|Dw|^2\right\}\mathrm{d}x & \text{if } w \in W^{1,2}(0,L), \\ \infty & \text{otherwise,} \end{cases} \tag{A.2}$$

$$E_0(w) = \begin{cases} c_0\mathcal{S}(w) & \text{if } w \in BV((0,L),\{\alpha,\beta\}), \\ \infty & \text{otherwise,} \end{cases} \tag{A.3}$$

and

$$G_L(w) = \int_0^L \left\{-\frac{c}{2}w^2 + \frac{c}{2}[(-\gamma^2D^2+1)^{-1/2}w]^2\right\}\mathrm{d}x. \tag{A.4}$$

It is proved by Modica [21], that E_ϵ Γ-converges to E_0 (with a slight change from $L^1(\Omega)$ to $L^2(0,1)$). The convergence is carried over to $\epsilon^{-1}F_\epsilon \to F_0$ by the continuity of G from X to R. □

Proposition A.2 does not immediately relate the minimizers of $\epsilon^{-1}F_\epsilon$ to the minimizers of F_0. To see the relation we first need a technical lemma.

LEMMA A.3. *Let $\{\epsilon_n\}$ be a sequence of positive numbers converging to 0, and $\{w_n\}$ a sequence of functions in X. If $\epsilon_n^{-1}F_{\epsilon_n}(w_n)$ is bounded above in n, then the sequence $\{w_n\}$ is relatively compact in X and its cluster points belong to $BV((0,1),\{\alpha,\beta\})$.*

Proof. We set

$$\phi(t) = \int_\alpha^t W^{1/2}(s)\,\mathrm{d}s. \tag{A.5}$$

Then (A.1) implies

$$|\phi(t)| \leqslant C + C|t|^{k/2+1}.$$

Set $v_n = \phi(w_n)$. By (A.1) and $k \geqslant 2$, we find,

$$|v_n| \leqslant C + C|w_n|^{k/2+1} \leqslant C + CW(w_n).$$

Following the above proof of the uniform lower bound for $\epsilon^{-1}F_\epsilon$, we obtain

$$\begin{aligned}\epsilon_n^{-1}F_{\epsilon_n}(w_n) &\geqslant \int_0^1 \left[\frac{1}{\epsilon_n}W(w_n) - \frac{c}{2}w_n^2\right]\mathrm{d}x \\ &= \frac{1}{2\epsilon_n}\int_0^1 W(w_n)\,\mathrm{d}x + \int_0^1 \left[\frac{1}{2\epsilon_n}W(w_n) - \frac{c}{2}w_n^2\right]\mathrm{d}x \\ &\geqslant \frac{1}{2\epsilon_n}\int_0^1 W(w_n)\,\mathrm{d}x - \frac{cC^2}{2}.\end{aligned}$$

Or, in other words,

$$\int_0^1 W(w_n)\,\mathrm{d}x \leqslant 2F_{\epsilon_n}(w_n) + cC^2\epsilon_n, \tag{A.6}$$

which means that $\{v_n\}$ is bounded in $L^1(0, 1)$.

On the other hand,

$$\begin{aligned}\epsilon_n^{-1}F_{\epsilon_n}(w_n) &\geqslant \frac{1}{2}\int_0^1 \left[\epsilon_n|Dw_n|^2 + \frac{1}{\epsilon_n}W(w_n)\right]\mathrm{d}x \\ &\quad + \int_0^1 \frac{1}{2\epsilon_n}\left[W(w_n) - \frac{c}{2}w_n^2\right]\mathrm{d}x \\ &\geqslant \int_0^1 (W(w_n))^{1/2}|Dw_n|\,\mathrm{d}x + \int_0^1 \frac{1}{2\epsilon_n}\left[W(w_n) - \frac{c}{2}w_n^2\right]\mathrm{d}x \\ &\geqslant \frac{1}{\sqrt{2}}\int_0^1 |Dv_n|\,\mathrm{d}x - \frac{cC^2}{2},\end{aligned}$$

and $\{v_n\}$ is bounded in $W^{1,1}(0, 1)$. The Sobolev imbedding theorem asserts that $\{v_n\}$ is relatively compact in $L^1(0, 1)$.

Now consider $w_n = \phi^{-1}(v_n)$. The formulas (A.1) and (A.5) imply that ϕ^{-1} is continuous, increasing and that

$$|\phi^{-1}(t)| \leqslant C + C|t|^{2/(k+2)}, \qquad |\phi^{-1}(t)|^2 \leqslant C + C|t|. \tag{A.7}$$

To prove that $\{w_n\}$ is relatively compact we show that every subsequence of $\{w_n\}$ has a L^2-convergent further subsequence.

Let $\{w_{n_l}\}$ be a subsequence of $\{w_n\}$. Then there are a subsequence of $\{v_{n_l} = \phi(w_{n_l})\}$, denoted by $\{v_{n_{l_m}}\}$, and $v \in L^1(0, 1)$ such that $v_{n_{l_m}} \to v$ in $L^1(0, 1)$ and $v_{n_{l_m}} \to v$ a.e. Then $w_{n_{l_m}} \to \phi^{-1}(v)$ a.e. Applying Vitali's convergence theorem [15, p. 203] to $v_{n_{l_m}}$, we find that for every $\varepsilon > 0$ there is $\eta > 0$, such that for every measurable set E, $|E| < \eta$ implies $\int_E |v_{n_{l_m}}| \,\mathrm{d}x < \varepsilon$ for all n. Then (A.7) implies

$$\int_E |w_{n_{l_m}}|^2 \,\mathrm{d}x \leqslant \int_E (C + C|v_{n_{l_m}}|)\mathrm{d}x < C\eta + C\varepsilon.$$

Now Vitali's theorem applied to $\{w_{n_{l_m}}\}$ asserts that $w_{n_{l_m}} \to \phi^{-1}(v)$ in $L^2(0, 1)$.

Let w be a cluster point of $\{w_n\}$, which means that there exists a subsequence $\{w_{n_l}\}$ such that $w_{n_l} \to w$ in $L^2(0, 1)$ as $l \to \infty$. Fatou's lemma, the boundedness of $\epsilon_n^{-1} F_{\epsilon_n}(w_n)$ and (A.6) imply that

$$0 \leqslant \int_0^1 W(w)\,\mathrm{d}x \leqslant \lim_{l\to\infty} \int_0^1 W(w_{n_l})\,\mathrm{d}x = 0.$$

Now for a.e. $x \in (0, 1)$, $w(x) = \alpha$ or β. If we consider $v_{n_l} = \phi(w_{n_l})$, then the boundedness of $\{v_{n_l}\}$ in $W^{1,1}(0, 1)$, proved earlier, implies that $\phi(w)$, the L^1-limit of $\{v_{n_l}\}$, is a BV function [10, Theorem 1, p. 172]. Since $\phi(w)$ only takes two values, $\phi(\alpha)$ and $\phi(\beta)$, we obtain

$$\phi(w) = \phi(\alpha) + \frac{\phi(\beta) - \phi(\alpha)}{\beta - \alpha}(w - \alpha).$$

Hence, w is also a BV function. □

From Proposition A.2 and Lemma A.3 we deduce how the minimizers of $\epsilon^{-1}F_\epsilon$ and F_0 are related.

PROPOSITION A.4. *Let $\{\epsilon_n\}$ be a positive sequence converging to* 0, *and w_{ϵ_n} be a minimizer of F_{ϵ_n} in X. Then $\{w_{\epsilon_n}\}$ is relatively compact in X and each of its cluster points is a minimizer of F_0 in X.*

References

1. J. Ball and R. James, *The Mathematics of Microstructure*. Birkhäuser, to appear.
2. J. Ball, R. James, R. Pego and P. Swart, On the dynamics of fine structure. *J. Nonlinear Sci.* **1** (1991) 17–70.
3. P. Bates and A. Chmaj, An integrodifferential model for phase transitions: Stationary solutions in higher space dimensions. *J. Stat. Phys.* **95** (1999) 1119–1139.
4. P. Bates, P. Fife, X. Ren and X. Wang, Traveling waves in a convolution model for phase transitions. *Arch. Rational Mech. Anal.* **138** (1997) 105–136.
5. D. Brandon, T. Liu and R. Rogers, Phase transitions and hysteresis in nonlocal and order parameter models. *Meccanica* **30**(5) (1995) 541–565.
6. J. Carr, M. Gurtin and M. Slemrod, Structured phase transitions on a finite interval. *Arch. Rational Mech. Anal.* **86** (1984) 317–351.
7. A. Chmaj and X. Ren, Homoclinic solutions of an integral equation: Existence and stability. *J. Differential Equations* **155** (1999) 17–43.

8. G. Dal Maso, *An Introduction to Γ-convergence. Progress in Nonlinear Differential Equations and Applications*. Birkhäuser, Boston (1993).
9. J. Ericksen, Equilibrium of bars. *J. Elasticity* **5** (1975) 191–201.
10. L. Evans and R. Gariepy, *Measure Theory and Fine Properties of Functions*. CRC Press, Boca Raton, FL (1992).
11. R. FitzHugh, *Biological Engineering*, H. Schwan (ed.), McGrow-Hill, New York (1969).
12. R. Fosdick and D. Mason, Single phase energy minimizers for materials with nonlocal spatial dependence. *Quart. Appl. Math.* **54**(1) (1996) 161–195.
13. R. Fosdick and D. Mason, On a model of nonlocal continuum mechanics Part II: Structure, Asymptotics, and Computations. *J. Elasticity* **48** (1997) 51–100.
14. R. Fosdick and D. Mason, On a model of nonlocal continuum mechanics, Part I: Existence and regularity. *SIAM J. Appl. Math.* **58**(4) (1998) 1278–1306.
15. E. Hewitt and K. Stromberg, *Real and Abstract Analysis*. Springer, Berlin (1965).
16. S. Kartha, D. Krumhansl, J. Sethna and L. Wickman, Disorder driven pretransitional tweed in martensitic transformations. *Phys. Rev. B* **52** (1995) 803–822.
17. A. Khachaturian, *Theory of Structural Deformations in Solids*. Wiley, New York (1983).
18. M. Killough, A diffusion interface approach to the development of microstructure in martensite, PhD Thesis, New York University (1998).
19. R. Kohn and S. Müller, Surface energy and microstructure in coherent phase transitions. *Comm. Pure. Appl. Math.* **47** (1994) 405–435.
20. M. Luskin, On the computation of crystalline microstructure. *Acta Numerica* **5** (1996) 191–258.
21. L. Modica, The gradient of phase transitions and the minimal interface criterion. *Arch. Rational Mech. Anal.* **98** (1987) 357–383.
22. S. Müller, Singular perturbations as a selection criterion for periodic minimizing sequences. *Cal. Var. Partial Diff. Equations* **1** (1993) 169–204.
23. S. Müller, *Variational Models for Microstructure and Phase Transitions*, Lecture Notes. Max-Plank-Institut, Leipzig (1998).
24. R. Peierls, The size of a dislocation. *Proc. Phys. Soc.* **52** (1940) 34–37.
25. M. Pitteri and G. Zanzotto, *Continuum Models of Phase Transitions and Twinning in Crystals*. CRC/Chapman&Hall, London, to appear.
26. X. Ren and M. Winter, Young measures in a nonlocal phase transition problem. *Proc. Roy. Soc. Edinburgh A* **127** (1997) 615–637.
27. R. Rogers and L. Truskinovsky, Discretization and hysteresis. *Physica B* **233** (1997) 370–375.
28. A. Roytburd, Martensitic transformation as a typical phase trnsformation in solids. *Solid State Phys.* **34** (1978) 317–390.
29. M. Slemrod, Admissibility criteria for propagating phase boundaries in a Van Der Waals fluid. *Arch. Rational Mech. Anal.* **81** (1983) 301–315.
30. S. Timoshenko, On the correction for shear of the differential equation for transverse vibrations of prizmatic bars. *Philos. Mag. Ser. 6* **41** (1921) 744.
31. L. Truskinovsky, Equilibrium phase boundaries. *Soviet. Phys. Dokl.* **27** (1982) 551–553.
32. L. Truskinovsky, About the normal growth approximation in the dynamical theory of phase transitions. *Cont. Mech. Thermodyn.* **6** (1993) 185–208.
33. L. Truskinovsky and G. Zanzotto, Finite scale microstructures and metastability in one-dimensional elasticity. *Meccanica* **30** (1995) 577–589.
34. L. Truskinovsky and G. Zanzotto, Ericksen's bar revisited: Energy wiggles. *J. Mech. Phys. Solids* **44**(8) (1996) 1371–1408.
35. A. Vainchtein, T. Healey, P. Rosakis and L. Truskinovsky, The role of the spinodal in one-dimensional phase transitions microstructures. *Phys. D* **115** (1998) 29–48.

Journal of Elasticity **59:** 357–367, 2000.

Optimal Design Problems for Elastic Bodies by Use of the Maximum Principle

WILLIAM H. WARNER
Department of Aerospace Engineering and Mechanics, University of Minnesota, Minneapolis, MN 55455, U.S.A. E-mail: warner@aem.umn.edu

Received 25 March 1999

For Roger Fosdick, amicus et collega.

Abstract. A class of optimal design problems for elastic bodies requires minimization of a nonconvex energy functional with respect to both kinematical variables and a design variable. Such functionals for rod, beam, sphere, and cylinder problems are special cases of a single functional. Minimization of this functional using Pontryagin's Maximum Principle to handle the nonconvexity and the inequality constraints leads to the choice of the design variable as taking on only its upper and lower bound values, agreeing with some previous results found for specific problems. A generalization to extensible beam-column problems is discussed.

Mathematics Subject Classifications (2000): 49 calculus of variations, optimal control, 74 mechanics of deformable solids.

Key words: Optimal design, elasticity, maximum principle.

1. Introduction; Problem Formulation

One class of optimal design problems for elastic bodies requires that an energy functional be minimized with respect to both a design variable and the ordinary kinematic variables that would be used in the traditional equilibrium analysis by minimum energy methods. These problems, generally nonconvex and involving inequality constraints, arise both in technical elasticity where rod, beam, and beam-column models are considered and in classical Hookean elasticity for spherically or cylindrically symmetric deformations. This work will show that a number of such problems where the design variable is bounded a priori and also subject to some overall integral constraint can be treated in a common fashion using the Maximum Principle, resulting in the determination that an optimal design (if it exists) must divide the body into alternating regions in which the design variable takes on only its minimum or maximum value. That is, in the parlance of the control theorist, the design must be a "bang-bang" control jumping in value at certain switching points.

In [1, 3] (see also [2]), Fosdick and Royer-Carfagni solved by function space methods two such problems in the optimal design of elastic bodies consisting of a mixture or alloy of two materials. In [1, 2], an elastic rod of constant cross-sectional area hanging under its own weight from a fixed support is given a prescribed end displacement at its lower end. Using the concentration density $c(x)$ of the stiffer material as a design variable subject to the inequality constraints $0 \leqslant c \leqslant 1$ and the volume fraction constraint

$$\int_0^L c(x)\,\mathrm{d}x = L\bar{c}, \quad 0 < \bar{c} < 1, \tag{1.1}$$

they ask for the distribution of $c(x)$ as well as that of the axial displacement $u(x)$ that minimizes the potential energy functional

$$\Pi = \int_0^L \left\{\frac{1}{2}E(c)Ae^2 - \rho(c)gAu\right\}\mathrm{d}x. \tag{1.2}$$

Here $e = \mathrm{d}u/\mathrm{d}x$ is the axial strain. The mass density of the mixture is the weighted average of the densities of the pure materials:

$$\rho(c) = \rho_0(1-c) + \rho_1 c > 0, \tag{1.3}$$

while the Young's modulus is assumed to satisfy

$$E(c) > 0, \qquad \frac{\mathrm{d}E}{\mathrm{d}c} > 0, \qquad \frac{\mathrm{d}^2}{\mathrm{d}c^2}\left(\frac{1}{E}\right) > 0. \tag{1.4}$$

The last inequality suffices to ensure that the strain energy density $W = EAe^2/2$ is nonconvex in the two variables c and e, and it is this feature that causes the mathematical complexity in the minimization problem.

In [3], they solved a similar problem for a sphere (hollow or solid) subject to uniform radial displacement on its outer boundary. With the usual assumptions of spherically symmetric deformation, the total strain energy to be minimized may be written

$$\Pi = \int_{R^*}^R \left\{\frac{(\lambda(c) + 2\mu(c))e^2}{2} + \frac{2\lambda eu}{r} + \frac{2(\lambda+\mu)u^2}{r^2}\right\}(4\pi r^2)\mathrm{d}r, \tag{1.5}$$

where $e = \mathrm{d}u/\mathrm{d}r$ and the mixture is assumed to be isotropic at each $c(r)$ with the Lamé moduli of elasticity for the mixture assumed to be linear in c:

$$\lambda(c) = \lambda_0(1-c) + \lambda_1 c, \qquad \mu(c) = \mu_0(1-c) + \mu_1 c. \tag{1.6}$$

Nonconvexity in c and the strain holds here also, as discussed in [3]. The admissible $c(r)$ are piecewise continuous, satisfy the inequality constraints $0 \leqslant c \leqslant 1$, and the volume fraction constraint

$$\int_{R^*}^R c(r)(4\pi r^2)\mathrm{d}r = \overline{V}\bar{c}, \quad 0 < \bar{c} < 1, \tag{1.7}$$

where $\overline{V}$ is the volume of the sphere.

For both of these problems the authors found that the optimal solution for the concentration variable consisted of piecewise constant distributions at the two bounds 0 and 1, with sometimes two and sometimes three regions depending on the parameters of the problem. Moreover, since $c(r)$ must be piecewise constant, the strain and displacement solutions had the form of well-known classical solutions in each subregion and could be fitted together to satisfy the appropriate continuity and boundary conditions for each problem.

The basic method they used depends on the construction of the lower convex envelope to the energy density and the determination by function space methods that the solution to the original problem must be the same as the solution using the envelope functional as a replacement problem.

In [4], I solved two design problems for the rod hanging under its own weight. I used the cross-sectional area $A(x)$ as design variable and took E and ρ as constants. The area satisfied upper and lower bound constraints:

$$A_1 \leqslant A(X) \leqslant A_2. \tag{1.8}$$

In one problem, the total volume was minimized subject to equilibrium equations and stress bounds as constraints; in the other, the potential energy was minimized subject to the inequality constraints on A and to the prescription of total volume:

$$\int_0^L A(X)\,\mathrm{d}X = V_0 = \overline{A}L, \quad A_1 < \overline{A} < A_2. \tag{1.9}$$

This second problem is equivalent to Fosdick and Royer–Carfagni's rod problem with a different scaling of the design variable. The body force $B = \rho g A$ is linear in the design variable in each case. The stiffness function $S = EA$ is now also linear in A; but functions linear in the design variable satisfy the nonconvexity condition stated above. Moreover, with this choice of design variable, my coefficients automatically satisfied the condition $S_1 B_1 < S_2 B_2$ equivalent to the condition $\rho_0 E_0 < \rho_1 E_1$ assumed to hold in [1] and so my solution was the same in form as that of [1] though with different physical interpretation. (In [5], I present the solution for the other ordering in this inequality and point out the difficulties in defining a solution when equality holds.) But mine was produced by treating the problem as a "control problem of Lagrange with inequality constraints" and applying the Pontryagin–Hestenes Maximum Principle as stated in [6, Chapter 6, Theorem 4.1].

I next looked at the sphere problem of [3] as a control problem and at the Euler–Bernouilli beam under the tranverse load of its own weight. The circular cylinder in plane strain can also be treated this way with an energy density similar to the sphere. The energy functional for the Euler–Bernouilli beam is

$$\Pi = \int_0^L \left\{ \frac{1}{2} E I \kappa^2 - \rho g A v \right\} \mathrm{d}x, \tag{1.10}$$

where $v(x)$ is the transverse deflection of the centerline, $\kappa(x)$ is its curvature, and the bending stiffness $S = EI$ involves the usual second moment of area. The total strain energy for the cylinder is

$$\Pi = \int_{R^*}^{R} \left\{ \frac{(\lambda + 2\mu)e^2}{2} + \frac{\lambda eu}{r} + \frac{(\lambda + 2\mu)u^2}{2r^2} \right\} (2\pi Lr)\,\mathrm{d}r, \tag{1.11}$$

where cylindrical polar coordinates are used and L is the length of the cylinder.

These four functionals and the design problems associated with them are all special cases of the following minimization problem. Suppose that there are N state variables $\mathbf{u} = \{u_i(z)\}$, $i = 1, 2, \ldots, N$, and two control variables $\varepsilon(z)$ and $\eta(z)$. "Kinematically admissible" sets of these are defined as follows. The u_i (which I shall call "generalized displacements") must be continuous on an interval $a < z < b$; ε (the "generalized strain") and η (the "design" variable) are piecewise continuous. Some or all of the values of the u_i are prescribed at $z = a$ and $z = b$, with at least one of these nonzero. The $\mathbf{u}$ and ε are related by the ordinary differential equations

$$\frac{\mathrm{d}u_i}{\mathrm{d}z} = f_i(z, \mathbf{u}) + \varepsilon(z) g_i(z, \mathbf{u}). \tag{1.12}$$

These are linear in ε and hold at all points of continuity of $\varepsilon(z)$. The problem is to minimize the functional

$$\Psi = \int_a^b \Phi(z, \mathbf{u}, \varepsilon, \eta)\,\mathrm{d}z \tag{1.13}$$

with respect to admissible $(\mathbf{u}, \varepsilon, \eta)$ and the further constraints on $\eta(z)$

$$\phi_1 = \eta_1 - \eta(z) \leqslant 0, \qquad \phi_2 = \eta(z) - \eta_2 \leqslant 0, \tag{1.14a}$$

$$\int_a^b \eta J\,\mathrm{d}z = \bar{\eta} \int_a^b J\,\mathrm{d}z, \qquad \eta_1 < \bar{\eta} < \eta_2. \tag{1.14b}$$

Here Φ satisfies conditions of convexity in ε (the "ordinary equilibrium" problem) and nonconvexity in the pair (ε, η):

$$\frac{\partial^2 \Phi}{\partial \varepsilon^2} > 0, \qquad \frac{\partial^2 \Phi}{\partial \varepsilon^2}\frac{\partial^2 \Phi}{\partial \eta^2} - \left(\frac{\partial^2 \Phi}{\partial \varepsilon \partial \eta} \right)^2 < 0; \tag{1.15}$$

$J(z) > 0$ (J can be zero at $z = a = 0$ only) and for all the problems given above

$$\Phi = J(z) \left\{ \frac{1}{2} S(\eta, z)\varepsilon^2 + P(\eta, z, \mathbf{u})\varepsilon + Q(\eta, z, \mathbf{u}) \right\} \tag{1.16}$$

with $S > 0$, $\partial S/\partial \eta > 0$, $\partial^2/\partial \eta^2 (1/S) > 0$ and P, Q at most linear in η.

2. Necessary Conditions for an Optimizer

The Maximum Principle [6, Chapter 6, Theorem 4.1] for control problems of Lagrange with inequality constraints states that if an optimizer $(\mathbf{u}, \varepsilon, \eta)^*$ exists then there must also exist multipliers

$$\lambda_0 \geqslant 0, \lambda, \qquad \mathbf{p} = \{p_i(z)\}, \quad i = 1, \dots, N, \quad \mu_1(z), \mu_2(z) \tag{2.1a}$$

not vanishing simultaneously on the closed interval $a \leqslant z \leqslant b$ and a "pre-Hamiltonian" function $H(z, \mathbf{u}, \mathbf{p}, \varepsilon, \eta, \mu_1, \mu_2)$ defined by

$$H = \sum_{k=1}^{N} p_k\big(f_k(z, \mathbf{u}) + \varepsilon g_k(z, \mathbf{u})\big) - \lambda_0 \Phi - \lambda \eta J - \mu_1 \phi_1 - \mu_2 \phi_2 \tag{2.1b}$$

such that the following conditions hold.

(i) The multipliers $\mu_1(z), \mu_2(z)$ are continuous on each interval of continuity of $\varepsilon^*(z), \eta^*(z)$. They are nonnegative. Each is zero wherever the corresponding inequality is strict. (Moreover, since η cannot equal both bounds simultaneously it must be true that the μs cannot be nonzero at the same place.)

(ii) The multipliers $\mathbf{p}(z)$ are continuous and satisfy, with $\mathbf{u}^*(z), \varepsilon^*(z), \eta^*(z)$ and the two μ's, the given differential equation constraints

$$\frac{\mathrm{d}u_i}{\mathrm{d}z} = \frac{\partial H}{\partial p_i} = f_i(z, \mathbf{u}) + \varepsilon g_i(z, \mathbf{u}) \tag{2.2a}$$

and the additional equations

$$\frac{\mathrm{d}p_i}{\mathrm{d}z} = -\frac{\partial H}{\partial u_i} = -\sum_{k=1}^{N} p_k \left(\frac{\partial f_k}{\partial u_i} + \varepsilon \frac{\partial g_k}{\partial u_i}\right) + \lambda_0 \frac{\partial \Phi}{\partial u_i}, \tag{2.2b}$$

$$\frac{\partial H}{\partial \varepsilon} = \sum_{k=1}^{N} p_k g_k - \lambda_0 \frac{\partial \Phi}{\partial \varepsilon} = 0, \tag{2.2c}$$

$$\frac{\partial H}{\partial \eta} = \mu_1 - \mu_2 - \lambda J(z) - \lambda_0 \frac{\partial \Phi}{\partial \eta} = 0 \tag{2.2d}$$

on each interval of continuity of $\varepsilon^*(z), \eta^*(z)$. Boundary conditions $p_i(a) = 0$, $p_i(b) = 0$ arising from the standard transversality relations will hold for those i for which the corresponding component of $\mathbf{u}$ is not prescribed.

(iii) The function H evaluated on a minimizer is continuous on the closed interval $a \leqslant z \leqslant b$ and satisfies the relation

$$\frac{\mathrm{d}H}{\mathrm{d}z} = \frac{\partial H}{\partial z} = \sum_{k=1}^{N} p_k \left(\frac{\partial f_k}{\partial z} + \varepsilon \frac{\partial g_k}{\partial z}\right) - \lambda_0 \frac{\partial \Phi}{\partial z} - \lambda \eta \frac{\mathrm{d}J}{\mathrm{d}z} \tag{2.3}$$

on each interval of continuity of $\varepsilon^*(z), \eta^*(z)$.

(iv) The supremum inequality

$$\begin{aligned} &H(z, \mathbf{u}^*, \mathbf{p}^*, \varepsilon, \eta, 0, 0) \\ &\quad = \sum_{k=1}^{N} p_k^*\big(f_k(z, \mathbf{u}^*) + \varepsilon g_k(z, \mathbf{u}^*)\big) - \lambda_0 \Psi(z, \mathbf{u}^*, \varepsilon, \eta) - \lambda \eta J \\ &\quad \leqslant H(z, \mathbf{u}^*, \mathbf{p}^*, \varepsilon^*, \eta^*, 0, 0) = \sup H \equiv \overline{H}(z) \end{aligned} \tag{2.4}$$

holds for all admissible controls $\varepsilon(z)$ and $\eta(z)$ on the left side and for $\mathbf{u}^*(z)$, $\varepsilon^*(z)$, $\eta^*(z)$, together with the solution values of the multipliers $\mathbf{p}^*(z)$, λ_0, λ on both sides.

3. The Optimal $\eta(z)$ Must Be Piecewise Constant

Since the solutions $\mathbf{u}^*(z)$, etc. are not known a priori to use in the Maximum-H inequality, candidate optimizers are generated by replacing the function on the left of the inequality (2.4) by

$$\widetilde{H}(\varepsilon, \eta) = \sum_{k=1}^{N} p_k\big(f_k(z, \mathbf{u}) + \varepsilon g_k(z, u)\big) - \lambda_0 \Phi - \lambda \eta J. \tag{3.1}$$

Its supremum will be found with respect to the control variables ε and η regarding $\mathbf{u}(z)$ and $\mathbf{p}(z)$ as known. This procedure only determines combinations that give $\widetilde{H}$ a local supremum and some other condition must be used to test any candidate for possible global optimality.

The program to be followed from here consists of the following steps.

1. Show that the choice $\lambda_0 = 0$ would lead to all multipliers vanishing, which is not allowed. Then λ_0 can be taken as 1 since it scales all other multipliers and H. This standard argument parallels that given in [4, Appendix B.1], and will not be presented here.
2. Show that finding the supremum of the function $\widetilde{H}(\varepsilon, \eta)$ leads to the choice of ε for any η to be given by the "elastic law" (2.2c) and to the choice of $\eta(z)$ as either η_1 or η_2 if conditions (2.2c) are satisfied. Imposing the integral constraint leads to the result that the fractional measures ξ_1 and ξ_2 at each of the bounds must satisfy

$$\xi_2 = 1 - \xi_1 = \bar{c} = \frac{\bar{\eta} - \eta_1}{\eta_2 - \eta_1}. \tag{3.2}$$

This will complete the goal set for this work. To go on to find further properties of the specific solutions, one must leave this general form and return to the specific problems; these solutions will appear in other papers. However, an outline of the procedure to be followed is easily stated.

3. The next step for each problem is to construct the solution form for an interval where η is constant. By use of these forms one may find the number of segments with alternating η-values in any optimal sequence if the problem is well set.

4. Construction of all candidates for global optimizer and their associated $\overline{H}$, $\mu(z)$, and Π quantities is carried out next. Comparison of the $\overline{H}$ values for the candidates and some of the properties of the multiplier functions $\mu_1(z)$, $\mu_2(z)$ leads usually (but not always) to a single global optimizer for each choice of end conditions and the other fixed parameters. Comparison of the values of the functional Π for the candidates usually is needed to provide the proof of global optimality.

Turn now to finding the supremum of (3.1) with respect to ε and η regarding all else as known. At any η, the first and second derivatives of $\widetilde{H}(\varepsilon, \eta)$ with respect to ε are

$$\frac{\partial \widetilde{H}}{\partial \varepsilon} = \frac{\partial H}{\partial \varepsilon} = \sum_{k=1}^{N} p_k g_k - \frac{\partial \Phi}{\partial \varepsilon},$$
$$\frac{\partial^2 \widetilde{H}}{\partial \varepsilon^2} = -\frac{\partial^2 \Phi}{\partial \varepsilon^2} < 0.$$

Thus for each η the supremum of $\widetilde{H}(\varepsilon, \eta)$ occurs for $\varepsilon = \varepsilon^*(\eta)$ where the first derivative is zero:

$$\sum_{k=1}^{N} p_k g_k - \frac{\partial \Phi}{\partial \varepsilon} = 0, \tag{3.3}$$

where the first derivative is zero. Substituting this into (3.1) one obtains

$$\begin{aligned} H^*(\eta) &= \widetilde{H}(\varepsilon^*(\eta), \eta) \\ &= \sum_{k=1}^{N} p_k f_k - \lambda \eta J + \left\{ \sum_{k=1}^{N} p_k g_k \varepsilon - \Phi \right\} \Bigg|_{\varepsilon=\varepsilon^*}. \end{aligned} \tag{3.4}$$

Differentiating this with respect to η leads to

$$\frac{\mathrm{d}H^*}{\mathrm{d}\eta} = \left.\frac{\partial \widetilde{H}(\varepsilon, \eta)}{\partial \varepsilon}\right|_{\varepsilon=\varepsilon^*} \frac{\mathrm{d}\varepsilon^*}{\mathrm{d}\eta} - \lambda J - \left.\frac{\partial \Phi}{\partial \eta}\right|_{\varepsilon=\varepsilon^*} = -\lambda J - \left.\frac{\partial \Phi}{\partial \eta}\right|_{\varepsilon=\varepsilon^*},$$
$$\frac{\mathrm{d}^2 H^*}{\mathrm{d}\eta^2} = -\left.\frac{\partial^2 \Phi}{\partial \varepsilon \partial \eta}\right|_{\varepsilon=\varepsilon^*} \frac{\mathrm{d}\varepsilon^*}{\mathrm{d}\eta} - \left.\frac{\partial^2 \Phi}{\partial \eta^2}\right|_{\varepsilon=\varepsilon^*}.$$

But differentiating the identity

$$\sum_{k=1}^{N} p_k g_k - \left.\frac{\partial \Phi}{\partial \varepsilon}\right|_{\varepsilon=\varepsilon^*} \equiv 0$$

with respect to η gives

$$\frac{\mathrm{d}\varepsilon^*}{\mathrm{d}\eta} = -\left.\frac{\partial^2 \Phi}{\partial \eta \partial \varepsilon}\right|_{\varepsilon=\varepsilon^*} \Bigg/ \left.\frac{\partial^2 \Phi}{\partial \varepsilon^2}\right|_{\varepsilon=\varepsilon^*},$$

so that the second derivative of H^* becomes

$$\frac{\mathrm{d}^2 H^*}{\mathrm{d}\eta^2} = \left\{ \left(\frac{\partial^2 \Phi}{\partial \eta \partial \varepsilon} \right)^2 - \frac{\partial^2 \Phi}{\partial \varepsilon^2} \frac{\partial^2 \Phi}{\partial \eta^2} \right\} \Big/ \left. \frac{\partial^2 \Phi}{\partial \varepsilon^2} \right|_{\varepsilon = \varepsilon^*} > 0. \tag{3.5}$$

The right-hand side of this relation is positive and so the supremum will occur at the endpoints of the η-interval.

Finally, since the integral constraint cannot be satisfied by η at either of its extreme values, any candidate must have at least two regions with at least one interior value $\hat{z}$ at which they meet. If $G_\alpha, \alpha = 1, 2$, is the union of the parts of the interval $[a, b]$ where $\eta = \eta_\alpha$, then satisfaction of the integral constraint requires

$$\begin{aligned} \eta_1 \int_{G_1} J(z)\,\mathrm{d}z + \eta_2 \int_{G_2} J(z)\,\mathrm{d}z &= \bar{\eta} \int_a^b J(z)\,\mathrm{d}z, \\ \int_{G_1} J(z)\,\mathrm{d}z + \int_{G_2} J(z)\,\mathrm{d}z &= \int_a^b J(z)\,\mathrm{d}z, \end{aligned}$$

from which we find the fractional measures at each value to be

$$\xi_1 = \frac{\int_{G_1} J(z)\,\mathrm{d}z}{\int_a^b J(z)\,\mathrm{d}z} = \frac{\eta_2 - \bar{\eta}}{\eta_2 - \eta_1}, \qquad \xi_2 = \frac{\int_{G_2} J(z)\,\mathrm{d}z}{\int_a^b J(z)\,\mathrm{d}z} = \frac{\bar{\eta} - \eta_1}{\eta_2 - \eta_1},$$

which is the same as (3.2).

Thus any candidate for optimizer in any problem with a functional that fits this pattern must have the design variable piecewise constant with the measures at each value known. This does not say that an optimizer exists or, if it does, that it is unique.

4. Extensions of the Result; the Beam-Column as Example

The computation raises a number of important questions for generalization. Can the strain energy density be more complicated than that used here which arises essentially from isotropic linear elastic forms? Do the side differential equation constraints have to be linear in the strain variable? Can there be more than one strain variable? Can the differential equations be implicit and not explicit in the derivatives of $\mathbf{u}$? How does one fit known nonzero boundary loads into the scheme?

One example where we can show that we can extend the result to more than one strain variable and to other forms of differential equations of constraint is the extensible beam-column problem related to the Euler buckling result. Combine the energies of Problems 1 and 2. At $x = 0$, set $u = v = 0$. For a pinned end, the bending moment $M = 0$ also; this will arise as a "natural boundary condition" or "dual function condition from transversality". At $x = L$, again take a pinned end with $v = 0$ as prescribed and $M = 0$ arising from transversality. For the axial displacement either set $u(L) = \Delta$ or else prescribe the end load F (> 0 for the strut in tension, < 0 for the compressive column problem) in the x-direction. In the latter case, a term $-Fu(L)$ must be added to the potential energy. Therefore,

$$\Pi = \int_0^L \left\{ \frac{1}{2} EI\kappa^2 + \frac{1}{2} EAe^2 - qv - pu \right\} \mathrm{d}x - Fu(L) \tag{4.1}$$

subject to the boundary conditions stated above and to differential equations of constraint (and, in the design problem, to the inequality and integral constraints given before).

There are a number of different choices for the differential equation constraints:

(a) those of Problems 1 and 2:

$$\frac{\mathrm{d}u}{\mathrm{d}x} = e, \qquad \frac{\mathrm{d}^2 v}{\mathrm{d}x^2} = \frac{\mathrm{d}\theta}{\mathrm{d}x} = \kappa; \tag{4.2a}$$

(b) those of Problem 2 for v, θ, and κ but with the first nonlinear term in the axial extension:

$$e = \frac{\mathrm{d}u}{\mathrm{d}x} + \frac{1}{2}\left(\frac{\mathrm{d}v}{\mathrm{d}x}\right)^2 = \frac{\mathrm{d}u}{\mathrm{d}x} + \frac{1}{2}\theta^2; \tag{4.2b}$$

(c) the same as the last with the a priori assumption of inextensibility:

$$e = 0 \quad \text{so that} \quad \frac{\mathrm{d}u}{\mathrm{d}x} = -\frac{1}{2}\theta^2; \tag{4.2c}$$

and

(d) both the nonlinear e relation (with or without a priori inextensibility) and a nonlinear curvature relation

$$\kappa = \frac{v''}{[1 + v'^2]^{3/2}}$$

or

$$\frac{\mathrm{d}\theta}{\mathrm{d}x} = \kappa\left[1 + \theta^2\right]^{3/2}. \tag{4.2d}$$

For (a), standard methods give complete decoupling between the equilibrium equations for the flexural and extensional problems. The dual variables in the Pontryagin-Hamiltonian approach corresponding to v, θ, and u are the shear force Q, bending moment M, and the axial force N if $\mathbf{u}(L)$ is prescribed or $N - F$ if the end force is prescribed. This last comes most easily from replacing the end force potential by its integral equivalent:

$$Fu(L) = F\big(u(L) - u(0)\big) = \int_0^L F\frac{\mathrm{d}u}{\mathrm{d}x}\,\mathrm{d}x = \int_0^L Fe\,\mathrm{d}x. \tag{4.3}$$

Introduction of the design variable and its associated constraints does not disturb these conditions but adds to the conditions that the supremum with respect to κ and e of $\widetilde{H}(\kappa, e, \eta)$ requires that the curvature and axial extension satisfy the standard $M = EI\kappa$ and $N = EAe$ elastic laws. The expected supremum condition with respect to η is that the second derivative of the $H^*(\eta)$ function is positive if the nonconvexity conditions on EI and EA hold and p, q are at most linear in η:

$$\frac{\mathrm{d}^2 H^*}{\mathrm{d}\eta^2} = \frac{M^2}{2}\frac{\partial^2}{\partial\eta^2}\left(\frac{1}{EI}\right) + \frac{N^2}{2}\frac{\partial^2}{\partial\eta^2}\left(\frac{1}{EA}\right) - \frac{\partial^2 p}{\partial\eta^2}u - \frac{\partial^2 q}{\partial\eta^2}v. \tag{4.4}$$

For the conditions of (b) the same conclusions concerning the interpretation of the dual variables occurs as in (a). For the end load, the integral representation of the potential is now

$$Fu(L) = \int_0^L F\frac{\mathrm{d}u}{\mathrm{d}x}\,\mathrm{d}x = \int_0^L F\left(e - \frac{1}{2}\theta^2\right)\mathrm{d}x. \tag{4.5}$$

Those familiar with the traditional presentation of the energy approach to the Euler column buckling problem will recognize the second term as the "energy of axial compression" when $F < 0$ which appears in that analysis. The effect of the nonlinearity is to change the moment equilibrium equation per unit length to

$$\frac{\mathrm{d}M}{\mathrm{d}x} = -Q + N\theta = -Q + N\frac{\mathrm{d}v}{\mathrm{d}x}, \tag{4.6}$$

again familiar from the Euler development. With a design variable η also included in the development, we are led to the same argument as for (a) to establish the conditions under which the design is "bang-bang".

For (c), e is no longer a control variable, leaving only κ and η. N does not satisfy an elastic law but arises as an internal force of constraint maintaining $e = 0$ just as the shear force Q maintains the no shear strain constraint of cross-sections remaining normal to the deformed center line. This a priori inextensibility is usually made in Euler buckling treatments with $p = q = 0$; it is not usually shown that this is in fact the condition that is most critical for the vanishing of the second variation in the usual approach to stability of equilibrium of a column.

For (d), it is probably simpler to go to the elastica formulation and use the arclength s in the deformed configuration so that $\mathrm{d}\theta/\mathrm{d}s = \kappa$; but the design variable result will still hold. All the details here have not yet been worked out, however.

References

1. R. Fosdick and G. Royer-Carfagni, The static state of a two-phase solid mixture in a stressed elastic bar. *Internat. J. Solids Struct.* **33** (1996) 2267–2281.
2. R. Fosdick, G. Royer-Carfagni and H. Weinberger, A note on the optimal state of a binary solid mixture in a stressed elastic bar. *Meccanica* **31** (1996) 519–525.
3. R. Fosdick and G. Royer-Carfagni, Alloy separation of a binary mixture in a stressed elastic sphere. *J. Elasticity* **42** (1996) 49–77.

4. W.H. Warner, Optimal design of elastic rods under axial gravitational load using the maximum principle. *Internat. J. Solids Struct.* **37** (2000) 2709–2726.
5. W.H. Warner, Optimal design of elastic rods: Extension of a minimum energy solution, *Internat. J. Solids Struct.*, accepted for publication (2000).
6. M. Hestenes, *Calculus of Variations and Optimal Control Theory*, Wiley (1966); Robert E. Krieger Publishing Co., Huntington, New York (1980).

Journal of Elasticity **59:** 369–392, 2000.

The Mullins Effect in a Pure Shear

MILLARD F. BEATTY
Department of Engineering Mechanics, University of Nebraska-Lincoln, Lincoln, NE 68588-0526, U.S.A. E-mail: mbeatty@unlserve.unl.edu

Received 31 March 1999

Dedicated to Roger Fosdick, in esteem and friendship, on the occasion of his sixtieth birthday.

Abstract. The Mullins effect in a rubberlike material subjected to a pure shear deformation is studied in the context of a recent theory of stress-softening for incompressible materials proposed by Beatty and Krishnaswamy. Some general technical results characterizing the mechanical response are presented. These show that the theory delivers results consistent with the overall behavior expected of a Mullins material, but usually exhibited in uniaxial extension or equibiaxial stretch experiments. The extent of stress-softening in a pure shear is shown to be much less than that due to an equibiaxial deformation, and only slightly greater than the degree of stress-softening induced by an uniaxial deformation, all to the same stretch. The Mullins effect in an equivalent simple shear deformation, even one having a rather large angle of shear, is small. The simple shear is the least damaging deformation among all of those mentioned here. Some graphical results, based on a special class of stress-softening materials applied to two parent material models – the familiar Mooney–Rivlin and a certain biotype material model, illustrate the general conclusions obtained for arbitrary Mullins materials. The inflation of a biomaterial membrane preconditioned in a pure shear deformation demonstrates the familiar stress-softening phenomenon observed in the inflation of a balloon.

1. Introduction

When a rubber specimen is loaded uniaxially, unloaded, and then reloaded, the load required to produce a given stretch during reloading is smaller than that required to produce the same stretch during the primary loading phase. This inelastic *stress-softening* phenomenon is known as the *Mullins effect* [1]. The mechanical response of a stress-softening material depends upon its deformation history. For a *Mullins material* this dependence is special; the material has a *selective memory* of only the *maximum previous strain* experienced during its deformation history. The magnitude of the strain is defined by

$$m \equiv \|\mathbf{B}\| = \sqrt{\mathbf{B}\cdot\mathbf{B}} = \sqrt{\mathrm{tr}\mathbf{B}^2} \tag{1.1}$$

in terms of the left Cauchy–Green deformation tensor $\mathbf{B} \equiv \mathbf{F}\mathbf{F}^{\mathrm{T}}$, and where tr denotes the trace operation. Note that in the undistorted state for which $\mathbf{B} = \mathbf{1}$, we have $m = \sqrt{3}$; otherwise, $m > \sqrt{3}$ for all isochoric deformations.

In a primary deformation of a virgin Mullins material, the maximum previous strain is the current value of m. When the material is unloaded, it remembers only the maximum previous strain $M = m_{\max}$ it experienced in the primary deformation, but it does not remember how it got there. If the material is reloaded again, its response, which is distinct from the virgin material response, is ideally elastic and for the same deformation $\mathbf{F}$ the material retraces its unloading response so long as the strain intensity $m \leqslant M$. The new elastic material is said to be *stress-softened*, or *preconditioned*. At this maximum previous strain, the stress-softened Mullins material recalls its virgin material response under additional loading for which $m > M$. When the material is unloaded again, its response follows a different elastic response that recalls only its maximum previous strain, and so on. The response of a stress-softened Mullins material is characterized as an ideally elastic material for all subsequent deformations of any sort so long as $m \leqslant M$. In this phenomenological description, for simplicity, we have supposed that the activation of stress-softening occurs immediately upon loading of the virgin material.

Based on a microstructural damage model proposed by Mullins and Tobin [2], the virgin Mullins material is described as an amorphous mixture of hard phase and soft phase microstructures. We may think of the hard phase as filler bonds of the vulcanizate, or perhaps as molecular chain clusters and chain entanglements, and the soft phase as the rubbery phase. The total amount of hard phase generally is much less than the amount of soft phase initially present in the microstructure. As the virgin material is deformed, a volume fraction α of the *initial portion* of hard phase at each material point $\mathbf{X}$ is converted immediately to an equivalent portion of new soft phase at $\mathbf{X}$, so the remaining volume fraction of the original portion of hard phase is $(1 - \alpha)$. Initially, $\alpha = 0$; and for a fully softened material, $\alpha = 1$. The current value of the softening fraction α changes only when loading occurs; otherwise, it remains fixed at its greatest previous value determined by the maximum previous strain. Of course, for the same deformation, the extent of stress-softening will vary from one Mullins material to another; and hence the volume fraction α is determined by a constitutive equation that depends only on the current strain intensity. We shall return to this later.

The foregoing characterization of a Mullins material neglects additional inelastic effects of hysteresis, permanent set, and any accompanying residual stress, and also ignores anisotropy induced by preconditioning. These matters are discussed by Johnson and Beatty [3–5]. We shall ignore these additional complications and focus on the main effect of stress-softening described above. We note, however, that Wineman and Rajagopal [6, 7] have developed a set of integral type constitutive equations to model deformation-induced microstructural damage effects, such as those induced by scission of molecular bonds that reform to create new microstructure. Subsequently, Wineman and Huntley [8], and Huntley et al. [9, 10] showed that this theory leads to permanent set and residual stress effects in applications modelled on a neo-Hookean type constitutive relation. Some parallel ideas based on the Mullins–Tobin microstructural damage model described earlier are used in a recent paper by Beatty and Krishnaswamy [11] to develop a simpler, yet general continuum theory of stress-softening in incompressible materials that neglects other inelastic effects. A companion theory for stress-softening in compressible materials is presented in [12]. Additional references dealing with both experimental and theoretical aspects of the Mullins effect may be found in [11]. In this work and in two further papers [13, 14] general technical and graphical results deduced from the theory are presented for uniaxial extension, simple shear, equibiaxial deformation, balloon inflation, and torsion, as well as compound deformations in which one deformation is followed by a different one. In the present paper, this theory is applied to study the effect of stress-softening in a pure shear deformation.

A brief description of the constitutive equations for a Mullins material is presented in the next section. The pure shear deformation is described in Section 3 where some general technical results characterizing the mechanical response of a Mullins material are derived. The effects of stress-softening on the orthogonal stress components in the pure shear and on elastic moduli, compared with those for the virgin material, are presented. The results in every instance are consistent with the overall behavior typical of a Mullins material, but usually described in the context of simple uniaxial or equibiaxial stretch experiments. The extent of stress-softening in a pure shear is shown to be much less than that due to an equibiaxial deformation, and only slightly greater than that induced by an uniaxial deformation, all to the same stretch. The Mullins effect in an equivalent simple shear deformation, even a rather large amount of shear, is small; the greater effect occurs in a pure shear. In fact, a simple shear generally is the least damaging deformation among all of the aforementioned deformations. Based on a special class of Mullins materials described in Section 4, some graphical results provided in Section 5 for a Mooney–Rivlin parent material model and in Section 6 for a certain biotype parent material model illustrate the general conclusions derived here for arbitrary Mullins materials. In Section 7, the inflation of a biomaterial membrane preconditioned in a pure shear deformation demonstrates the familiar stress-softening phenomenon observed in the inflation of a balloon. It is shown that the inflation pressure is considerably reduced by preconditioning of the virgin material in a pure shear. We begin with a sketch of the constitutive theory of stress-softening materials presented in [11].

2. Constitutive Equations for a Mullins Material

The strain intensity m of an assigned deformation is defined by (1.1); and hence we may say more precisely that $\dot{m} = \mathbf{B} \cdot \dot{\mathbf{B}}/m > 0$ defines *loading* from a *softening point* at $m = m_{\max}$; $\dot{m} = 0$ identifies *neutral loading*; and $\dot{m} < 0$ describes *unloading* from a softening point, after which $\sqrt{3} \leqslant m < m_{\max}$ defines an *elastic point*. When loading occurs, $m_{\max}$ is the current value of m. Recalling our earlier discussion of the Mullins–Tobin model in which α is the volume fraction of the initial portion of hard phase that is transformed immediately to soft phase, we define the *softening function* by

$$\alpha = \begin{cases} \hat{\alpha}(m) & \text{during loading,} \\ \hat{\alpha}(m_{\max}) = const & \text{otherwise,} \end{cases} \tag{2.1}$$

and for which

$$\hat{\alpha}(\sqrt{3}) = 0, \qquad 0 \leqslant \hat{\alpha}(m) \leqslant \hat{\alpha}(m_{\max}) \leqslant 1. \tag{2.2}$$

Notice that the constitutive equation (2.1) involves an isotropic scalar-valued function of the deformation tensor $\mathbf{B}$, relative to an undistorted state for which $\mathbf{F} = \mathbf{1}$. Hence, for any assigned deformation $\mathbf{F}$ from an undistorted state of a virgin material point $\mathbf{X}$, the softening function has the same value α for every direction at $\mathbf{X}$. The constitutive equation (2.1) thus characterizes isotropic stress-softening of the virgin material.

The first relation in (2.2) specifies that stress-softening commences immediately upon initiation of loading of the virgin material. It is possible, however, to admit initiation of stress-softening only after m has reached a critical activation value m_a, say, for which $\hat{\alpha}(m_a) = 0$ holds and such that the material response is ideally elastic for all $m \leqslant m_a$. This feature is included in the Wineman–Ragagopal theory [6, 7]. Here we consider a simpler model for which $m_a = \sqrt{3}$. The case $\alpha = 1$ in $(2.2)_2$ corresponds to a fully softened material. In practice though, the material fails well before full softening can occur; or, perhaps no further microstructural damage is induced by loading beyond a certain strain intensity m^*. In either case, we may further limit the model so that $\hat{\alpha}(m) \leqslant \alpha^* < 1$, where α^* is the "ultimate" total fraction of the original portion of hard phase that has been transformed to soft phase. Furthermore, since softening in the material increases with increasing values of m in the domain $\mathfrak{D} = [\sqrt{3}, m_{\max}]$, $\hat{\alpha}(m)$ must be a monotone increasing function on $\mathfrak{D}$, so that $\mathrm{d}\hat{\alpha}(m)/\mathrm{d}m > 0$ at each value of $m \in \mathfrak{D}$. We are now prepared to introduce the constitutive equation for the Cauchy stress in a stress-softening material.

A Mullins material is an inelastic, stress-softening material that possesses memory of only the maximum previous strain $m = m_{\max}$ experienced in its deformation history. Otherwise, for a fixed value $m_{\max} = M$ of the maximum previous strain, the material behavior for $m < M$ is ideally elastic. Thus, in accordance with (2.1), for each fixed value of $\alpha \in [0, 1]$, the stress-softened material may be regarded as an elastic material. In this sense, the Mullins material is characterized by an infinite family of elastic materials parameterized by α. We confine attention to materials that are isotropic relative to the undistorted, virgin state. Accordingly, we consider the general isotropic stress-softening model characterized by

$$\mathbf{T} = \begin{cases} -p\mathbf{1} + \hat{\beta}_1\big(I_1, I_2; \hat{\alpha}(m)\big)\mathbf{B} + \hat{\beta}_{-1}\big(I_1, I_2; \hat{\alpha}(m)\big)\mathbf{B}^{-1} & \text{during loading,} \\ -q\mathbf{1} + \hat{\beta}_1\big(I_1, I_2; \hat{\alpha}(m_{\max})\big)\mathbf{B} + \hat{\beta}_{-1}\big(I_1, I_2; \hat{\alpha}(m_{\max})\big)\mathbf{B}^{-1} & \text{otherwise,} \end{cases} \tag{2.3}$$

in which $\mathbf{T}$ is the Cauchy stress tensor, $\hat{\beta}_\Gamma$ are material response functions, and p and q are undetermined pressures reflecting the incompressibility constraint. These pressures must have the same value at $m = m_{\max}$ at which the stress in $(2.3)_{1,2}$ is continuous; elsewhere, they generally are distinct.

The *virgin material* response is characterized by the constitutive equation $(2.3)_1$, and the elastic, *stress-softened material* response is described by $(2.3)_2$. When the virgin material is unloaded from a state of maximum previous strain reached in $(2.3)_1$, the stress-softened material response follows equation $(2.3)_2$. In fact, as long as $m < m_{\max}$, the material response follows $(2.3)_2$ for decreasing as well as increasing strain. If now we begin from the undeformed state of the newly softened material and deform it differently, the response must follow the second constitutive law $(2.3)_2$ for all $m \leqslant M = m_{\max}$, the maximum previous strain induced by the primary deformation. The response of the newly softened material is ideally elastic up to the point at which $m = M$. When the material is further loaded in continuing the second deformation of the stress-softening material from its maximum previous strain at the softening point, the response is described anew by the virgin material constitutive equation $(2.3)_1$ for $m > M$.

Thus far our constitutive theory for stress-softening materials consists of the stress response equations (2.3) and an equation for the isotropic softening function (2.1) during loading. More precisely, *the material model characterized by* (2.3) *is called a Mullins material if and only if the response functions* $\hat{\beta}_1$ *and* $-\hat{\beta}_{-1}$ *are monotone decreasing functions of the stress-softening variable* $\alpha = \hat{\alpha}(m)$ *for all* $m \in \mathfrak{D}$. Hence, the stress-softening material response functions must satisfy

$$\begin{aligned} \hat{\beta}_1\big(I_1, I_2; \hat{\alpha}(m)\big) &> \hat{\beta}_1\big(I_1, I_2; \hat{\alpha}(M)\big), \\ -\hat{\beta}_{-1}\big(I_1, I_2; \hat{\alpha}(m)\big) &> -\hat{\beta}_{-1}\big(I_1, I_2; \hat{\alpha}(M)\big), \end{aligned} \tag{2.4}$$

or equivalently,

$$\frac{\partial \hat{\beta}_1(I_1, I_2; \hat{\alpha}(m))}{\partial \hat{\alpha}} < 0, \qquad -\frac{\partial \hat{\beta}_{-1}(I_1, I_2; \hat{\alpha}(m))}{\partial \hat{\alpha}} < 0, \tag{2.5}$$

for all isochoric deformations $\mathbf{B}$ with $m < M = m_{\max}$.

The empirical inequalities (see, e.g., [15, 16]) applied to the response functions in (2.3) require

$$\hat{\beta}_1\big(I_1, I_2; \hat{\alpha}(m)\big) > 0, \qquad \hat{\beta}_{-1}\big(I_1, I_2; \hat{\alpha}(m)\big) \leqslant 0 \tag{2.6}$$

for all isochoric deformations $\mathbf{B}$ with $m \leqslant M = m_{\max}$. Equality in $(2.6)_2$ holds when and only when $\hat{\beta}_{-1} \equiv 0$ in (2.3). Notice that the foregoing definition of a Mullins material is independent of the auxiliary empirical inequalities. We shall see, however, that both the empirical inequalities (2.6) and the equivalent monotone properties (2.4) and (2.5) of the stress-softened response functions are useful in characterizing the physical behavior of a Mullins material subjected to specific deformations. We next apply the theory to study the response of a Mullins material in a pure shear deformation.

3. Stress-softening in a Pure Shear

Let x_i and X_A denote the respective rectangular Cartesian components of the position vectors $\mathbf{x}$ and $\mathbf{X}$ of a particle in the current and reference configurations of a body in a common reference frame $\varphi = \{O; \mathbf{e}_k\}$ with origin at point O and orthonormal basis $\mathbf{e}_k$. The deformation describing an isochoric, pure shear is given by

$$x_1 = \lambda X_1, \qquad x_2 = X_2, \qquad x_3 = \lambda^{-1} X_3, \tag{3.1}$$

where the axial stretch $\lambda \in (0, \infty)$. In accordance with (3.1), a rectangular sheet or block is stretched along the $\mathbf{e}_1$ axis an amount $\lambda_1 = \lambda$, while its sides perpendicular to the $\mathbf{e}_2$ axis are held fixed so that the stretch $\lambda_2 = 1$. The tensor $\mathbf{B}$ in the corresponding tensor product basis $\mathbf{e}_{jk} \equiv \mathbf{e}_j \otimes \mathbf{e}_k$ has the form

$$\mathbf{B} = \lambda^2 \mathbf{e}_{11} + \mathbf{e}_{22} + \lambda^{-2} \mathbf{e}_{33}. \tag{3.2}$$

The principal invariants I_k of $\mathbf{B}$ are thus given by

$$I_1 = I_2 = 1 + \lambda^2 + \frac{1}{\lambda^2}, \qquad I_3 = \det \mathbf{B} = 1. \tag{3.3}$$

By (1.1), the magnitude of $\mathbf{B}$ is

$$m = \sqrt{1 + \lambda^4 + \frac{1}{\lambda^4}}; \tag{3.4}$$

and hence the deformation rate is

$$\dot{m} = \frac{2}{m}\left(\lambda^3 - \frac{1}{\lambda^5}\right)\dot{\lambda}. \tag{3.5}$$

The loading criterion applied to (3.5) in an elongation for which $\lambda > 1$ reduces to $\dot{\lambda} > 0$ for loading and to $\dot{\lambda} < 0$ for unloading. For compression, the previous inequalities are reversed. If $\lambda \neq 1$, then in neutral loading for which $\dot{m} = 0$, $\dot{\lambda} = 0$ corresponds to holding λ fixed at its pre-existing value. Thus, as anticipated, the loading criterion coincides with familiar intuitive notions in a pure shear deformation.

3.1. GENERAL RESULTS FOR STRESS-SOFTENING IN A PURE SHEAR

We now show that our general constitutive equations (2.3), together with (2.4) and (2.6), yield technical results typical of the kind of behavior expected of stress-softening materials in a pure shear deformation. The results are similar to those demonstrated in [11–13] for simple extension, simple shear, equibiaxial deformation, and inflation.

The pure shear can be sustained by plane biaxial Cauchy stresses $T_{11} = t_1$, $T_{22} = t_2$, with all other $T_{jk} = 0$. From $(2.3)_1$ and (3.2), we see that the equilibrium equations are satisfied trivially for constant p. Hence, with $T_{33} = 0$, elimination of the parameter p from the system $(2.3)_1$ yields the biaxial Cauchy stress components in a pure shear of a general isotropic, virgin Mullins material under loading:

$$t_1 = \left(\lambda^2 - \frac{1}{\lambda^2}\right)\left(\hat{\beta}_1(I_1, I_2; \hat{\alpha}(m)) - \hat{\beta}_{-1}(I_1, I_2; \hat{\alpha}(m))\right), \tag{3.6}$$

$$t_2 = \left(1 - \frac{1}{\lambda^2}\right)\left(\hat{\beta}_1(I_1, I_2; \hat{\alpha}(m)) - \lambda^2\hat{\beta}_{-1}(I_1, I_2; \hat{\alpha}(m))\right), \tag{3.7}$$

in which we recall (3.3) and (3.4). The component t_1 is the *axial stress*, and t_2 is the *transverse, restraining stress*. In consequence of the empirical inequalities (2.6), we see from (3.6) and (3.7) that $t_1 > 0$, $t_2 > 0$ when $\lambda > 1$ (a simple tension), and $t_1 < 0$, $t_2 < 0$ when $\lambda < 1$ (a simple compression), as expected intuitively in a pure shear.

Fixing $m = M \equiv m_{\max}$ and unloading from this state in the sense described earlier, we may obtain directly from (3.6) and (3.7) the corresponding axial and transverse Cauchy stress components $T_{11} = \tau_1$, $T_{22} = \tau_2$ in the elastic stress-softened material described by $(2.3)_2$; namely,

$$\tau_1 = \left(\lambda^2 - \frac{1}{\lambda^2}\right)\left(\hat{\beta}_1(I_1, I_2; \hat{\alpha}(M)) - \hat{\beta}_{-1}(I_1, I_2; \hat{\alpha}(M))\right), \tag{3.8}$$

$$\tau_2 = \left(1 - \frac{1}{\lambda^2}\right)\left(\hat{\beta}_1(I_1, I_2; \hat{\alpha}(M)) - \lambda^2\hat{\beta}_{-1}(I_1, I_2; \hat{\alpha}(M))\right), \tag{3.9}$$

in which, by (3.4),

$$M = \sqrt{1 + \Lambda^4 + \frac{1}{\Lambda^4}}, \tag{3.10}$$

$\Lambda \equiv \lambda_{\max}$ being the maximum previous stretch attained prior to unloading. Of course, the equilibrium equations are satisfied trivially. Indeed, the stress-stretch equations (3.8) and (3.9) are exactly the relations obtained by similar construction from $(2.3)_2$ and for which the stresses are continuous with $t_1 = \tau_1$, $t_2 = \tau_2$ at $m = M$ in (3.6) and (3.7). In view of (2.6), it follows from (3.6) through (3.9), that $t_1 = t_2 = \tau_1 = \tau_2 = 0$, when and only when $\lambda = 1$. Otherwise, in view of the monotonicity conditions (2.4), we have

$$\frac{t_1(\lambda)}{\tau_1(\lambda)} \geqslant 1, \qquad \frac{t_2(\lambda)}{\tau_2(\lambda)} \geqslant 1, \tag{3.11}$$

equality holding only at $\lambda = \Lambda$. *Therefore, at each value of the stretch $\lambda \in (1, \Lambda)$ in a pure shear, the biaxial Cauchy stress intensities in the virgin material are always greater than the corresponding stress intensities in the elastic stress-softened material. Moreover, for every Mullins material for which $\hat{\beta}_{-1} = 0$, the ratios of the biaxial Cauchy stresses in the virgin material and in the elastic, stress-softened* *material subject to a pure shear deformation satisfy the same universal relation, namely,*

$$\frac{t_1}{t_2} = \frac{\tau_1}{\tau_2} = 1 + \lambda^2. \tag{3.12}$$

Thus, for this class of Mullins materials, both the axial stress t_1 in the virgin material and τ_1 in the stress-softened material exceed the respective transverse restraining stress t_2 and τ_2 by exactly the same extent. Further, the corresponding orthogonal Cauchy stress intensities in the virgin material exceed those in the softened material so that, by (3.11) and (3.12), $t_1(\lambda)/\tau_1(\lambda) = t_2(\lambda)/\tau_2(\lambda) \geqslant 1$. Two special material models in this class appear later.

3.1.1. *Additional Stress Component Relations for Pure Shear*

The difference of the principal stresses (3.6) and (3.7) for a general isotropic stress-softening virgin material during loading is given by

$$t_1 - t_2 = (\lambda^2 - 1)\left(\hat{\beta}_1(I_1, I_2; \hat{\alpha}(m)) - \frac{1}{\lambda^2}\hat{\beta}_{-1}(I_1, I_2; \hat{\alpha}(m))\right). \tag{3.13}$$

In consequence of the empirical inequalities (2.6), it follows from (3.13) that $t_1 > t_2$ when $\lambda > 1$ (a pure shear by tension), and $t_1 < t_2$ when $\lambda < 1$ (a pure shear by compression). The same remarks hold for the principal stresses (3.8) and (3.9) in the general elastic stress-softened material for which the principal stress difference $\tau_1 - \tau_2$ may be read from (3.13) with $m = M \equiv m_{\max}$ fixed. In view of the monotonicity conditions (2.4), it follows that

$$\frac{t_1 - t_2}{\tau_1 - \tau_2} \geqslant 1, \tag{3.14}$$

equality holding only at $\lambda = \Lambda$. Therefore, *at each value of the stretch $\lambda \in (1, \Lambda)$ in a pure shear, the principal stress difference in the virgin material is always greater than the corresponding principal stress difference in the softened material; and hence in a pure shear of any Mullins material, $|t_1 - t_2| > |\tau_1 - \tau_2|$.* Henceforward, we shall focus on the tension example, noting that similar results may be stated for compression.

We also observe a connection of the results with the Mullins–Tobin damage model described earlier. We recall (2.4) and (2.5) for a stress-softening material and observe from (3.6) and (3.7) that the biaxial Cauchy stresses in the virgin material, considered as functions of $\hat{\alpha}(m)$, decrease as the current value of m increases. *Hence, for a Mullins material in a pure shear, the intensities of the biaxial Cauchy stresses t_k in the virgin material decrease as hard phase is transformed to soft phase with increasing m; that is, the biaxial stresses must be monotone decreasing functions of $\hat{\alpha}(m)$; $\partial t_k/\partial\hat{\alpha}(m) < 0$, $k = 1, 2$.*

On the basis of a somewhat similar argument applied to the Cauchy stresses τ_k in an elastic stress-softened material, we can further characterize geometrical properties of the stress–stretch response curves for stress-softened materials in a pure shear. It is seen from (3.8) and (3.9) that over an appropriate range of stretch, each stress component decreases with increasing softening $\hat{\alpha}(M)$, and that the τ_k are monotone decreasing functions of $\hat{\alpha}(M)$. This means that for a maximum previous strain $M_2 > M_1$, we have $\tau_k(\lambda, M_2) < \tau_k(\lambda, M_1)$ for $\lambda \in (1, \Lambda_1]$; that is, the stress-stretch curve $\tau_{k2} = \tau_k(\lambda, M_2)$ is everywhere situated below the curve $\tau_{k1} = \tau_k(\lambda, M_1)$ for $\lambda \in (1, \Lambda_1]$, except at their common origin at $\lambda = 1$, where $\tau_{k1} = \tau_{k2} = 0$. These geometrical attributes of stress-softening material response in a pure shear also can be viewed in terms of effective material moduli.

3.1.2. *Effective Moduli for Mullins Materials in a Pure Shear*

It may be useful for experimental purposes to examine the tangent moduli $\mathrm{d}t_k/\mathrm{d}\lambda$ obtained from (3.6) and (3.7) in the tension case. Suppose that the derivatives of the material response terms are finite in the undistorted state where $\lambda = 1$. We can then show that the respective axial and transverse effective Young's moduli $E_{10} \equiv \mathrm{d}t_1/\mathrm{d}\lambda|_{\lambda=1}$ and $E_{20} \equiv \mathrm{d}t_2/\mathrm{d}\lambda|_{\lambda=1}$ of the virgin material stress–stretch graphs at $\lambda = 1$ in a pure shear are given by

$$E_{10} = 2E_{20} = 4\mu_0 = \frac{4}{3}E_0, \tag{3.15}$$

where $\mu_0 \equiv \hat{\beta}_1(3, 3, 0)) - \hat{\beta}_{-1}(3, 3, 0)$ is the shear modulus of the virgin material in its natural state, and $E_0 = 3\mu_0$ is the corresponding Young's slope of the virgin material in a simple tension experiment. Similarly, for a fixed value of the maximum previous strain, (3.8) and (3.9) yield the corresponding effective Young's moduli $E_{kM} \equiv \mathrm{d}\tau_k/\mathrm{d}\lambda|_{\lambda=1}$ for the elastic stress-softened material: $E_{1M} = 2E_{2M} = 4\mu_M = \frac{4}{3}E_M$, in which $\mu_M \equiv \hat{\beta}_1(3, 3; \hat{\alpha}(M))) - \hat{\beta}_{-1}(3, 3; \hat{\alpha}(M))$ is the shear modulus of the elastic stress-softened material in its undeformed state, and $E_M = 3\mu_M$ is the corresponding Young's modulus in a simple tensile test. *For a general Mullins material, actually any incompressible, isotropic elastic material, the effective axial modulus is twice as great as the effective transverse modulus in a pure shear.*

The properties (3.15) and their stress-softened counterparts thus relate the effective biaxial moduli in a pure shear of the virgin and elastic stress-softened materials to their classical moduli in a simple tensile test. Recalling (2.4), we see that the relation $E_0 > E_M > 0$ holds for all Mullins materials in simple uniaxial extension; therefore, for the virgin and stress-softened materials, we have $E_{10} > E_{1M}$ and $E_{20} > E_{2M}$. *Consequently, in an undeformed stress-free state, the slope of each biaxial stress–stretch graph for the current stress-softened material in a pure shear will be less than that of the virgin material and also less than those of any previously softened material for which $m < M$.*

At a softening point for which $\lambda = \Lambda$, bearing in mind that $\hat{\alpha}(m)$ is a monotone increasing function of m, we may see that the slope of the elastic stress-softened material curve always exceeds that of the virgin material. Therefore, *at $\lambda = \Lambda$,* *each tangent modulus* $E_{kT} \equiv \mathrm{d}t_k/\mathrm{d}\lambda|_{\lambda=\Lambda}$ *of the virgin material in a pure shear must be smaller than the corresponding tangent modulus* $E_{k\tau} \equiv \mathrm{d}\tau_k/\mathrm{d}\lambda|_{\lambda=\Lambda}$ *of the stress-softened material whose maximum previous stretch is* $\lambda = \Lambda$.

The foregoing remarks on stress component relations, effective moduli, and geometry of the stress–stretch response curves identify some technical implications of the definition of a stress-softening material applied in a pure shear. Although some of the results may appear intuitively evident, here they follow easily as natural consequences of our general constitutive theory for Mullins materials. Notice that all of the foregoing analytical *predictions* coincide with the ideal phenomenological description of stress-softening presented earlier. Therefore, these preliminary results, obtained from the general constitutive model in a pure shear deformation, should provide confidence in its use in less familiar applications.

We know from [13] that the respective effective Young's moduli in an equibiaxial stretch of the virgin and stress-softened materials are given by $E_{E0} = 2E_0$ and $E_{EM} = 2E_M$ in terms of those in a simple uniaxial stretch. The foregoing relations thus connect the various technical moduli E_0 in an uniaxial stretch, E_{E0} in an equibiaxial stretch, and E_{10} and E_{20} in a pure shear of the virgin material: $E_{E0} = 2E_0 = \frac{3}{2}E_{10} = 3E_{20}$, parallel relations holding for each elastic stress-softened material. Comparisons of the extent of stress-softening in a simple uniaxial stretch, an equibiaxial stretch, and an equivalent simple shear with the degree of softening in a pure shear are presented next.

3.2. COMPARISON OF SOFTENING IN PURE SHEAR AND UNIAXIAL EXTENSION

It is shown in [11] that the strain intensity m_{u} of an incompressible material in a simple uniaxial deformation with stretch λ is given by

$$m_{\mathrm{u}} = \sqrt{\lambda^4 + \frac{2}{\lambda^2}}. \tag{3.16}$$

We consider a pure shear deformation with the *same* axial stretch $\lambda \neq 1$, and denote the corresponding strain intensity by m_{p}. With the aid of (3.4), we have $m_{\mathrm{p}}^2 - m_{\mathrm{u}}^2 = (1-\lambda^{-2})^2 > 0$ for all $\lambda \neq 1$. It thus follows that $m_{\mathrm{p}} > m_{\mathrm{u}}$ for each value of $\lambda \neq 1$. In view of the monotone nature of the softening function, but otherwise independent of its specific functional form, it is seen that for the same axial stretch $\lambda \neq 1$, $\hat{\alpha}(m_{\mathrm{p}}) > \hat{\alpha}(m_{\mathrm{u}})$. Therefore, we have the following general result relating a pure shear to an uniaxial stretch. *The maximum previous strain intensity in a pure shear is greater than it is in an uniaxial deformation with the same stretch, equality holding only in the undeformed state; therefore, the degree of softening is greater in a pure shear than it is in an uniaxial deformation with the same stretch. Thus, in this sense, a greater degree of microstructural damage occurs in the pure shear deformation. Further, the stretch in a simple uniaxial deformation required to produce the same maximum previous strain,* $M_{\mathrm{u}} = M_{\mathrm{p}}$, *hence, the same degree of softening* $\hat{\alpha}(M_p) = \hat{\alpha}(M_u)$, *is greater than the axial stretch required in a pure shear.*

It may be seen that the strain intensity m_u in an uniaxial elongation, over a physically reasonable range of stretch, differs very little from the strain intensity (3.4) for a pure shear with the *same* stretch; but the variation is greater in a compressional deformation. Specifically, for $\lambda = 10$, $\Delta m \equiv m_p - m_u = 4.4 \times 10^{-3}$, which is quite small, whereas for $\lambda = 0.5$, $\Delta m = 1.291$ is more substantial.

3.3. COMPARISON OF SOFTENING IN PURE SHEAR AND IN EQUIBIAXIAL STRETCH

For an incompressible material [13], the strain intensity m_e in an equibiaxial deformation with stretch λ is given by

$$m_e = \sqrt{2\lambda^4 + \frac{1}{\lambda^8}}. \tag{3.17}$$

We consider a pure shear deformation with the *same* axial stretch $\lambda \neq 1$, and use (3.4) to show that $m_e^2 - m_p^2 = \lambda^{-8}(\lambda^4 + 1)(\lambda^4 - 1)^2 > 0$ for all $\lambda \neq 1$. Hence, $m_e > m_p$ for each value of $\lambda \neq 1$. Monotonicity of the softening function thus implies that for the same axial stretch $\lambda \neq 1$, $\hat{\alpha}(m_e) > \hat{\alpha}(m_p)$. This yields the following general result relating a pure shear to an equibiaxial stretch. *The maximum previous strain intensity in an equibiaxial deformation is greater than it is in a pure shear deformation with the same stretch, equality holding only in the undeformed state; hence, the degree of softening is greater in an equibiaxial deformation than it is in a pure shear deformation. Thus, in this sense, a greater degree of microstructural damage occurs in the equibiaxial deformation. Further, the axial stretch in a pure shear deformation required to produce the same maximum previous strain* $M_p = M_e$, *hence, the same degree of softening* $\hat{\alpha}(M_p) = \hat{\alpha}(M_e)$, *is greater than that required in an equibiaxial deformation.* The same conclusions hold for the equibiaxial, membrane inflation problem described in [13].

The difference $\Delta m \equiv m_e - m_p$ in the strain intensity for the same stretch in an equibiaxial deformation and a pure shear, in both elongation and compression, can vary substantially. For example, at $\lambda = 6$, $\Delta m = 14.9$ in extension; and for $\lambda = 0.50$, $\Delta m = 11.873$ in compression.

Collecting the foregoing results here and in the previous section, we conclude that, for the *same* stretch $\lambda \neq 1$,

$$m_e(\lambda) > m_p(\lambda) > m_u(\lambda); \tag{3.18}$$

and hence, also

$$\hat{\alpha}(m_e) > \hat{\alpha}(m_p) > \hat{\alpha}(m_u). \tag{3.19}$$

Therefore, among the three classes of deformations studied above, the greater degree of stress-softening is induced in an equibiaxial deformation of the material, while the least occurs an equal uniaxial stretch, and the latter is only slightly less than the degree of softening induced in a pure shear with the same stretch. The fact that the softening effect in an equibiaxial deformation exceeds that in an uniaxial deformation is reported in [13]; and the interesting application to the well known stress-softening phenomenon in a balloon inflation is described there. We shall return to the ballon inflation effect later in connection with a pure shear of a biomaterial membrane.

3.4. COMPARISON OF SOFTENING IN PURE SHEAR AND SIMPLE SHEAR

Here we compare the equivalent amount of shear $\kappa = \tan\psi$ required in a simple shear of angle ψ to produce the same strain intensity due to a pure shear with axial stretch λ. It is shown in [11] that the effect of stress-softening in a simple shear deformation, even a rather large shear, is small compared to the effect due to even a fairly small, simple uniaxial stretch. In view of our previous general result relating softening in a pure shear and in a simple uniaxial deformation, it is to be expected that a greater degree of softening will occur in a pure shear. To illustrate this, we recall that the principal invariants in a simple shear are given by $I_1 = I_2 = 3 + \kappa^2$. Thus, by (3.3), the equivalent amount of shear in a simple shear is related to the axial stretch in a pure shear in accordance with $\kappa^2 = \lambda^2 + \lambda^{-2} - 2$, that is,

$$\kappa = \frac{1}{\lambda}(\lambda^2 - 1), \quad \text{or} \quad \psi = \tan^{-1}\left(\frac{1}{\lambda}(\lambda^2 - 1)\right). \tag{3.20}$$

Suppose the virgin material is subjected to pure shear for which the maximum previous stretch is $\lambda = \Lambda$, and unloaded to its undeformed state. Then the relation (3.20) determines the greatest angle of shear $\psi = \Psi$ to which the new elastic stress-softened material may be subjected in an equivalent simple shear deformation without inducing further microstructural damage. Or conversely. The relation (3.20) demonstrates that for physically meaningful finite axial stretches and angles of shear, the greater effect of stress-softening is most likely produced in a pure shear, whether or not it is the primary deformation. It can be shown from (3.20) that for *small* values of the maximum previous engineering strain $\varepsilon_{\max} \equiv \Lambda - 1$, the corresponding angle of shear (in degrees) required to effect the same maximum previous strain $m_{\max} = M$ is given by $\psi_{\max} = \Psi = (360/\pi)\varepsilon_{\max}$. So, for $\varepsilon_{\max} = \pm 0.2$, say, we find $\psi_{\max} = \pm 22.92°$, for example, which is substantial. One finds from (3.20) that to effect by a simple shear deformation the same amount stress-softening $\hat{\alpha}(M)$ that would be induced by an axial stretch $\Lambda = 3.069$ corresponding to the ultimate axial stretch measured in the Rivlin–Saunders pure shear tests in [18], for example, would require an angle of shear of $\Psi = 70°$! Clearly not practicable. *Thus the bulk extent of stress-softening has already occurred in the primary pure shear of the material, and even a gross simple shear will not alter the elastic response of the stress-softened material. Hence, the effect of stress-softening in a simple shear deformation, even a rather large shear, is small; the larger effect of softening occurs in the pure shear. In fact, a simple shear generally is the least damaging deformation among all of those considered above.*

The foregoing results are established for all stress-softening materials in the class (2.3) for an unspecified monotone softening function. To be more specific in the description of stress-softening effects in a pure shear, one must consider particular models.

4. Simplified Constitutive Models

We shall now introduce simplified analytical models for both the stress-softening function and the stress–stretch constitutive equations for a Mullins material. These are useful for graphical illustrations of analytical results. First, a specific constitutive equation for the softening function is introduced. We then focus on a simplified Mullins material model.

4.1. A SPECIAL SOFTENING FUNCTION

An example constitutive equation for the softening function $\hat{\alpha}(m)$ useful in graphical illustrations is provided by

$$\hat{\alpha}(m) = 1 - \mathrm{e}^{-b(m-\sqrt{3})}, \tag{4.1}$$

where b is a positive material constant called the *softening rate parameter* [11]. This monotone increasing function of m clearly satisfies (2.2). Based on this constitutive equation, we may compare numerically the extent of softening for the different kinds of deformations described above.

Suppose, for example, that three samples of virgin material are respectively subjected to the same stretch $\lambda = 6$ in an uniaxial, an equibiaxial, and a pure shear deformation. Then for a softening rate $b = 0.02$, say, (3.16), (3.4), and (3.17) yield the respective strain intensities $m_{\mathrm{u}} = 36.001$, $m_{\mathrm{p}} = 36.014$, $m_{\mathrm{e}} = 50.912$; and the corresponding softening fractions given by (4.1) are $\hat{\alpha}_{\mathrm{u}}(36.001) = 49.61\%$, $\hat{\alpha}_{\mathrm{p}}(36.014) = 49.62\%$, $\hat{\alpha}_{\mathrm{e}}(50.912) = 62.60\%$. In the context of our Mullins–Tobin type stress-softening model, the last value shows, for example, that in a virgin stress-softening material subjected to an equibiaxial stretch $\lambda = 6$, 62.6% of the hard phase is transformed to soft phase. It is seen that these sample values agree with (3.18) and (3.19) for the same stretch.

We may look at these examples differently. Suppose that the virgin material is subjected an uniaxial stretch $\lambda_{\mathrm{u}} = 6$, so that $m_{\mathrm{u}} = 36.001$ and $\hat{\alpha}_{\mathrm{u}} = 49.61\%$. We can now determine the amount of axial stretch that will produce the same degree of stress-softening in a pure shear of another sample of the same virgin material. We set $m = m_{\mathrm{u}}$ in (3.4) to obtain the slightly smaller axial stretch $\lambda_{\mathrm{p}} = 5.9989$ for the pure shear. Similarly, with $m_{\mathrm{e}} = m_{\mathrm{u}}$ in (3.17), we find a much smaller amount of equibiaxial stretch $\lambda_{\mathrm{e}} = 5.0454$ is needed to produce the same degree of stress-softening in the same virgin material so that $\hat{\alpha}_{\mathrm{e}} = \hat{\alpha}_{\mathrm{u}}$. Thus, all of the foregoing examples are consistent with the general analytical description of stress-softening behavior described previously.

Although the form of the function (4.1), as remarked in [11], is suggested by the nature of experimental results related to the strain amplification function studied in [5], no experimental investigations that identify a specific relation for the stress-softening function are known. Therefore, the model (4.1) is introduced for analytical convenience. Other functions are possible. A second example is described in [11]. Though future experimental work may suggest an altogether different stress-softening constitutive equation, the general analytical results presented above are not affected by a specific choice of the softening function $\hat{\alpha}(m)$, and it is unlikely that the main effects of stress-softening illustrated by use of the analytical model (4.1) shall be too different.

4.2. A SPECIAL STRESS-SOFTENING MATERIAL MODEL

The constitutive equations and the technical applications presented so far are general. The precise dependence of the response functions on the volume fraction $\hat{\alpha} \in [0, 1]$ at the maximum previous strain of the material is unspecified, as are the response functions themselves. It is useful in illustration of some further results to consider within the general class (2.3) a simplified Mullins material model for which the response functions are defined by

$$\hat{\beta}_\Gamma\big(I_1, I_2; \hat{\alpha}(m)\big) = \big(1 - \hat{\alpha}(m)\big)\aleph_\Gamma(I_1, I_2). \tag{4.2}$$

We require that the new response functions $\aleph_\Gamma = \aleph_\Gamma(I_1, I_2)$, $\Gamma = 1, -1$ satisfy the empirical inequalities so that (2.6) hold for response functions $\hat{\beta}_\Gamma$ of the stress-softening material, provided $\hat{\alpha} \neq 1$, which is the fully softened case in (2.2). It is evident that the response functions $\hat{\beta}_1$ and $-\hat{\beta}_{-1}$ are monotone decreasing functions of $\hat{\alpha}(m)$ that satisfy the conditions in (2.4) for a Mullins material. When loading does not occur, m is fixed at its maximum previous value, $m_{\max} = M$; and the material response is ideally elastic with new response functions $\hat{\beta}_\Gamma(I_1, I_2; \hat{\alpha}(m_{\max})) = (1 - \hat{\alpha}(m_{\max}))\aleph_\Gamma(I_1, I_2)$ whose absolute values are reduced in consequence of the microstructural material transformation from hard to soft phase.

With (4.2), the constitutive equation (2.3) is then reduced to

$$\mathbf{T} = \begin{cases} \big(1 - \hat{\alpha}(m)\big)\mathbf{T}_0 & \text{during loading,} \\ \big(1 - \hat{\alpha}(M)\big)\mathbf{T}_0 & \text{otherwise,} \end{cases} \tag{4.3}$$

in which the Cauchy stress $\mathbf{T}_0$ is determined by

$$\mathbf{T}_0 = -p_0\mathbf{1} + \aleph_1\mathbf{B} + \aleph_{-1}\mathbf{B}^{-1}, \tag{4.4}$$

where p_0 is an undetermined pressure. Therefore, the simplified constitutive theory for a Mullins material expressed in (4.3) consists of the stress response of some *parent elastic material* (4.4) and an equation for the monotone increasing softening function $\hat{\alpha}(m)$. We note that equations similar to (4.3) hold also for the first Piola–Kirchhoff (or engineering) stress $\mathbf{P} = \mathbf{T}\mathbf{F}^{-\mathrm{T}}$ with $\mathbf{P}_0 = \mathbf{T}_0\mathbf{F}^{-\mathrm{T}}$.

The special constitutive model (4.3) shows that the ratios of the physical stress components T_{jk} in the virgin material to the corresponding components τ_{jk} in the stress-softened material (subjected to the same deformation) are determined by the stress-softening function alone. That is, in accordance with

$$\frac{T_{jk}}{\tau_{jk}} = \frac{1 - \hat{\alpha}(m)}{1 - \hat{\alpha}(M)} \geqslant 1, \quad j, k = 1, 2, 3, \text{ no sum}, \tag{4.5}$$

where we recall (2.2) with $m_{\max} = M$. Equality in (4.5) holds when and only when $m = M$. The rule (4.5) shows that, *at the same value of the deformation tensor, the magnitude of the stress component in the softened material is always smaller than the magnitude of the corresponding stress component in the virgin material.* This is one of the predominant features of the Mullins effect exhibited in experiments and characterized in the general analysis presented earlier. Notice that the more general results obtained in (3.11) also follow from (4.5).

For an isotropic, incompressible, hyperelastic parent material, the strain energy density W per unit reference volume depends on $\mathbf{B}$ only through the values of its invariants I_1, I_2: $W = \widehat{W}(I_1, I_2)$. The response functions $\aleph_\Gamma(I_1, I_2)$ are then given by

$$\aleph_1 = 2W_1, \qquad \aleph_{-1} = -2W_2, \tag{4.6}$$

where $W_\alpha \equiv \partial \widehat{W}/\partial I_\alpha$. We next identify specific parent material models for use in applications.

4.3. SPECIAL PARENT MATERIAL MODELS

We shall study stress-softening behavior of two classes of parent elastic materials. The first is the familiar Mooney–Rivlin rubberlike solid for which the strain energy function is

$$W(I_1, I_2) = \frac{\mu}{2}\left[(I_1 - 3) + \gamma(I_2 - 3)\right]; \tag{4.7}$$

and hence, by (4.6)

$$\aleph_1 = \mu, \qquad \aleph_{-1} = -\gamma\mu, \tag{4.8}$$

where $\mu > 0$ and $\gamma \geqslant 0$ are material constants. These constants are related to Young's modulus E_0 and to the shear modulus μ_0 in the undistorted state, in accordance with $E_0/3 = \mu_0 = \mu(1 + \gamma)$. The neo-Hookean model is obtained with $\gamma = 0$.

A recent study by Emery et al. [17] has shown that the "strain softening" response of ventricular myocardium to cyclic cavity pressure loading is similar to the stress-softening response exhibited by elastomers in uniaxial extension. Based on the general one-dimensional model proposed by Johnson and Beatty [3], the authors introduce a certain exponential left ventricle pressure-volume relation for strain softening that closely approximates the experimentally observed softening response of myocardium, an effect that depends on the maximum previous pressure. The consistent changes observed in the pressure-volume relation with the history of the maximum previous pressure show that left ventricular stiffness decreases as the maximum previous pressure increases. It is shown in the experiments that when the tissue is loaded to a pressure P^*, unloaded and then reloaded to a pressure P_1, unloaded and reloaded to a pressure P_2 for which $P_1 < P_2 < P^*$, the pressure-volume curves follow the primary unloading curve from the maximum previous pressure P^*; hence, the stress-softened material response is elastic. The response changes, however, when the pressure exceeds P^*. Although our current study is not related to the experiments in [17], we shall see later that similar stress-softening response is predicted by our biomaterial model in a pure shear deformation. Hence, for our second parent elastic material we choose a hyperelastic biomaterial for which the strain energy density is

$$W(I_1, I_2) = \frac{\mu_0}{2\gamma}\{e^{\gamma(I_1-3)} - 1\}, \tag{4.9}$$

where $\mu_0 > 0$ is the constant shear modulus in the undistorted state and γ is another non-negative material constant. From (4.6), the elastic response functions are found to be

$$\aleph_1 = \mu_0 e^{\gamma(I_1-3)}, \qquad \aleph_{-1} = 0. \tag{4.10}$$

The case $\gamma = 0$ describes a neo-Hookean material. Notice that for both the neo-Hookean and biomaterial models, $\hat{\beta}_{-1} = 0$; and hence the universal relation (3.12) holds for these special parent materials.

5. Graphical Results for the Mooney–Rivlin Parent Material

The softening material response for the Mooney–Rivlin parent material (4.8) with the softening function (4.1) is obtained from (3.6) through (3.9). With the aid of (4.2), the normalized biaxial stresses $t_k^* \equiv t_k/\mu_0$ in a plane sheet of the virgin material under loading are described by

$$\begin{aligned} t_1^* &= \left(\lambda^2 - \frac{1}{\lambda^2}\right)e^{-b(m-\sqrt{3})}, \\ t_2^* &= \left(1 - \frac{1}{\lambda^2}\right)\left(\frac{1+\gamma\lambda^2}{1+\gamma}\right)e^{-b(m-\sqrt{3})} \quad \text{when } \dot{\lambda} > 0, \end{aligned} \tag{5.1}$$

where we recall (3.4). For a fixed value $M = m_{\max}$ of the maximum previous strain in (3.10), $\tau_k^* \equiv t_k^*|_{m=M}$ gives the biaxial stress–stretch response in the stress-softened Mooney–Rivlin material. For $\gamma = 0$, we may obtain results for the neo-Hookean parent material model; but we shall not present these here. Notice, however, that the axial stress t_1^* in (5.1) is independent of γ; it thus holds for every

Mooney–Rivlin parent material having the same softening rate b, hence also for every neo-Hookean parent material.

The choice of parameters for our first illustration is motivated by the Rivlin–Saunders pure shear experiments on a thin sheet of natural, gum rubber described in [18]. The data show that the normalized axial engineering stress $P_1^* = t_1^*/\lambda$ is a monotone increasing function of the stretch in the interval $\lambda \in [1, 3.069]$, where $\lambda_{\max} = 3.069$ is the greatest axial stretch to which the material was subjected in their pure shear test. The range of deformation was limited to avoid effects of strain induced crystallization. From (3.10), this stretch corresponds to a maximum previous strain $M = M_p = 9.4723$. It follows from $(5.1)_1$ that P_1^* is monotone over this range of stretch provided that $b < 0.0558$. Thus, for $b = 0.055$, say, the extent of softening in a pure shear with a maximum previous stretch $\Lambda_p = 3.069$, in accordance with the constitutive model (4.1), is $\hat{\alpha} = 34.7\%$. To effect the same maximum previous strain M_p in a simple uniaxial extension with stretch Λ_u requires that $\sqrt{\Lambda_u^4 + 2/\Lambda_u^2} = M_p$, by (3.16). We thus find for the model (4.1) that the same degree of softening may be effected by only a slightly larger uniaxial stretch $\Lambda_u = 3.076$. Therefore, using the Rivlin–Saunders data as a guide, we see that *the extent of stress-softening in a simple uniaxial extension is comparable to, but slightly less than that induced by a pure shear*. This agrees with the general result (3.19).

Of course, monotone response of the axial stress in the pure shear need not imply monotone response of the transverse, restraining stress over the same range of stretch. Notice from (5.1) that the normalized transverse, restraining engineering stresses $P_2^* = t_2^*$ and $\pi_2^* \equiv P_2^*|_{m=M} = \tau_2^*$ depend on γ, whereas the normalized axial stresses P_1^* and $\pi_1^* \equiv \tau_1^*/\lambda$ do not. It is shown in [11] that, for $\gamma = \{0.01, 0.33\}$ and an ultimate stretch $\lambda_{\text{ult}} = 6$, an appropriate softening rate parameter for monotone increasing stress–stretch behavior exhibited in uniaxial experiments in [3] is $b = 0.013$, so we adopt the same value here. Notice for this example that the axial stress response over the Rivlin–Saunders range will remain monotone as before.

For a given softening rate, the softening fractions determined by (4.1) are the same for every Mooney–Rivlin type softening material. Graphs of the normalized axial stress–stretch functions for the virgin and stress-softened Mooney–Rivlin parent material based on $(5.1)_1$, unloaded from softening points at $\lambda_{\max} = \{3, 4, 5, 6\}$ are shown in Figure 1. The corresponding percentages of softening $\hat{\alpha}(m_{\max})$ are noted. The effects resulting from use of different normalized axial stress components vary. The normalized Cauchy stresses t_1^* and τ_1^* are shown in Figure 1(a); and the corresponding normalized engineering stresses P_1^* and π_1^* are shown in Figure 1(b). The normalized transverse stress components for two values of γ are mapped in Figure 2. The corresponding values of the stress-softening function are noted. Notice from Figures 1(a), 2(a) and 2(b) that $t_1^* > t_2^* = P_2^*$ and $\tau_1^* > \tau_2^* = \pi_2^*$, as established generally in remarks following (3.13).

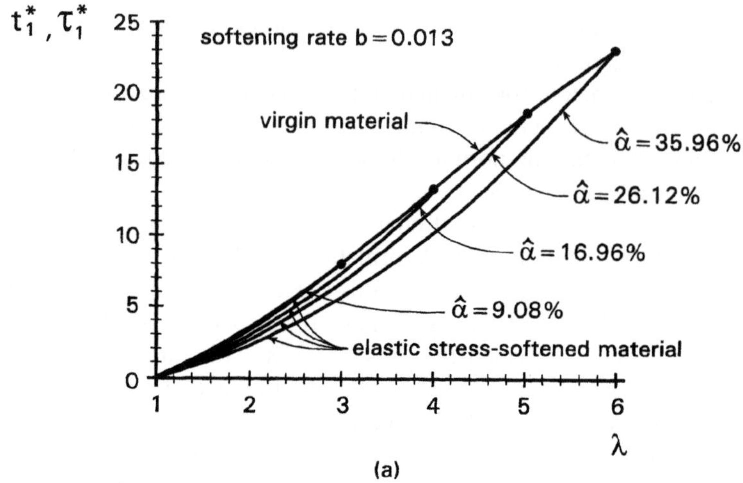

(a)

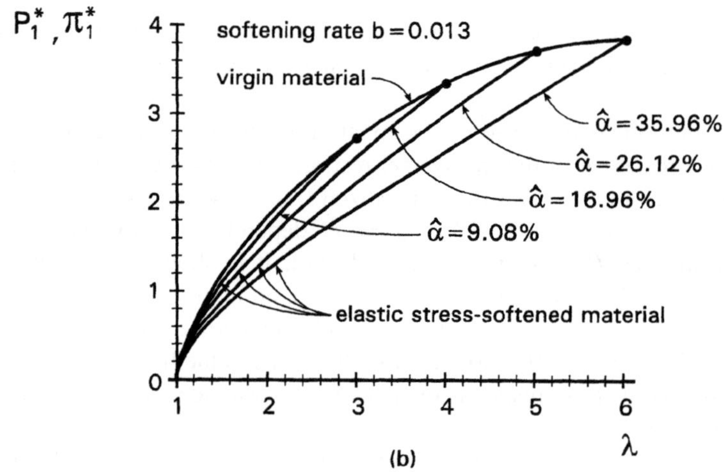

(b)

Figure 1. Normalized axial stress–stretch plots for all Mooney–Rivlin parent materials with $b = 0.013$. (a) Normalized axial Cauchy stresses, (b) normalized axial engineering stresses.

The nature of the results for the same softening rate parameter are quite different. The transverse stress for $\gamma = 0.01$ in Figure 2(a) rises to a maximum $P^*_{\max} = 0.894$ at $\lambda = 3.963$, whereas the transverse stress response for $\gamma = 0.33$ in Figure 2(b) has a much greater intensity and is monotone increasing for a range extending to an ultimate stretch $\lambda_{\text{ult}} = 6$. Results for the neo-Hookean model are similar. Of course, the effects will vary with the nature of the parent material and the values of the various material parameters.

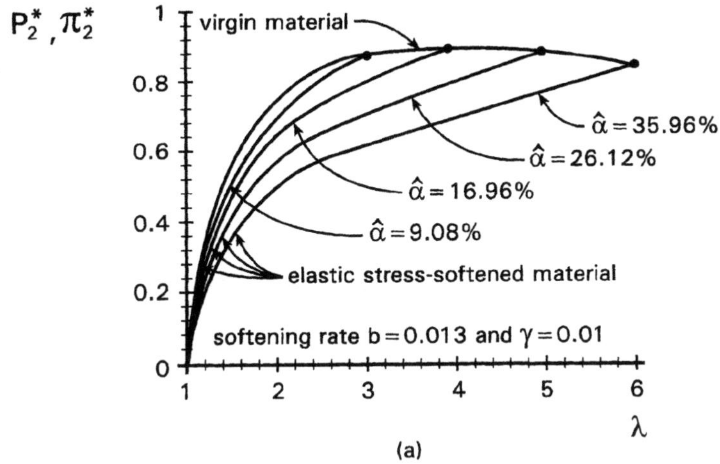

(a)

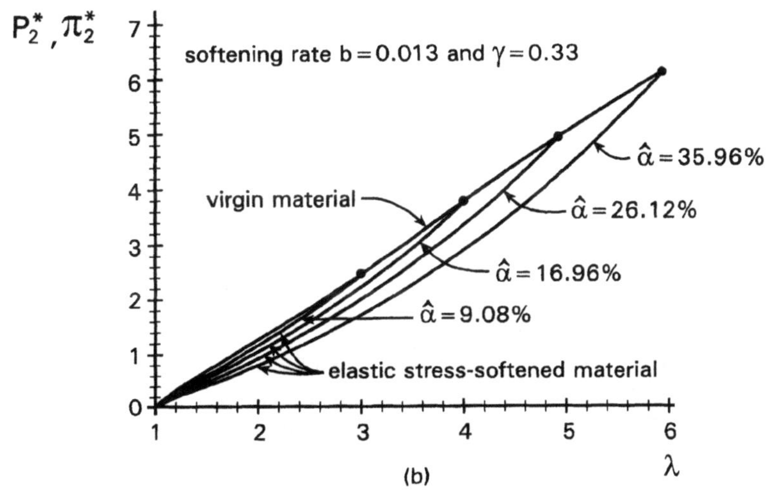

(b)

Figure 2. Normalized transverse stress–stretch plots for Mooney–Rivlin parent materials with $b = 0.013$ and (a) $\gamma = 0.01$, (b) $\gamma = 0.33$.

6. Graphical Results for the Parent Biomaterial

The normalized biaxial stresses $t_k^* = t_k/\mu_0$ for the virgin material response of a parent biomaterial (4.10) with the softening function (4.1) in a pure shear deformation are obtained from (3.6), (3.7), and (4.2). These yield

$$t_1^* = \left(\lambda^2 - \frac{1}{\lambda^2}\right) e^{\gamma(I_1-3)-b(m-\sqrt{3})}, \qquad t_2^* = \left(1 - \frac{1}{\lambda^2}\right) e^{\gamma(I_1-3)-b(m-\sqrt{3})}, \tag{6.1}$$

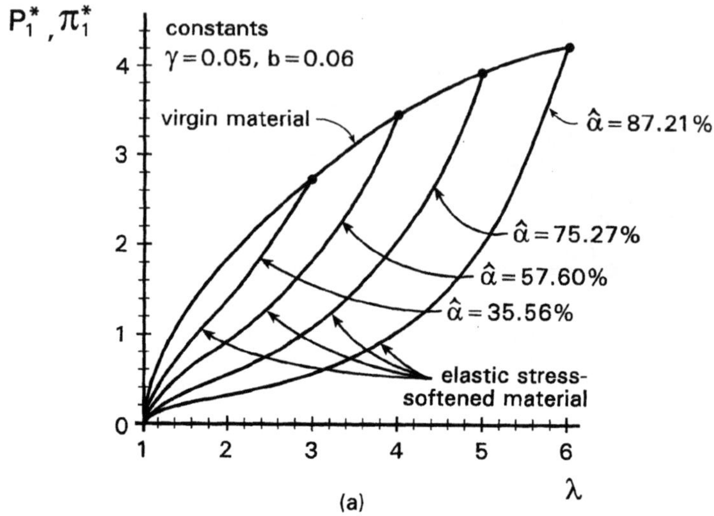

(a)

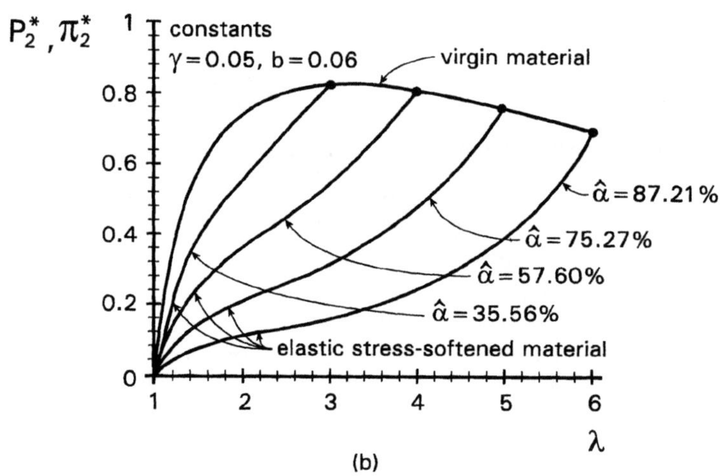

(b)

Figure 3. Normalized engineering stress–strech graphs for a parent biomaterial with $b = 0.06$ and $\gamma = 0.05$. (a) Axial stress response, (b) transverse stress response.

in which we recall $(3.3)_1$ and (3.4). The corresponding normalized biaxial stress-stretch response in the elastic stress-softened biomaterial is given by $\tau_k^* \equiv t_k^*|_{m=M}$, for a fixed value of M in (3.10). Again, with $\gamma = 0$, we may recover results for the neo-Hookean model.

We wish to preserve the monotone character of the axial load in the Rivlin–Saunders pure shear experiments [18]. An appropriate choice of b now depends on the value of the positive material constant γ in (6.1). For $\gamma = 0.01$, we may take

$b = 0.02$, for example. In this case, graphs of the stress–stretch response obtained from (6.1) for the biomaterial are similar to those shown in Figures 1 and 2 for the Mooney–Rivlin parent material, though the degree of stress-softening of the biomaterial is greater; specifically, $\hat{\alpha}(36.014) = 49.62\%$ at the ultimate stretch $\lambda_{\mathrm{ult}} = 6$. Of course, the effect varies with the values of the material constants. The effect for $\gamma = 0.05$ and $b = 0.06$, though similar, is more striking. Graphs of the normalized axial engineering stress–stretch relations $P_1^* = t_1^*/\lambda$ and $P_2^* = t_2^*$ for the virgin material response and the corresponding relations $\pi_k^* = P_k^*|_{m=M}$ for the elastic stress-softened biomaterial based on (6.1) are shown in Figure 3. The extent of stress-softening varies only with the value of b, but the material response also depends on γ. For each value of the maximum previous stretch $\Lambda = \{3, 4, 5, 6\}$, it is seen that the degree of stress-softening is considerably larger than the corresponding fraction determined for both the Mooney–Rivlin example and aforementioned biomaterial having $\gamma = 0.01$. Currently, the ultimate softening fraction is $\hat{\alpha}(36.014) = 87.21\%$; yet the normalized stress levels shown in Figures 2 and 3 are not too different. The maximum stress in Figure 3 is $P_{\max}^* = 0.8193$ at $\lambda = 3.229$. It is emphasized that the biomaterial parameters in these examples are provided only for illustration; they are not designed to mimic any specific biotissue.

7. Preconditioning Effect of a Pure Shear on Balloon Inflation

Here we recall the softening effect encountered when one inflates a balloon. At first, the effort required to expand the balloon is considerable; then suddenly the task becomes easier, and the balloon expands with less effort, less air pressure. Upon inflating this balloon a second time, the effort required is noticeably smaller than that needed initially. It is also observed that one can ease the primary effort required to inflate the balloon by stretching it several times in an uniaxial extension. Since a pure shear has a slightly stronger effect on stress-softening, it is expected that preconditioning the material by a pure shear will result in a similar membrane inflation effect. Of course, we conjecture that this preconditioning phenomenon is just the Mullins effect at work. This was demonstrated recently by Beatty and Krishnaswamy [13]. Here we describe the preconditioning effect of a pure shear on the inflation of a spherical membrane. We consider for illustration a biomaterial membrane, but the effect is the same for other models.

With this familiar experience in mind, let us imagine that we have a virgin stress-softening spherical membrane that we first subject to a pure shear with a maximum axial stretch $\Lambda_{\mathrm{p}} = 6$, say, after which the material is unloaded to an undeformed, stress-free reference configuration. The corresponding maximum previous strain M_{p} is given by (3.10). Now suppose that following this preconditioning of the virgin material, the membrane is inflated to the same maximum previous strain $M_{\mathrm{e}} = M_{\mathrm{p}}$ for which the equibiaxial stretch λ is given by (3.17). Then the equibiaxial stretch satisfies the relation

$$2\lambda^{12} - M_{\mathrm{p}}^2\lambda^8 + 1 = 0. \tag{7.1}$$

For a maximum axial stretch $\Lambda_p = 6$, the unique equibiaxial extensional root is $\lambda = \Lambda_e = 5.046$. Thus, by our preconditioning the material in a pure shear with axial stretch $\Lambda_p = 6$, the same maximum previous strain in an equibiaxial deformation requires a stretch of only $\Lambda_e = 5.046$. Because this value differs little from the value $\Lambda_e = 5.045$ fixed by M_u for an uniaxial stretch $\Lambda_u = 6$, it is to be expected that the preconditioning effect should be very much the same for both the uniaxial case reported in [13] and for the pure shear studied here. It is nonetheless useful to revisit this result in a different setting. So, let us see what effect the preconditioning by a pure shear has on the pressure response in the inflation of a spherical biomaterial membrane.

It is shown in [13] that the normalized inflation pressure response for the virgin biomaterial is described by

$$P_I^* = \frac{1}{\lambda}\left(1 - \frac{1}{\lambda^6}\right) e^{\gamma(2\lambda^2+\lambda^{-4}-3)} e^{-b(m_e(\lambda)-\sqrt{3})} \tag{7.2}$$

for all $m_e(\lambda) \leqslant M_{ult}$, the ultimate previous strain. Here we recall (3.17). We consider the stress-softened (preconditioned) material for which the maximum previous strain is M_p due to a pure shear. The normalized inflation pressure function for the preconditioned biomaterial provided in [13] is then given by

$$p_I^* = \frac{1}{\lambda}\left(1 - \frac{1}{\lambda^6}\right) e^{\gamma(2\lambda^2+\lambda^{-4}-3)} e^{-b(M_p-\sqrt{3})}. \tag{7.3}$$

When $M_{ult} \geqslant M_p$, the biomaterial recalls its virgin material response in (7.2) starting from $m_e(\lambda) = M_p$ up to M_{ult}. For illustration of the results in Figure 4, we adopt the same data used in [13].

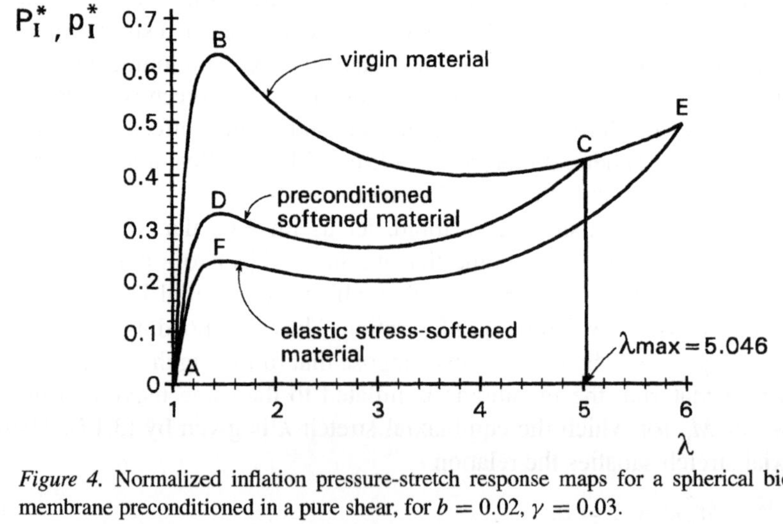

Figure 4. Normalized inflation pressure-stretch response maps for a spherical biomaterial membrane preconditioned in a pure shear, for $b = 0.02$, $\gamma = 0.03$.

The elastic preconditioned inflation response is described by the curve ADC. This is the predicted response curve (7.3) following preconditioning by a pure shear with axial stretch to $\Lambda_p = 6$ for which the maximum previous strain is $M_p = 36.014$. This describes the elastic stress-softened response of the biomaterial membrane for both loading and unloading so long as the magnitude of the strain does not exceed the maximum previous equibiaxial strain $M_e = M_p$ at the equibiaxial stretch $\lambda = \Lambda_e = 5.046$. The material remembers this maximum previous strain condition, but it does not recall the deformation path by which it got there; and hence, when the equibiaxial stretch exceeds this value, the material recalls its virgin description and the pressure-stretch response in Figure 4 traces the curve CE until the ultimate stretch $\lambda_{ult} = 6$ is reached at the softening point E. If the membrane is unloaded from E, in accordance with (7.2) in which m_e is replaced by $M_e = 50.912$, the response is again elastic and traces the path AFE in either direction so long as $m_e < M_e$ at E. For the biomaterial response shown in Figure 4, the virgin yield pressure at B is $P_I^*(1.4034) = 0.6284$, while the stress-softened yield pressure at D following preconditioning by a pure shear with stretch $\Lambda_p = 5.046$ is $p_I^*(1.4334) = 0.3237$, or 51.5% smaller. Notice in Figure 4 the additional recovery effect as the virgin material pressure falls to a minimum pressure $p_I^*(3.968) = 0.3928$ and then grows again toward C. Additional details of the pressure response curves may be found in [13]. *This example shows that the yield pressure of the biomaterial preconditioned by a pure shear is substantially smaller than the yield pressure of the virgin biomaterial. The preconditioning effect physically experienced in the inflation of a toy balloon is evident – the subsequent inflation pressure is considerably reduced by preconditioning of the virgin material in a pure shear as well as in an uniaxial stretch, to nearly the same extent.* Precisely the same effect holds for a Mooney–Rivlin parent material.

Acknowledgement

This work was funded by Grant No. CMS-9634817 from the National Science Foundation.

References

1. L. Mullins, Effect of stretching on the properties of rubber. *J. Rubber Res.* **16** (1947) 275–289.
2. L. Mullins and N. Tobin, Theoretical model for the elastic behavior of filler-reinforced vulcanized rubbers. *Rubber Chem. Tech.* **30** (1957) 551–571.
3. M.A. Johnson and M.F. Beatty, The Mullins effect in uniaxial extension and its influence on the transverse vibration of a rubber string. *Contin. Mech. Thermodyn.* **5** (1993) 83–115.
4. M.A. Johnson and M.F. Beatty, The Mullins effect in equibiaxial extension and its influence on the inflation of a balloon. *Internat. J. Engrg. Sci.* **33** (1995) 223–245.
5. M.A. Johnson and M.F. Beatty, A constitutive equation for the Mullins effect in stress controlled uniaxial extension experiments. *Contin. Mech. Thermodyn.* **5** (1993) 301–318.
6. A.S. Wineman and K.R. Rajagopal, On a constitutive theory for materials undergoing microstructural changes. *Arch. Mech.* **42** (1990) 53–75.

7. K.R. Rajagopal and A.S. Wineman, A constitutive equation for nonlinear solids which undergo deformation induced microstructural changes. *Internat. J. Plasticity* **8** (1992) 385–395.
8. A.S. Wineman and H.E. Huntley, Numerical simulation of the effect of damaged induced softening on the inflation of a circular rubber membrane. *Internat. J. Solids Struct.* **31** (1994) 3295–3313.
9. H.E. Huntley, A.S. Wineman and K.R. Rajagopal, Chemorheological relaxation, residual stress and permanent set arising in radial deformation of an elastomeric hollow sphere. *Math. Mech. Solids* **1** (1996) 267–299.
10. H.E. Huntley, A.S. Wineman and K.R. Rajagopal, Stress softening, strain localization and permanent set in the circumferential shear of an incompressible elastomeric cylinder. *IMA J. Appl. Math.* **59** (1997) 309–338.
11. M.F. Beatty and S. Krishnaswamy, A theory of stress-softening in incompressible isotropic materials. *J. Mech. Phys. Solids* **48** (2000) 1931–1965.
12. S. Krishnaswamy and M.F. Beatty, The Mullins effect in compressible solids. *Internat. J. Engrg. Sci.* **38** (2000) 1397–1414.
13. M.F. Beatty and S. Krishnaswamy, The Mullins effect in equibiaxial extension. *Z. angew. Mat. Phys.*, to appear.
14. S. Krishnaswamy and M.F. Beatty, Damage induced stress-softening in the combined torsion, extension and inflation of a cylindrical tube. *Quart. J. Mech. Appl. Math.*, to appear.
15. M.F. Beatty, Introduction to nonlinear elasticity. In: M.M. Carroll and M.A. Hayes (eds), *Nonlinear Effects in Fluids and Solids*. Plenum, New York, London (1996) pp. 13–104.
16. C. Truesdell and W. Noll, *The Nonlinear Field Theories of Mechanics*, Flügge's Handbuch der Physik III/3 Springer, Berlin (1965).
17. J.L. Emery, J.H. Omen and A.D. McCulloch, Strain softening in rat left ventricular myocardium. *J. Biomech. Engrg.* **119** (1997) 6–13.
18. R.S. Rivlin and D.W. Saunders, Large elastic deformations of isotropic materials. VII. Experiments on the deformation of rubber. *Phil. Trans. Roy. Soc. London A* **243** (1951) 251–288.

Journal of Elasticity **59:** 393–398, 2000.

The Compatibility Constraint in Linear Elasticity

P. PODIO-GUIDUGLI
Dipartimento di Ingegneria Civile, Università di Roma "Tor Vergata", Via di Tor Vergata 110, I-00133 Roma, Italy. E-mail: ppg@uniroma2.it

Received 20 August 1999

Dedicated to Roger Fosdick on the occasion of his sixtieth birthday.

Abstract. It is shown that the St.Venant–Beltrami equation of compatibility can be regarded as an internal constraint on the admissible strains, a constraint maintained by reactive stress fields that are collectively characterized in a global manner. It is also shown that such reactive stresses have an important role in a formal integration of the mixed boundary-value problem of linear elasticity. The main tools – Beltrami's map, Donati's theorem, and Cesàro's formula – have been available for a century or so; they are here used in a form as close to original as possible.

Mathematics Subject Classifications (2000): 73C02, 73C35.

Key words: linear elasticity, compatibility, mixed boundary-value problems.

1. Introduction

Let Ω be an open, bounded, and simply-connected region of the ordinary three-dimensional point space, with smooth boundary $\partial\Omega$, a single closed surface. A symmetric-valued tensor field $\mathbf{E}$ of class $C^N(\Omega)$ $(N \geqslant 2)$ is said to satisfy the St.Venant–Beltrami *equation of compatibility* if

$$\operatorname{curl}\operatorname{curl}\mathbf{E} = \mathbf{0} \quad \text{in } \Omega. \tag{1.1}$$

Each compatible field $\mathbf{E}$ is associated (uniquely, to within an infinitesimal rigid displacement field) to a displacement field $\mathbf{u}$ of class $C^{N+1}(\Omega)$ such as to satisfy the *strain-displacement relation*

$$\nabla\mathbf{u} + \nabla\mathbf{u}^{\mathrm{T}} = 2\mathbf{E}. \tag{1.2}$$

In this paper I consider relation (1.1) in the context of linear elasticity, and propose to regard it as an internal constraint of an unusual type, the *compatibility constraint*. To open the way to this interpretation, I rephrase and marginally update certain results of Beltrami (1892) and Donati (1890, 1894) (*vid.* [3, Sections 17 and 18]), two exponents of last century's great Italian school of elasticity.

The reactive stress fields arising in response to the compatibility constraint, which has a *global* nature, depend in general on the shape and the size of Ω, but have no algebraic specialty, as is instead the rule with the usual *pointwise* internal constraints (Section 2). Since those reactive stress fields are divergenceless and vanish near the boundary of Ω, they do not affect the equilibrium equations (Section 3); yet, their presence is crucial to a seemingly new procedure of formal integration for the mixed boundary-value problem of linear elasticity (Section 4), as is a constructive formula by Cesàro (1906) (*vid.* [3, Section 14]), another Italian elasticist of the same category as Beltrami and Donati.

2. Pointwise, Partwise, and Global Internal Constraints

In this section, with a view toward putting the notion of a compatibility constraint into perspective, I briefly summarize the standard concept of an internal constraint, as well as other, more or less well-known, related concepts.

According to the standard Cauchy-type model, a continuous body is partitionable into subbodies of arbitrary shape and size. Consistent with this assumption, all constitutive statements – internal constraints included – take the *pointwise* form of assertions about the typical material point of a body class.

For example, the state of a linearly elastic material point is specified by assigning the value of three fields: the displacement $\mathbf{u}$, the strain $\mathbf{E}$, and the stress $\mathbf{S}$. A material point is *constrained* whenever $\mathbf{E}$ (or $\mathbf{u}$, and hence $\mathbf{E}$, to guarantee consistency with the strain–displacement relation (1.2)) is required to belong to a given *constraint manifold* $\mathcal{C}$, a subspace of the space Sym of all symmetric tensors. This confinement condition is called an *internal constraint*, and thought of as maintained by a "frictionless" *reactive stress* $\mathbf{S}^R$, i.e., a stress that expends no power in any admissible motion. Since $\mathcal{C}$ is flat, the *reaction manifold*

$$\mathcal{R} = \{\mathbf{S}^R \in \mathrm{Sym} \mid \mathbf{S}^R \cdot \mathbf{E} = 0, \ \mathbf{E} \in \mathcal{C}\} \tag{2.1}$$

(the collection of all admissible reactive stresses) turns out to be flat as well. Dually, given a flat reaction manifold $\mathcal{R}$, the constraint manifold can be defined to be

$$\mathcal{C} = \{\mathbf{E} \in \mathrm{Sym} \mid \mathbf{S}^R \cdot \mathbf{E} = 0, \ \mathbf{S}^R \in \mathcal{R}\}; \tag{2.2}$$

the direct sum $\mathcal{C} \oplus \mathcal{R}$ exhausts Sym.* Finally, for the *active stress* the value $\mathbf{S}^A(\mathbf{E})$ at $\mathbf{E}$ of a linear constitutive mapping defined over $\mathcal{C}$, a state of a $\mathcal{C}$-constrained material point is any triplet $\{\mathbf{u}, \mathbf{E}, \mathbf{S}\}$ with $\mathbf{E} \in \mathcal{C}$ and $\mathbf{S} = \mathbf{S}^A(\mathbf{E}) + \mathbf{S}^R$, $\mathbf{S}^R \in \mathcal{R}$.

More general notions of internal constraint have been proposed (see the discussion in [1, Section XII.12]), and for good reasons. To mention one, the standard pointwise characterization identifies certain reactions, but not all those one can think and make use of.

* E.g., a material point is *incompressible* if $\mathrm{div}\,\mathbf{u} = 0$. Since $\mathrm{div}\,\mathbf{u} = \mathrm{tr}\,\mathbf{E}(\mathbf{u})$ for $\mathbf{E}(\mathbf{u}) = 1/2(\nabla\mathbf{u} + \nabla\mathbf{u}^{\mathrm{T}})$, and since by the definition of trace $\mathrm{tr}\,\mathbf{E} = \mathbf{1} \cdot \mathbf{E}$, with $\mathbf{1}$ the unit tensor, the incompressibility constraint manifold is $\mathcal{C} = \{\mathbf{E} \in \mathrm{Sym} \mid \mathbf{1} \cdot \mathbf{E} = 0\}$ and hence the associated reaction manifold is $\mathcal{R} = \mathrm{span}\{\mathbf{1}\}$; dually, given such an $\mathcal{R}$, one finds that $\mathcal{C}$ consists of all traceless symmetric tensors.

Generally speaking, reactive stress fields compensate for the difficulty inherent to finding a solution of a constrained form to a given boundary-value problem; they do so by helping to satisfy the field and boundary equilibrium conditions; the richer the reaction manifold where one can pick for this purpose, the better.

When an internal constraint, rather than in terms of admissible strains, is posited in terms of admissible displacements, it may happen that – at variance with the standard case exemplified in footnote* – the corresponding strains have no algebraic distinction: the pointwise characterization in (2.1) would then call for a null reactive stress. One can think, however, of reactive stress fields satisfying weaker orthogonality conditions than in (2.1), and hence generally being different from zero at a typical point of Ω. If a body's part-collection should not include subbodies of arbitrary shape and size, a *partwise* notion of internal constraint becomes appropriate, namely,

$$\int_{\Pi} \mathbf{S}^R \cdot \operatorname{sym}(\nabla \mathbf{u}) = 0 \tag{2.3}$$

for each part $\Pi \subset \Omega$ and each admissible displacement field $\mathbf{u}$.‡ If instead the body is arbitrarily partitionable, one is led to postulate a *global* constraint condition, such as

$$\int_{\Omega} \mathbf{S}^R \cdot \operatorname{sym}(\nabla \mathbf{u}) = 0 \tag{2.4}$$

for each admissible displacement field $\mathbf{u}$.

When a constraint manifold consists of strain fields over the body that need not satisfy any pointwise algebraic condition, partwise or global constraint conditions are appropriate being completely similar to (2.3) or (2.4), but quantified over the admissible strains rather than the symmetric gradients of the admissible displacements; the induced reaction manifolds consist of algebraically generic fields as well.

‡ This is the case for the "thin" bodies studied in structural mechanics. Consider, e.g., a *plate-like* body Ω whose reference shape has the form of a right cylinder of cross section $\mathcal{P}$, axis parallel to $\mathbf{z}$, and thickness 2ε: the parts of Ω are themselves cylinders of axis $\mathbf{z}$ and length 2ε; only their cross section $\mathcal{p} \in \mathcal{P}$ is arbitrary. If equilibrium displacements of the constrained form

$$\mathbf{u}(x,\zeta) = \mathbf{u}_0(x) + \zeta \mathbf{u}_1(x), \qquad (x,\zeta) \in \mathcal{P} \times (-\varepsilon, +\varepsilon),$$

are sougth, then (2.3) allows for whatever reaction fields $\mathbf{S}^R$ over Ω agree with the conditions

$$\mathbf{N}^R = \mathbf{0}, \quad \mathbf{M}^R(\mathbf{1} - \mathbf{z} \otimes \mathbf{z}) = \mathbf{0} \quad \text{in } \mathcal{P},$$

where

$$\mathbf{N}^R(x) = \int_{-\varepsilon}^{+\varepsilon} \mathbf{S}^R(x,\zeta)\,\mathrm{d}\zeta, \qquad \mathbf{M}^R(x) = \int_{-\varepsilon}^{+\varepsilon} \zeta \mathbf{S}^R(x,\zeta)\,\mathrm{d}\zeta$$

(see [2, 4] for more details).

All these constraint conditions are expressed as generalizations of the standard notion (2.1) of a reaction manifold; by duality, corresponding generalizations of the standard notion (2.2) of a constraint manifold are easily achieved. The compatibility constraint (1.1) will be quickly seen to admit an expression of the latter type, relation (3.4) below, which has a global form, because a compatible body is presumed to be arbitrarily partitionable.

3. The Compatibility Constraint

Let $\mathcal{T}$ denote the collection of all test fields over Ω, i.e., of all symmetric-valued tensor fields $\mathbf{T}$ of class $C^\infty(\Omega)$ that vanish near $\partial\Omega$ (a field ϕ defined over $\overline{\Omega}$ is said to vanish near $\partial\Omega$ if there is a neighborhood $\mathcal{N}$ of $\partial\Omega$ such that $\phi = 0$ on $\mathcal{N} \cap \overline{\Omega}$ [3, Section 7]).

Beltrami's map

$$\mathbf{B}(\mathbf{T}) = \operatorname{curl}\operatorname{curl}\mathbf{T} \tag{3.1}$$

transforms $\mathcal{T}$ into the subcollection $\mathbf{B}(\mathcal{T})$ of all its divergenceless elements; moreover, each sufficiently smooth solution of the homogeneous equilibrium equation

$$\operatorname{div}\mathbf{S} = \mathbf{0} \tag{3.2}$$

has a Beltrami representation $\mathbf{S} = \operatorname{curl}\operatorname{curl}\mathbf{A}$ in terms of some smooth, symmetric-valued field $\mathbf{A}$. Now, a version of the fundamental lemma of the calculus of variations allows one to write (1.1) as

$$\int_\Omega \mathbf{T}\cdot\operatorname{curl}\operatorname{curl}\mathbf{E} = 0, \quad \mathbf{T} \in \mathcal{T}, \tag{3.3}$$

or, equivalently, as

$$\int_\Omega \mathbf{R}\cdot\mathbf{E} = 0, \quad \mathbf{R} \in \mathbf{B}(\mathcal{T}). \tag{3.4}$$

Donati's theorem states that each symmetric-valued tensor field $\mathbf{E}$ of class $C^N(\Omega)$ ($N \geqslant 2$) that satisfies (3.4) also satisfies the equation of compatibility (1.1).

In the light of the developments in the previous section, I define the compatibility manifold to be the collection $\mathfrak{C}$ of all symmetric-valued tensor fields $\mathbf{E}$ of class $C^N(\Omega)$ ($N \geqslant 2$) that satisfy (3.4), and interpret $\mathcal{R} = \mathbf{B}(\mathcal{T})$ as the collection of all reactive stress fields $\mathbf{S}^R$ associated with the compatibility constraint. Remarkably, such reactive fields depend in general on the shape and size of Ω. Thus, in particular, one should expect a *scale effect* here, just as in fracture mechanics; it would make no sense to talk about compatible or incompatible material elements, but only of *compatible or incompatible material bodies*. Variants of (3.4) in the form of variational inequalities would conceivably fit a theory of materials that do not support tension, such as mansonry-like materials.

4. Integration of the Mixed Problem of Linear Elasticity

Let an equilibrium problem in linear elasticity be posed, and assume that the equilibrium stress $\mathbf{S}$ has the form

$$\mathbf{S} = \mathbf{S}^R + \mathbf{S}^A, \quad \mathbf{S}^R \in \mathcal{R}. \tag{4.1}$$

Then, $\mathbf{S}^A$ must satisfy

$$\operatorname{div} \mathbf{S}^A + \mathbf{b}_o = \mathbf{0} \quad \text{in } \Omega \tag{4.2}$$

and

$$\mathbf{S}^A \mathbf{n} = \mathbf{s}_o \quad \text{on } \partial_1 \Omega, \tag{4.3}$$

where $\mathbf{b}_o$ and $\mathbf{s}_o$ are the given volume and surface loads, and $\partial_1 \Omega \subset \partial \Omega$. A completeness result due to Schaefer and Gurtin [3, Section 17] suggests the following representation for $\mathbf{S}^A$:

$$\mathbf{S}^A(\mathbf{h}) = \nabla \mathbf{h} + \nabla \mathbf{h}^{\mathrm{T}} - (\operatorname{div} \mathbf{h})\mathbf{1}, \tag{4.4}$$

in terms of a vector field $\mathbf{h}$ such as to verify the equations that obtain when (4.4) is substituted into (4.2) and (4.3), namely,

$$-\Delta \mathbf{h} = \mathbf{b}_o \qquad \text{in } \Omega, \tag{4.5}$$

$$2\partial_{\mathbf{n}} \mathbf{h} - \operatorname{curl} \mathbf{h} \times \mathbf{n} - (\operatorname{div} \mathbf{h})\mathbf{n} = \mathbf{s}_o \quad \text{on } \partial_1 \Omega \tag{4.6}$$

(here $\partial_{\mathbf{n}} \mathbf{h} = (\nabla \mathbf{h})\mathbf{n}$).

One is then led to look for a strain field

$$\mathbf{E} = \mathbb{K}[\mathbf{S}^A(\mathbf{h})] + \mathbf{D}, \tag{4.7}$$

with $\mathbb{K}$ the compliance tensor of the material and $\mathbf{D}$ a *discrepancy strain*, such as to satisfy (3.4). The choice of $\mathbf{E}$ would be further restricted by first constructing the displacement field $\mathbf{u}(x)$ at each point $x \in \overline{\Omega}$ by means of *Cesàro's formula*

$$\mathbf{u}(x) = \int_{x_o}^{x} \mathbf{U}(y, x)\, \mathrm{d}(y - x_o), \tag{4.8}$$

where

$$U_{ij}(y, x) = E_{ij}(y) + \big[E_{ij,k}(y) - E_{kj,i}(y)\big](x - y)_k, \tag{4.9}$$

and then imposing the remaining boundary condition

$$\mathbf{u} = \mathbf{u}_o \quad \text{on } \partial_2 \Omega \tag{4.10}$$

(here, as customary, $\partial \Omega_2 \cup \partial \Omega_1 = \partial \Omega$ and $\partial \Omega_2 \cap \partial \Omega_1 = \emptyset$).

Acknowledgements

The support of M.U.R.S.T., Progetto Nazionale "Termomeccanica dei Continui Classici e dei Materiali Nuovi" and, later, of Progetto Cofinanziato 1998 "Modelli Matematici per la Scienza dei Materiali" is gratefully acknowledged.

A preliminary version of this work was presented at the 5th Meeting on "Unilateral Problems in Structural Analysis", Ferrara (Italy), June 12–14, 1997; a fuller version at the Symposium on "Advances in the Continuum Mechanics and Thermodynamics of Material Behavior" in honor of Professor Roger Fosdick, held during the ASME Joint Applied Mechanics and Materials Conference, Blacksburg (VA, USA), June 27–30, 1999. At the end of the latter presentation J.F. Pommaret made the interesting observation that the reactive stress plays the role of a lagrangian multiplier when the quadratic strain energy of linear elasticity is minimized over compatible strains.

References

1. S.S. Antman, *Nonlinear Problems of Elasticity*. Springer, New York (1995).
2. A. DiCarlo, P. Podio-Guidugli and W.O. Williams, Shells with thickness distension. *Internat. J. Solids Struct.* **38**(6) (2001) 7405–7429.
3. M.E. Gurtin, The linear theory of elasticity. In: *Handbuch der Physik*, Bd. VI a/2. Springer, Berlin/Heidelberg/New York (1972).
4. P. Podio-Guidugli, Local and global constraints. In: *The 3rd Meeting on Current Ideas in Mechanics and Related Fields*, Segovia (June 1995).

Journal of Elasticity **59:** 399–436, 2000.

Pressurized Shape Memory Thin Films

RICHARD D. JAMES[1] and RAFFAELLA RIZZONI[2]
[1]*Department of Aerospace Engineering and Mechanics, 107 Akerman Hall, University of Minnesota, Minneapolis, MN 55455, U.S.A. E-mail: james@umn.edu*
[2]*Dipartimento di Ingegneria, Università di Ferrara, Via Saragat 1, 44100 Ferrara, Italy. E-mail: rizzoni@ing.unife.it*

Received 29 December 1999; in revised form 22 February 2000

Dedicated to Roger Fosdick on the occasion of his 60th birthday.

Abstract. We study the behavior of a martensitic thin film with a hydrostatic pressure applied underneath the film. The problem is formulated in 3-D for a single crystal film of thickness h, and a Cosserat membrane theory is derived by Γ-convergence techniques in the limit $h \to 0$. The membrane theory is further simplified using a second Γ-convergence argument based on hard moduli. The resulting theory supports energy minimizing "tunnels": structures having the shape of part of a cylinder cut by a plane parallel to its axis, obtained by releasing the film from the substrate along a strip with a certain orientation. As the temperature is raised (at fixed pressure) the energy minimizing shape collapses gradually to the substrate, accompanied by a martensite-to-austenite phase transformation. During this process the tunnel supports a microstructure consisting of fine bands of austenite parallel to the axis of the tunnel, alternating with bands of a single variant of martensite. Formulas for the associated volume–temperature–pressure relation are given: in these the latent heat of transformation plays an important role.

1. Introduction

In recent years it has become possible to derive rigorously from 3-D nonlinear elasticity special theories for thin structures, without adopting an *ansatz* for the deformation (As an incomplete selection of these derivations, we list Acerbi et al. [1], Bhattacharya and James [11] and Le Dret and Raoult [20, 21]). If the starting theory is general 3-D finite elasticity, then it can be stated that these derivations give the definitive plate–shell–thin film theories. They therefore settle the long-standing question of which, among the many such theories available in the literature, is the appropriate theory for a thin body. There remains a lot to be done: currently, except for one formal argument, the only results are for the membrane theory. Bending theories evidently are obtained from a higher-order Γ-convergence argument.

These developments also allow one to approach with a certain measure of confidence the analysis of thin bodies of nonclassical materials, such as thin films of martensitic materials (cf. [11, 25]), which undergo a diffusionless phase transformation and have free energies with energy wells. Martensitic films can change shape when their temperature is changed, and some materials that undergo a reversible martensitic transformation display the shape memory effect. This effect is: the specimen is deformed by loads at low temperature (causing rearrangements of the different variants of martensite), the loads are removed but the specimen remains deformed (because all the variants have the same free energy density), the specimen is heated, causing a phase transformation from martensite to austenite, and it returns to its original shape. It is then cooled through the transformation with no macroscopic change of shape (because the martensite variants are able to form microstructures with this property), and the cycle can begin again. By alternately heating and loading a specimen, it can be used as an actuator. As discussed by Krulevitch [19], the NiTi shape memory material is among the actuator materials with the largest value of the work output per cycle per volume. Here, "volume" refers to the volume of the actuator. The "per volume" part of this formula is one reason that this material might perform well at small scales. Another reason is that the slow response of bulk shape memory actuators, caused by the necessity to cool them, is greatly improved by the rapid rate of heat transfer possible at small scales, especially in a thin film heated and cooled on its faces.

There are two simple overall design principles for these actuators that are sharply delineated by the Γ-convergence arguments. First, the membrane theory emerges at order h (unlike bending, which emerges at h^3). Therefore, to take advantage of the large work output *per volume* of the shape–memory materials, one should design the actuator to work in membrane mode. Second, the film should be released from the substrate, assuming it is an ordinary elastic material, so that the highly constraining effect of the substrate is eliminated.

There is another potentially attractive feature of the use of martensitic films for microactuators in MEMS (= micro-electro-mechanical systems). The techniques of microelectronics (e.g., molecular beam epitaxy) open up the possibility of making single crystal films. These could avoid the "fighting between the grains" that is associated with transformation in bulk polycrystals, which reduces the effective transformation strain much below its maximum value in single crystals. Oriented single crystal films could be released from the substrate on certain well-defined regions and undergo a large deformation, large work output, relatively high frequency shape memory effect. These ideas have motivated theoretical (Bhattacharya and James [11]) and experimental (Dong et al. [15]) studies. From the former have emerged some structures – "tents" and "tunnels" – that are energy minimizing under zero stress and exhibit large deformations.

As explained above, to produce the shape memory effect, loads have to be applied. Ideally, these are the actual loads that the actuator must do work against, but, in any case, the so-called bias stresses must be applied to produce the change of shape in the martensitic state in the first place. The idea explored in this paper is that the loads are produced by a pressure on the film. This could be done in the following way. The film could be released from the substrate on a well-defined region by back-etching the substrate (the judicious use of an etch-stop might be relevant here). Then a pressure could be applied from either above (advantageous, to prevent peeling of the film) or below. This scheme is suited to potential applications like pumps and valves in which the pressure is produced by the working fluid. The key questions are: what structure? what volume–temperature–pressure relation is predicted?

The membrane theory of a stiff, single phase material is the theory of isometric mappings of a plane. The deformed shapes are developable surfaces. A spherical bubble is not one, but a piece of a cylinder is. Moreover, according to the isoperimetric inequality, a circular cylinder holds the greatest volume for its area, among prismatic shapes. Thus, it is natural to consider a film released on a strip, so that under pressure, it bulges into a cylindrical shape. By raising the temperature, the film is made to undergo the shape memory effect and the film collapses to a flat shape. This is the situation studied in this paper. We speculate that a complex system of such tunnels could be patterned onto a chip and, by selective heating, bubbles of different fluids could be pumped around, mixed, reacted, etc.

The plan of this paper is as follows. We formulate the 3-D problem of a thin film of martensitic material of certain shape, acted upon by a pressure underneath, in Section 2. We include bulk and interfacial energies. The presence of the pressure necessitates that we impose slightly stronger growth conditions than is usual to prevent the film from blowing up. In Section 3 we rescale the pressure $p^h = Ph$, $P = \text{const}$, and do the Γ-convergence argument, which is a modification of that of Bhattacharya and James [11]. The energy that emerges is a nonlinear membrane theory with an additive contribution pV where V is the volume enclosed between the membrane and a suitable plane (Section 3). To further simplify this energy, we do a second Γ-convergence argument in Section 4 based on the presence of hard moduli, a situation that is expected to be relevant for martensitic materials with large transformation strain. For the resulting *constrained theory* microstructures are replaced by Young measures having support on the energy wells, and the macroscopic deformation gradient is recovered as the center of mass of the Young measure. This theory is very easy to use and it supports tunnels with circular cross-section. They are proved to be energy minimizing among cylindrical deformations (Section 5) for suitable materials, and probably they are energy minimizing in general under their own boundary conditions. The dependence of the volume enclosed by the tunnel on pressure and temperature is found in Section 5 (Figures 5 and 6). An unexpected behavior is observed: instead of the tunnel collapsing suddenly to the flat shape as the temperature is raised, the collapse is more gradual (Figure 7), and complete collapse within the constrained theory is only possible with infinitely large temperature. During collapse, the tunnel exhibits a microstructure of axial bands of austenite and one variant of martensite (Figure 8). These results need to be explored for special materials, which we postpone to later work.

This research has been conducted in parallel with a related computational study [9].

2. The 3-dimensional problem

We assume that an origin $\mathbf{O}$ and an orthonormal basis, $\{\mathbf{e}_1, \mathbf{e}_2, \mathbf{e}_3\}$, have been chosen in the 3-dimensional Euclidean space. We identify the point $\mathbf{X}$ with the vector $\mathbf{X} - \mathbf{O}$, whose components will be indicated by X_1, X_2, X_3. In the reference configuration, the film is assumed to occupy the cylindrical region

$$\Omega^h := \left\{(X_1, X_2, X_3) \in \mathbb{R}^3: (X_1, X_2) \in S,\ X_3 \in (0, h)\right\}, \tag{2.1}$$

where S is an open bounded subset of $\mathbb{R}^2$ with Lipschitz boundary, and h is the thickness of the film. A deformation of the film is a mapping $\mathbf{y}: \Omega^h \mapsto \mathbb{R}^3$. Recalling that the film is attached to a substrate except for the released region S, we model the presence of the substrate by imposing the linear boundary condition

$$\mathbf{y} = \mathbf{y}_A(\mathbf{X}) := \mathbf{A}\mathbf{X}, \quad \mathbf{X} \in \partial S \times (0, h), \tag{2.2}$$

where $\mathbf{A}$ is a constant 3×3 matrix. The material of the film is supposed to be homogeneous and thermoelastic. At the temperature $\theta \in (0, +\infty)$, the strain energy of the film corresponding to the deformation $\mathbf{y}$ is given by

$$\int_{\Omega^h} \phi(\nabla\mathbf{y}; \theta)\, \mathrm{d}\mathbf{X}, \tag{2.3}$$

where $\phi: M^{3\times3} \times (0, +\infty) \mapsto [0, +\infty)$ is the strain energy density, and $M^{3\times3}$ is the set of all 3×3 matrices. The function ϕ is assumed to be smooth in both its arguments and to satisfy the following growth hypothesis: there exist positive constants c_1, c_2, c_3, c_4, and $3 < q < 6$ such that

$$c_1\, |\mathbf{F}|^3 - c_2 \leqslant \phi(\mathbf{F}; \theta) \leqslant c_3\, |\mathbf{F}|^q - c_4, \tag{2.4}$$

for all matrices $\mathbf{F}$ in $M^{3\times3}$, and for θ in $(0, +\infty)$. Furthermore, ϕ is assumed to be frame indifferent, i. e., it is assumed to satisfy the condition

$$\phi(\mathbf{QF}; \theta) = \phi(\mathbf{F}; \theta), \tag{2.5}$$

for all $\mathbf{F} \in M^{3\times3}$ and $\theta \in [0, +\infty)$, and for all proper rotations

$$\mathbf{Q} \in SO(3) := \left\{\mathbf{Q} \in M^{3\times3} | \mathbf{QQ}^T = \mathbf{I}, \det\mathbf{Q} = 1\right\}. \tag{2.6}$$

Stable deformations of the film, subject to a hydrostatic pressure p^h acting on its lower surface, are assumed to correspond to the deformations $\mathbf{y} \in W^{2,2}(\Omega^h; \mathbb{R}^3)$ which minimize the total energy

$$\begin{aligned} E^h(\mathbf{y}; \theta) \ := \ & \int_{\Omega^h} \left(\phi(\nabla\mathbf{y}; \theta) + \kappa|\nabla^2\mathbf{y}|^2\right) \mathrm{d}\mathbf{X} \\ & - \frac{p^h}{3} \int_{S\times\{0\}} \mathbf{y} \cdot (\mathbf{y}_{,1} \wedge \mathbf{y}_{,2})\, \mathrm{d}X_1\, \mathrm{d}X_2, \end{aligned} \tag{2.7}$$

in the class of deformations

$$\mathcal{H} := \{\mathbf{y} \in W^{2,2}(\Omega^h; \mathbb{R}^3) | \mathbf{y} = \mathbf{A}\mathbf{X},\ \mathbf{X} \in \partial S \times (0, h)\}. \tag{2.8}$$

In the expression (2.7) of the total energy, the notation $\mathbf{y}_{,i}$ denotes the vector $\nabla\mathbf{y}\mathbf{e}_i$, for $i = 1, 2$. The term $|\nabla^2\mathbf{y}|^2$, which denotes the full $3 \times 3 \times 3$ matrix of second derivatives, not just the Laplacean, penalizes the formation of interfaces, κ being a small positive constant; the norm here is $\sqrt{\mathbf{y}_{,ij} \cdot \mathbf{y}_{,ij}}$. As discussed below, the term

$$V(\mathbf{y}) := \frac{1}{3}\int_{S\times\{0\}} \mathbf{y} \cdot (\mathbf{y}_{,1} \wedge \mathbf{y}_{,2})\, \mathrm{d}X_1\, \mathrm{d}X_2 \tag{2.9}$$

represents the volume enclosed between the plane $\mathbf{y}_A(S \times \{0\})$ and the deformed lower surface of the film. Physically, the last term of (2.7) may be interpreted as the energy of a gas under the film whose pressure can be assumed constant. This situation is realized in practice by micromachining a hole in the substrate and by pressurizing the film using a reservoir of gas whose volume is much larger than typical changes of volume due to film deformation. We note that if the volume of this reservoir is on the order of volume changes produced by the deformations of the film, then the last term of (2.7) would have to be replaced by the general expression for the free energy of such a gas, accounting for its compressibility. In the latter case our predictions concerning stability could be substantially changed [18].

The volume functional (2.9) evaluated at a deformation $\mathbf{y}$ of the film gives a reasonable expression for the volume of the region $\mathcal{V}_\mathbf{y}$ enclosed between the plane $\mathbf{y}_A(S \times \{0\})$ and the deformed lower surface of the film $\mathbf{y}(S \times \{0\})$. This can be seen through the following heuristic calculation. Let $|\mathcal{V}_\mathbf{y}|$ denote the three-dimensional Lebesgue measure of $\mathcal{V}_\mathbf{y}$, and let $\mathbf{Z} = (Z_1, Z_2, Z_3)$ be a point belonging to $\mathcal{V}_\mathbf{y}$. We have

$$|\mathcal{V}_\mathbf{y}| = \int_{\mathcal{V}_\mathbf{y}} \mathrm{d}\mathbf{Z} = \frac{1}{3}\int_{\mathcal{V}_\mathbf{y}} \mathrm{div}\mathbf{Z}\, \mathrm{d}\mathbf{Z}. \tag{2.10}$$

Using the divergence theorem, we get

$$|\mathcal{V}_\mathbf{y}| = \frac{1}{3}\int_{\mathbf{y}(S\times\{0\})} \mathbf{Z} \cdot \mathbf{n}(\mathbf{Z})\, \mathrm{d}a + \frac{1}{3}\int_{\mathbf{y}_A(S\times\{0\})} \mathbf{Z} \cdot \mathbf{n}(\mathbf{Z})\, \mathrm{d}a, \tag{2.11}$$

where, in each integral, $\mathbf{n}(\mathbf{Z})$ is the unit outward normal at the point $\mathbf{Z}$ to the surface on which the integral is defined. Since $\mathbf{Z} \cdot \mathbf{n}(\mathbf{Z}) = 0$ at each point $\mathbf{Z}$ of $\mathbf{y}_A(S \times \{0\})$, the second integral is zero. Taking $\mathbf{y}^o$, the trace of $\mathbf{y}$ on the surface $S \times \{0\}$, as a parametric representation of the surface $\mathbf{y}(S \times \{0\})$, equation (2.11) may be rewritten as

$$\begin{aligned} |\mathcal{V}_\mathbf{y}| &= \frac{1}{3}\int_S \mathbf{y}^o \cdot (\mathbf{y}^o{}_{,1} \wedge \mathbf{y}^o{}_{,2})\, \mathrm{d}X_1\, \mathrm{d}X_2 \\ &= \frac{1}{3}\int_{S\times\{0\}} \mathbf{y} \cdot (\mathbf{y}_{,1} \wedge \mathbf{y}_{,2})\, \mathrm{d}X_1\, \mathrm{d}X_2 = V(\mathbf{y}). \end{aligned} \tag{2.12}$$

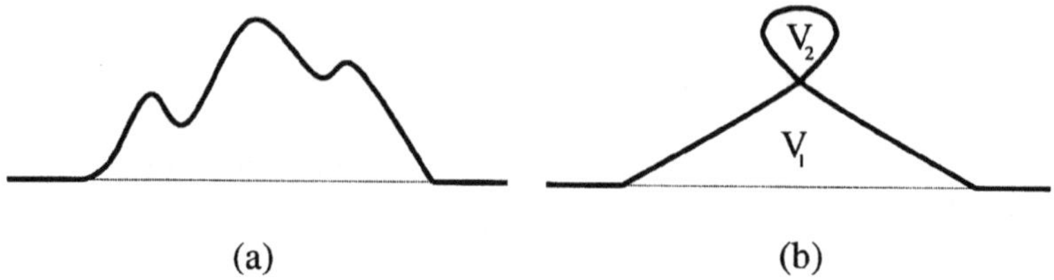

Figure 1. Globally invertible (a) and not globally invertible (b) deformation of the film.

We note that the expression (2.9) for the volume under the film has some pathologies related to global invertibility, illustrated in Figure 1 for the case $\mathbf{A} = \mathbf{I}$. In Figure 1(a), the volume under the film is expected to be reasonably modeled by (2.9). However, in Figure 1(b) the formula (2.9) gives the volume $V_1 - V_2$, which could not possibly be occupied by a fluid. Without Dirichlet boundary conditions on $\mathbf{y}$, we do not know how to impose global invertibility assumptions on the minimization of E^h, so we are not able to rule out situations as shown in Figure 1(b). However, we do not expect such configurations to arise from energy minimization.

The volume functional (2.9) is defined for the deformations $\mathbf{y}$ of $\mathcal{H}$. This can be seen as follows. By Sobolev's embedding theorem [2, Theorem 5.4], each element of $W^{2,2}(\Omega^h; \mathbb{R}^3)$ has trace on $S \times \{0\}$ belonging to $C^0(\overline{S}; \mathbb{R}^3) \cap W^{1,2}(S; \mathbb{R}^3)$, with $\overline{S}$ denoting the closure of S. By the continuity of the trace, there exists a positive constant C such that

$$|V(\mathbf{y})| \leqslant C \int_S \left|\mathbf{y}^{\circ}{}_{,1} \wedge \mathbf{y}^{\circ}{}_{,2}\right| \mathrm{d}X_1 \, \mathrm{d}X_2. \tag{2.13}$$

The functional

$$L(\mathbf{y}^{\circ}) := \int_S \left|\mathbf{y}^{\circ}{}_{,1} \wedge \mathbf{y}^{\circ}{}_{,2}\right| \mathrm{d}X_1 \, \mathrm{d}X_2 \tag{2.14}$$

gives the Lebesgue area of the surface parametrized by $\mathbf{y}^{\circ}$ [12]. It is known that

$$\int_S \left|\mathbf{y}^{\circ}{}_{,1} \wedge \mathbf{y}^{\circ}{}_{,2}\right| \mathrm{d}X_1 \, \mathrm{d}X_2 \leqslant \frac{1}{2} \int_S \left(|\mathbf{y}^{\circ}{}_{,1}|^2 + |\mathbf{y}^{\circ}{}_{,2}|^2\right) \mathrm{d}X_1 \, \mathrm{d}X_2, \tag{2.15}$$

and the inequality holds only when $\mathbf{y}^{\circ}$ satisfies the conditions $|\mathbf{y}^{\circ}{}_{,1}| = |\mathbf{y}^{\circ}{}_{,2}|$, $(\mathbf{y}^{\circ}{}_{,1} \cdot \mathbf{y}^{\circ}{}_{,2}) = 0$ [16]. From (2.13), (2.15), and the regularity of the trace, it follows that $V(\mathbf{y})$ is defined in $\mathcal{H}$. Moreover, the volume functional (2.9) is weakly continuous, up to a subsequence, in $W^{2,2}(\Omega^h; \mathbb{R}^3)$. Indeed, consider a sequence $\mathbf{y}_n$ weakly converging to $\mathbf{y}$ in $W^{2,2}(\Omega^h; \mathbb{R}^3)$, by Sobolev's Theorem there exists a subsequence, not relabeled, such that

$$\mathbf{y}^{\circ}{}_n \rightharpoonup \mathbf{y}^{\circ} \quad \text{in } W^{1,q}(S; \mathbb{R}^3), \tag{2.16}$$

for $3 < q \leqslant 4$. This together with the weak continuity of minors [3], implies

$$(\mathbf{y}^{\circ}{}_{n,1} \wedge \mathbf{y}^{\circ}{}_{n,2}) \cdot \mathbf{e}_j \rightharpoonup (\mathbf{y}^{\circ}{}_{,1} \wedge \mathbf{y}^{\circ}{}_{,2}) \cdot \mathbf{e}_j \quad \text{in } L^{q/2}(S; \mathbb{R}), \tag{2.17}$$

for $j = 1, 2, 3$. Furthermore, by the Rellich–Kondrachov Theorem [2, Theorem 6.2], there exists a further subsequence, not relabeled, such that

$$\mathbf{y}^{\mathrm{o}}{}_{n} \rightarrow \mathbf{y}^{\mathrm{o}} \quad \text{in } C^0(\overline{S}; \mathbb{R}^3), \tag{2.18}$$

which, together with (2.17), implies $V(\mathbf{y}_n) \rightarrow V(\mathbf{y})$.

The fundamental estimate for dealing with the volume functional is the following isoperimetric inequality, whose proof can be found in [24, 26]. With an abuse of notation, with the same symbol V we shall denote the volume functional defined for deformations of $\mathcal{H}$, as in (2.9), and the volume functional defined for parametric surfaces in $C^0(\overline{S}; \mathbb{R}^3) \cap W^{1,2}(S; \mathbb{R}^3)$.

PROPOSITION 2.1. *Let* $\mathbf{f}, \mathbf{g} \in W^{1,2}(S; \mathbb{R}^3) \cap C^0(\overline{S}; \mathbb{R}^3)$ *be two parametrizations of two 3-dimensional surfaces such that* $\mathbf{f} = \mathbf{g}$ *on* ∂S. *Then,*

$$\left|V(\mathbf{f}) - V(\mathbf{g})\right|^2 \leqslant \frac{1}{36\pi}\left|L(\mathbf{f}) + L(\mathbf{g})\right|^3. \tag{2.19}$$

For the proof of the existence of minimizers of the total energy E^h, we shall need the following lemmas, which are applications of the isoperimetric inequality.

LEMMA 2.2. *Let* $t > 0$ *and let* $\mathbf{f} \in W^{1,3}(S; \mathbb{R}^3) \cap C^0(\overline{S}; \mathbb{R}^3)$ *be such that*

$$\mathbf{f} = \mathbf{A}\mathbf{e}_1 X_1 + \mathbf{A}\mathbf{e}_2 X_2 + \mathbf{A}\mathbf{e}_3 t, \quad (X_1, X_2) \in \partial S, \tag{2.20}$$

with $\mathbf{A}$ *a given* 3×3 *matrix. Then, denoting by* $|S|$ *the area of* S, *we have*

$$|V(\mathbf{f})| \leqslant \frac{|S|^{1/2}}{12\sqrt{\pi}} \|\nabla \mathbf{f}\|^3_{L^3(S)} + c, \tag{2.21}$$

with

$$c = \frac{\sqrt{2}}{6\sqrt{\pi}} |S|^{3/2} |\mathbf{A}\mathbf{e}_1 \wedge \mathbf{A}\mathbf{e}_2|^{3/2} + \frac{t}{3}|S||\det \mathbf{A}|.$$

Proof. By the triangle inequality, the isoperimetric inequality (2.19), Jensen's inequality [14, Theorem 2.2] and the inequality (2.15), we obtain

$$\begin{aligned}
\left|V(\mathbf{f})\right| &\leqslant \left|V(\mathbf{f}) - V(\mathbf{g})\right| + \left|V(\mathbf{g})\right| \leqslant \frac{1}{6\sqrt{\pi}}\left|L(\mathbf{f}) + L(\mathbf{g})\right|^{3/2} + \left|V(\mathbf{g})\right| \\
&\leqslant \frac{\sqrt{2}}{6\sqrt{\pi}}\left(L(\mathbf{f})^{3/2} + L(\mathbf{g})^{3/2}\right) + \left|V(\mathbf{g})\right| \\
&\leqslant \frac{1}{12\sqrt{\pi}} \|\nabla \mathbf{f}\|^3_{L^2(S)} + \frac{\sqrt{2}}{6\sqrt{\pi}} L(\mathbf{g})^{3/2} + \left|V(\mathbf{g})\right|.
\end{aligned} \tag{2.22}$$

Hence, in view of the imbedding of L^3 into L^2 [2, Theorem 2.8], we have

$$|V(\mathbf{f})| \leqslant \frac{|S|^{1/2}}{12\sqrt{\pi}} \|\nabla \mathbf{f}\|^3_{L^3(S)} + \frac{\sqrt{2}}{6\sqrt{\pi}} L(\mathbf{g})^{3/2} + \left|V(\mathbf{g})\right|. \tag{2.23}$$

We now choose

$$\mathbf{g} = \mathbf{A}\mathbf{e}_1 X_1 + \mathbf{A}\mathbf{e}_2 X_2 + \mathbf{A}\mathbf{e}_3 t. \tag{2.24}$$

Simple calculations give (2.21). □

We now extend the bound given in Lemma 2.2 to the interior of the film.

LEMMA 2.3. *For any* $\mathbf{y} \in \mathcal{H}$ *we have*

$$\begin{aligned}\left|V(\mathbf{y})\right| \leqslant & \left(\frac{|S|^{1/2}}{12\sqrt{\pi} h} + \frac{1}{3\sqrt{3}}\right)\|\nabla \mathbf{y}\|^3_{L^3(\Omega^h)} + \frac{\sqrt{2}}{6\sqrt{\pi}}|S|^{3/2}|\mathbf{A}\mathbf{e}_1 \wedge \mathbf{A}\mathbf{e}_2|^{3/2} \\ & + \frac{h}{2}|S||\det\mathbf{A}|. \end{aligned} \tag{2.25}$$

Proof. Fix t in $(0, h)$ and let $\mathbf{y}^t$ be the trace of $\mathbf{y}$ on the surface $S \times \{t\}$. Then $\mathbf{y}^t$ satisfies the hypothesis of Lemma 2.2 and thus it satisfies the inequality (2.21). In Appendix A, we prove that

$$V(\mathbf{y}) = V\left(\mathbf{y}^t\right) + \frac{2}{3} t |S| \det\mathbf{A} - \int_{S\times(0,t)} \det\nabla\mathbf{y}\, \mathrm{d}\mathbf{X}, \tag{2.26}$$

which, in view of (2.21), gives

$$\begin{aligned}\left|V(\mathbf{y})\right| \leqslant & \frac{|S|^{1/2}}{12\sqrt{\pi}}\|\nabla \mathbf{y}^t\|^3_{L^3(S)} + \frac{\sqrt{2}}{6\sqrt{\pi}}|S|^{3/2}|\mathbf{A}\mathbf{e}_1 \wedge \mathbf{A}\mathbf{e}_2|^{3/2} \\ & + t|S||\det\mathbf{A}| + \int_{S\times(0,t)} |\det\nabla\mathbf{y}|\, \mathrm{d}\mathbf{X}. \end{aligned} \tag{2.27}$$

Adding $|\mathbf{y}_{,3}(X_1, X_2, t)|^2$ to the argument of the square root in the first term of the right-hand side, and using the positivity of the integrand of the last term, we get

$$\begin{aligned}\left|V(\mathbf{y})\right| \leqslant & \frac{|S|^{1/2}}{12\sqrt{\pi}} \int_{S\times\{t\}} \left(\sum_{i=1}^{3} \left|\mathbf{y}_{,i}(X_1, X_2, t)\right|^2\right)^{3/2} \mathrm{d}X_1\, \mathrm{d}X_2 \\ & + \frac{\sqrt{2}}{6\sqrt{\pi}}|S|^{3/2}|\mathbf{A}\mathbf{e}_1 \wedge \mathbf{A}\mathbf{e}_2|^{3/2} + t|S||\det\mathbf{A}| \\ & + \int_{\Omega^h} |\det\nabla\mathbf{y}|\, \mathrm{d}\mathbf{X}. \end{aligned} \tag{2.28}$$

Integrating over $t \in (0, h)$ and applying Fubini's theorem, we find

$$\begin{aligned}h\left|V(\mathbf{y})\right| \leqslant & \frac{|S|^{1/2}}{12\sqrt{\pi}}\|\nabla \mathbf{y}\|^3_{L^3(\Omega^h)} + \frac{\sqrt{2}}{6\sqrt{\pi}} h|S|^{3/2}|\mathbf{A}\mathbf{e}_1 \wedge \mathbf{A}\mathbf{e}_2|^{3/2} \\ & + \frac{h^2}{2}|S||\det\mathbf{A}| + h\int_{\Omega^h} |\det\nabla\mathbf{y}|\, \mathrm{d}\mathbf{X}. \end{aligned} \tag{2.29}$$

We now divide by h and use the inequality

$$|\det\nabla\mathbf{y}| \leqslant \frac{1}{3\sqrt{3}}|\nabla\mathbf{y}|^3, \tag{2.30}$$

[23, Section 2.3], to obtain the result. □

The following proposition establishes, for suitable values of the pressure and for a fixed value of the thickness h, the existence of minimizers of the total energy functional in the class of deformations $\mathcal{H}$. In the argument, the temperature θ shall be held fixed, so we suppress it from the notation.

PROPOSITION 2.4. *Assume that* $\inf_{\mathbf{y}\in\mathcal{H}} E^h(\mathbf{y}) < \infty$. *If*

$$p^h \leqslant c_1 \Big/ \frac{|S|^{1/2}}{12\sqrt{\pi}h} + \frac{1}{3\sqrt{3}} =: p^h_{cr}, \tag{2.31}$$

there exists at least a minimizer of E^h *in* $\mathcal{H}$.

Proof. From the growth assumptions (2.4) and from the bound on the volume (2.25), we obtain

$$\begin{aligned} E^h(\mathbf{y}) \geqslant\ & \kappa\|\nabla^2\mathbf{y}\|^2_{L^2(\Omega^h)} + \left(c_1 - p^h\left(\frac{|S|^{1/2}}{12\sqrt{\pi}h} + \frac{1}{3\sqrt{3}}\right)\right)\|\nabla\mathbf{y}\|^3_{L^3(\Omega^h)} \\ & - c_2|\Omega^h| - \frac{\sqrt{2}}{6\sqrt{\pi}}|S|^{3/2}|\mathbf{Ae}_1 \wedge \mathbf{Ae}_2|^{3/2} - \frac{h}{2}|S||\det\mathbf{A}|. \end{aligned} \tag{2.32}$$

In view of the assumption (2.31), the second term on the right-hand side is non-negative and thus we can write

$$\begin{aligned} E^h(\mathbf{y}) \geqslant\ & \kappa\|\nabla^2\mathbf{y}\|^2_{L^2(\Omega^h)} - c_2|\Omega^h| - \frac{\sqrt{2}}{6\sqrt{\pi}}|S|^{3/2}|\mathbf{Ae}_1 \wedge \mathbf{Ae}_2|^{3/2} \\ & - \frac{h}{2}|S||\det\mathbf{A}|. \end{aligned} \tag{2.33}$$

According to the generalized Poincaré inequality, [22, Theorem 3.6.4], there exist two positive constants C_1 and C_2 such that

$$\|\mathbf{y}\|^p_{L^p(\Omega^h)} + \|\nabla\mathbf{y}\|^p_{L^p(\Omega^h)} \leqslant C_1\|\nabla^2\mathbf{y}\|^p_{L^p(\Omega^h)} + C_2 \tag{2.34}$$

for each p in $[1, \infty)$ and for each $\mathbf{y}$ in $\mathcal{H}$. Combining (2.33) and (2.34) written for $p = 2$, we obtain

$$\begin{aligned} E^h(\mathbf{y}) \geqslant\ & \min\left\{\frac{\kappa}{2}; \frac{\kappa}{2C_1}\right\}\|\mathbf{y}\|^2_{W^{2,2}(\Omega^h)} - C_2 - c_2|\Omega^h| \\ & - \frac{\sqrt{2}}{6\sqrt{\pi}}|S|^{3/2}|\mathbf{Ae}_1 \wedge \mathbf{Ae}_2|^{3/2} - \frac{h}{2}|S||\det\mathbf{A}|. \end{aligned} \tag{2.35}$$

Let now $\mathbf{y}_n$ be a minimizing sequence for E^h. Because the infimum of E^h is bounded by hypothesis, from the bound (2.35) it follows that the sequence $\mathbf{y}_n$ is uniformly bounded in $W^{2,2}(\Omega^h; \mathbb{R}^3)$. We can then extract a subsequence, not relabeled, such that

$$\mathbf{y}_n \rightharpoonup \bar{\mathbf{y}} \quad \text{in } W^{2,2}(\Omega^h; \mathbb{R}^3). \tag{2.36}$$

The limit $\bar{\mathbf{y}}$ belongs to $\mathcal{H}$. Indeed, by the compact embedding of $W^{2,2}(\Omega^h; \mathbb{R}^3)$ into $C^0(\bar{\Omega}^h; \mathbb{R}^3)$, [2, Theorem 6.2], we can extract a further subsequence uniformly converging to $\bar{\mathbf{y}}$ in the closure of Ω^h, $\bar{\Omega}^h$. Therefore, $\bar{\mathbf{y}}$ satisfies the boundary condition (2.2). Using the convexity of the second gradient term and the continuity of ϕ, together with the bound (2.4), we also have

$$\liminf_{n\to\infty} \int_{\Omega^h} \left(\phi(\nabla \mathbf{y}_n) + \kappa\left|\nabla^2 \mathbf{y}_n\right|^2\right) \mathrm{d}\mathbf{X} \geqslant \int_{\Omega^h} \left(\phi(\nabla \bar{\mathbf{y}}) + \kappa\left|\nabla^2 \bar{\mathbf{y}}\right|^2\right) \mathrm{d}\mathbf{X}, \tag{2.37}$$

[14, Theorem 3.4]. From (2.37) and from the weak continuity of the volume functional in $W^{2,2}(\Omega^h; \mathbb{R}^3)$, it follows that the energy functional E^h is weakly lower semicontinuous in $\mathcal{H}$. Therefore, $\bar{\mathbf{y}}$ is a minimizer. □

If $p^h > p^h_{cr}$ and the lower bound of the strain energy density (2.4) holds with equality, then the infimum of the energy in $W^{2,2}(\Omega^h; \mathbb{R}^3)$ is expected to be $-\infty$, corresponding to rupture of the film. We also note that, since the constant which multiplies $\|\nabla \mathbf{y}\|_{L^3}$ in the bound on the volume (2.25) is not optimal, p^h_{cr} provides only a lower bound for the pressure at rupture. The presence of the thickness h and of the area $|S|$ of the film in the expression of p^h_{cr} introduces a scale effect. Indeed, if we keep $|S|$ fixed and let h increase, i.e., if we consider a thicker and thicker film, then, as the physical intuition suggests, p^h_{cr} increases and thus the film can sustain larger and larger value of the pressure. On the other hand, if we keep h fixed and increase $|S|$, i.e., if we consider a film of fixed thickness which becomes wider and wider, then p^h_{cr} decreases and thus the film sustains smaller and smaller values of the pressure.

Another interesting point is that, to get the lower bound on the energy (2.35) and thus to guarantee the existence of a continuous minimizer, not only the condition on the pressure is needed, but it also necessary that the exponent which appears in the growth condition (2.4) be not less than three. This issue arises from the competition between the loading potential, which, as it can be seen from (2.25), is dominated by the cube of $\|\nabla \mathbf{y}\|_{L^3(\Omega^h)}$, and the strain energy. In this respect, an analogy can be established with the phenomenon of cavitation [4].

3. Thin Film Theory

In the previous section, for each value of thickness h, we proved the existence of at least one minimizer, say $\mathbf{y}^h$, of the total energy functional E^h in the class of admissible deformations $\mathcal{H}$. We now consider the behavior of the minimizers $\mathbf{y}^h$

as $h \to 0^+$. The approach which we follow was introduced by Bhattacharya and James in [11] for a film whose total energy is given by (2.7) with $p^h = 0$. In the following, we adopt the notation ∇_p for the gradient in the plane of the film:

$$\nabla_p \mathbf{y}^h := \mathbf{y}^h_{,1} \otimes \mathbf{e}_1 + \mathbf{y}^h_{,2} \otimes \mathbf{e}_2, \tag{3.1}$$

and the notation $\mathbf{y}^h_{,1} \,|\mathbf{y}^h_{,2}|\, \mathbf{y}^h_{,3}$ for $\mathbf{y}^h_{,1} \otimes \mathbf{e}_1 + \mathbf{y}^h_{,2} \otimes \mathbf{e}_2 + \mathbf{y}^h_{,3} \otimes \mathbf{e}_3$. We begin by considering an equivalent minimization problem set on a cylindrical domain of fixed height. This can be obtained through the change of variables

$$Z_1 = X_1, \quad Z_2 = X_2, \quad Z_3 = \frac{X_3}{h}, \tag{3.2}$$

which maps the reference configuration of the film, Ω^h, into the configuration

$$\Omega_1 := \{(Z_1, Z_2, Z_3) \in \mathbb{R}^3 \colon (Z_1, Z_2) \in S,\ Z_3 \in (0,1)\}. \tag{3.3}$$

Then, if $\mathbf{y} \in \mathcal{H}$ is an admissible deformation of the film, the rescaled deformation

$$\mathbf{y} = \mathbf{y}(\mathbf{X}(\mathbf{Z})) =: \tilde{\mathbf{y}}(\mathbf{Z}), \tag{3.4}$$

belongs to the set

$$\mathcal{H}_1 := \{\tilde{\mathbf{y}} \in W^{2,2}(\Omega_1; \mathbb{R}^3) | \tilde{\mathbf{y}} = (\mathbf{A}\mathbf{e}_1|\mathbf{A}\mathbf{e}_1|h\mathbf{A}\mathbf{e}_3)\mathbf{Z},\ \mathbf{Z} \in \partial S \times (0,1)\}. \tag{3.5}$$

We accordingly rescale the total energy by setting

$$\begin{aligned} E_1^h(\tilde{\mathbf{y}}) := \frac{1}{h} E^h(\tilde{\mathbf{y}}) &= \int_{\Omega_1} \kappa\Big(|\nabla_p^2 \tilde{\mathbf{y}}|^2 + \frac{2}{h^2}|\nabla_p \tilde{\mathbf{y}}_{,3}|^2 + \frac{1}{h^4}|\tilde{\mathbf{y}}_{,33}|^2\Big)\, \mathrm{d}\mathbf{Z} \\ &+ \int_{\Omega_1} \phi\Big(\tilde{\mathbf{y}}_{,1}|\tilde{\mathbf{y}}_{,2}|\frac{1}{h}\tilde{\mathbf{y}}_{,3}\Big)\, \mathrm{d}\mathbf{Z} - \frac{p^h}{3h}\int_{S\times\{0\}} \tilde{\mathbf{y}} \cdot (\tilde{\mathbf{y}}_{,1} \wedge \tilde{\mathbf{y}}_{,2})\, \mathrm{d}Z_1\, \mathrm{d}Z_2. \end{aligned} \tag{3.6}$$

The existence of minimizers of E_1^h in $\mathcal{H}_1$ follows from the existence of minimizers of E^h in $\mathcal{H}$ (Proposition 2.4), and from the fact that if $\mathbf{y}^h$ is a minimizer of E^h in $\mathcal{H}$, then $\tilde{\mathbf{y}}^h$ is a minimizer of E_1^h in $\mathcal{H}_1$. Note also that because

$$\|\nabla \mathbf{y}\|^3_{L^3(\Omega^h)} = h \Big\|(\tilde{\mathbf{y}}_{,1}|\tilde{\mathbf{y}}_{,2}|\frac{1}{h}\tilde{\mathbf{y}}_{,3})\Big\|^3_{L^3(\Omega_1)}, \tag{3.7}$$

then, by (2.25) any $\tilde{\mathbf{y}} \in \mathcal{H}_1$ satisfies the inequality

$$|V(\tilde{\mathbf{y}})| \leqslant \Big(\frac{|S|^{1/2}}{12\sqrt{\pi}} + \frac{h}{3\sqrt{3}}\Big) \Big\|(\tilde{\mathbf{y}}_{,1}|\tilde{\mathbf{y}}_{,2}|\frac{1}{h}\tilde{\mathbf{y}}_{,3})\Big\|^3_{L^3(\Omega_1^h)} + \widetilde{D}, \tag{3.8}$$

with

$$\widetilde{D} := \frac{\sqrt{2}}{6\sqrt{\pi}} |S|^{3/2} |\mathbf{A}\mathbf{e}_1 \wedge \mathbf{A}\mathbf{e}_2|^{3/2} + \frac{h}{2}|S||\det \mathbf{A}|.$$

For the asymptotic approach, it is also essential that the magnitude of the applied pressure be scaled appropriately in the thickness. From (3.6) (see also (2.31)), it turns out that the right order of magnitude is $p^h = Ph$, where P is a constant independent of h.

THEOREM 3.1. *Assume that*

$$P < \frac{12\sqrt{\pi}c_1}{|S|^{1/2}}. \tag{3.9}$$

Then, the family of minimizers $\tilde{\mathbf{y}}^h \in \mathcal{H}_1$ has a subsequence, not relabeled, such that

$$\left.\begin{aligned} &\nabla_p^2\tilde{\mathbf{y}}^h \to \nabla^2\hat{\mathbf{y}},\\ &\frac{1}{h}\nabla_p\tilde{\mathbf{y}}^h_{,3} \to \nabla\hat{\mathbf{b}},\\ &\frac{1}{h^2}\tilde{\mathbf{y}}^h_{,33} \to 0 \end{aligned}\right\} \quad \text{in } L^2, \tag{3.10}$$

where $\hat{\mathbf{y}} \in W^{2,2}(S;\mathbb{R}^3)$ and $\hat{\mathbf{b}} \in W^{1,2}(S;\mathbb{R}^3)$ are vector fields independent of Z_3. The couple $(\hat{\mathbf{y}}, \hat{\mathbf{b}})$ minimizes the limit energy

$$\begin{aligned} E(\mathbf{y},\mathbf{b}) := \int_S \Big\{&\kappa\big\{|\nabla^2\mathbf{y}|^2 + 2|\nabla\mathbf{b}|^2\big\} + \phi(\mathbf{y}_{,1}|\mathbf{y}_{,2}|\mathbf{b})\\ &- \frac{P}{3}\mathbf{y}\cdot(\mathbf{y}_{,1}\wedge\mathbf{y}_{,2})\Big\}\mathrm{d}Z_1\,\mathrm{d}Z_2, \end{aligned} \tag{3.11}$$

among all couples $(\mathbf{y},\mathbf{b}) \in W^{2,2}(S;\mathbb{R}^3) \times W^{1,2}(S;\mathbb{R}^3)$ satisfying the boundary conditions

$$\left.\begin{aligned} &\mathbf{y} = \mathbf{A}\mathbf{e}_1 Z_1 + \mathbf{A}\mathbf{e}_2 Z_2,\\ &\mathbf{b} = \mathbf{A}\mathbf{e}_3 \end{aligned}\right\} \quad (Z_1, Z_2) \in \partial S. \tag{3.12}$$

Proof. The lower bound in (2.4) and the inequality (3.8) give for the rescaled energy the following lower bound :

$$\begin{aligned} E_1^h(\tilde{\mathbf{y}}^h) \geqslant\ &\kappa\left\{\|\nabla_p^2\tilde{\mathbf{y}}^h\|^2_{L^2(\Omega_1)} + 2\left\|\frac{1}{h}\nabla_p\tilde{\mathbf{y}}^h_{,3}\right\|^2_{L^2(\Omega_1)} + \left\|\frac{1}{h^2}\tilde{\mathbf{y}}^h_{,33}\right\|^2_{L^2(\Omega_1)}\right\}\\ &+\left\{c_1 - P\left(\frac{|S|^{1/2}}{12\sqrt{\pi}} + \frac{h}{3\sqrt{3}}\right)\right\}\left\|\left(\tilde{\mathbf{y}}_{,1}|\tilde{\mathbf{y}}_{,2}|\frac{1}{h}\tilde{\mathbf{y}}_{,3}\right)\right\|^3_{L^3(\Omega_1^h)}\\ &-\widetilde{D} + c_2|\Omega_1|, \end{aligned} \tag{3.13}$$

from which, in view of the assumption (3.9), we get

$$\begin{aligned} E_1^h(\tilde{\mathbf{y}}^h) \geqslant\ &\kappa\left\{\left\|\nabla_p^2\tilde{\mathbf{y}}^h\right\|^2_{L^2(\Omega_1)} + 2\left\|\frac{1}{h}\nabla_p\tilde{\mathbf{y}}^h_{,3}\right\|^2_{L^2(\Omega_1)} + \left\|\frac{1}{h^2}\tilde{\mathbf{y}}^h_{,33}\right\|^2_{L^2(\Omega_1)}\right\}\\ &-D + c_2|\Omega_1|, \end{aligned} \tag{3.14}$$

for h sufficiently small, with

$$D = \frac{\sqrt{2}}{6\sqrt{\pi}}|S|^{3/2}|\mathbf{A}\mathbf{e}_1 \wedge \mathbf{A}\mathbf{e}_2|^{3/2}.$$

Because $\tilde{\mathbf{y}}^h$ is a minimizer, we can test it against the affine deformation $\mathbf{y} = (\mathbf{A}\mathbf{e}_1|\mathbf{A}\mathbf{e}_2|h\mathbf{A}\mathbf{e}_3)\mathbf{Z}$ of $\mathcal{H}_1$ to obtain the upper bound

$$E_1^h(\tilde{\mathbf{y}}^h) \leqslant |\Omega_1|\phi\big(\mathbf{A}\mathbf{e}_1|\mathbf{A}\mathbf{e}_2|\mathbf{A}\mathbf{e}_3\big). \tag{3.15}$$

Combining (3.14) with (3.15) and setting $c := |\Omega_1|\phi(\mathbf{A}\mathbf{e}_1|\mathbf{A}\mathbf{e}_2|\mathbf{A}\mathbf{e}_3)/\kappa + D$, we get

$$\|\nabla_p^2\tilde{\mathbf{y}}^h\|_{L^2(\Omega_1)}^2 \leqslant c, \tag{3.16}$$

$$2\left\|\frac{1}{h}\nabla_p\tilde{\mathbf{y}}^h_{,3}\right\|_{L^2(\Omega_1)}^2 \leqslant c, \tag{3.17}$$

$$\left\|\frac{1}{h^2}\tilde{\mathbf{y}}^h_{,33}\right\|_{L^2(\Omega_1)}^2 \leqslant c. \tag{3.18}$$

Because we assume $h \leqslant 1$, we also have

$$2\|\nabla_p\tilde{\mathbf{y}}^h_{,3}\|_{L^2(\Omega_1)}^2 \leqslant c, \tag{3.19}$$

$$\|\tilde{\mathbf{y}}^h_{,33}\|_{L^2(\Omega_1)}^2 \leqslant c. \tag{3.20}$$

By (3.16), (3.19), and (3.20), we see that

$$\|\nabla^2\tilde{\mathbf{y}}^h\|_{L^2(\Omega_1)}^2 \leqslant c, \tag{3.21}$$

and this bound, together with the Poincaré inequality (2.34) written for $\tilde{\mathbf{y}}^h$ and $p = 2$, gives in turn

$$\|\tilde{\mathbf{y}}^h\|_{L^2(\Omega_1)}^2 + \|\nabla\tilde{\mathbf{y}}^h\|_{L^2(\Omega_1)}^2 \leqslant C_1c + C_2. \tag{3.22}$$

Therefore, the family of minimizers $\tilde{\mathbf{y}}^h$ is uniformly bounded in $W^{2,2}$, and thus it contains a subsequence, not relabed, such that

$$\tilde{\mathbf{y}}^h \rightharpoonup \hat{\mathbf{y}} \quad \text{in } W^{2,2}\big(\Omega_1; \mathbb{R}^3\big). \tag{3.23}$$

In view of (3.17) and (3.18), we also have

$$\left\|\nabla\left(\frac{1}{h}\tilde{\mathbf{y}}^h_{,3}\right)\right\|_{L^2(\Omega_1)}^2 = \left\|\frac{1}{h}\nabla_p\tilde{\mathbf{y}}^h_{,3}\right\|_{L^2(\Omega_1)}^2 + \left\|\frac{1}{h^2}\tilde{\mathbf{y}}^h_{,33}\right\|_{L^2(\Omega_1)}^2 \leqslant \frac{3}{2}c, \tag{3.24}$$

which, together with the Poincaré inequality written in the form

$$\int_{\Omega_1}\left|\frac{1}{h}\tilde{\mathbf{y}}^h_{,3}\right|^2 \mathrm{d}\mathbf{Z} \leqslant \overline{C}_1\left\{\int_{\Omega_1}\left|\frac{1}{h}\nabla\tilde{\mathbf{y}}^h_{,3}\right|^2 \mathrm{d}\mathbf{Z} + \left|\int_{\partial S\times(0,1)}\frac{1}{h}\tilde{\mathbf{y}}^h_{,3}\,\mathrm{d}a\right|^2\right\} \tag{3.25}$$

[13, Theorem 6.1-8], implies

$$\left\| \frac{1}{h} \tilde{\mathbf{y}}^h_{,3} \right\|^2_{L^2(\Omega_1)} \leqslant \overline{C}_1 \left[\frac{3}{2} c + |\partial S|^2 |A\mathbf{e}_3|^2 \right], \tag{3.26}$$

where $|\partial S|$ denotes the length of the boundary curve ∂S of S. In view of (3.24) and (3.26), we conclude that $(1/h)\tilde{\mathbf{y}}^h_{,3}$ is uniformly bounded in $W^{1,2}(\Omega_1; \mathbb{R}^3)$. Therefore, there exists a sequence, not relabeled, such that

$$\frac{1}{h} \tilde{\mathbf{y}}^h_{,3} \rightharpoonup \hat{\mathbf{b}} \quad \text{in} W^{1,2}(\Omega_1; \mathbb{R}^3), \tag{3.27}$$

and thus, up to a further subsequence, $\tilde{\mathbf{y}}^h_{,3}$ converges to zero almost everywhere in Ω_1. Because Ω_1 is convex in the Z_3 direction, the limit $\hat{\mathbf{y}}$ is independent of Z_3. Note also that, by (3.23) and by Rellich's Theorem, there exists a subsequence of $\tilde{\mathbf{y}}^h$, not relabeled, uniformly converging to $\hat{\mathbf{y}}$ in the closure of Ω_1. Therefore, $\hat{\mathbf{y}}$ satisfies the boundary condition $(3.12)_1$. From (3.18) we also get that $\tilde{\mathbf{y}}^h_{,33}$ converges to zero almost everywhere in Ω_1, and again from the convexity of Ω_1 in the Z_3 direction, the limit $\hat{\mathbf{b}}$ turns out to be independent of Z_3. Finally, by the trace theorem [17, 4.3], $\hat{\mathbf{b}}$ satisfies the boundary condition $(3.12)_2$. We now write

$$\nabla^2_p \tilde{\mathbf{y}}^h = \nabla^2_p \hat{\mathbf{y}} + e^h_p, \qquad e^h_p \rightharpoonup 0 \quad \text{in } L^2, \tag{3.28}$$

$$\frac{1}{h} \nabla_p \tilde{\mathbf{y}}^h_{,3} = \nabla \hat{\mathbf{b}} + e^h_3, \quad e^h_3 \rightharpoonup 0 \quad \text{in } L^2. \tag{3.29}$$

Let $n \mapsto \hat{\mathbf{b}}_n \in C^\infty(\overline{S})$ be an approximating smooth sequence strongly converging to $\hat{\mathbf{b}}$ in $W^{1,2}(\Omega_1; \mathbb{R}^3)$ such that $\hat{\mathbf{b}}_n(Z_1, Z_2) = \mathbf{A}\mathbf{e}_3$ for $(Z_1, Z_2) \in \partial S$ and $\hat{\mathbf{b}}_{n,3}(\mathbf{Z}) = 0$ for each $\mathbf{Z}$ in Ω_1. Because $\tilde{\mathbf{y}}^h$ is a minimizer of E^h_1, we can test it against the deformation $\tilde{\mathbf{y}}^h_n := \hat{\mathbf{y}} + h\hat{\mathbf{b}}_n Z_3$, which, in virtue of the properties enjoyed by $\hat{\mathbf{b}}_n$, belongs to the set $\mathcal{H}_1$. Using (3.28) and (3.29), we get

$$\begin{aligned}
&\int_{\Omega_1} \kappa \Big\{ \left|\nabla^2_p \hat{\mathbf{y}}\right|^2 + \left|e^h_p\right|^2 + 2\nabla^2_p \hat{\mathbf{y}} \cdot e^h_p + 2\left(\left|\nabla \hat{\mathbf{b}}\right|^2 + \left|e^h_3\right|^2 + 2\nabla \hat{\mathbf{b}} \cdot e^h_3\right) \\
&\qquad + \left|\frac{1}{h^2} \tilde{\mathbf{y}}^h_{,33}\right|^2 \Big\} + \phi\left(\tilde{\mathbf{y}}^h_{,1} | \tilde{\mathbf{y}}^h_{,2} | \frac{1}{h} \tilde{\mathbf{y}}^h_{,3} \right) \mathrm{d}\mathbf{Z} \\
&\qquad - \frac{P}{3} \int_{S\times\{0\}} \tilde{\mathbf{y}}^h \cdot (\tilde{\mathbf{y}}^h_{,1} \wedge \tilde{\mathbf{y}}^h_{,2}) \mathrm{d}Z_1 \, \mathrm{d}Z_2 \\
&\leqslant \int_{\Omega_1} \kappa \left\{ \left|\nabla^2_p \hat{\mathbf{y}}\right|^2 + h^2 Z_3^2 \left|\nabla^2_p \hat{\mathbf{b}}_n\right|^2 + 2h^2 Z_3^2 \nabla^2_p \hat{\mathbf{y}} \cdot \nabla^2 \hat{\mathbf{b}}_n + 2\left|\nabla \hat{\mathbf{b}}_n\right|^2 \right\} \\
&\qquad + \phi(\hat{\mathbf{y}}_{,1} + hZ_3 \hat{\mathbf{b}}_{n,1} | \hat{\mathbf{y}}_{,2} + hZ_3 \hat{\mathbf{b}}_{n,2} | \hat{\mathbf{b}}_n) \, \mathrm{d}\mathbf{Z} \\
&\qquad - \frac{P}{3} \int_S \hat{\mathbf{y}} \cdot (\hat{\mathbf{y}}_{,1} \wedge \hat{\mathbf{y}}_{,2}) \, \mathrm{d}Z_1 \, \mathrm{d}Z_2.
\end{aligned} \tag{3.30}$$

After simplifying the first term on both sides, fix n and take the lim sup as $h \to 0^+$. Because $\hat{\mathbf{b}}_n$ is smooth, we can simplify the second and the third term on the right-hand side. Moreover, in view of (3.28) and (3.29), the third and sixth term on the left-hand side converge to zero. Using the upper bound in (2.4), the Lebesgue theorem and the weak continuity of the volume functional in $W^{2,2}$, we reduce to

$$\limsup_{h\to 0^+} \int_{\Omega_1} \kappa\left\{|e_p^h|^2 + 2|e_3^h|^2 + \left|\frac{1}{h^2}\tilde{\mathbf{y}}^h_{,33}\right|^2\right\} \mathrm{d}\mathbf{Z}$$
$$\leqslant \int_{\Omega_1} 2\kappa\{|\nabla\hat{\mathbf{b}}_n|^2 - |\nabla\hat{\mathbf{b}}|^2\}\,\mathrm{d}\mathbf{Z}$$
$$+ \int_{\Omega_1} \big(\phi(\hat{\mathbf{y}}_{,1}|\hat{\mathbf{y}}_{,2}|\hat{\mathbf{b}}_n) - \phi(\hat{\mathbf{y}}_{,1}|\hat{\mathbf{y}}_{,2}|\hat{\mathbf{b}})\big)\,\mathrm{d}\mathbf{Z}. \tag{3.31}$$

Let now $n \to \infty$ and use again the upper bound in (2.4). Then the “sup” can be dropped in (3.31) and we have improved the convergence in (3.28) and (3.29) to strong. This also shows that the limit energy is given by (3.11) evaluated at $(\hat{\mathbf{y}}, \hat{\mathbf{b}})$.

To establish the minimum principle, we choose the deformation $\hat{\mathbf{y}}^h := \hat{\mathbf{y}}(Z_1, Z_2) + h\hat{\mathbf{b}}(Z_1, Z_2)Z_3$ as a test function, with $\hat{\mathbf{y}}$ and $\hat{\mathbf{b}} \in C^\infty(\overline{S}; \mathbb{R}^3)$ satisfying the boundary conditions (3.12). Repeating the argument from (3.30) to (2.30) gives the minimum principle for smooth competitors. The minimum principle for competitors in $W^{2,2}(S; \mathbb{R}^3) \times W^{1,2}(S; \mathbb{R}^3)$ follows by approximation. □

The limit energy (3.11) turns out to depend upon two independent vector fields, $\hat{\mathbf{y}}$ and $\hat{\mathbf{b}}$, which describe the deformation of the middle surface of the film S, and the deformation in the direction perpendicular to the film, respectively. Therefore, Theorem 3.1 provides a 2-dimensional Cosserat theory, with $\hat{\mathbf{b}}$ the Cosserat director. The energy (3.11) is a membrane energy supplemented by an interfacial energy term (the term multiplying κ). The latter has a similar form as a bending energy, but its physical origins are the same as the analogous term in the 3-dimensional energy (2.7). That is, it is intended to model energy associated with lattice curvature, arising from lattice radii of curvature that are comparable to atomic spacing, which occur in the present case when there are interfaces between variants or phases. Thus κ should not be considered as a classical bending modulus.

Because of this interpretation, κ is expected to be much smaller than a typical modulus that describes the growth of ϕ away from its energy wells. It is therefore known from many studies that the presence of κ merely smooths interfaces, and this could also be verified by an elementary Γ-convergence argument. Hence, in the following section we drop the term multiplying κ and study the pressurized membrane energy alone.

4. Martensitic Thin Films

Martensitic crystals display a diffusionless solid-to-solid phase transformation between a symmetric high temperature phase (austenite) and different symmetry-related variants of a low temperature phase (martensite). For temperatures above the transformation temperature θ_{cr}, the austenite phase is the stable phase, while for temperatures below θ_{cr}, the martensite is stable. At the transformation temperature θ_{cr}, both phases are stable. To model the change of phase, we follow Ball and James [7, 8] in introducing a nonconvex strain energy density ϕ with energy wells at the matrices

$$\mathcal{A} := SO(3), \tag{4.1}$$

for the austenite phase, and

$$\mathcal{M} := \{\mathbf{F} \in M^{3\times 3} \mid \exists \mathbf{R} \in SO(3), \exists \mathbf{U} \in \{\mathbf{U}_1, \mathbf{U}_2, \ldots \mathbf{U}_n\}: \mathbf{F} = \mathbf{RU}\}, \tag{4.2}$$

for the martensite variants. $SO(3)$ is the set of all proper rotations, and $\mathbf{U}_i$, $i = 1, 2, \ldots, N$, are distinct positive definite symmetric matrices representing the transformation strains of the variants of martensite from the austenite, taken as reference configuration; they can be determined by measurements of the lattice parameters of the material. For the austenite and the martensite phases, we introduce nonconvex strain energy densities ϕ_a, ϕ_m, which are continuous non negative scalar functions defined over $M^{3\times 3}$ and such that ϕ_a is minimized on $\mathcal{A}$ and the minimum value is zero, while ϕ_m is minimized on $\mathcal{M}$ and the minimum value is zero. The energies ϕ_a, ϕ_m, are also supposed to satisfy the following growth hypotheses: there exist positive constants c_{a1}, c_{a2}, c_{a3}, c_{a4}, c_{m1}, c_{m2}, c_{m3}, c_{m4} and $3 < q < 6$ such that

$$c_{a1}|\mathbf{F}|^3 - c_{a2} \leqslant \phi_a(\mathbf{F}) \leqslant c_{a3}|\mathbf{F}|^q - c_{a4}, \tag{4.3}$$

$$c_{m1}|\mathbf{F}|^3 - c_{m2} \leqslant \phi_m(\mathbf{F}) \leqslant c_{m3}|\mathbf{F}|^q - c_{m4}, \tag{4.4}$$

for each $\mathbf{F} \in M^{3\times 3}$. These hypotheses are consistent with the growth assumptions (2.4).

Introducing a $\chi \in C^0(M^{3\times 3}; [0, 1])$ such that

$$\chi(\mathbf{F}) = 0 \quad \Leftrightarrow \quad \mathbf{F} \in \mathcal{A}, \tag{4.5}$$

$$\chi(\mathbf{F}) = 1 \quad \Leftrightarrow \quad \mathbf{F} \in \mathcal{M}, \tag{4.6}$$

we assume for the strain energy density ϕ the simple form

$$\phi(\mathbf{F}; \theta) = \chi(\mathbf{F})\big(\phi_m(\mathbf{F}) + l_m(\theta)\big) + (1 - \chi(\mathbf{F}))\big(\phi_a(\mathbf{F}) + l_a(\theta)\big). \tag{4.7}$$

The expression (4.7) is a simple but realistic way of modeling the exchange of stability between austenite and martensite. The terms $l_m(\theta)$ and $l_a(\theta)$, which are related to the latent heat of transformation, are positive material constants depending continuously upon the temperature θ and such that

$$\left.\begin{array}{ll} l_a(\theta) > l_m(\theta) & \text{if } \theta < \theta_{cr}, \\ l_a(\theta) < l_m(\theta) & \text{if } \theta > \theta_{cr}, \\ l_a(\theta_{cr}) = l_m(\theta_{cr}). & \end{array}\right\} \tag{4.8}$$

In view of the definition (4.7) and of the assumptions (4.1), (4.2),(4.5), (4.6), and (4.8), ϕ has a multi-well structure, the terms $l_m(\theta)$, $l_a(\theta)$ corresponding to the heights of the martensite and of the austenite wells, respectively. In particular, if $\theta < \theta_{cr}$, $\phi(\cdot;\theta)$ attains the absolute minimum at all matrices belonging to the set of martensite wells $\mathcal{M}$. Indeed, because χ has values in [0, 1] and because ϕ_a and ϕ_m are positive, we have

$$\inf_{\mathbf{F}\in M^{3\times 3}} \phi(\mathbf{F};\theta) \geqslant \min\{l_a(\theta); l_m(\theta)\}. \tag{4.9}$$

If $\theta < \theta_{cr}$, the right-hand side coincides with $l_m(\theta)$, and since $\phi(\mathbf{F};\theta) = l_m(\theta)$ for all $\mathbf{F} \in \mathcal{M}$, then ϕ is minimized at $\mathcal{M}$. Analogously, it can be shown that if $\theta > \theta_{cr}$, then $\phi(\cdot;\theta)$ is minimized at $\mathcal{A}$ and $\phi(\cdot;\theta_{cr})$ is minimized at $\mathcal{A} \cup \mathcal{M}$.

Experiments indicate that, at equilibrium, the deformation gradient stays very close to the wells, even though it could not be precisely at the minima, because of the presence of the term $-PV(\mathbf{y})$. For many martensitic materials with "hard moduli", this suggests that the equilibrium microstructures can be approximately described by deformations whose gradients satisfy the constraint of lying on the wells. The idea of this approach, called the *constrained theory of martensite* and first proposed in [6], is to study the asymptotic behavior of the sequence of total energy functionals

$$E^n(\mathbf{y},\mathbf{b}) := \int_S \left(\phi_n(\mathbf{y}_{,1}|\mathbf{y}_{,2}|\mathbf{b}) - \frac{P}{3}\mathbf{y}\cdot(\mathbf{y}_{,1}\wedge\mathbf{y}_{,2})\right) \mathrm{d}Z_1\,\mathrm{d}Z_2, \tag{4.10}$$

as $n \to \infty$, with

$$\phi_n(\mathbf{F};\theta) := \chi(\mathbf{F})\big(n\phi_m(\mathbf{F}) + l_m(\theta)\big) + \big(1-\chi(\mathbf{F})\big)\big(n\phi_a(\mathbf{F}) + l_a(\theta)\big). \tag{4.11}$$

The interfacial energy term $\kappa\{|\nabla_p^2\mathbf{y}|^2 + 2|\nabla_p\mathbf{b}|^2\}$ has been neglected in (4.10). Indeed, if the film is large enough, the elastic energy is much larger than the interfacial energy. In this respect, each element of the sequence (4.10) provides a reasonable approximation of the expression (3.11) of the energy of a very thin film with strain energy density ϕ_n growing more and more steeply away from the wells. Each element of the sequence (4.10) is assumed to be defined on the set of functions

$$\begin{aligned}\mathcal{K} := \big\{&(\mathbf{y},\mathbf{b}) \in W^{1,q}(S;\mathbb{R}^3) \times L^q(S;\mathbb{R}^3) | \mathbf{y} = \mathbf{A}\mathbf{e}_1 Z_1 + \mathbf{A}\mathbf{e}_2 Z_2,\\ &\mathbf{b} = \mathbf{A}\mathbf{e}_3,\ (Z_1, Z_2) \in \partial S\big\},\end{aligned} \tag{4.12}$$

where q is the exponent which appears in (4.3), (4.4). This set turns out to be the "natural" domain of the energy functionals in (4.10). Indeed, in view of the upper bounds in (4.3), (4.4) and of the bound on the volume functional (2.21), each couple $(\mathbf{y},\mathbf{b})$ in $\mathcal{K}$ has energy $E^n(\mathbf{y},\mathbf{b})$ finite. As shown by the following theorem, the result of the constrained theory is a new simplified variational problem whose solutions are searched among fine mixtures; these are mathematically described by families of Young measures with supports contained in the set $\mathcal{A} \cup \mathcal{M}$. We recall the essential property of the Young measure [5]. Let $C_0(M^{3\times 3};\mathbb{R})$ denote the continuous functions on $M^{3\times 3}$ with compact support. Given a sequence $\mathbf{F}^n \in L^1(S; M^{3\times 3})$, we may find a family of probability measures $(\nu_{\mathbf{Z}}), \mathbf{Z} \in S$ and a subsequence of $\mathbf{F}^n$, not relabeled, such that, for any $\psi \in C_0(M^{3\times 3};\mathbb{R})$,

$$\psi(\mathbf{F}^n) \stackrel{*}{\rightharpoonup} \int_{M^{3\times 3}} \psi(\mathbf{F})\,\mathrm{d}\nu_{\mathbf{Z}}(\mathbf{F}) \quad \text{in } L^\infty(S;\mathbb{R}). \tag{4.13}$$

Young measures are useful tools for the analysis of the microstructure [7, 8]. The family of measures $(\nu_{\mathbf{Z}}), \mathbf{Z} \in S$ characterizes the local limit distribution of the values $\mathbf{F}^n$ as $n \to \infty$. If the sequence $\mathbf{F}^n$ is thought of as representing a sequence $(\mathbf{y}^n_{,1}|\mathbf{y}^n_{,2}|\mathbf{b}^n)$ and if the measures $\nu_{\mathbf{Z}}$ are supported on the set $\mathcal{M} \cup \mathcal{A}$, then the $\nu_{\mathbf{Z}}$ turn out to describe the local proportions of phases and the microstructure of the material.

THEOREM 4.1. *Assume that there exists a constant C, independent of n, such that*

$$\inf_{(\mathbf{y},\mathbf{b})\in\mathcal{K}} E^n(\mathbf{y},\mathbf{b}) \leqslant C < +\infty, \tag{4.14}$$

and a sequence $(\mathbf{y}^n, \mathbf{b}^n)$ in $\mathcal{K}$ such that

$$E^n\left(\mathbf{y}^n,\mathbf{b}^n\right) \leqslant \inf_{(\mathbf{y},\mathbf{b})\in\mathcal{K}} E^n(\mathbf{y},\mathbf{b}) + \frac{1}{n}. \tag{4.15}$$

Then, there exists a subsequence, not relabeled, such that

$$\mathbf{y}^n \rightharpoonup \hat{\mathbf{y}} \quad \text{in } W^{1,3}\left(S;\mathbb{R}^3\right), \tag{4.16}$$

$$\mathbf{b}^n \rightharpoonup \hat{\mathbf{b}} \quad \text{in } L^3\left(S;\mathbb{R}^3\right), \tag{4.17}$$

and the family of Young measures, $(\hat{\nu}_{\mathbf{Z}}), \mathbf{Z} \in S$, generated by the sequence $(\mathbf{y}^n_{,1}|\mathbf{y}^n_{,2}|\mathbf{b}^n)$ is such that the couple $(\hat{\nu}_{\mathbf{Z}}, \hat{\mathbf{y}})$ is a minimizer of the limit energy

$$e(\nu_{\mathbf{Z}};\mathbf{y}) := \int_S \left\{ \left(l_m(\theta) - l_a(\theta)\right) \int_{\mathcal{M}} \mathrm{d}\nu_{\mathbf{Z}}(\mathbf{F}) - \frac{P}{3}\mathbf{y}\cdot(\mathbf{y}_{,1}\wedge\mathbf{y}_{,2}) \right\} \mathrm{d}Z_1\,\mathrm{d}Z_2 + l_a(\theta)|S|, \tag{4.18}$$

among all couples $(\nu_{\mathbf{Z}}, \mathbf{y})$ such that $(\nu_{\mathbf{Z}}), \mathbf{Z} \in S$, is a family of Young measures with supports in the set $\mathcal{M} \cup \mathcal{A}$ and $\mathbf{y} \in W^{1,3}(S;\mathbb{R}^3)$ satisfies the constraint

$$\mathbf{y}_{,\alpha}(\mathbf{Z}) = \int_{\mathcal{M}\cup\mathcal{A}} \mathbf{F}\mathbf{e}_\alpha\,\mathrm{d}\nu_{\mathbf{Z}}(\mathbf{F}), \quad \alpha = 1, 2, \tag{4.19}$$

at almost every $\mathbf{Z} \in S$, and the boundary condition

$$\mathbf{y} = \mathbf{A}\mathbf{e}_1 Z_1 + \mathbf{A}\mathbf{e}_2 Z_2, \quad \mathbf{Z} \in \partial S. \tag{4.20}$$

REMARKS. We note that (4.14) places a restriction on the boundary conditions in (4.12). This could be quantified, but we don't do it here. In the next section, we give an example (the tunnel) in which (4.14) holds with the boundary conditions (5.1). In Theorem 4.1, condition (4.15) is to allow for the possibility of nonattainment of the minimum. Also, Young measure refers in this paper to Young measures arising from a sequence that is bounded in $\mathcal{K}$.

Proof. In view of the hypotheses (4.14) and (4.15) and recalling that $n \geqslant 1$, we have

$$E^n(\mathbf{y}^n, \mathbf{b}^n) \leqslant C + 1. \tag{4.21}$$

Substituting to E^n its expression (4.10) with ϕ_n given by (4.11) and dividing by n, we obtain

$$\begin{aligned}&\int_S \Big\{ \chi(\mathbf{y}^n_{,1}|\mathbf{y}^n_{,2}|\mathbf{b}^n)\phi_m(\mathbf{y}^n_{,1}|\mathbf{y}^n_{,2}|\mathbf{b}^n) + (1-\chi(\mathbf{y}^n_{,1}|\mathbf{y}^n_{,2}|\mathbf{b}^n))\phi_a(\mathbf{y}^n_{,1}|\mathbf{y}^n_{,2}|\mathbf{b}^n)\\ &\quad + \frac{1}{n}\big[\chi(\mathbf{y}^n_{,1}|\mathbf{y}^n_{,2}|\mathbf{b}^n) l_m(\theta) + (1-\chi(\mathbf{y}^n_{,1}|\mathbf{y}^n_{,2}|\mathbf{b}^n)) l_a(\theta)\big]\Big\} \mathrm{d}Z_1\,\mathrm{d}Z_2\\ &\quad - \frac{P}{n} V(\mathbf{y}^n) \leqslant \frac{C+1}{n}.\end{aligned} \tag{4.22}$$

Now we use the bound on the volume functional (2.21), written for $t = 0$, and the lower bounds in (4.3) and (4.4) to get

$$\begin{aligned}&\int_S \Big\{ \big[c_{m1}\chi(\mathbf{y}^n_{,1}|\mathbf{y}^n_{,2}|\mathbf{b}^n) + c_{a1}(1-\chi(\mathbf{y}^n_{,1}|\mathbf{y}^n_{,2}|\mathbf{b}^n))\big]|(\mathbf{y}^n_{,1}|\mathbf{y}^n_{,2}|\mathbf{b}^n)|^3\\ &\quad - \big[c_{m2}\chi(\mathbf{y}^n_{,1}|\mathbf{y}^n_{,2}|\mathbf{b}^n) + c_{a2}(1-\chi(\mathbf{y}^n_{,1}|\mathbf{y}^n_{,2}|\mathbf{b}^n))\big]\\ &\quad + \frac{1}{n}\big[\chi(\mathbf{y}^n_{,1}|\mathbf{y}^n_{,2}|\mathbf{b}^n) l_m(\theta) + (1-\chi(\mathbf{y}^n_{,1}|\mathbf{y}^n_{,2}|\mathbf{b}^n)) l_a(\theta)\big]\Big\} \mathrm{d}Z_1\,\mathrm{d}Z_2\\ &\quad - \frac{P}{n}\left[\frac{|S|^{1/2}}{12\sqrt{\pi}}\|\nabla\mathbf{y}^n\|^3_{L^3(S)} + c\right] \leqslant \frac{C+1}{n}.\end{aligned} \tag{4.23}$$

Using the inequality $|(\mathbf{y}^n_{,1}|\mathbf{y}^n_{,2}|\mathbf{b}^n)|^3 \geqslant |\nabla\mathbf{y}^n|^3 + |\mathbf{b}^n|^3$ and recalling that $0 \leqslant \chi \leqslant 1$, we also have

$$\begin{aligned}&\left(\min\{c_{a1}; c_{m1}\} - \frac{P|S|^{1/2}}{12n\sqrt{\pi}}\right)\|\nabla\mathbf{y}^n\|^3_{L^3(S)} + \min\{c_{a1}; c_{m1}\}\|\nabla\mathbf{b}^n\|^3_{L^3(S)}\\ &\quad + |S|\left(\frac{1}{n}\min\{l_m(\theta); l_a(\theta)\} - \max\{c_{m2}; c_{a2}\}\right) - \frac{P}{n}c \leqslant \frac{C+1}{n}.\end{aligned} \tag{4.24}$$

Therefore, for n sufficiently large, there exists a constant $\overline{C}$, independent of n, such that

$$\|\nabla\mathbf{y}^n\|^3_{L^3(S)} + \|\nabla\mathbf{b}^n\|^3_{L^3(S)} \leqslant \overline{C}. \tag{4.25}$$

By the weak compactness in Sobolev's spaces, it then follows that

$$\nabla \mathbf{y}^n \rightharpoonup \nabla \hat{\mathbf{y}} \quad \text{in } L^3(S;\mathbb{R}^3), \tag{4.26}$$

$$\mathbf{b}^n \rightharpoonup \hat{\mathbf{b}} \quad \text{in } L^3(S;\mathbb{R}^3), \tag{4.27}$$

up to a subsequence. In view of (4.26) and of the Poincaré inequality (2.34) written for $\mathbf{y}^n$ and $p = 3$, $\mathbf{y}^n$ is uniformly bounded in $W^{1,3}(S;\mathbb{R}^3)$, and thus (4.16) holds up to a subsequence. Besides, by Rellich's theorem, there exists a further subsequence, not relabeled, such that

$$\mathbf{y}^n \to \hat{\mathbf{y}} \quad \text{in } C^0(\overline{S};\mathbb{R}^3). \tag{4.28}$$

Therefore, the limit $\hat{\mathbf{y}}$ satisfies the boundary condition (4.20). From (4.26) and from the bound on the volume (2.21), we also get

$$\frac{1}{n} V(\mathbf{y}^n) \to 0, \tag{4.29}$$

which, in view of (4.21) and of the positivity of ϕ_m, ϕ_a, χ and $1-\chi$, in turn implies

$$\int_S \chi(\mathbf{y}^n_{,1}|\mathbf{y}^n_{,2}|\mathbf{b}^n)\phi_m(\mathbf{y}^n_{,1}|\mathbf{y}^n_{,2}|\mathbf{b}^n)\mathrm{d}Z_1\,\mathrm{d}Z_2 \to 0, \tag{4.30}$$

$$\int_S (1-\chi(\mathbf{y}^n_{,1}|\mathbf{y}^n_{,2}|\mathbf{b}^n))\phi_a(\mathbf{y}^n_{,1}|\mathbf{y}^n_{,2}|\mathbf{b}^n)\mathrm{d}Z_1\,\mathrm{d}Z_2 \to 0. \tag{4.31}$$

Therefore, by the fundamental property of the Young measures (4.13), there exists another subsequence of $(\mathbf{y}^n_{,1}|\mathbf{y}^n_{,2}|\mathbf{b}^n)$, generating a family of Young measures $(\hat{\nu}_\mathbf{Z})$, $\mathbf{Z} \in S$, such that

$$\int_S \int_{M^{3\times 3}} \chi(\mathbf{F})\phi_m(\mathbf{F}) d\hat{\nu}_\mathbf{Z}(\mathbf{F})\,\mathrm{d}Z_1\,\mathrm{d}Z_2 = 0, \tag{4.32}$$

$$\int_S \int_{M^{3\times 3}} (1-\chi(\mathbf{F}))\phi_a(\mathbf{F}) d\hat{\nu}_\mathbf{Z}(\mathbf{F})\,\mathrm{d}Z_1\,\mathrm{d}Z_2 = 0. \tag{4.33}$$

These imply that the support of $\hat{\nu}_\mathbf{Z}$ is contained in $\mathcal{A} \cup \mathcal{M}$ for almost every $\mathbf{Z} \in S$ [8, Lemma 3.3]. Using (4.30), (4.31), we now construct a further "rare" subsequence, not relabeled, such that

$$\begin{aligned}\frac{C+1}{n^2} \geqslant \int_S \{&\chi(\mathbf{y}^n_{,1}|\mathbf{y}^n_{,2}|\mathbf{b}^n)\phi_m(\mathbf{y}^n_{,1}|\mathbf{y}^n_{,2}|\mathbf{b}^n) \\ &+ (1-\chi(\mathbf{y}^n_{,1}|\mathbf{y}^n_{,2}|\mathbf{b}^n))\phi_a(\mathbf{y}^n_{,1}|\mathbf{y}^n_{,2}|\mathbf{b}^n)\}\,\mathrm{d}Z_1\,\mathrm{d}Z_2.\end{aligned} \tag{4.34}$$

This subsequence has the same Young measure, $(\hat{\nu}_\mathbf{Z})$, $\mathbf{Z} \in S$, but it is such that

$$\begin{aligned}n \int_S \{&\chi(\mathbf{y}^n_{,1}|\mathbf{y}^n_{,2}|\mathbf{b}^n)\phi_m(\mathbf{y}^n_{,1}|\mathbf{y}^n_{,2}|\mathbf{b}^n) \\ &+ (1-\chi(\mathbf{y}^n_{,1}|\mathbf{y}^n_{,2}|\mathbf{b}^n))\phi_a(\mathbf{y}^n_{,1}|\mathbf{y}^n_{,2}|\mathbf{b}^n)\}\mathrm{d}Z_1\,\mathrm{d}Z_2 \to 0.\end{aligned} \tag{4.35}$$

Note that by (4.26), (4.28) and the weak continuity of minors [3], we have

$$\int_S \mathbf{y}^n \cdot (\mathbf{y}^n_{,1} \wedge \mathbf{y}^n_{,2}) \mathrm{d}Z_1 \, \mathrm{d}Z_2 \to \int_S \hat{\mathbf{y}} \cdot (\hat{\mathbf{y}}_{,1} \wedge \hat{\mathbf{y}}_{,2}) \mathrm{d}Z_1 \, \mathrm{d}Z_2, \tag{4.36}$$

and this, together with (4.30), (4.31) and the fact that the support of $(\hat{\nu}_{\mathbf{Z}})$, $\mathbf{Z} \in S$ is contained in $\mathcal{A} \cup \mathcal{M}$, implies that the limit of $E^n(\mathbf{y}^n, \mathbf{b}^n)$ is given by (4.18) evaluated at $(\hat{\mathbf{y}}, \hat{\nu}_{\mathbf{Z}})$. To establish the minimum principle, we consider a family of Young measures $(\nu_{\mathbf{Z}})$, $\mathbf{Z} \in S$, supported on $\mathcal{A} \cup \mathcal{M}$, arising from a sequence $(\bar{\mathbf{y}}^n_{,1} | \bar{\mathbf{y}}^n_{,2} | \bar{\mathbf{b}}^n)$, with $(\bar{\mathbf{y}}^n, \bar{\mathbf{b}}^n) \in \mathcal{K}$, such that (without loss of generality)

$$\nabla \bar{\mathbf{y}}^n \rightharpoonup \mathbf{y} \quad \text{in } L^3(S), \tag{4.37}$$

$$\bar{\mathbf{b}}^n \rightharpoonup \mathbf{b} \quad \text{in } L^3(S), \tag{4.38}$$

$$n \int_S \{ \chi (\bar{\mathbf{y}}^n_{,1} | \bar{\mathbf{y}}^n_{,2} | \bar{\mathbf{b}}^n) \phi_m (\bar{\mathbf{y}}^n_{,1} | \bar{\mathbf{y}}^n_{,2} | \bar{\mathbf{b}}^n) + (1 - \chi (\bar{\mathbf{y}}^n_{,1} | \bar{\mathbf{y}}^n_{,2} | \bar{\mathbf{b}}^n)) \phi_a (\bar{\mathbf{y}}^n_{,1} | \bar{\mathbf{y}}^n_{,2} | \bar{\mathbf{b}}^n) \} \mathrm{d}Z_1 \, \mathrm{d}Z_2 \to 0. \tag{4.39}$$

From (4.15) we find

$$\begin{aligned} E^n(\mathbf{y}^n, \mathbf{b}^n) &= \int_S \Big\{ \chi (\mathbf{y}^n_{,1} | \mathbf{y}^n_{,2} | \mathbf{b}^n) \big[n\phi_m (\mathbf{y}^n_{,1} | \mathbf{y}^n_{,2} | \mathbf{b}^n) + l_m(\theta) \big] \\ &\quad + (1 - \chi (\mathbf{y}^n_{,1} | \mathbf{y}^n_{,2} | \mathbf{b}^n)) \big[n\phi_a (\mathbf{y}^n_{,1} | \mathbf{y}^n_{,2} | \mathbf{b}^n) + l_a(\theta) \big] \\ &\quad - \frac{P}{3} \mathbf{y}^n \cdot (\mathbf{y}^n_{,1} \wedge \mathbf{y}^n_{,2}) \Big\} \mathrm{d}Z_1 \, \mathrm{d}Z_2 \\ &\leqslant \inf_{(\mathbf{y},\mathbf{b}) \in \mathcal{K}} E^n(\mathbf{y}, \mathbf{b}) + \frac{1}{n} \leqslant E^n(\bar{\mathbf{y}}^n, \bar{\mathbf{b}}^n) + \frac{1}{n} \\ &= \int_S \Big\{ \chi (\bar{\mathbf{y}}^n_{,1} | \bar{\mathbf{y}}^n_{,2} | \bar{\mathbf{b}}^n) \big[n\phi_m (\bar{\mathbf{y}}^n_{,1} | \bar{\mathbf{y}}^n_{,2} | \bar{\mathbf{b}}^n) + l_m(\theta) \big] \\ &\quad + (1 - \chi (\bar{\mathbf{y}}^n_{,1} | \bar{\mathbf{y}}^n_{,2} | \bar{\mathbf{b}}^n)) \big[n\phi_a (\bar{\mathbf{y}}^n_{,1} | \bar{\mathbf{y}}^n_{,2} | \bar{\mathbf{b}}^n) + l_a(\theta) \big] \\ &\quad - \frac{P}{3} \bar{\mathbf{y}}^n \cdot (\bar{\mathbf{y}}^n_{,1} \wedge \bar{\mathbf{y}}^n_{,2}) \Big\} \mathrm{d}Z_1 \, \mathrm{d}Z_2 + \frac{1}{n}. \end{aligned} \tag{4.40}$$

Taking the limit as $n \to \infty$ and using (4.35), (4.36), (4.37), (4.38) and the weak continuity of minors, we obtain the minimum principle. □

According to Theorem 4.1, the behavior of the film is governed by the limit energy (4.18), defined for couples $(\nu_{\mathbf{Z}}; \mathbf{y})$, where $(\nu_{\mathbf{Z}})$, $\mathbf{Z} \in S$, is a family of Young measures supported on $\mathcal{M} \cup \mathcal{A}$, and $\mathbf{y}$ is a vector field satisfying the boundary conditions (4.20). The vector field $\mathbf{y}$ and the measures $\nu_{\mathbf{Z}}$ are related through the constraint (4.19). The vector field $\hat{\mathbf{b}}$ does not appear explicitly in the expression of the limit energy (4.18). However, because the family of Young measures $(\hat{\nu}_{\mathbf{Z}})$, $\mathbf{Z} \in S$, is generated by the sequence $(\mathbf{y}^n_{,1}|\mathbf{y}^n_{,2}|\mathbf{b}^n)$, we have

$$\hat{\mathbf{b}} = \int_{\mathcal{A}\cup\mathcal{M}} \mathbf{F}\mathbf{e}_3 \, d\hat{\nu}_{\mathbf{Z}}(\mathbf{F}) \tag{4.41}$$

at almost every $\mathbf{Z} \in S$. The energy (4.18) is much easier to study than the original energy.

5. Tunnels

Thin film deformations involving gradients only from the martensite and the austenite wells are studied in [10, 11]. A particularly interesting deformation, especially in connection with the possible applications in the design of microactuators, is the *tunnel deformation*, sketched in Figure 2. To illustrate this deformation, we consider a martensitic film released on the rectangular region $S = (0, l_1) \times (0, l_2)$, for which we adopt boundary conditions more general than (4.20). In particular, we assume

$$\begin{cases} \mathbf{y}(Z_1, Z_2) = Z_1\mathbf{e}_1 + Z_2\mathbf{e}_2 & \text{for } Z_1 \in [0, l_1],\ Z_2 = 0, l_2, \\ \mathbf{y}(Z_1, Z_2) \cdot \mathbf{e}_1 = Z_1 & \text{for } Z_1 = 0, l_1,\ Z_2 \in [0, l_2]. \end{cases} \tag{5.1}$$

These conditions model the situation of a rectangular film attached to the substrate only along the edges parallel to the direction of $\mathbf{e}_1$. The edges parallel to the direction of $\mathbf{e}_2$ are restricted to move on planes perpendicular to the plane of the film.

To ensure the existence of the tunnel deformation, it is necessary to make suitable assumptions on the set of the martensite wells $\mathcal{M}$ [11]. In particular, we assume that the conditions

$$\mathbf{e}_3 \cdot \mathrm{Adj}\big(\mathbf{U}^2 - \mathbf{I}\big)\mathbf{e}_3 = 0, \tag{5.2}$$

$$\mathrm{tr}\mathbf{U}^2 - \mathbf{e}_3 \cdot \mathbf{U}^2\mathbf{e}_3 - 2 > 0 \tag{5.3}$$

hold for some symmetric matrix $\mathbf{U} \in \mathcal{M}$. These, in turn, are satisfied if and only if there exist a rotation $\mathbf{Q} \in SO(3)$ and a vector $\mathbf{e}$ such that

$$(\mathbf{QU} - \mathbf{I})\mathbf{e} = 0, \quad \mathbf{e} \cdot \mathbf{e}_3 = 0, \quad |\mathbf{e}| = 1, \tag{5.4}$$

$$\mathbf{n} \cdot \mathbf{U}^2\mathbf{e} = 0, \quad \text{where } \mathbf{n} = \mathbf{e} \wedge \mathbf{e}_3, \tag{5.5}$$

$$|\mathbf{Un}| > 1, \tag{5.6}$$

[11, Proposition 5.2]. Condition (5.4) says that an interface between the austenite and a variant of martensite described by $\mathbf{U}$ can be formed in the direction of $\mathbf{e}$. Equation (5.5) is a condition of vanishing shear, while the inequality (5.6) says that the film is stretched in the direction perpendicular to the interface. If we orient the film so that the two directions of $\mathbf{e}$ and of $\mathbf{n}$ coincide with the directions of $\mathbf{e}_1$ and of $\mathbf{e}_2$, respectively, then the conditions from (5.4) to (5.6) become

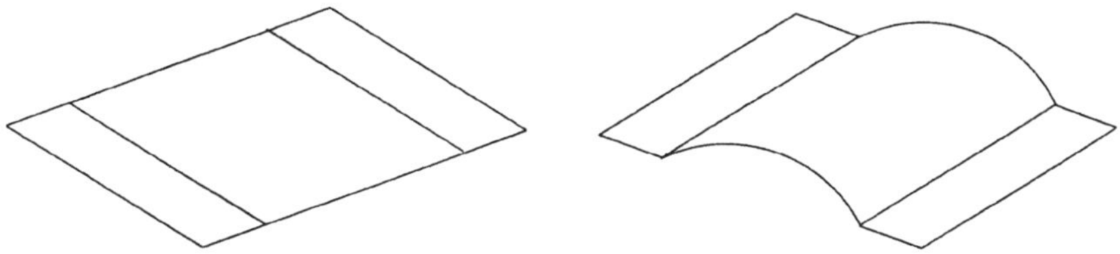

Figure 2. The flat and the tunnel configurations.

$$\mathbf{QUe}_1 = \mathbf{e}_1, \tag{5.7}$$

$$\mathbf{e}_1 \cdot \mathbf{U}^2\mathbf{e}_2 = 0. \tag{5.8}$$

$$|\mathbf{Ue}_2| > 1. \tag{5.9}$$

In the the constrained theory, the tunnel deformation is described by the sequence $(\mathbf{y}^n, \mathbf{b}^n)$ with $\mathbf{y}^n$ the sequence of cylindrical deformations

$$\mathbf{y}^n(Z_1, Z_2) = Z_1\mathbf{e}_1 + \hat{u}(Z_2)\mathbf{e}_2 + \hat{v}(Z_2)\mathbf{e}_3, \tag{5.10}$$

with

$$\hat{u}(Z_2) := |\mathbf{Ue}_2| \int_0^{Z_2} \cos\left(\alpha - \frac{2\alpha}{l_2}t\right)\mathrm{d}t, \tag{5.11}$$

$$\hat{v}(Z_2) := |\mathbf{Ue}_2| \int_0^{Z_2} \sin\left(\alpha - \frac{2\alpha}{l_2}t\right)\mathrm{d}t, \tag{5.12}$$

and with α the solution in $(0, 2\pi)$ to the equation

$$\sin\alpha = \frac{\alpha}{|\mathbf{Ue}_2|}. \tag{5.13}$$

The sequence of Cosserat directors $\mathbf{b}^n$ is given by $\mathbf{R}(Z_2)\overline{\mathbf{R}}\mathbf{QUe}_3$, $\mathbf{R}(Z_2)$ and $\overline{\mathbf{R}}$ being the rotation matrices

$$\begin{aligned}\mathbf{R}(Z_2) &:= \mathbf{e}_1 \otimes \mathbf{e}_1 + \cos\left(\alpha - \frac{2\alpha}{l_2}Z_2\right)[\mathbf{e}_2 \otimes \mathbf{e}_2 + \mathbf{e}_3 \otimes \mathbf{e}_3] \\ &\quad + \sin\left(\alpha - \frac{2\alpha}{l_2}Z_2\right)[-\mathbf{e}_2 \otimes \mathbf{e}_3 + \mathbf{e}_3 \otimes \mathbf{e}_2],\end{aligned} \tag{5.14}$$

$$\begin{aligned}\overline{\mathbf{R}} &:= \mathbf{e}_1 \otimes \mathbf{e}_1 + \frac{\mathbf{QUe}_2 \cdot \mathbf{e}_2}{|\mathbf{Ue}_2|}[\mathbf{e}_2 \otimes \mathbf{e}_2 + \mathbf{e}_3 \otimes \mathbf{e}_3] \\ &\quad - \frac{\mathbf{QUe}_2 \cdot \mathbf{e}_3}{|\mathbf{Ue}_2|}[-\mathbf{e}_2 \otimes \mathbf{e}_3 + \mathbf{e}_3 \otimes \mathbf{e}_2].\end{aligned} \tag{5.15}$$

In the plane $(\mathbf{e}_2, \mathbf{e}_3)$, the couple $(\hat{u}, \hat{v})$ describes the circular arch of length $l_2|\mathbf{Ue}_2|$, starting at the point $(0, 0)$, ending at $(0, l_2)$, and lying in the positive half-plane. The family of Young measures arising from the sequence $(\mathbf{y}^n_{,1}|\mathbf{y}^n_{,2}|\mathbf{b}^n)$ is simply a Dirac mass centered at $\mathbf{R}(Z_2)\overline{\mathbf{R}}\mathbf{QU}$:

$$\nu_{\mathbf{Z}} = \delta_{\mathbf{R}(Z_2)\bar{\mathbf{R}}\mathbf{QU}}. \tag{5.16}$$

It is easy to see that the tunnel deformation is energy minimizing in the constrained theory when $\theta \leqslant \theta_{cr}$ and no pressure is applied under the film. In this case, the total energy (4.18) reduces to

$$e(\nu_{\mathbf{Z}}; \mathbf{y}) = \int_S \left\{ \big(l_m(\theta) - l_a(\theta)\big) \int_{\mathcal{M}} \mathrm{d}\nu_{\mathbf{Z}}(\mathbf{F}) \right\} \mathrm{d}Z_1\, \mathrm{d}Z_2 + l_a(\theta)|S|, \tag{5.17}$$

to be minimized among the Young measures $\nu_{\mathbf{Z}}$ such that

$$\operatorname{supp} \nu_{\mathbf{Z}} \subset \mathcal{A} \quad \text{if } \theta > \theta_{cr}, \tag{5.18}$$

$$\operatorname{supp} \nu_{\mathbf{Z}} \subset \mathcal{M} \quad \text{if } \theta < \theta_{cr}, \tag{5.19}$$

$$\operatorname{supp} \nu_{\mathbf{Z}} \subset \mathcal{A} \cup \mathcal{M} \quad \text{if } \theta = \theta_{cr}, \tag{5.20}$$

and whose center of mass satisfies the constraint (4.19) and the boundary conditions (5.1). If $\theta \leqslant \theta_{cr}$, then, in view of (5.19), the couple $(\nu_{\mathbf{Z}}; \mathbf{y})$ given by (5.16) and (5.10) is minimizing. If $\theta > \theta_{cr}$, then, from (5.18), minimizers involve only the austenite phase, and therefore the couple $(\delta_{\mathbf{I}}, \mathbf{Z})$, with $\delta_{\mathbf{I}}$ the Dirac mass centered at the identity $\mathbf{I}$ and arising from a sequence of "flat" deformations, is minimizing. The reversible, temperature activated change of stability between the tunnel and the flat configurations, both sketched in Figure 2, makes it possible to employ the film as an actuator [10, 11].

Let us turn to the case $P \neq 0$. Now the total energy (4.18) is the sum of the bulk energy (5.17) and the free energy of the gas $-PV(\mathbf{y})$, and thus minimizing deformations $\mathbf{y}$ involve gradients from the wells which maximize the volume $V(\mathbf{y})$. In this respect, the tunnel deformation is a good candidate to be a minimizer. In the next subsection, we prove that the tunnel deformation is a minimizer in certain ranges of pressure and temperature. In the proof, we restrict ourselves to consider only cylindrical deformations $\mathbf{y}$, but we believe that our results also hold under weaker hypotheses on $\mathbf{y}$.

5.1. CYLINDRICAL DEFORMATIONS

Let us consider deformation $(\nu_{\mathbf{Z}}, \mathbf{y})$ of the constrained theory with $\nu_{\mathbf{Z}}$ independent of Z_1 and with $\mathbf{y}$ of the type (5.10). Again we assume $\mathcal{M}$ containing a matrix with positive strain so that

$$\gamma := \max\left\{|\mathbf{F}\mathbf{e}_2| : \mathbf{F} \in \mathcal{M}\right\} > 1. \tag{5.21}$$

We assume the film made by a good "tunnel material", so that the maximizer of (5.21) satisfies (5.4) and (5.5). Using the constraint (4.19) and the kinematic assumption (5.10), we find

$$u(Z_2) = \int_{\mathcal{M}} \mathbf{F}\mathbf{e}_2 \cdot \mathbf{e}_2\, \mathrm{d}\nu_{Z_2}(\mathbf{F}), \tag{5.22}$$

$$v(Z_2) = \int_{\mathcal{M}} \mathbf{F}\mathbf{e}_2 \cdot \mathbf{e}_3\, \mathrm{d}\nu_{Z_2}(\mathbf{F}), \tag{5.23}$$

at almost every $Z_2 \in (0, l_2)$. Besides, each deformation of the type (5.10) automatically satisfies the boundary condition $(5.1)_2$, while $(5.1)_1$ gives

$$u(0) = v(0) = 0, \tag{5.24}$$

$$u(l_2) = l_2, \qquad v(l_2) = 0. \tag{5.25}$$

In view of (5.10), (5.24), and (5.25), the volume of the gas under the film has the expression

$$V(\mathbf{y}) = \int_0^{l_2} u'(Z_2)v(Z_2)\,\mathrm{d}Z_2, \tag{5.26}$$

where the prime denotes the first derivative. Therefore, the energy (4.18) reduces to

$$\begin{aligned} e(\nu_{\mathbf{Z}}; \mathbf{y}) &= l_1\left\{\big(l_m(\theta) - l_a(\theta)\big)l_2\bar{\lambda} + l_a(\theta)l_2 - P\int_0^{l_2} u'(Z_2)v(Z_2)\,\mathrm{d}Z_2\right\} \\ &=: \mathcal{E}(\bar{\lambda}; u, v), \end{aligned} \tag{5.27}$$

where

$$\bar{\lambda} := \frac{1}{l_2}\int_0^{l_2}\int_{\mathcal{M}} \mathrm{d}\nu_{Z_2}(\mathbf{F})\,\mathrm{d}Z_2 \in [0, 1] \tag{5.28}$$

indicates the average volume fraction of martensite along the direction of $\mathbf{e}_2$. Using the constraints (5.22), (5.23), the triangle inequality, the definitions (5.21) and (5.28), and recalling that ν_{Z_2} is a probability measure, we get

$$\begin{aligned} \int_0^{l_2} \sqrt{(u')^2 + (v')^2}\,\mathrm{d}Z_2 &\leqslant \int_0^{l_2}\left|\int_{\mathcal{M}} \mathbf{F}\mathbf{e}_2\,\mathrm{d}\nu_{Z_2}\right|\mathrm{d}Z_2 \\ &\leqslant \int_0^{l_2}\int_{\mathcal{M}} |\mathbf{F}\mathbf{e}_2|\,\mathrm{d}\nu_{Z_2}\,\mathrm{d}Z_2 \leqslant l_2(\gamma\bar{\lambda} + 1 - \bar{\lambda}). \end{aligned} \tag{5.29}$$

This inequality provides an upper bound on the length of the curve describing the deformed configuration of the cross-section of the film. Ignoring other possible compatibility conditions between $\bar{\lambda}$ and (u, v) arising from the constraints (5.22), (5.23), we minimize the energy (5.27) with $(\bar{\lambda}; u, v)$ satisfying the constraint (5.29) and the boundary conditions (5.24), (5.25), thereby giving a lower bound for $\mathcal{E}(\bar{\lambda}; u, v)$. We begin by keeping $\bar{\lambda}$ fixed and by minimizing over (u, v). This means that we seek the curve lying in the $(\mathbf{e}_2, \mathbf{e}_3)$ plane, joining the origin with the point $(0, l_2)$, having length not greater than $l_2(\gamma\bar{\lambda} + 1 - \bar{\lambda})$ and enclosing the largest area. In Appendix 2, we prove that the circular arch of length $l_2(\gamma\bar{\lambda}+1-\bar{\lambda})$, parametrized by the couple (U, V) with

$$U(Z_2) := (\gamma\bar{\lambda} + 1 - \bar{\lambda})\int_0^{Z_2} \cos\left(\alpha - \frac{2\alpha}{l_2}t\right)\mathrm{d}t, \tag{5.30}$$

$$V(Z_2) := (\gamma\bar{\lambda} + 1 - \bar{\lambda})\int_0^{Z_2} \sin\left(\alpha - \frac{2\alpha}{l_2}t\right)\mathrm{d}t, \tag{5.31}$$

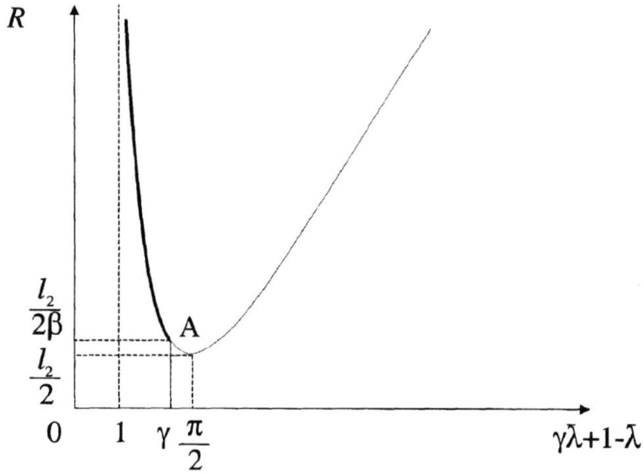

Figure 3. Radius R of the cross-section of the tilm versus $\gamma\bar{\lambda}+1-\bar{\lambda}$.

and α the solution in $(0, 2\pi)$ to the equation

$$\sin\alpha = \frac{\alpha}{(\gamma\bar{\lambda}+1-\bar{\lambda})}, \tag{5.32}$$

encloses the largest area, given by

$$\int_0^{l_2} U'(Z_2)V(Z_2)\,\mathrm{d}Z_2 = \begin{cases} \frac{1}{2}l_2R(\gamma\bar{\lambda}+1-\bar{\lambda}) - \frac{l_2}{2}\sqrt{R^2-\frac{l_2^2}{4}}, & \text{for } 1 < (\gamma\bar{\lambda}+1-\bar{\lambda}) \leqslant \frac{\pi}{2}, \\ \frac{1}{2}l_2R(\gamma\bar{\lambda}+1-\bar{\lambda}) + \frac{l_2}{2}\sqrt{R^2-\frac{l_2^2}{4}}, & \text{for } (\gamma\bar{\lambda}+1-\bar{\lambda}) > \frac{\pi}{2}. \end{cases} \tag{5.33}$$

In the last equation, R, which denotes the radius of the arch parametrized by (U, V), satisfies the implicit relation

$$\sin\left((\gamma\bar{\lambda}+1-\bar{\lambda})\frac{l_2}{2R}\right) = \frac{l_2}{2R}, \tag{5.34}$$

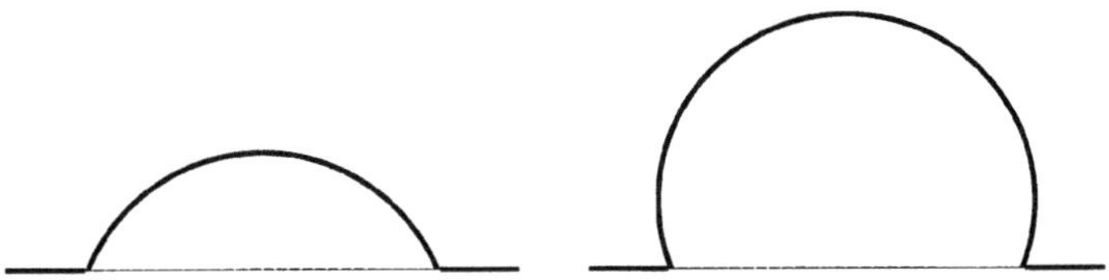

Figure 4. Two possible tunnel configurations.

plotted in Figure 3. The presence of the two expression in (5.33) refers to the two cases shown in Figure 4. Substituting (5.33) into the expression of the total energy (5.27), we get the lower bound

$$\mathcal{E}(\bar{\lambda}; u, v) \geqslant \begin{cases} l_1 l_2 \Big\{ \big(l_m(\theta) - l_a(\theta)\big)\bar{\lambda} + l_a(\theta) - \dfrac{P}{2}(\gamma\bar{\lambda} + 1 - \bar{\lambda})R \\ \quad + \dfrac{P}{2}\sqrt{R^2 - \dfrac{l_2^2}{4}} \Big\}, \quad \text{for } 1 < (\gamma\bar{\lambda} + 1 - \bar{\lambda}) \leqslant \dfrac{\pi}{2}, \\ l_1 l_2 \Big\{ \big(l_m(\theta) - l_a(\theta)\big)\bar{\lambda} + l_a(\theta) - \dfrac{P}{2}(\gamma\bar{\lambda} + 1 - \bar{\lambda})R \\ \quad - \dfrac{P}{2}\sqrt{R^2 - \dfrac{l_2^2}{4}} \Big\}, \quad \text{for } (\gamma\bar{\lambda} + 1 - \bar{\lambda}) > \dfrac{\pi}{2}. \end{cases} \tag{5.35}$$

Because the right-hand side turns out to depend only upon $\bar{\lambda}$, the bound can be further improved by minimizing with respect to $\bar{\lambda}$. Let $\bar{\lambda}_{\min} \in [0, 1]$ denote the minimizer. Now we show that there exists a family of Young measures which achieves the lower bound (5.35). This family arises from the sequence $(\mathbf{y}^n_{,1}|\mathbf{y}^n_{,2}|\mathbf{b}^n)$ with

$$\mathbf{y}^n(Z_1, Z_2) = Z_1\mathbf{e}_1 + u_n(Z_2)\mathbf{e}_2 + v_n(Z_2)\mathbf{e}_3, \tag{5.36}$$

$$u_n(Z_2) := \int_0^{Z_2} r_n(t)\cos(\beta_n t + \alpha_n)\,\mathrm{d}t, \tag{5.37}$$

$$v_n(Z_2) := \int_0^{Z_2} r_n(t)\sin(\beta_n t + \alpha_n)\,\mathrm{d}t, \tag{5.38}$$

and with r_n the piecewise constant periodic function with period l_2/n such that

$$r_n(t) := \begin{cases} \gamma & \text{for } p\frac{l_2}{n} \leqslant t \leqslant p\frac{l_2}{n} + \bar{\lambda}_{\min}\frac{l_2}{n}, \\ 1 & \text{for } p\frac{l_2}{n} + \bar{\lambda}_{\min}\frac{l_2}{n} < t \leqslant (p+1)\frac{l_2}{n}, \end{cases} \tag{5.39}$$

p being an integer between 0 and $n - 1$. The sequence of Cosserat directors $\mathbf{b}^n$ is defined by

$$\mathbf{b}^n(Z_2) = \mathbf{R}(Z_2)\overline{\mathbf{R}}\mathbf{Q}\mathbf{U}\mathbf{e}_3 \quad \text{wherever } r_n(Z_2) = \gamma, \tag{5.40}$$

$$\mathbf{b}^n(Z_2) = \mathbf{R}(Z_2)\mathbf{e}_3 \quad \text{wherever } r_n(Z_2) = 1. \tag{5.41}$$

Here, $\mathbf{R}(Z_2)$ and $\overline{\mathbf{R}}$ are the rotations (5.14), (5.15), $\mathbf{U}$ a solution to the maximum problem in (5.21), and $\mathbf{Q}$ a rotation such that $\mathbf{QU}$ satisfies the condition (5.7)–(5.9) for building up a tunnel.

By construction, $u_n(0) = v_n(0) = 0$. In Appendix 3, we prove the existence of $\alpha_n^*, \beta_n^* \in [0, 2\pi]$ for which the boundary conditions (5.25) are also satisfied. We also show that

$$\alpha_n^* \to \alpha, \tag{5.42}$$

$$\beta_n^* \to -\frac{2\alpha}{l_2}, \tag{5.43}$$

up to a subsequence, with α as in (5.32). Using (5.42), (5.43), the weak convergence of r_n and $\mathbf{b}^n$ to their averages [14, Theorem 1.5], and the definitions (5.14), (5.15), we find

$$\left(\mathbf{y}^n_{,1}|\mathbf{y}^n_{,2}|\mathbf{b}^n\right) \overset{*}{\rightharpoonup} \bar{\lambda}_{\min}\mathbf{R}(Z_2)\overline{\mathbf{R}}\mathbf{QU} + \left(1 - \bar{\lambda}_{\min}\right)\mathbf{R}(Z_2) \quad \text{in } L^\infty. \tag{5.44}$$

To compute the Young measure generated by $(\mathbf{y}^n_{,1}|\mathbf{y}^n_{,2}|\mathbf{b}^n)$, we note that for any function $\psi \in C_0(M^{3\times 3}; \mathbb{R})$

$$\psi\left(\mathbf{y}^n_{,1}|\mathbf{y}^n_{,2}|\mathbf{b}^n\right) \overset{*}{\rightharpoonup} \bar{\lambda}_{\min}\psi(\mathbf{R}(Z_2)\overline{\mathbf{R}}\mathbf{QU}) + (1 - \bar{\lambda}_{\min})\psi\left(\mathbf{R}(Z_2)\right) \quad \text{in } L^\infty \tag{5.45}$$

[23, Corollary 3.3]. Thus, in view of (4.13), the family of Young measures arising from $(\mathbf{y}^n_{,1}|\mathbf{y}^n_{,2}|\mathbf{b}^n)$ is given by

$$\hat{\nu}_{Z_2} = \bar{\lambda}_{\min}\delta_{\mathbf{R}(Z_2)\overline{\mathbf{R}}\mathbf{QU}} + (1 - \bar{\lambda}_{\min})\delta_{\mathbf{R}(Z_2)}. \tag{5.46}$$

Because

$$\left|\int_{\mathcal{M}} \mathbf{F}\mathbf{e}_2 \, \mathrm{d}\hat{\nu}_{Z_2}\right| = \bar{\lambda}_{\min}\gamma + 1 - \bar{\lambda}_{\min}, \tag{5.47}$$

the family of Young measures $\hat{\nu}$ achieves the lower bound in (5.35).

Let us now turn to evaluating $\bar{\lambda}_{\min}$. We differentiate the right-hand side of (5.35) with respect to $\bar{\lambda}$ and use (5.34) to get

$$\frac{\mathrm{d}\mathcal{E}(\bar{\lambda}; U, V)}{\mathrm{d}\bar{\lambda}} = l_1 l_2\left[(l_m(\theta) - l_a(\theta)) - P(\gamma - 1)R\right], \tag{5.48}$$

which vanishes at the unique solution

$$R = \frac{(l_m(\theta) - l_a(\theta))}{P(\gamma - 1)}. \tag{5.49}$$

While there is a unique stationary point R, the corresponding value of $\bar{\lambda}$ may not be unique (i.e., see Figure 3 with $\gamma > \pi/2$). The minimizing values of $\bar{\lambda}$ are the following:

- Case 1.

$$\gamma \leqslant \pi/2, \quad \text{and} \quad P < 2\beta \frac{(l_m(\theta) - l_a(\theta))}{l_2(\gamma - 1)} =: P_1,$$

with β the solution in $(0, \pi/(2\gamma))$ to the equation

$$\sin(\gamma\beta) = \beta. \tag{5.50}$$

There is a unique minimizer $\bar{\lambda}_{\min}$ given by

$$\bar{\lambda}_{\min} = \frac{2(l_m(\theta) - l_a(\theta))}{l_2 P(\gamma - 1)^2} \arcsin\left[\frac{l_2 P(\gamma - 1)}{2(l_m(\theta) - l_a(\theta)}\right] - \frac{1}{(\gamma - 1)}. \tag{5.51}$$

- Case 2. $\gamma \leqslant \pi/2$, and $P \geqslant P_1$. The unique global minimizer is $\bar{\lambda}_{\min} = 1$.
- Case 3.

$$\gamma > \pi/2, \quad \text{and} \quad P < 2\xi \frac{(l_m(\theta) - l_a(\theta))}{l_2(\gamma - 1)} =: P_2,$$

with $\xi \in (0, 1)$ the solution to the equation

$$\frac{1}{\xi^2} \arcsin \xi + \frac{1}{\xi}\sqrt{1 - \xi^2} = \frac{2\gamma}{\xi} - \frac{1}{\beta}\left[\gamma + \sqrt{1 - \beta^2}\right], \tag{5.52}$$

where β is now the solution in $(\pi/(2\gamma); 1)$ to (5.50). The unique global minimizer is again given by (5.51).
- Case 4. $\gamma > \pi/2$, and $P = P_2$. There are two global minimizer, one at $\bar{\lambda}$ given by (5.51) and one at $\bar{\lambda}_{\min} = 1$.
- Case 5. $\gamma > \pi/2$, and $P > P_2$. The unique global minimizer is $\bar{\lambda}_{\min} = 1$.

In Case 3, there is also a relative minimum at $\bar{\lambda} = 1$ if

$$2\beta \frac{(l_m(\theta) - l_a(\theta))}{l_2(\gamma - 1)} < P < P_2, \tag{5.53}$$

while in Case 5 there is a relative minimum at $\bar{\lambda}$ given by (5.51) if

$$P_2 < P < 2\frac{(l_m(\theta) - l_a(\theta))}{l_2(\gamma - 1)}. \tag{5.54}$$

To give a physical interpretation of these results, in all cases we fix the temperature θ above the transformation temperature and the material (so γ is fixed), and we increase the pressure starting from an appropriate value. In Case 1, a fine mixture of austenite and martensite is globally stable. The volume fraction of martensite $\bar{\lambda}_{\min}$ increases as P increases starting from zero. Correspondingly, the length of the cross-section, which has the form of a circular arch, increases and the film encloses

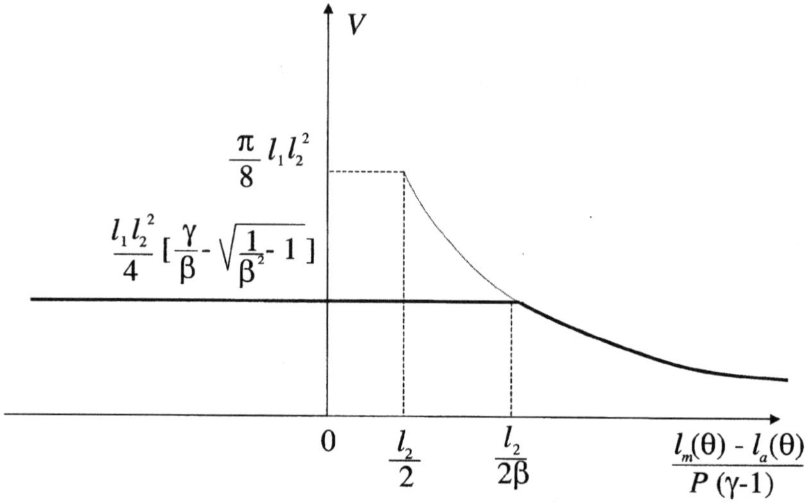

Figure 5. Volume V enclosed by the film versus the ratio $(l_m(\theta) - l_a(\theta))/(P(\gamma - 1))$ in the case $\gamma \in (1, \pi/2]$.

more and more volume. From (5.33), (5.49), and (5.51), we find that this volume is given by

$$V = \frac{l_1 l_2^2}{4}\left[\frac{4(l_m(\theta) - l_a(\theta))^2)}{P l_2^2(\gamma - 1)^2} \arcsin\left(\frac{(P l_2(\gamma - 1))}{2(l_m(\theta) - l_a(\theta))}\right) - \sqrt{\frac{4(l_m(\theta) - l_a(\theta))^2)}{P l_2^2(\gamma - 1)^2} - 1}\right], \tag{5.55}$$

which corresponds to the curve plotted in Figure 5. For $P = P_1$ (see Case 2), we have $\bar{\lambda}_{\min} = 1$ and thus the austenite has completely transformed. At this point the length of the cross-section reaches its maximum value $l_2\gamma$ and the film encloses the volume

$$V = \frac{l_1 l_2^2}{4}\left[\frac{\gamma}{\beta} - \frac{1}{\beta}\sqrt{1 - \beta^2}\right], \tag{5.56}$$

with β defined as in Case 1. If the pressure is further increased from P_1, the length of the cross-section and the volume enclosed remain constant.

In Case 3, as the pressure increases, a mixture of austenite and martensite with increasing volume fraction (5.51) is globally stable. The volume enclosed by the film is still given by the relation (5.55) plotted now in Figure 6. For P as in (5.53), the martensite becomes a metastable configuration and at $P = P_2$, (see Case 4), both the martensite and the mixture are globally stable. If P is further increased (Case 5), then the mixture becomes metastable while the martensite becomes globally stable. The presence of metastable states introduces the possibility of a hysteresis loop in Figure 6.

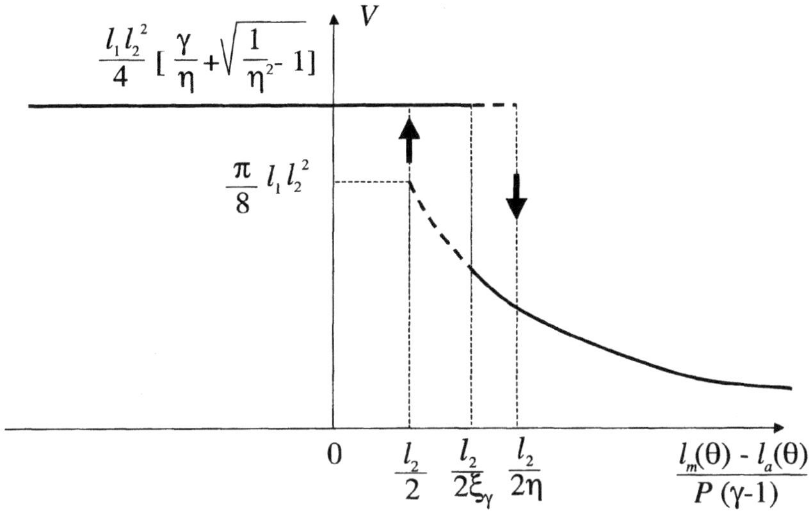

Figure 6. Volume V enclosed by the film versus the ratio $(l_m(\theta) - l_a(\theta))/(P(\gamma - 1))$ in the case $\gamma \in (\pi/2, +\infty)$. Dashed curve corresponds to relative minimizers.

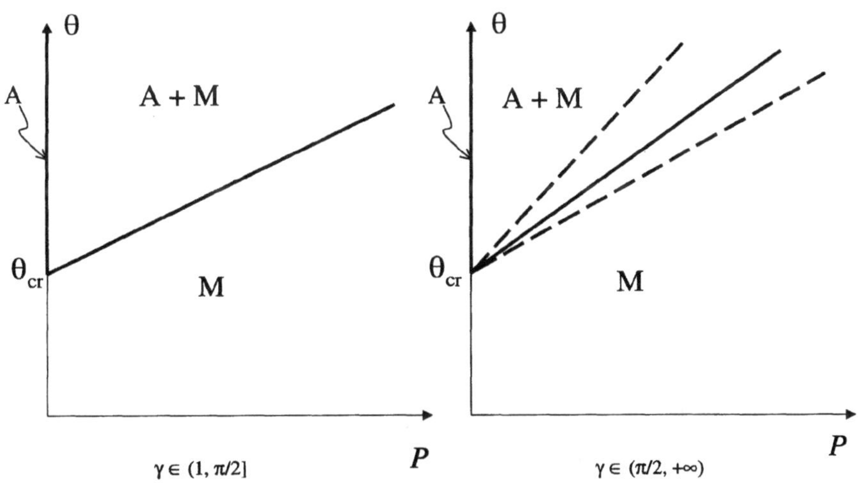

Figure 7. Pressure–temperature phase diagram for the material of the film. The cases drawn correspond to linear relations $l_m(\theta)$, $l_a(\theta)$. The sector enclosed by dashed lines contains metastable states. Note that some martensite is present even at high temperatures.

If we assume $l_m(\theta)$ and $l_a(\theta)$ to depend linearly upon the temperature, we can summarize these results in the pressure-temperature phase diagram depicted in Figure 7, in which the two cases $\gamma \leqslant \pi/2$, and $\gamma > \pi/2$ are drawn separately. For values of the pressure and of the temperature lying in the regions marked with M, the tunnel deformation with $\mathbf{y}$ given by (5.10), (5.30), (5.31) with $\bar{\lambda}_{\min} = 1$, and $\mathbf{b} = \mathbf{R}(Z_2)\overline{\mathbf{R}}\mathbf{QUe}_3$ is globally stable. At each point of the film, the material

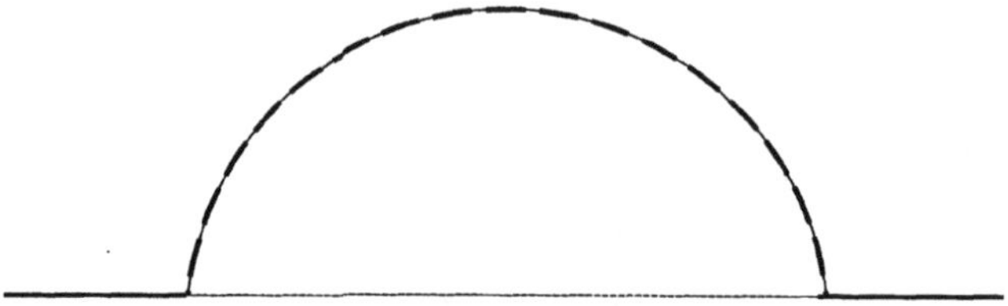

Figure 8. Microstructure of the film at equilibrium.

is in the variant of martensite $\mathbf{U}$. For values of pressure and temperature lying in the regions marked with $A + M$, the macroscopic deformation of the film is now given by the tunnel deformation (5.10), (5.30), (5.31) evaluated at $\bar{\lambda} = \bar{\lambda}_{\min}$, and by $\mathbf{b} = (\bar{\lambda}_{\min}\mathbf{R}(Z_2)\bar{\mathbf{R}}\mathbf{QU} + (1 - \bar{\lambda}_{\min}))\mathbf{R}(Z_2))\mathbf{e}_3$; correspondingly, the material is a fine mixture of austenite and of the variant of martensite $\mathbf{U}$. The family of Young measures (5.46) describes the microstructure of the material. Recalling the construction of the sequence generating the measures (5.46), the microstructure is found to consist of martensitic regions alternated with austenitic region, both regions having the shape of thin strips parallel to the axis of the tunnel, as depicted in Figure 8.

Acknowledgement

The authors thank AFOSR/MURI F49620-98-1-0433 for supporting this research. The work also benefitted from the support of ONR N00014-95-1-1145 and -91-J-4034, ARO DA/DAAG55-98-1-0335, NSF DMS-9505077 and DMS 0074043, and the Isaac Newton Institute for Mathematical Sciences. RR also thanks the Italian M.U.R.S.T. for the financial support provided through the project "Modelli matematici per la scienza dei materiali" Co.fin. (1998).

Appendix A

We now give a proof of the identity (2.26).

LEMMA A.1. *Let* $\mathbf{y} \in \mathcal{H}$ *and* $t > 0$. *Then*

$$\frac{1}{3}\int_{S\times\{t\}} \mathbf{y}\cdot(\mathbf{y}_{,1}\wedge\mathbf{y}_{,2})\,\mathrm{d}X_1\,\mathrm{d}X_2 - \frac{1}{3}\int_{S\times\{0\}} \mathbf{y}\cdot(\mathbf{y}_{,1}\wedge\mathbf{y}_{,2})\,\mathrm{d}X_1\,\mathrm{d}X_2 + \frac{2}{3}t|S|\det\mathbf{A} = \int_{S\times(0,t)} \det\nabla\mathbf{y}\,\mathrm{d}\mathbf{X}. \tag{A.1}$$

Proof. Let $\mathbf{y}^n \in C^\infty(\Omega^h;\mathbb{R}^3)$ be a smooth sequence approximating $\mathbf{y}$ in $\mathcal{H}$. Denoting with ϵ_{ijk} the Ricci tensor, we have

$$\int_{S\times(0,t)} \det\nabla\mathbf{y}^n\,\mathrm{d}\mathbf{X} = \int_{S\times(0,t)} \frac{1}{6}\,\epsilon_{ijk}\epsilon_{pqr}y^n_{i,p}y^n_{j,q}y^n_{k,r}\,\mathrm{d}\mathbf{X}. \tag{A.2}$$

By integrating by parts and applying the divergence theorem, we obtain

$$\int_{S\times(0,t)} \det\nabla\mathbf{y}^n \, d\mathbf{X} = -\int_{S\times(0,t)} \frac{1}{6}\epsilon_{ijk}\epsilon_{pqr}\left(y^n_{i,p}y^n_{j,q}\right)_{,r} y^n_k \, d\mathbf{X} + \int_{\partial(S\times(0,t))} \frac{1}{6}\epsilon_{ijk}\epsilon_{pqr} y^n_{i,p} y^n_{j,q} y^n_k N_r \, da, \tag{A.3}$$

where $\mathbf{N} = N_1\mathbf{e}_1 + N_2\mathbf{e}_2 + N_3\mathbf{e}_3$ is the outward unit normal to $\partial(S \times (0,t))$. The first term on the right hand side of (A.3) is zero because it is the inner product of symmetric and antisymmetric tensors. Hence,

$$\begin{aligned}\int_{S\times(0,t)} \det\nabla\mathbf{y}^n \, d\mathbf{X} &= \int_{\partial(S\times(0,t))} \frac{1}{3}(\mathrm{Adj}\nabla\mathbf{y}^n)\mathbf{y}^n \cdot \mathbf{N} \, da,\\ &= -\int_{S\times\{0\}} \frac{1}{3}(\mathrm{Adj}\nabla\mathbf{y}^n)\mathbf{y}^n \cdot \mathbf{e}_3 \, dX_1 \, dX_2\\ &\quad + \int_{S\times\{t\}} \frac{1}{3}(\mathrm{Adj}\nabla\mathbf{y}^n)\mathbf{y}^n \cdot \mathbf{e}_3 \, dX_1 \, dX_2\\ &\quad + \int_{\partial S\times\{0,t\}} \frac{1}{3}(\mathrm{Adj}\nabla\mathbf{y}^n)\mathbf{y}^n \cdot \mathbf{N} \, da.\end{aligned} \tag{A.4}$$

Let $\mathbf{X}(s)$ be a parametrization of the boundary of S, with $\mathbf{X}(0) = \mathbf{X}(1)$. Then, $(\mathbf{X}(s), X_3)$ is a parametrization of the surface $\partial S \times \{0, t\}$, which is assumed to orient the surface $\partial S \times \{0, t\}$ so that the vector

$$\frac{\mathbf{X}_{,s} \wedge \mathbf{e}_3}{|\mathbf{X}_{,s} \wedge \mathbf{e}_3|} \tag{A.5}$$

is the outward pointing unit normal. Using the identities

$$(\mathrm{Adj}\mathbf{A})^{\mathrm{T}}(\mathbf{b} \wedge \mathbf{c}) = \mathbf{Ab} \wedge \mathbf{Ac},$$
$$(\mathrm{Adj}\mathbf{A})\mathbf{A} = (\det\mathbf{A})\mathbf{I},$$

and the boundary conditions (2.2), we obtain

$$\begin{aligned}&\int_{\partial S\times\{0,t\}} (\mathrm{Adj}\nabla\mathbf{y}^n(\mathbf{X}))\mathbf{y}^n(\mathbf{X}) \cdot \mathbf{N}(\mathbf{X}) \, da\\ &= \int_0^t \int_0^1 (\mathrm{Adj}\nabla\mathbf{y}^n(\mathbf{X}(s), X_3))\mathbf{y}^n(\mathbf{X}(s), X_3) \cdot (\mathbf{X}(s)_{,s} \wedge \mathbf{e}_3) \, ds \, dX_3\\ &= \int_0^t \int_0^1 \mathbf{y}^n(\mathbf{X}(s), X_3) \cdot \left(\mathrm{Adj}\nabla\mathbf{y}^n(\mathbf{X}(s), X_3)\right)^{\mathrm{T}} (\mathbf{X}(s)_{,s} \wedge \mathbf{e}_3) \, ds \, dX_3\\ &= \int_0^t \int_0^1 \mathbf{y}^n(\mathbf{X}(s), X_3) \cdot \left(\nabla\mathbf{y}^n(\mathbf{X}(s), X_3)\mathbf{X}(s)_{,s} \wedge \nabla\mathbf{y}^n(\mathbf{X}(s), X_3)\mathbf{e}_3\right) ds \, dX_3\\ &= \int_0^t \int_0^1 \mathbf{A}(\mathbf{X}(s), X_3) \cdot \left(\mathbf{A}\mathbf{X}(s)_{,s} \wedge \mathbf{A}\mathbf{e}_3\right) ds \, dX_3\end{aligned}$$

$$= \int_0^t \int_0^1 \mathbf{A}\big(\mathbf{X}(s), X_3\big) \cdot (\mathrm{Adj}\mathbf{A})^{\mathrm{T}}\big(\mathbf{X}(s)_{,s} \wedge \mathbf{e}_3\big)\, \mathrm{d}s\, \mathrm{d}X_3$$

$$= \int_0^t \int_0^1 (\mathrm{Adj}\mathbf{A})\mathbf{A}\big(\mathbf{X}(s), X_3\big) \cdot \big(\mathbf{X}(s)_{,s} \wedge \mathbf{e}_3\big)\, \mathrm{d}s\, \mathrm{d}X_3$$

$$= \det\mathbf{A} \int_0^t \int_0^1 (\mathbf{X}(s), X_3) \cdot \big(\mathbf{X}(s)_{,s} \wedge \mathbf{e}_3\big)\, \mathrm{d}s\, \mathrm{d}X_3. \tag{A.6}$$

By the divergence theorem,

$$\int_0^1 \big(\mathbf{X}(s), X_3\big) \cdot \big(\mathbf{X}(s)_{,s} \wedge \mathbf{e}_3\big)\, \mathrm{d}s = \int_S \mathrm{div}\mathbf{X}\, \mathrm{d}\mathbf{X} = 2|S|, \tag{A.7}$$

which, substituted into (A.6), gives

$$\int_{\partial S \times \{0,t\}} \big(\mathrm{Adj}\nabla\mathbf{y}^n\big)\mathbf{y}^n \cdot \mathbf{N}\, \mathrm{d}a = 2t|S|\det A. \tag{A.8}$$

This, together with (A.4) and the identity $(\mathrm{Adj}\nabla\mathbf{y})\mathbf{y}\cdot\mathbf{e}_3 = \mathbf{y}\cdot(\mathbf{y}_{,1}\wedge\mathbf{y}_{,2})$, gives (A.1) for the approximating sequence $\mathbf{y}^n$. Letting $n \to \infty$, by the continuity in $W^{2,2}$ of the volume functional and of the last term in (A.1), we obtain (A.1) for $\mathbf{y}$. □

Appendix B

We prove a convenient 2-dimensional version of the isoperimetric inequality.

LEMMA B.2. *Let $\gamma > 1$ and $\bar\lambda \in [0, 1]$ be given. Then, for any couple $(u, v) \in (W^{1,3}(0, l_2))^2$ satisfying the boundary conditions (5.24), (5.25) and the constraint (5.29),*

$$\int_0^{l_2} u'(Z_2)v(Z_2)\, \mathrm{d}Z_2 \leqslant \begin{cases} \dfrac{1}{2} l_2 R(\gamma\bar\lambda + 1 - \bar\lambda) - \dfrac{l_2}{2}\sqrt{R^2 - \dfrac{l_2^2}{4}}, \\ \qquad \text{for } 1 < \gamma\bar\lambda + 1 - \bar\lambda \leqslant \dfrac{\pi}{2}, \\ \dfrac{1}{2} l_2 R(\gamma\bar\lambda + 1 - \bar\lambda) + \dfrac{l_2}{2}\sqrt{R^2 - \dfrac{l_2^2}{4}}, \\ \qquad \text{for } \gamma\bar\lambda + 1 - \bar\lambda > \dfrac{\pi}{2}. \end{cases} \tag{B.1}$$

Proof. We recall that for any two planar curves parametrized by the couples (u, v) and (ϕ, ψ) in $(W^{1,3}(0, l_2))^2$ satisfying the same boundary conditions, the following 2-dimensional version of the isoperimetric inequality (2.19) holds:

$$\left|\int_0^{l_2} \big(u'(Z_2)v(Z_2) - \phi'(Z_2)\psi(Z_2)\big)\,\mathrm{d}Z_2\right| \leqslant \frac{1}{4\pi}\left[\int_0^{l_2} \left(\sqrt{(u')^2(Z_2)+(v')^2(Z_2)} + \sqrt{(\phi')^2(Z_2)+(\psi')^2(Z_2)}\right)\mathrm{d}Z_2\right]^2 \tag{B.2}$$

[24], from which, using the triangle inequality, we have

$$\int_0^{l_2} u'(Z_2)v(Z_2)\,\mathrm{d}Z_2 \leqslant \int_0^{l_2} \phi'(Z_2)\psi(Z_2)\,\mathrm{d}Z_2 + \frac{1}{4\pi}\left[\int_0^{l_2}\left(\sqrt{(u')^2(Z_2)+(v')^2(Z_2)} + \sqrt{(\phi')^2(Z_2)+(\psi')^2(Z_2)}\right)\mathrm{d}Z_2\right]^2. \tag{B.3}$$

Let now us choose

$$\phi(Z_2) = \frac{\pi-\alpha}{\sin\alpha}\int_0^{Z_2} \cos\left[\frac{2}{l_2}(\pi-\alpha)s - (\pi-\alpha)\right]\mathrm{d}s, \tag{B.4}$$

$$\psi(Z_2) = \frac{\pi-\alpha}{\sin\alpha}\int_0^{Z_2} \sin\left[\frac{2}{l_2}(\pi-\alpha)s - (\pi-\alpha)\right]\mathrm{d}s, \tag{B.5}$$

with α the solution in $(0, 2\pi)$ to the equation (5.32). The couple (ϕ, ψ) is a parametric representation of a circular arch joining the origin with the point $(0, l_2)$, lying in the negative halfplane, and having length

$$\int_0^{l_2} \sqrt{(\phi')^2(Z_2)+(\psi')^2(Z_2)}\,\mathrm{d}Z_2 = 2\pi R - l_2(\gamma\bar{\lambda} + 1 - \bar{\lambda}), \tag{B.6}$$

with the radius of the arch R satisfying (5.34). The (algebraic) area enclosed by (ϕ, ψ) is given by

$$\int_0^{l_2} \phi'(Z_2)\psi(Z_2)\,\mathrm{d}Z_2 = \begin{cases} \frac{1}{2}l_2R(\gamma\bar{\lambda}+1-\bar{\lambda}) - \pi R^2 - \frac{l_2}{2}\sqrt{R^2 - \frac{l_2^2}{4}}, & \text{if } \gamma \in \left(1, \frac{\pi}{2}\right], \\ \frac{1}{2}l_2R(\gamma\bar{\lambda}+1-\bar{\lambda}) - \pi R^2 + \frac{l_2}{2}\sqrt{R^2 - \frac{l_2^2}{4}}, & \text{if } \gamma \in \left(\frac{\pi}{2}, +\infty\right). \end{cases} \tag{B.7}$$

Substituting (B.6) and (B.7) into the isoperimetric inequality (B.3) gives (B.1). □

A straightforward calculation shows that the upper value in (B.1) is achieved by the circular arch parametrized by the couple (U, V) with U and V given by (5.30), (5.31).

Appendix C

LEMMA C.3. *Let $\gamma > 1$ and $\bar{\lambda} \in [0, 1]$ be given. Then, the system of equations*

$$\begin{cases} \int_0^{l_2} r_n(t)\cos(\beta_n t + \alpha_n)\,dt = l_2, \\ \int_0^{l_2} r_n(t)\sin(\beta_n t + \alpha_n)\,dt = 0, \end{cases} \tag{C.1}$$

with r_n defined as in (5.39), admits at least a solution (α_n^*, β_n^*) for each integer n. Moreover,

$$\alpha_n^* \to \alpha, \tag{C.2}$$

$$\beta_n^* \to -\frac{2\alpha}{l_2}, \tag{C.3}$$

up to a subsequence, with α the solution in $(0, 2\pi)$ to (5.32).

Proof. We change variables to reduce the system (C.1) to the form

$$\begin{cases} \int_0^1 q_n(t)\cos(\beta_n l_2 s + \alpha_n)\,ds = 1, \\ \int_0^1 q_n(t)\sin(\beta_n l_2 s + \alpha_n)\,ds = 0, \end{cases} \tag{C.4}$$

with $q_n(s) := r_n(l_2 s)$. The system (C.4) is equivalent to

$$\begin{cases} \cos\alpha_n \displaystyle\int_0^1 q_n(s)\cos(\beta_n l_2 s)\,ds - \sin\alpha_n \int_0^1 q_n(s)\sin(A_n s)\,ds = 1, \\ \cos\alpha_n \displaystyle\int_0^1 q_n(s)\sin(\beta_n l_2 s)\,ds + \sin\alpha_n \int_0^1 q_n(s)\cos(\beta_n l_2 s)\,ds = 0. \end{cases} \tag{C.5}$$

Take the square, sum and use the trigonometric identity $\sin^2\alpha_n + \cos^2\alpha_n = 1$ to get

$$1 = \left[\int_0^1 q_n(s)\cos(\beta_n l_2 s)\,ds\right]^2 + \left[\int_0^1 q_n(s)\sin(\beta_n l_2 s)\,ds\right]^2 =: f^n(\beta_n). \tag{C.6}$$

Note that $f^n(0) = 0$. Besides, because

$$\lim_{n\to\infty} f^n(2\pi) = 0, \tag{C.7}$$

there exists an integer, say n_0, such that for each $n > n_0$

$$f^n(2\pi) < \frac{1}{2}. \tag{C.8}$$

Therefore, using the continuity of f^n, we conclude that there exists a $\beta_n^* \in [0, 2\pi]$ which solves (C.6). Besides, in view of (C.6), we may define α_n^* in $[0, 2\pi]$ such that

$$\cos\alpha_n^* = \int_0^1 q_n(s)\cos(\beta_n l_2 s)\,\mathrm{d}s. \tag{C.9}$$

From (C.6) and (C.9), we find

$$\left[\int_0^1 q_n(s)\sin(\beta_n l_2 s)\,\mathrm{d}s\right]^2 = \sin^2\alpha_n^*. \tag{C.10}$$

Since there are two solutions of (C.10), without loss of generality we choose

$$\left[\int_0^1 q_n(s)\sin(\beta_n l_2 s)\,\mathrm{d}s\right] = -\sin\alpha_n^*. \tag{C.11}$$

On using (C.10), (C.11), and (C.6), the equations in (C.5) are identically satisfied for α_n^*, β_n^*. Besides, because $\alpha_n^*, \beta_n^* \in [0, 2\pi]$, we have

$$\alpha_n^* \to \alpha, \tag{C.12}$$
$$\beta_n^* \to \beta, \tag{C.13}$$

up to a subsequence. The limit problem associated with (C.1)

$$\begin{cases} (\gamma\bar{\lambda} + 1 - \bar{\lambda})\displaystyle\int_0^{l_2} \cos(\beta t + \alpha)\,\mathrm{d}t = l_2, \\ (\gamma\bar{\lambda} + 1 - \bar{\lambda})\displaystyle\int_0^{l_2} \sin(\beta t + \alpha)\,\mathrm{d}t = 0, \end{cases} \tag{C.14}$$

is equivalent to

$$\begin{cases} \beta = -\dfrac{2\alpha}{l_2}, \\ \sin\alpha = \dfrac{\alpha}{(\bar{\lambda}\gamma + 1 - \bar{\lambda})}. \end{cases} \tag{C.15}$$

This system admits a unique solution in $(0, 2\pi)$, since $(\gamma\bar{\lambda} + 1 - \bar{\lambda}) > 1$ by hypothesis. □

References

1. E. Acerbi, G. Buttazzo and D. Percivale, A variational definition for the strain energy of an elastic string. *J. Elasticity* **25** (1991) 137–148.
2. R.A. Adams, *Sobolev Spaces*. Academic Press, New York (1975).
3. J.M. Ball, Convexity conditions and existence theorems in nonlinear elasticity. *Arch. Rational Mech. Anal.* **63** (1977) 337–403.

4. J.M. Ball, Discontinuous equilibrium solutions and cavitation in nonlinear elasticity. *Phil. Trans. Roy. Soc. London* **A 306** (1982) 557–611.
5. J.M. Ball, A version of the fundamental theorem for Young measures, PDE's and continuum models of phase transitions. In: M. Rascle, D. Serre and M. Slemrod (eds), *Lecture Notes in Physics*, Vol. 344. Springer, Berlin (1989) pp. 207–215.
6. J.M. Ball, C. Chu and R.D. James, Hysteresis during stress-induced variant rearrangement. *J. Phys. IV* **5**, colloque C8 (1995) 245–251.
7. J.M. Ball and R.D. James, Fine phase mixtures as minimizers of energy. *Arch. Rational Mech. Anal.* **100** (1987) 13–52.
8. J.M. Ball and R.D. James, Proposed experimental tests of a theory of fine microstructure and the two-well problem. *Phil. Trans. Roy. Soc. London* **A 338** (1992) 389–450.
9. P. Bělik, Brule and Luskin, On numerical modeling of deformations of pressurized martensitic thin films. Preprint (1999).
10. K. Bhattacharya and R.D. James, A theory of shape-memory thin films with applications. In: E.P. George, R. Gotthardt, K. Otsuka, S. Trolier-McKinstry and M. Wun-Fogle, (eds), *Materials for Smart Systems II*, MRS Symposium Proceedings Series, Vol. 459. Materials Research Society, Pittsburgh (1997) pp. 311–316.
11. K. Bhattacharya and R.D. James, A theory of thin films of martensitic materials with applications to microactuators. *J. Mech. Phys. Solids* **47** (1999) 531–576.
12. L. Cesari, Surface area. *Ann. of Math.* **35** (1956).
13. P.G. Ciarlet, *Mathematical Elasticity. Vol. I: Three-Dimensional Elasticity*. North-Holland, Amsterdam (1988).
14. B. Dacorogna, *Direct Methods in the Calculus of Variations*. Springer, Berlin (1989).
15. J.W. Dong, L.C. Chen, R.D. James, S. McKernan and C.J. Palmstrøm, MBE growth of ferromagnetic single crystal (001) Ni_2MnGa on (001) GaAs. *Appl. Phys. Lett.* **75** (1999) 1443–1445.
16. J. Douglas, Solution of the problem of Plateau. *Trans. Amer. Math. Soc.* **33** (1931) 263–321.
17. L.C. Evans and R.F. Gariepy, *Measure Theory and Fine Properties of Functions*. CRC Press, Boca Raton, FL (1992).
18. W. Kitsche, I. Müller and P. Strehlow, Simulation of pseudo–elastic behavior in a system of rubber balloons. In: S. Antman, J.L. Ericksen, D. Kinderlehrer and I. Müller (eds), *Metastability and Incompletely Posed Problems*, IMA Volumes in Mathematics and its Applications, Vol. 3 (1987) 213–239.
19. P. Krulevitch, A.P. Lee, P.B. Ramsey, J.C. Trevino, J. Hamilton and M.A. Northrup, Thin film shape memory alloy microactuators. *J. MEMS* **5** (1996) 270–282.
20. H. Le Dret and A. Raoult, The nonlinear membrane model as variational limit of nonlinear 3-dimensional elasticity. *J. Math. Pures Appl.* **74** (1995) 549–578.
21. H. Le Dret and A. Raoult, The membrane shell model: a variational asymptotic derivation. *J. Nonlinear Sci.* **6** (1996) 59–84.
22. C.B. Morrey, *Multiple Integrals in the Calculus of Variations*. Springer, Berlin (1966).
23. S. Müller, Variational models for microstructures and phase transitions, Lectures delivered at the C.I.M.E. summer school. In: *Calculus of Variations and Geometric Evolution Problems*. Cetraro (1996).
24. T. Rado, The isoperimetric inequality and the Lebesgue measure definition of surface area. *J. Nonlinear Sci.* **6** (1947) 59–84.
25. Y.C. Shu, Heterogeneous thin films of martensitic materials. *Arch. Rational Mech. Anal.* (1998), submitted.
26. H.C. Wente, An existence theorem for surfaces of constant mean curvature. *J. Math. Anal. Appl.* **26** (1969) 318–344.

MIX
Papier aus verantwortungsvollen Quellen
Paper from responsible sources
FSC® C105338

If you have any concerns about our products,
you can contact us on
ProductSafety@springernature.com

In case Publisher is established outside the EU,
the EU authorized representative is:
Springer Nature Customer Service Center GmbH
Europaplatz 3, 69115 Heidelberg, Germany

Printed by Libri Plureos GmbH
in Hamburg, Germany